EMBEDDED SYSTEMS:

REAL-TIME INTERFACING TO ARM® CORTEX™-M MICROCONTROLLERS

Volume 2

Fifth Edition,

December 2017

Jonathan W. Valvano

Fifth edition
6th Printing
December 2017

ARM and uVision are registered trademarks of ARM Limited.
Cortex and Keil are trademarks of ARM Limited.
Stellaris and Tiva are registered trademarks Texas Instruments.
Code Composer Studio is a trademark of Texas Instruments.
All other product or service names mentioned herein are the trademarks of their respective owners.

In order to reduce costs, this college textbook has been self-published. For more information about my classes, my research, and my books, see http://users.ece.utexas.edu/~valvano/

For corrections and comments, please contact me at: valvano@mail.utexas.edu. Please cite this book as: J. W. Valvano, Embedded Systems: Real-Time Interfacing to ARM® Cortex™-M Microcontrollers, http://users.ece.utexas.edu/~valvano/, ISBN: 978-1463590154, 2017.

Copyright © 2017 Jonathan W. Valvano
All rights reserved. No part of this work covered by the copyright herein may be reproduced, transmitted, stored, or used in any form or by any means graphic, electronic, or mechanical, including but not limited to photocopying, recording, scanning, digitizing, taping, web distribution, information networks, or information storage and retrieval, except as permitted under Section 107 or 108 of the 1976 United States Copyright Act, without the prior written permission of the publisher.
ISBN-13: 978-1463590154
ISBN-10: 1463590156

Table of Contents

Preface to Third Edition vii
Preface to Fourth Edition vii
Preface to Fifth Edition vii
Preface viii
Acknowledgements ix

1. Introduction to Embedded Systems 1
   1.1. Computer Architecture 2
   1.2. Embedded Systems 10
   1.3. The Design Process 16
   1.4. Digital Logic and Open Collector 29
   1.5. Digital Representation of Numbers 40
   1.6. Ethics 56
   1.7. Exercises 58
   1.8. Lab Assignments 62

2. ARM Cortex-M Processor 63
   2.1. Cortex™-M Architecture 64
   2.2. Texas Instruments TM4C I/O pins 71
   2.3. ARM® Cortex™-M Assembly Language 81
   2.4. Parallel I/O ports 101
   2.5. Phase-Lock-Loop 111
   2.6. SysTick Timer 114
   2.7. Choosing a Microcontroller 118
   2.8. Exercises 120
   2.9. Lab Assignments 121

3. Software Design 123
   3.1. Attitude 124
   3.2. Quality Programming 126

3.3. Software Style Guidelines 127
3.4. Modular Software 142
3.5. Finite State Machines 155
3.6. Threads 166
3.7. First In First Out Queue 169
3.8. Memory Management and the Heap 177
3.9. Introduction to Debugging 180
3.10. Exercises 192
3.11. Lab Assignments 194

4. Hardware-Software Synchronization 195
4.1. Introduction 196
4.2. Timing 201
4.3. Petri Nets 206
4.4. Kahn Process Networks 209
4.5. Edge-triggered Interfacing 211
4.6. Configuring Digital Output Pins 214
4.7. Blind-cycle Interfacing 215
4.8. Busy-Wait Synchronization 227
4.9. UART Interface 232
4.10. Keyboard Interface 241
4.11. Exercises 246
4.12. Lab Assignments 248

5. Interrupt Synchronization 251
5.1. Multithreading 252
5.2. Interthread Communication and Synchronization 255
5.3. Critical Sections 263
5.4. NVIC on the ARM® Cortex-M Processor 267
5.5. Edge-triggered Interrupts 272
5.6. Interrupt-Driven UART 275
5.7. Periodic Interrupts using SysTick 279
5.8. Low-Power Design 283

    5.9. Debugging Profile..284
    5.10. Exercises..285
    5.11. Lab Assignments..289
6. Time Interfacing...291
    6.1. Input Capture or Input Edge Time Mode 291
    6.2. Output Compare or Periodic Timer..305
    6.3. Pulse Width Modulation 309
    6.4. Frequency Measurement 312
    6.5. Binary Actuators..316
    6.6. Integral Control of a DC Motor 329
    6.7. Exercises..331
    6.8. Lab Assignments..333
7. Serial Interfacing..335
    7.1. Introduction to Serial Communication 336
    7.2. RS232 Interfacing 342
    7.3. RS422/USB/RS423/RS485 Balanced Differential Lines 347
    7.4. Logic Level Conversion 352
    7.5. Synchronous Transmission and Receiving using the SSI 353
    7.6. Inter-Integrated Circuit (I^2C) Interface..364
    7.7. Introduction to Universal Serial Bus (USB).....................................377
    7.8. Exercises..384
    7.9. Lab Assignments..387
8. Analog Interfacing 389
    8.1. Resistors and Capacitors 389
    8.2. Op Amps 392
    8.3. Analog Filters..408
    8.4. Digital to Analog Converters 411
    8.5. Analog to Digital Converters 423
    8.6. Exercises..436
    8.7. Lab Assignments..438
9. System-Level Design...439

9.1. Design for Manufacturability 439
9.2. Power 441
9.3 Tolerance 454
9.4. Design for Testability 456
9.5. Printed Circuit Board Layout and Enclosures 457
9.6. Exercises 460
9.7. Lab Assignments 460

10. Data Acquisition Systems 461

10.1. Introduction 461
10.2. Transducers 466
10.3. Discrete Calculus 478
10.4. Data Acquisition System Design 480
10.5. Analysis of Noise 486
10.6. Data Acquisition Case Studies 495
10.7. Exercises 507
10.8. Lab Assignments 511

11. Introduction to Communication Systems 513

11.1. Fundamentals 513
11.2. Communication Systems Based on the UARTs 517
11.3. Wireless Communication 521
11.4. Internet of Things 526
11.5. Exercises 544
11.6. Lab Assignments 545

Appendix 1. Glossary 546

Appendix 2. Solutions to Checkpoints 563

Index 571

Reference Material 583

Preface to Third Edition

There are a new features added to this third edition. The new development platform based on the TM4C123 is called Tiva LaunchPad. This new microcontroller runs at 80 MHz, include single-precision floating point, have two 12-bit ADCs, and support DMA and USB. A wonderful feature of these new boards is their low cost. As of August 2017, the boards are available on TI.com as part number EK-TM4C123GXL for $12.99. Although this edition now focuses on the M4, the concepts still apply to the M3, and the web site associated with this book has example projects based on the LM3S811, LM3S1968, and LM3S8962.

Preface to Fourth Edition

This fourth edition includes the new TM4C1294-based LaunchPad. Most of the code in the book is specific for the TM4C123-based LaunchPad. There are now two lost-cost development platforms called Tiva LaunchPad. The EK-TM4C123GXL LaunchPad retails for $12.99, and the EK-TM4C1294XL Connected LaunchPad retails for $19.99. The various LM3S, and TM4C microcontrollers are quite similar, so this book along with the example code on the web can be used for any of these microcontrollers. Compared to the TM4C123, the new TM4C1294 microcontroller runs faster, has more RAM, has more ROM, includes Ethernet, and has more I/O pins. This fourth edition switches the syntax from C to the industry-standard C99, adds a line-tracking robot, designs an integral controller for a DC motor

Preface to Fifth Edition

This fifth edition has expanded sections on line-tracking finite state machines, profiling, frequency measurements, low-power design, hibernation, measurement of supply current, and the Internet of Things.

This fifth edition focuses on the TM4C123 and MSP432E401Y/TM4C1294-based LaunchPads. Most of the code in the book is specific for the TM4C123-based LaunchPad. However, the book website includes corresponding example projects for the LM3S811, LM3S1968, LM4F120, MSP432E401Y, and TM4C1294, which are ARM® Cortex™-M microcontrollers from Texas Instruments. There are now two lost-cost development platforms called Tiva LaunchPad. The EK-TM4C123GXL LaunchPad retails for $12.99, and the MSP432E401Y/TM4C1294 LaunchPads retail for $19.99. The various LM3S, LM4F and TM4C microcontrollers are quite similar, so this book along with the example code on the web can be used for any of these microcontrollers. This fifth edition 5th printing includes information on the new MSP432E401Y. The web site now includes projects configured for both Keil uVision and Code Composer Studio.

Preface

Embedded systems are a ubiquitous component of our everyday lives. We interact with hundreds of tiny computers every day that are embedded into our houses, our cars, our toys, and our work. As our world has become more complex, so have the capabilities of the microcontrollers embedded into our devices. The ARM® Cortex™-M family represents a new class of microcontrollers much more powerful than the devices available ten years ago. The purpose of this book is to present the design methodology to train young engineers to understand the basic building blocks that comprise devices like a cell phone, an MP3 player, a pacemaker, antilock brakes, and an engine controller.

This book is the second in a series of three books that teach the fundamentals of embedded systems as applied to the ARM® Cortex™-M family of microcontrollers. The three books are primarily written for undergraduate electrical and computer engineering students. They could also be used for professionals learning the ARM platform. The first book <u>Embedded Systems: Introduction to ARM Cortex-M Microcontrollers</u> is an introduction to computers and interfacing focusing on assembly language and C programming. This second book focuses on interfacing and the design of embedded systems. The third book <u>Embedded Systems: Real-Time Operating Systems for ARM Cortex-M Microcontrollers</u> is an advanced book focusing on operating systems, high-speed interfacing, control systems, and robotics.

An embedded system is a system that performs a specific task and has a computer embedded inside. A system is comprised of components and interfaces connected together for a common purpose. This book presents components, interfaces and methodologies for building systems. Specific topics include the architecture of microcontrollers, design methodology, verification, hardware/software synchronization, interfacing devices to the computer, timing diagrams, real-time operating systems, data collection and processing, motor control, analog filters, digital filters, real-time signal processing, wireless communication, and the internet of things.

In general, the area of embedded systems is an important and growing discipline within electrical and computer engineering. The educational market of embedded systems has been dominated by simple microcontrollers like the PIC, the 9S12, and the 8051. This is because of their market share, low cost, and historical dominance. However, as problems become more complex, so must the systems that solve them. A number of embedded system paradigms must shift in order to accommodate this growth in complexity. First, the number of calculations per second will increase from millions/sec to billions/sec. Similarly, the number of lines of software code will also increase from thousands to millions. Thirdly, systems will involve multiple microcontrollers supporting many simultaneous operations. Lastly, the need for system verification will continue to grow as these systems are deployed into safety critical applications. These changes are more than a simple growth in size and bandwidth. These systems must employ parallel programming, high-speed synchronization, real-time operating systems, fault tolerant design, priority interrupt handling, and networking. Consequently, it will be important to provide our students with these types of design experiences. The ARM platform is both low cost and provides the high performance features required in future embedded systems. Although the ARM market share is currently not huge, its share will grow. Furthermore, students trained on the ARM will be equipped to design systems across the complete spectrum from simple to complex. The purpose of writing these three books at this time is to bring engineering education into the 21st century.

This book employs many approaches to learning. It will not include an exhaustive recapitulation of the information in data sheets. First, it begins with basic fundamentals, which allows the reader to solve new problems with new technology. Second, the book presents many detailed design examples. These examples illustrate the process of design. There are multiple structural components that assist learning. Checkpoints, with answers in the back, are short easy to answer questions providing immediate feedback while reading. Simple homework, with answers to the odd questions on the web, provides more detailed learning opportunities. The book includes an index and a glossary so that information can be searched. The most important learning experiences in a class like this are of course the laboratories. Each chapter has suggested lab assignments. More detailed lab descriptions are available on the web. Specifically, look at the lab assignments for EE445L and EE445M.

There is a web site accompanying this book **http://users.ece.utexas.edu/~valvano/arm**. Posted here are ARM Keil™ uVision® projects for each the example programs in the book. Code Composer Studio™ versions are also available for most examples. You will also find data sheets and Excel spreadsheets relevant to the material in this book.

These three books will cover embedded systems for ARM® Cortex™-M microcontrollers with specific details on the LM3S811, LM3S1968, LM3S8962, LM4F120, TM4C123, and TM4C1294/MSP432E401Y. Most of the topics can be run on the low-cost TM4C123. Ethernet examples can be run on the LM3S8962 and MSP432E401Y. In these books the terms **LM3S** and **TM4C** will refer to any of the Texas Instruments ARM® Cortex™-M based microcontrollers. Although the solutions are specific for the **LM3S** and **TM4C** families, it will be possible to use these books for other ARM derivatives.

Acknowledgements

I owe a wonderful debt of gratitude to Daniel Valvano. He wrote and tested most of the software examples found in this book. Secondly, he created and maintains the example web site, **http://users.ece.utexas.edu/~valvano/arm.**

Many shared experiences contributed to the development of this book. First I would like to acknowledge the many excellent teaching assistants I have had the pleasure of working with. Some of these hard-working, underpaid warriors include Pankaj Bishnoi, Rajeev Sethia, Adson da Rocha, Bao Hua, Raj Randeri, Santosh Jodh, Naresh Bhavaraju, Ashutosh Kulkarni, Bryan Stiles, V. Krishnamurthy, Paul Johnson, Craig Kochis, Sean Askew, George Panayi, Jeehyun Kim, Vikram Godbole, Andres Zambrano, Ann Meyer, Hyunjin Shin, Anand Rajan, Anil Kottam, Chia-ling Wei, Jignesh Shah, Icaro Santos, David Altman, Nachiket Kharalkar, Robin Tsang, Byung Geun Jun, John Porterfield, Daniel Fernandez, Deepak Panwar, Jacob Egner, Sandy Hermawan, Usman Tariq, Sterling Wei, Seil Oh, Antonius Keddis, Lev Shuhatovich, Glen Rhodes, Geoffrey Luke, Karthik Sankar, Tim Van Ruitenbeek, Raffaele Cetrulo, Harshad Desai, Justin Capogna, Arindam Goswami, Jungho Jo, Mehmet Basoglu, Kathryn Loeffler, Evgeni Krimer, Nachiappan Valliappan, Razik Ahmed, Sundeep Korrapati, Song Zhang, Zahidul Haq, Matthew Halpern, Cruz Monrreal II, Pohan Wu, Saugata Bhattacharyya, Omar Baca, Aditya Saraf, Mahesh Srinivasan, Victoria Bill, Alex Hsu, Jason He, Youngchun Kim, Dylan Zika, Cody Horton, Lavanya Venkatesan, and Corey Cormier. These teaching assistants have contributed greatly to the contents of this book and particularly to its laboratory assignments. Since 1981, I estimate I have taught embedded systems to over 5000 students. My students have recharged my energy each semester with their enthusiasm, dedication, and quest for knowledge. I have decided not to acknowledge them all individually. However, they know I feel privileged to have had this opportunity.

Next, I appreciate the patience and expertise of my fellow faculty members here at the University of Texas at Austin. From a personal perspective Dr. John Pearce provided much needed encouragement and support throughout my career. In addition, Drs. John Cogdell, John Pearce, and Francis Bostick helped me with analog circuit design. The book and accompanying software include many finite state machines derived from the digital logic examples explained to me by Dr. Charles Roth. Over the last few years, I have enjoyed teaching embedded systems with Drs. Ramesh Yerraballi, Mattan Erez, Andreas Gerstlauer, Vijay Janapa Reddi, Nina Telang, and Bill Bard. Bill has contributed to both the excitement and substance of our laboratory based on this book. With pushing from Bill and TAs Robin, Glen, Lev, and John, we have added low power, PCB layout, systems level design, surface mount soldering, and wireless communication to our lab experience. You can see descriptions and photos of our EE445L design competition at **http://users.ece.utexas.edu/~valvano/**. Many of the suggestions and corrections from Chris Shore and Drew Barbier of ARM about Volume 1 applied equally to this volume. Paul Nossaman, from Texas Instruments, helped explain some of the power circuits presented in Chapter 9. Austin Blackstone created and debugged the Code Composer Studio™ versions of the example programs posted on the web. Austin also taught me how to run the CC3100 Wifi examples on the LaunchPad.

Sincerely, I appreciate the valuable lessons of character and commitment taught to me by my parents and grandparents. I recall how hard my parents and grandparents worked to make the world a better place for the next generation. Most significantly, I acknowledge the love, patience and support of my wife, Barbara, and my children, Ben Daniel and Liz. In particular, Dan designed and tested most of the TM4C software presented in this book.

By the grace of God, I am truly the happiest man on the planet, because I am surrounded by these fine people. Good luck.

Jonathan W. Valvano

The true engineering experience occurs not with your eyes and ears, but rather with your fingers and elbows. In other words, engineering education does not happen by listening in class or reading a book; rather it happens by designing under the watchful eyes of a patient mentor. So, go build something today, then show it to someone you respect!

This book is dedicated to my mom and dad, who demonstrated to me by their lives that "Not all of us can do great things. But we can do small things with great love".

1. Introduction to Embedded Systems

Chapter 1 objectives are to:
- Review computer architecture
- Introduce embedded systems
- Present a process for design
- Discuss practical aspects of digital logic, including open collector
- Review how numbers are represented in binary
- Define ethics

The overall objective of this book is to teach the design of embedded systems. It is effective to learn new techniques by doing them. But, the dilemma in teaching a laboratory-based topic like embedded systems is that there is a tremendous volume of details that first must be learned before hardware and software systems can be designed. The approach taken in this book is to learn by doing, starting with very simple problems and building up to more complex systems later in the book.

In this chapter we begin by introducing some terminology and basic components of a computer system. In order to understand the context of our designs, we will overview the general characteristics of embedded systems. It is in these discussions that we develop a feel for the range of possible embedded applications. Next we will present a template to guide us in design. We begin a project with a requirements document. Embedded systems interact with physical devices. Often, we can describe the physical world with mathematical models. If a model is available, we can then use it to predict how the embedded system will interface with the real world. When we write software, we mistakenly think of it as one dimensional, because the code looks sequential on the computer screen. Data flow graphs, call graphs, and flow charts are multidimensional graphical tools to understand complex behaviors. Because courses taught using this book typically have a lab component, we will review some practical aspects of digital logic and interfacing signals to the microcontroller.

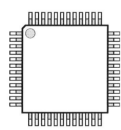

Next, we show multiple ways to represent numbers in the computer. Choosing the correct format is necessary to implement efficient and correct solutions. Fixed-point numbers are the typical way embedded systems represent non-integer values. Floating-point numbers, typically used to represent non-integer values on a general purpose computer, will also be presented.

Because embedded systems can be employed in safety critical applications, it is important for engineers be both effective and ethical. Throughout the book we will present ways to verify the system is operating within specifications.

1.1. Computer Architecture

1.1.1. Computers, microprocessors, memory, and microcontrollers

A **computer** combines a processor, random access memory (RAM), read only memory (ROM), and input/output (I/O) ports. The common bus in Figure 1.1 defines the von Neumann architecture, where instructions are fetched from ROM on the same bus as data fetched from RAM. **Software** is an ordered sequence of very specific instructions that are stored in memory, defining exactly what and when certain tasks are to be performed. The **processor** executes the software by retrieving and interpreting these instructions one at a time. A **microprocessor** is a small processor, where small refers to size (i.e., it fits in your hand) and not computational ability. For example, Intel Xeon, AMD FX and Sun SPARC are microprocessors. An ARM® Cortex™-M microcontroller includes a processor together with the bus and some peripherals. A **microcomputer** is a small computer, where again small refers to size (i.e., you can carry it) and not computational ability. For example, a desktop PC is a microcomputer.

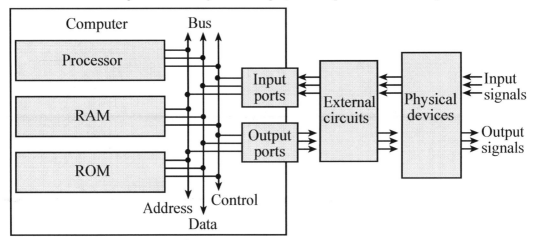

Figure 1.1. The basic components of a computer system include processor, memory and I/O.

A very small microcomputer, called a **microcontroller**, contains all the components of a computer (processor, memory, I/O) on a single chip. As shown in Figure 1.2, the Atmel ATtiny, the Texas Instruments MSP430, and the Texas Instruments TM4C123 are examples of microcontrollers. Because a microcomputer is a small computer, this term can be confusing because it is used to describe a wide range of systems from a 6-pin ATtiny4 running at 1 MHz with 512 bytes of program memory to a personal computer with state-of-the-art 64-bit multi-core processor running at multi-GHz speeds having terabytes of storage.

The computer can store information in **RAM** by writing to it, or it can retrieve previously stored data by reading from it. Most RAMs are **volatile**; meaning if power is interrupted and restored the information in the RAM is lost. Most microcontrollers have **static RAM** (SRAM) using six metal-oxide-semiconductor field-effect transistors (MOSFET) to create each memory bit. Four transistors are used to create two cross-coupled inverters that store the binary information, and the other two are used to read and write the bit.

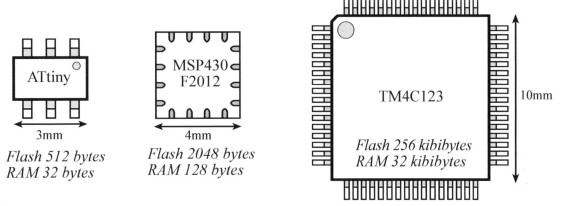

Figure 1.2. A microcontroller is a complete computer on a single chip.

Information is programmed into **ROM** using techniques more complicated than writing to RAM. From a programming viewpoint, retrieving data from a ROM is identical to retrieving data from RAM. ROMs are nonvolatile; meaning if power is interrupted and restored the information in the ROM is retained. Some ROMs are programmed at the factory and can never be changed. A Programmable ROM (PROM) can be erased and reprogrammed by the user, but the erase/program sequence is typically 10000 times slower than the time to write data into a RAM. PROMs used to need ultraviolet light to erase, and then we programmed them with voltages. Now, most PROMs now are electrically erasable (EEPROM), which means they can be both erased and programmed with voltages. We cannot program ones into the ROM. We first erase the ROM, which puts ones into its storage memory, and then we program the zeros as needed. **Flash ROM** is a popular type of EEPROM. Each flash bit requires only two MOSFET transistors. The input (gate) of one transistor is electrically isolated, so if we trap charge on this input, it will remain there for years. The other transistor is used to read the bit by sensing whether or not the other transistor has trapped charge. In regular EEPROM, you can erase and program individual bytes. Flash ROM must be erased in large blocks. On many of TM4C microcontrollers, we can erase the entire ROM or just a 1024-byte block. Because flash is smaller than regular EEPROM, most microcontrollers have a large flash into which we store the software. For all the systems in this book, we will store instructions and constants in flash ROM and place variables and temporary data in static RAM.

Checkpoint 1.1: What are the differences between a microcomputer, a microprocessor and a microcontroller?

Checkpoint 1.2: Which has a higher information density on the chip in bits per mm^2: static RAM or flash ROM? Assume all MOSFETs are approximately the same size in mm^2.

Observation: Memory is an object that can transport information across time.

The external devices attached to the microcontroller provide functionality for the system. An **input port** is hardware on the microcontroller that allows information about the external world to be entered into the computer. The microcontroller also has hardware called an **output port** to send information out to the external world. Most of the pins shown in Figure 1.2 are input/output ports.

An **interface** is defined as the collection of the I/O port, external electronics, physical devices, and the software, which combine to allow the computer to communicate with the external world. An example of an input interface is a switch, where the operator toggles the switch, and the software can recognize the switch position. An example of an output interface is a light-emitting diode (LED), where the software can turn the light on and off, and the operator can see whether or not the light is shining. There is a wide range of possible inputs and outputs, which can exist in either digital or analog form. In general, we can classify I/O interfaces into four categories

Parallel - binary data are available simultaneously on a group of lines

Serial - binary data are available one bit at a time on a single line

Analog - data are encoded as an electrical voltage, current, or power

Time - data are encoded as a period, frequency, pulse width, or phase shift

Checkpoint 1.3: What are the differences between an input port and an input interface?

Checkpoint 1.4: List three input interfaces available on a personal computer.

Checkpoint 1.5: List three output interfaces available on a personal computer.

In this book, numbers that start with 0x (e.g., 0x64) are specified in **hexadecimal**, which is base 16 (0x64 = $6*16^1+4*16^0$ = 100). Some assemblers start hexadecimal numbers with $ (e.g., $64). Other assembly languages add an "H" at the end to specify hexadecimal (e.g., 64H or 64h). Yale Patt's LC3 assembler uses just the "x" (e.g., x64).

In a system with **memory mapped I/O**, as shown in Figure 1.1, the I/O ports are connected to the processor in a manner similar to memory. I/O ports are assigned addresses, and the software accesses I/O using reads and writes to the specific I/O addresses. The software inputs from an input port using the same instructions as it would if it were reading from memory. Similarly, the software outputs from an output port using the same instructions as it would if it were writing to memory. A **bus** is defined as a collection of signals, which are grouped for a common purpose. The bus has three types of signals: address signals, data signals, and control signals. Together, the bus directs the data transfer between the various modules in the computer. There are five buses on ARM® Cortex™-M processor, as illustrated in Figure 1.3. The address specifies which module is being accessed, and the data contains the information being transferred. The control signals specify the direction of transfer, the size of the data, and timing information. The **ICode bus** is used to fetch instructions from flash ROM. All ICode bus fetches contain 32 bits of data, which may be one or two instructions. The **DCode bus** can fetch data or debug information from flash ROM. The **system bus** can read/write data from RAM or I/O ports. The **private peripheral bus** (PPB) can access some of the common peripherals like the interrupt controller. The multiple-bus architecture allows simultaneous bus activity, greatly improving performance over single-bus architectures. For example, the processor can simultaneously fetch an instruction out of flash ROM using the ICode bus while it writes data into RAM using the system bus. From a software development perspective, the fact that there are multiple buses is transparent. This means we write code like we would on any computer, and the parallel operations occur automatically. The TM4C123 has 256 kibibytes (2^{18} bytes) of flash ROM and 32768 bytes of RAM. The TM4C1294/ MSP432E401Y has 1024 kibibytes (2^{20} bytes) of flash ROM and 256 kibibytes of RAM. The RAM begins at 0x2000.0000, and the flash ROM begins at 0x0000.0000.

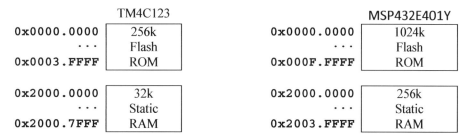

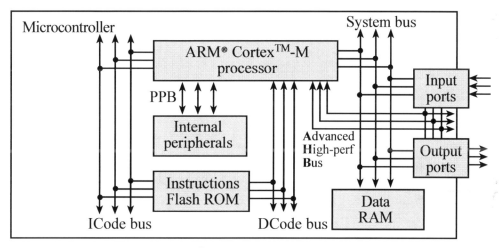

Figure 1.3. Harvard architecture of an ARM® Cortex™-M-based microcontroller.

The Cortex-M4 series includes an additional bus called the Advanced High-Performance Bus (AHB or AHPB). This bus improves performance when communicating with high-speed I/O devices like USB. In general, the more operations that can be performed in parallel, the faster the processor will execute. In summary:

ICode bus	Fetch opcodes from ROM
DCode bus	Read constant data from ROM
System bus	Read/write data from RAM or I/O, fetch opcode from RAM
PPB	Read/write data from internal peripherals like the NVIC
AHB	Read/write data from high-speed I/O and parallel ports

Instructions and data are accessed the same way on a von Neumann machine. The Cortex™-M processor is a Harvard architecture because instructions are fetched on the ICode bus and data accessed on the system bus. The address signals on the ARM® Cortex™-M processor include 32 lines, which together specify the memory address (0x0000.0000 to 0xFFFF.FFFF) that is currently being accessed. The address specifies both which module (input, output, RAM, or ROM) as well as which cell within the module will communicate with the processor. The data signals contain the information that is being transferred and also include 32 bits. However, on the system bus it can also transfer 8-bit or 16-bit data. The control signals specify the timing, the size, and the direction of the transfer. We call a complete data transfer a bus cycle. Two types of transfers are allowed, as shown in Table 1.1. In most systems, the processor always controls the address (where to access), the direction (read or write), and the control (when to access.)

Type	Address Driven by	Data Driven by	Transfer
Read Cycle	Processor	RAM, ROM or Input	Data copied to processor
Write Cycle	Processor	Processor	Data copied to output or RAM

Table 1.1. Simple computers generate two types of bus cycles.

A **read cycle** is used to transfer data into the processor. During a read cycle the processor first places the address on the address signals, and then the processor issues a read command on the control signals. The slave module (RAM, ROM, or I/O) will respond by placing the contents at that address on the data signals, and lastly the processor will accept the data and disable the read command.

The processor uses a **write cycle** to store data into memory or I/O. During a write cycle the processor also begins by placing the address on the address signals. Next, the processor places the information it wishes to store on the data signals, and then the processor issues a write command on the control signals. The memory or I/O will respond by storing the information into the proper place, and after the processor is sure the data has been captured, it will disable the write command.

The **bandwidth** of an I/O interface is the number of bytes/sec that can be transferred. If we wish to transfer data from an input device into RAM, the software must first transfer the data from input to the processor, then from the processor into RAM. On the ARM, it will take multiple instructions to perform this transfer. The bandwidth depends both on the speed of the I/O hardware and the software performing the transfer. In some microcontrollers like the TM4C123 and TM4C1294/MSP432E401Y, we will be able to transfer data directly from input to RAM or RAM to output using direct memory access (DMA). When using DMA the software time is removed, so the bandwidth only depends on the speed of the I/O hardware. Because DMA is faster, we will use this method to interface high bandwidth devices like disks and networks. During a **DMA read cycle** data flows directly from the memory to the output device. General purpose computers also support DMA allowing data to be transferred from memory to memory. During a **DMA write cycle** data flows directly from the input device to memory.

Input/output devices are important in all computers, but they are especially significant in an embedded system. In a computer system with **I/O-mapped I/O**, the control bus signals that activate the I/O are separate from those that activate the memory devices. These systems have a separate address space and separate instructions to access the I/O devices. The original Intel 8086 had four control bus signals MEMR, MEMW, IOR, and IOW. MEMR and MEMW were used to read and write memory, while IOR and IOW were used to read and write I/O. The Intel x86 refers to any of the processors that Intel has developed based on this original architecture. Even though we do not consider the personal computer (PC) an embedded system, there are embedded systems developed on this architecture. There are a number of single board embedded platforms based on the Intel Atom. The Intel x86 processors continue to implement this separation between memory and I/O. Rather than use the regular memory access instructions, the Intel x86 processor uses special **in** and **out** instructions to access the I/O devices. The advantages of I/O-mapped I/O are that software cannot inadvertently access I/O when it thinks it is accessing memory. In other words, it protects I/O devices from common software bugs, such as bad pointers, stack overflow, and buffer overflows. In contrast, systems with memory-mapped I/O are easier to design, and the software is easier to write.

1.1.2. ARM Cortex-M processor

The ARM Cortex-M processor has four major components, as illustrated in Figure 1.4. There are four **bus interface units** (BIU) that read data from the bus during a read cycle and write data onto the bus during a write cycle. Both the TM4C123 and TM4C1294/MSP432E401Y microcontrollers support DMA. The BIU always drives the address bus and the control signals of the bus. The **effective address register** (EAR) contains the memory address used to fetch the data needed for the current instruction. Cortex-M microcontrollers execute Thumb® instructions extended with Thumb-2 technology. An overview of these instructions will be presented in Chapter 2. The Cortex-M4F microcontrollers include a floating-point processor. However, in this book we will focus on integer and fixed-point arithmetic.

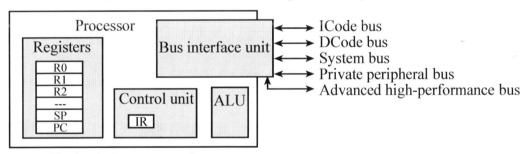

Figure 1.4. The four basic components of a processor.

The **control unit** (CU) orchestrates the sequence of operations in the processor. The CU issues commands to the other three components. The **instruction register** (IR) contains the operation code (or op code) for the current instruction. When extended with Thumb-2 technology, op codes are either 16 or 32 bits wide. In an embedded system the software is converted to machine code, which is a list of instructions, and stored in nonvolatile flash ROM. As instructions are fetched, they are placed in a **pipeline**. This allows instruction fetching to run ahead of execution. Instructions are fetched in order and executed in order. However, it can execute one instruction while fetching the next.

The **registers** are high-speed storage devices located in the processor (e.g., R0 to R15). Registers do not have addresses like regular memory, but rather they have specific functions explicitly defined by the instruction. Registers can contain data or addresses. The **program** counter (PC) points to the memory containing the instruction to execute next. On the ARM Cortex-M processor, the PC is register 15 (R15). In an embedded system, the PC usually points into nonvolatile memory like flash ROM. The information stored in nonvolatile memory (e.g., the instructions) is not lost when power is removed. The **stack pointer** (SP) points to the RAM, and defines the top of the stack. The stack implements last in first out (LIFO) storage. On the ARM® Cortex™-M processor, the SP is register 13 (R13). The stack is an extremely important component of software development, which can be used to pass parameters, save temporary information, and implement local variables. The **program status register** (PSR) contains the status of the previous operation, as well as some operating mode flags such as the interrupt enable bit. This register is called the flag register on the Intel computers.

The **arithmetic logic unit** (ALU) performs arithmetic and logic operations. Addition, subtraction, multiplication and division are examples of arithmetic operations. And, or, exclusive or, and shift are examples of logical operations.

Checkpoint 1.6: For what do the acronyms CU DMA BIU ALU stand?

In general, the execution of an instruction goes through four phases. First, the computer fetches the machine code for the instruction by reading the value in memory pointed to by the program counter (PC). Some instructions are 16 bits, while others are 32 bits. After each instruction is fetched, the PC is incremented to the next instruction. At this time, the instruction is decoded, and the effective address is determined (EAR). Many instructions require additional data, and during phase 2 the data is retrieved from memory at the effective address. Next, the actual function for this instruction is performed. During the last phase, the results are written back to memory. All instructions have a phase 1, but the other three phases may or may not occur for any specific instruction.

On the ARM Cortex-M processor, an instruction may read memory or write memory, but it does not both read and write memory in the same instruction. Each of the phases may require one or more bus cycles to complete. Each bus cycle reads or writes one piece of data. Because of the multiple bus architecture, most instructions execute in one or two cycles. For more information on the time to execute instructions, see Table 3.1 in the Cortex-M Technical Reference Manual. ARM is a **reduced instruction set computer** (RISC), which achieves high performance by implementing very simple instructions that run extremely fast.

Phase	Function	Bus	Address	Comment
1	Instruction fetch	Read	PC++	Put into IR
2	Data read	Read	EAR	Data passes through ALU
3	Operation	-	-	ALU operations, set PSR
4	Data store	Write	EAR	Results stored in memory

Table 1.2. Four phases of execution.

An instruction on a RISC processor does not have both a phase 2 data read cycle and a phase 4 data write cycle. In general, a RISC processor has a small number of instructions, instructions have fixed lengths, instructions execute in 1 or 2 bus cycles, there are only a few instructions (e.g., load and store) that can access memory, no one instruction can both read and write memory in the same instruction, there are many identical general purpose registers, and there are a limited number of addressing modes.

Conversely, processors are classified as **complex instruction set computers** (CISC), because one instruction is capable of performing multiple memory operations. For example, CISC processors have instructions that can both read and write memory in the same instruction. Assume `Data` is an 8-bit memory variable. The following Intel 8080 instruction will increment the 8-bit variable, requiring a read memory cycle, ALU operation, and then a write memory cycle.

```
          INR Data      ; Intel 8080
```

Other CISC processors like the 6800, 9S12, 8051, and Pentium also have memory increment instructions requiring both a phase 2 data read cycle and a phase 4 data write cycle. In general, a CISC processor has a large number of instructions, instructions have varying lengths, instructions execute in varying times, there are many instructions that can access memory, the processor can both read and write memory in one instruction, the processor has fewer and more specialized registers, and the processor has many addressing modes.

1.1.3. History

In 1968, two unhappy engineers named Bob Noyce and Gordon Moore left the Fairchild Semiconductor Company and created their own company, which they called Integrated Electronics (Intel). Working for Intel in 1971, Federico Faggin, Ted Hoff, and Stan Mazor invented the first single chip microprocessor, the Intel 4004. It was a four-bit processor designed to solve a very specific application for a Japanese company called Busicon. Busicon backed out of the purchase, so Intel decided to market it as a "general purpose" microprocessing system. The product was a success, which lead to a series of more powerful microprocessors: the Intel 8008 in 1974, the Intel 8080 also in 1974. Both the Intel 8008 and the Intel 8080 were 8-bit microprocessors that operated from a single +5V power supply using N-channel metal-oxide semiconductor (NMOS) technology.

Seeing the long term potential for this technology, Motorola released its MC6800 in 1974, which was also an 8-bit processor with about the same capabilities of the 8080. Although similar in computing power, the 8080 and 6800 had very different architectures. The 8080 used isolated I/O and handled addresses in a fundamentally different way than data. Isolated I/O defines special hardware signals and special instructions for input/output. On the 8080, certain registers had capabilities designed for addressing, while other registers had capabilities for specific for data manipulation. In contrast, the 6800 used memory-mapped I/O and handled addresses and data in a similar way. As we defined earlier, input/output on a system with memory-mapped I/O is performed in a manner similar to accessing memory.

During the 1980s and 1990s, Motorola and Intel traveled down similar paths. The microprocessor families from both companies developed bigger and faster products: Intel 8085, 8088, 80x86, ... and the Motorola 6809, 68000, 680x0... During the early 1980's another technology emerged, the microcontroller. In sharp contrast to the microprocessor family, which optimized computational speed and memory size requiring more power and larger physical sizes, the microcontroller devices minimized power consumption and physical size, striving for only modest increases in computational speed and memory size. Out of the Intel architecture came the 8051 family (www.semiconductors.philips.com), and out of the Motorola architecture came the 6805, 6811, and 6812 microcontroller family (www.freescale.com). Many of the same fundamental differences that existed between the original 8-bit Intel 8080 and Motorola 6800 have persisted over forty years of microprocessor and microcontroller developments. In 1999, Motorola shipped its 2 billionth MC68HC05 microcontroller. In 2004, Motorola spun off its microcontroller products as Freescale Semiconductor. Microchip is a leading supplier of 8-bit microcontrollers.

The first ARM processor was conceived in the 1983 by Acorn Computers, which at the time was one of the leaders of business computers in the United Kingdom. The first chips were delivered in 1985. At that time ARM referred to Acorn RISC Machine. In 1990, a new company ARM Ltd was formed with Acorn, Apple, and VLSI Technology as founding partners, changing the ARM acronym to Advanced RISC Machine. As a company, the ARM business model involves the designing and licensing of intellectual property (IP) rather than the manufacturing and selling of actual semiconductor chips. ARM has sold 600 processor licenses to more than 200 companies. Virtually every company that manufacturers integrated circuits in the computer field produces a variant of the ARM processor. ARM currently dominates the high-performance low-power embedded system market. ARM processors account for approximately 90% of all embedded 32-bit RISC processors and are used in consumer electronics, including PDAs, cell phones, music players, hand-held game consoles, and calculators. The ARM Cortex-A is used in applications processors, such as smartphones. The ARM Cortex-R is appropriate for real-time applications, and ARM Cortex-M targets microcontrollers. Examples of microcontrollers built using the ARM Cortex-M core are MSP432/TM4C by Texas Instruments, STM32 by STMicroelectronics, LPC17xx by NXP Semiconductors, TMPM330 by Toshiba, EM3xx by Ember, AT91SAM3 by Atmel, and EFM32 by Energy Micro. As of March 2017, over 100 billion ARM processors have shipped from over 950 companies.

What will the future unfold? One way to predict the future is to study the past. How embedded systems interact with humans has been and will continue to be critical. Improving the human experience has been the goal of many systems. Many predict the number of microcontrollers will soon reach into the trillions. As this happens, communication, security, energy, politics, resources, and economics will become increasingly important. When there are this many computers, it will be possible to make guesses about how to change, then let a process like evolution select which changes are beneficial. In fact, a network of embedded systems with tight coupling to the real world, linked together for a common objective, is now being called a **cyber-physical system** (CPS). The **internet of things** (IoT) will increase in importance.

One constant describing the history of computers is continuous change coupled with periodic monumental changes. Therefore, engineers must focus their education on fundamental principles rather than the voluminous details. They must embrace the concept of lifelong learning. Most humans are fundamentally good, but some are not. Therefore, engineers acting in an ethical manner can guarantee future prosperity of the entire planet.

1.2. Embedded Systems

An **embedded system** is an electronic system that includes a one or more microcontrollers that is configured to perform a specific dedicated application, drawn previously as Figure 1.1. To better understand the expression "embedded system," consider each word separately. In this context, the word embedded means "a computer is hidden inside so one can't see it." The word "system" refers to the fact that there are many components which act in concert achieving the common goal. As mentioned earlier, input/output devices characterize the embedded system, allowing it to interact with the real world.

The software that controls the system is programmed or fixed into flash ROM and is not accessible to the user of the device. Even so, software maintenance is still extremely important. Software maintenance is verification of proper operation, updates, fixing bugs, adding features, and extending to new applications and end user configurations. Embedded systems have these four characteristics.

First, embedded systems typically perform a single function. Consequently, they solve a limited range of problems. For example, the embedded system in a microwave oven may be reconfigured to control different versions of the oven within a similar product line. But, a microwave oven will always be a microwave oven, and you can't reprogram it to be a dishwasher. Embedded systems are unique because of the microcontroller's I/O ports to which the external devices are interfaced. This allows the system to interact with the real world.

Second, embedded systems are tightly constrained. Typically, system must operate within very specific performance parameters. If an embedded system cannot operate with specifications, it is considered a failure and will not be sold. For example, a cell-phone carrier typically gets 832 radio frequencies to use in a city, a hand-held video game must cost less than $50, an automotive cruise control system must operate the vehicle within 3 mph of the set-point speed, and a portable MP3 player must operate for 12 hours on one battery charge.

Third, many embedded systems must operate in real-time. In a **real-time system**, we can put an upper bound on the time required to perform the input-calculation-output sequence. A real-time system can guarantee a worst case upper bound on the response time between when the new input information becomes available and when that information is processed. Another real-time requirement that exists in many embedded systems is the execution of periodic tasks. A periodic task is one that must be performed at equal time intervals. A real-time system can put a small and bounded limit on the time error between when a task should be run and when it is actually run. Because of the real-time nature of these systems, microcontrollers in the TM4C family have a rich set of features to handle all aspects of time.

The fourth characteristic of embedded systems is their small memory requirements as compared to general purpose computers. There are exceptions to this rule, such as those which process video or audio, but most have memory requirements measured in thousands of bytes. Over the years, the memory in embedded systems has increased, but the gap memory size between embedded systems and general purpose computers remains. The original microcontrollers had thousands of bytes of memory and the PC had millions. Now, microcontrollers can have millions of bytes, but the PC has billions.

There have been two trends in the microcontroller field. The first trend is to make microcontrollers smaller, cheaper, and lower power. The Atmel ATtiny, Microchip PIC, and Texas Instruments MSP430 families are good examples of this trend. Size, cost, and power are critical factors for high-volume products, where the products are often disposable. On the other end of the spectrum is the trend of larger RAM and ROM, faster processing, and increasing integration of complex I/O devices, such as Ethernet, radio, graphics, and audio. It is common for one device to have multiple microcontrollers, where the operational tasks are distributed and the microcontrollers are connected in a local area network (LAN). These high-end features are critical for consumer electronics, medical devices, automotive controllers, and military hardware, where performance and reliability are more important than cost. However, small size and low power continue as important features for all embedded systems.

The RAM is volatile memory, meaning its information is lost when power is removed. On some embedded systems a battery powers the microcontroller. When in the off mode, the microcontroller goes into low-power sleep mode, which means the information in RAM is maintained, but the processor is not executing. The MSP430, MSP432, and ATtiny require less than one µA of current in sleep mode.

Checkpoint 1.7: What is an embedded system?

Checkpoint 1.8: What goes in the RAM on a smartphone?

Checkpoint 1.9: Why does your smartphone need so much flash ROM?

The computer engineer has many design choices to make when building a real-time embedded system. Often, defining the problem, specifying the objectives, and identifying the constraints are harder than actual implementations. In this book, we will develop computer engineering design processes by introducing fundamental methodologies for problem specification, prototyping, testing, and performance evaluation.

A typical automobile now contains an average of ten microcontrollers. In fact, upscale homes may contain as many as 150 microcontrollers and the average consumer now interacts with microcontrollers up to 300 times a day. The general areas that employ embedded systems encompass every field of engineering:

- Consumer electronics, wearables
- Communications
- Military
- Business
- Medical
- Home
- Automotive
- Industrial
- Shipping
- Computer components

In general, embedded systems have inputs, perform calculations, make decisions, and then produce outputs. The microcontrollers often must communicate with each other. How the system interacts with humans is often called the **human-computer interface** (HCI) or **man-machine interface** (MMI). To get a sense of what "embedded system" means we will present brief descriptions of four example systems.

Example 1.1: The goal of a **pacemaker** is to regulate and improve heart function. To be successful the engineer must understand how the heart works and how disease states cause the heart to fail. Its inputs are sensors on the heart to detect electrical activity, and its outputs can deliver electrical pulses to stimulate the heart. Consider a simple pacemaker with two sensors, one in the right atrium and the other in the right ventricle. The sensor allows the pacemaker to know if the normal heart contraction is occurring. This pacemaker has one right ventricular stimulation output. The embedded system analyzes the status of the heart deciding where and when to send simulation pulses. If the pacemaker recognizes the normal behavior of atrial contraction followed shortly by ventricular contraction, then it will not stimulate. If the pacemaker recognizes atrial contraction without a following ventricular contraction, then is will pace the ventricle shortly after each atrial contraction. If the pacemaker senses no contractions or if the contractions are too slow, then it can pace the ventricle at a regular rate. A pacemaker can also communicate via radio with the doctor to download past performance and optimize parameters for future operation. Some pacemakers can call the doctor on the phone when it senses a critical problem. Pacemakers are real-time systems because the time delay between atrial sensing and ventricular triggering is critical. Low power and reliability are important.

Example 1.2: The goal of a **smoke detector** is to warn people in the event of a fire. It has two inputs. One is a chemical sensor that detects the presence of smoke, and the other is a button that the operator can push to test the battery. There are also two outputs: an LED and the alarm. Most of the time, the detector is in a low-power sleep mode. If the test button is pushed, the detector performs a self-diagnostic and issues a short sound if the sensor and battery are ok. Once every 30 seconds, it wakes up and checks to see if it senses smoke. If it senses smoke, it will alarm. Otherwise it goes back to sleep. Advanced smoke detectors should be able to communicate with other devices in the home. If one sensor detects smoke, all alarms should sound. If multiple detectors in the house collectively agree there is really a fire, they could communicate with the fire department and with the neighboring houses. To design and deploy a collection of detectors, the engineer must understand how fires start and how they spread. Smoke detectors are not real-time systems. However, reliability and low power are important.

Example 1.3: The goal of a **motor controller** is to cause a motor to spin in a desired manner. Sometimes we control speed, as in the cruise control on an automobile. Sometimes we control position as in moving paper through a printer. In a complex robotics system, we may need to simultaneously control multiple motors and multiple parameters such as position, speed, and torque. Torque control is important for building a robot that walks. The engineer must understand the mechanics of how the motor interacts with its world and the behavior of the interface electronics. The motor controller uses sensors to measure the current state of the motor, such as position, speed, and torque. The controller accepts input commands defining the desired operation. The system uses actuators, which are outputs that affect the motor. A typical actuator allows the system to set the electrical power delivered to the motor. Periodically, the microcontroller senses the inputs and calculates the power needed to minimize the difference between measured and desired parameters. This needed power is output to the actuator. Motor controllers are real-time systems, because performance depends greatly on when and how fast the controller software runs. Accuracy, stability, and time are important.

Example 1.4: The goal of a **traffic controller** is to minimize waiting time and to save energy. The engineer must understand the civil engineering of how city streets are laid out and the behavior of human drivers as they interact with traffic lights and other drivers. The controller uses sensors to know the number of cars traveling on each segment of road. Pedestrians can also push walk buttons. The controller will accept input commands from the fire or police department to handle emergencies. The outputs are the traffic lights at each intersection. The controller collects sensor inputs and calculates the traffic pattern needed to minimize waiting time, while maintaining safety. Traffic controllers are not real-time systems, because human safety is not sacrificed if a request is delayed. In contrast, an air traffic controller must run in real time, because safety is compromised if a response to a request is delayed. The system must be able to operate under extreme conditions such as rain, snow, freezing temperature, and power outages. Computational speed and sensor/light reliability are important.

Checkpoint 1.10: There is a microcontroller embedded in an alarm clock. List three operations the software must perform.

When designing embedded systems we need to know how to interface a wide range of signals that can exist in digital, analog, or time formats.

Table 1.3 lists example products and the functions performed by their embedded systems. The microcontroller accepts inputs, performs calculations, and generates outputs.

Functions performed by the microcontroller

Consumer/Home:
Washing machine	Controls the water and spin cycles, saving water and energy
Wearables	Measures speed, distance, calories, heart rate, wireless communication
Remote controls	Accepts key touches, sends infrared pulses, learns how to interact with user
Clocks and watches	Maintains the time, alarm, and display
Games and toys	Entertains the user, joystick input, video output
Audio/video	Interacts with the operator, enhances performance with sounds and pictures
Set-back thermostats	Adjusts day/night thresholds saving energy

Communication:
Answering machines	Plays outgoing messages and saves incoming messages
Telephone system	Switches signals and retrieves information
Cellular phones	Interacts with touch screen, microphone, accelerometer, GPS, and speaker
Internet of things	Sends and receives messages with other computers around the world

Automotive:
Automatic braking	Optimizes stopping on slippery surfaces
Noise cancellation	Improves sound quality, removing noise
Theft deterrent devices	Allows keyless entry, controls alarm
Electronic ignition	Controls sparks and fuel injectors
Windows and seats	Remembers preferred settings for each driver
Instrumentation	Collects and provides necessary information

Military:
Smart weapons	Recognizes friendly targets
Missile guidance	Directs ordnance at the desired target
Global positioning	Determines where you are on the planet, suggests paths, coordinates troops
Surveillance	Collects information about enemy activities

Industrial/Business/Shipping:
Point-of-sale systems	Accepts inputs and manages money, keeps credit information secure
Temperature control	Adjusts heating and cooling to maintain temperature
Robot systems	Inputs from sensors, controls the motors improving productivity
Inventory systems	Reads and prints labels, maximizing profit, minimizing shipping delay
Automatic sprinklers	Controls the wetness of the soil maximizing plant growth

Medical:
Infant apnea monitors	Detects breathing, alarms if stopped
Cardiac monitors	Measures heart function, alarms if problem
Cancer treatments	Controls doses of radiation, drugs, or heat
Prosthetic devices	Increases mobility for the handicapped
Medical records	Collect, organize, and present medical information

Computer Components:
Mouse	Translates hand movements into commands for the main computer
USB flash drive	Facilitates the storage and retrieval of information
Keyboard	Accepts key strokes, decodes them, and transmits to the main computer

Table 1.3. Products involving embedded systems.

In contrast, a **general-purpose computer** system typically has a keyboard, disk, and graphics display and can be programmed for a wide variety of purposes. Typical general-purpose applications include word processing, electronic mail, business accounting, scientific computing, cloud computing, and web servers. General-purpose computers have the opposite of the four characteristics listed above. First, they can perform a wide and dynamic range of functions. Because the general-purpose computer has a removable disk or network interface, new programs can easily be added to the system. The user of a general-purpose computer does have access to the software that controls the machine. In other words, the user decides which operating system to run and which applications to launch. Second, they are loosely constrained. For example, the Java machine used by a web browser will operate on an extremely wide range of computer platforms. Third, general-purpose machines do not run in real-time. Yes, we would like the time to print a page on the printer to be fast, and we would like a web page to load quickly, but there are no guaranteed response times for these types of activities. In fact, the real-time tasks that do exist (such as sound recording, burning CD, and graphics) are actually performed by embedded systems built into the system. Fourth, general purpose computers employ billions, if not trillions of memory cells.

The most common type of general-purpose computer is the personal computer, which is based on the x86 architecture (below $3,000). Computers more powerful than the personal computer can be grouped in the workstation ($3,000 to $50,000 range) or the supercomputer categories (above $50,000). See the web site www.top500.org for a list of the fastest computers on the planet. These computers often employ multiple processors and have much more memory than the typical personal computer. The workstations and supercomputers are used for handling large amounts of information (business applications), running large simulations (weather forecasting), searching (www.google.com), or performing large calculations (scientific research). This book will not cover the general-purpose computer, although many of the basic principles of embedded systems do apply to all types of systems.

The I/O interfaces are a crucial part of an embedded system because they provide necessary functionality. Most personal computers have the same basic I/O devices (e.g., mouse, keyboard, video display, CD, USB, and hard drive.) In contrast, there is no common set of I/O that all embedded system have. The software together with the I/O ports and associated interface circuits give an embedded computer system its distinctive characteristics. A **device driver** is a set of software functions that facilitate the use of an I/O port. Another name for device driver is **application programmer interface** (API). In this book we will study a wide range of I/O ports supported by the TM4C microcontrollers. Parallel ports provide for digital input and outputs. Serial ports employ a wide range of formats and synchronization protocols. The serial ports can communicate with devices such as:

- Sensors
- Liquid Crystal Display (LCD) and light emitting diode (LED) displays
- Analog to digital converters (ADC) and digital to analog converters (DAC)
- Wireless devices like Bluetooth, ZigBee and Wifi.

Analog to digital converters convert analog voltages to digital numbers. Digital to analog converters convert digital numbers to analog voltages. The timer features include:

- Fixed rate periodic execution
- Square wave and Pulse Width Modulated outputs (PWM)
- Input capture used for period, frequency and pulse width measurement

1.3. The Design Process

1.3.1. Requirements document

Before beginning any project, it is a good idea to have a plan. The following is one possible outline of a requirements document. Although originally proposed for software projects, it is appropriate to use when planning an embedded system, which includes software, electronics, and mechanical components. IEEE publishes a number of templates that can be used to define a project (IEEE STD 830-1998). A requirements document states what the system will do. It does not state how the system will do it. The main purpose of a requirements document is to serve as an agreement between you and your clients describing what the system will do. This agreement can become a legally binding contract. Write the document so that it is easy to read and understand by others. It should be unambiguous, complete, verifiable, and modifiable.

The requirements document should not include how the system will be designed. This allows the engineer to make choices during the design to minimize cost and maximize performance. Rather it should describe the problem being solved and what the system actually does. It can include some constraints placed on the development process. Ideally, it is co-written by both the engineers and the non-technical clients. However, it is imperative that both the engineers and the clients understand and agree on the specifics in the document.

1. Overview
    1.1. Objectives: Why are we doing this project? What is the purpose?
    1.2. Process: How will the project be developed?
    1.3. Roles and Responsibilities: Who will do what? Who are the clients?
    1.4. Interactions with Existing Systems: How will it fit in?
    1.5. Terminology: Define terms used in the document.
    1.6. Security: How will intellectual property be managed?
2. Function Description
    2.1. Functionality: What will the system do precisely?
    2.2. Scope: List the phases and what will be delivered in each phase.
    2.3. Prototypes: How will intermediate progress be demonstrated?
    2.4. Performance: Define the measures and describe how they will be determined.
    2.5. Usability: Describe the interfaces. Be quantitative if possible.
    2.6. Safety: Explain any safety requirements and how they will be measured.
3. Deliverables
    3.1. Reports: How will the system be described?
    3.2. Audits: How will the clients evaluate progress?
    3.3. Outcomes: What are the deliverables? How do we know when it is done?

Observation: To build a system without a requirements document means you are never wrong, but never done.

1.3.2. Modeling

One of the common threads in the example embedded systems presented in Section 1.2 is the need to understand the behavior of the physical system with which the embedded system interacts. Sometimes this understanding is only human intuition. However, the design process will be much more successful if this understanding can be represented in mathematical form. Scientists strive to describe physical processes with closed-form mathematical equations. For example, Newton's second law for damped harmonic oscillators is

$$F(t) = m\frac{d^2x}{dt^2} + c\frac{dx}{dt} + kx$$

where x is the one-dimensional position of the object (m), t is time (s), F is the applied force (N), m is the mass of the object (kg), c is called the viscous damping coefficient (kg/s), and k is the spring constant (N/m). Another example is Maxwell–Faraday equation (or Faraday's law of induction)

$$\nabla \times \boldsymbol{E} = -\frac{\partial \boldsymbol{B}}{\partial t} \quad \text{or} \quad \oint_C \boldsymbol{E} \cdot d\boldsymbol{l} = -\frac{\partial \Phi_{B,S}}{\partial t}$$

where $\boldsymbol{E}$ is the electric field (V/m), $\boldsymbol{B}$ is the magnetic field (Wb/m^2), C is the closed curve along the boundary of surface S, $d\boldsymbol{l}$ is differential vector element of path length tangential to the path/curve (m), and $\Phi_{B,S}$ magnetic flux through any surface S (Wb). A third example is heat conduction

$$\nabla \cdot (k\nabla T) + q = \rho c \frac{\partial T}{\partial t} \quad \text{or} \quad \frac{\partial}{\partial x}\left(k\frac{\partial T}{\partial x}\right) + q = \rho c \frac{\partial T}{\partial t}$$

where k is thermal conductivity (W/m/°C), T is temperature (°C), x is one-dimensional distance(m), q is internal heat generations (W/m^3), ρ is density (kg/m^3), c is specific heat at constant pressure (W-s/kg/°C) and t is time (s). The system is **causal** if its output depends only on current and past inputs. Let $S(x)$ define the output of a model for an input x. A system is **linear** if $S(ax_1+bx_2) = aS(x_1)+bS(x_2)$. A linear time-invariant system (LTI) is a system that is both linear and time invariant.

Some of the difficulties in solving closed form equations such as these include multidimensional space, irregular boundaries, and non-constant properties. These difficulties can be overcome using computational methods such as the finite element method (FEM). Still many problems remain. Inaccuracies in property values cause errors in the computational method. The biggest problem however is in the equations themselves. Many important real life problems exhibit nonlinear behavior not described by scientific equations.

Observation: It is important to properly model the boundary conditions.

Observation: It is important to use accurate physical parameters when modelling.

Consequentially, engineers tend to use empirical models of the world with which the embedded system interacts. The parameters of an **empirical model** are determined by experimental measurement under conditions similar to how the system will be deployed. Typically the models are discrete in time, because the measurements are discrete in time. The models can be linear or nonlinear as needed. These models often have memory, meaning the outputs are a function of both the current inputs and previous inputs/outputs. One of the simplest measures of **stability** is called bounded-input bounded-output, which means if all input signals are bounded then all output signals will also be bounded. For example, **performance maps** are used in engine control to optimize performance. They are empirical equations relating control parameters (such as applied power) and measured parameters (such as shaft rotational speed) to desired output parameters (such as generated torque). Even if difficult, it is appropriate to develop an abstract model describing the interaction between embedded system and the real world. We will present some models when designing more complex systems later in the book.

1.3.3. Top-down design

In this section, we will present the **top-down design** process. The process is called top-down, because we start with the high-level designs and work down to low-level implementations. The basic approach is introduced here, and the details of these concepts will be presented throughout the remaining chapters of the book. As we learn software/hardware development tools and techniques, we can place them into the framework presented in this section. As illustrated in Figure 1.5, the development of a product follows an analysis-design-implementation-testing cycle. For complex systems with long life-spans, we traverse multiple times around the development cycle. For simple systems, a one-time pass may suffice. Even after a system is deployed, it can reenter the life cycle to add features or correct mistakes.

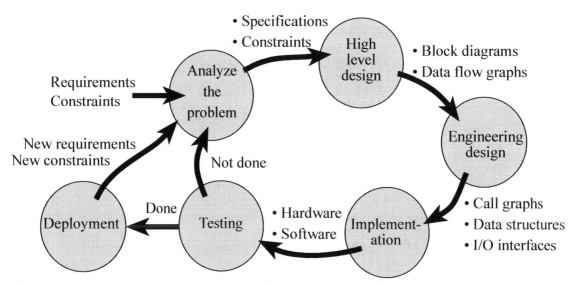

Figure 1.5. System development cycle or life-cycle. After the system is done it can be deployed.

During the **analysis phase**, we discover the requirements and constraints for our proposed system. We can hire consultants and interview potential customers in order to gather this critical information. A **requirement** is a specific parameter that the system must satisfy, describing what the system should do. We begin by rewriting the system requirements, which are usually written as a requirements document. In general, **specifications** are detailed parameters describing how the system should work. For example, a requirement may state that the system should fit into a pocket, whereas a specification would give the exact size and weight of the device. For example, suppose we wish to build a motor controller. During the analysis phase, we would determine obvious specifications such as range, stability, accuracy, and response time. The following measures are often considered during the analysis phase:

Safety: The risk to humans or the environment.
Accuracy: The difference between the expected truth and the actual parameter
Precision: The number of distinguishable measurements
Resolution: The smallest change that can be reliably detected
Response time: The time between a triggering event and the resulting action
Bandwidth: The amount of information processed per time
Signal to noise ratio: The quotient of the signal amplitude divided by the noise
Maintainability: The flexibility with which the device can be modified
Testability: The ease with which proper operation of the device can be verified
Compatibility: The conformance of the device to existing standards
Mean time between failure: The reliability of the device defining the life if a product
Size and weight: The physical space required by the system and its mass
Power: The amount of energy it takes to operate the system
Nonrecurring engineering cost (NRE cost): The one-time cost to design and test
Unit cost: The cost required to manufacture one additional product
Time-to-prototype: The time required to design build and test an example system
Time-to-market: The time required to deliver the product to the customer
Human factors: The degree to which our customers enjoy/like/appreciate the product

There are many parameters to consider and their relative importance may be difficult to ascertain. For example, in consumer electronics the human interface can be more important than bandwidth or signal to noise ratio. Often, improving the performance on one parameter can be achieved only by decreasing the performance of another. This art of compromise defines the tradeoffs an engineer must make when designing a product. A **constraint** is a limitation, within which the system must operate. The system may be constrained to such factors as cost, safety, compatibility with other products, use of specific electronic and mechanical parts as other devices, interfaces with other instruments and test equipment, and development schedule.

Checkpoint 1.11: What's the difference between a requirement and a specification?

When you write a paper, you first decide on a theme, and next you write an outline. In the same manner, if you design an embedded system, you define its specification (what it does), and begin with an organizational plan. In this section, we will present three graphical tools to describe the organization of an embedded system: data flow graphs, call graphs, and flowcharts. You should draw all three for every system you design.

During the **high-level design phase**, we build a conceptual model of the hardware/software system. It is in this model that we exploit as much abstraction as appropriate. The project is broken in modules or subcomponents. Modular design will be presented in Chapter 3. During this phase, we estimate the cost, schedule, and expected performance of the system. At this point we can decide if the project has a high enough potential for profit. A **data flow graph** is a block diagram of the system, showing the flow of information. Arrows point from source to destination. It is good practice to label the arrows with the information type and bandwidth. The rectangles represent hardware components and the ovals are software modules. We use data flow graphs in the high-level design, because they describe the overall operation of the system while hiding the details of how it works. Issues such as safety (e.g., Isaac Asimov's first Law of Robotics "*A robot may not harm a human being, or, through inaction, allow a human being to come to harm*") and testing (e.g., we need to verify our system is operational) should be addressed during the high-level design.

An example data flow graph for a motor controller is shown in Figure 1.6. Notice that the arrows are labeled with data type and bandwidth. The requirement of the system is to deliver power to a motor so that the speed of the motor equals the desired value set by the operator using a keypad. In order to make the system easier to use and to assist in testing, a liquid crystal display (LCD) is added. The sensor converts motor speed an electrical voltage. The amplifier converts this signal into the 0 to +3.3 V voltage range required by the ADC. The ADC converts analog voltage into a digital sample. The ADC routines, using the ADC and timer hardware, collect samples and calculate voltages. Next, this software uses a table data structure to convert voltage to measured speed. The user will be able to select the desired speed using the keypad interface. The desired and measured speed data are passed to the controller software, which will adjust the power output in such a manner as to minimize the difference between the measured speed and the desired speed. Finally, the power commands are output to the actuator module. The actuator interface converts the digital control signals to power delivered to the motor. The measured speed and speed error will be sent to the LCD module.

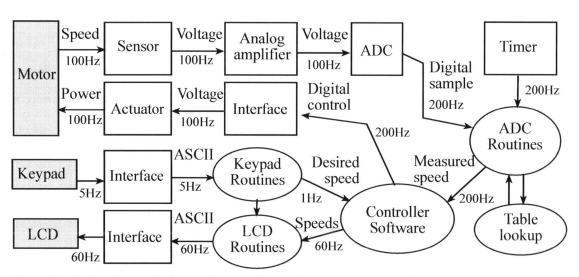

Figure 1.6. A data flow graph showing how signals pass through a motor controller.

The next phase is **engineering design**. We begin by constructing a preliminary design. This system includes the overall top-down hierarchical structure, the basic I/O signals, shared data structures and overall software scheme. At this stage there should be a simple and direct correlation between the hardware/software systems and the conceptual model developed in the high-level design. Next, we finish the top-down hierarchical structure, and build mock-ups of the mechanical parts (connectors, chassis, cables etc.) and user software interface. Sophisticated 3-D CAD systems can create realistic images of our system. Detailed hardware designs must include mechanical drawings. It is a good idea to have a second source, which is an alternative supplier that can sell our parts if the first source can't deliver on time. A **call graph** is a directed graph showing the calling relationships between software and hardware modules. If a function in module A calls a function in module B, then we draw an arrow from A to B. If a function in module A input/outputs data from hardware module C, then we draw an arrow from A to C. If hardware module C can cause an interrupt, resulting in software running in module A, then we draw an arrow from C to A. A hierarchical system will have a tree-structured call graph.

A call graph for this motor controller is shown in Figure 1.7. Again, rectangles represent hardware components and ovals show software modules. An arrow points from the calling routine to the module it calls. The I/O ports are organized into groups and placed at the bottom of the graph. A high-level call graph, like the one shown in Figure 1.7, shows only the high-level hardware/software modules. A detailed call graph would include each software function and I/O port. Normally, hardware is passive and the software initiates hardware/software communication, but as we will learn in Chapter 5, it is possible for the hardware to interrupt the software and cause certain software modules to be run. In this system, the timer hardware will cause the ADC software to collect a sample at a regular rate. The controller software calls the keypad routines to get the desired speed, calls the ADC software to get the motor speed at that point, determines what power to deliver to the motor and updates the actuator by sending the power value to the actuator interface. The controller software calls the LCD routines to display the status of the system. Acquiring data, calculating parameters, outputting results at a regular rate is strategic when performing digital signal processing in embedded systems.

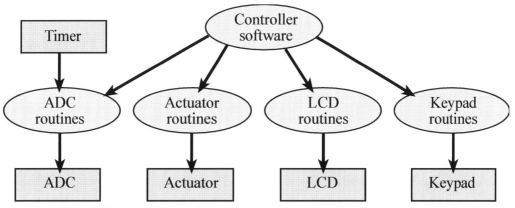

Figure 1.7. A call graph for a motor controller.

Checkpoint 1.12: What confusion could arise if two software modules were allowed to access the same I/O port? This situation would be evident on a call graph if the two software modules had arrows pointing to the same I/O port.

Observation: If module A calls module B, and B returns data, then a data flow graph will show an arrow from B to A, but a call graph will show an arrow from A to B.

Data structures include both the organization of information and mechanisms to access the data. Again safety and testing should be addressed during this low-level design.

The next phase is **implementation**. An advantage of a top-down design is that implementation of subcomponents can occur concurrently. The most common approach to developing software for an embedded system is to use a cross-assembler or **cross-compiler** to convert source code into the machine code for the target system. The machine code can then be loaded into the target machine. Debugging embedded systems with this simple approach is very difficult for two reasons. First, the embedded system lacks the usual keyboard and display that assist us when we debug regular software. Second, the nature of embedded systems involves the complex and real-time interaction between the hardware and software. These real-time interactions make it impossible to test software with the usual single-stepping and print statements.

The next technological advancement that has greatly affected the manner in which embedded systems are developed is **simulation**. Because of the high cost and long times required to create hardware prototypes, many preliminary feasibility designs are now performed using hardware/software simulations. A simulator is a software application that models the behavior of the hardware/software system. If both the external hardware and software program are simulated together, even although the simulated time is slower than the clock on the wall, the real-time hardware/software interactions can be studied.

During the initial iterations of the development cycle, it is quite efficient to implement the hardware/software using simulation. One major advantage of simulation is that it is usually quicker to implement an initial product on a simulator versus constructing a physical device out of actual components. Rapid prototyping is important in the early stages of product development. This allows for more loops around the analysis-design-implementation-testing cycle, which in turn leads to a more sophisticated product.

During the **testing** phase, we evaluate the performance of our system. First, we debug the system and validate basic functions. Next, we use careful measurements to optimize performance such as static efficiency (memory requirements), dynamic efficiency (execution speed), accuracy (difference between expected truth and measured), and stability (consistent operation.) Debugging techniques will be presented throughout the book. Testing is not performed at the end of project when we think we are done. Rather testing must be integrated into all phases of the design cycle. Once tested the system can be **deployed**.

Maintenance is the process of correcting mistakes, adding new features, optimizing for execution speed or program size, porting to new computers or operating systems, and reconfiguring the system to solve a similar problem. No system is static. Customers may change or add requirements or constraints. To be profitable, we probably will wish to tailor each system to the individual needs of each customer. Maintenance is not really a separate phase, but rather involves additional loops around the development cycle.

1.3.4. Flowcharts

In this section, we introduce the flowchart syntax that will be used throughout the book. Programs themselves are written in a linear or one-dimensional fashion. In other words, we type one line of software after another in a sequential fashion. Writing programs this way is a natural process, because the computer itself usually executes the program in a top-to-bottom sequential fashion. This one-dimensional format is fine for simple programs, but conditional branching and function calls may create complex behaviors that are not easily observed in a linear fashion. Even the simple systems have multiple software tasks. Furthermore, a complex application will require multiple microcontrollers. Therefore, we need a multi-dimensional way to visualize software behavior. **Flowcharts** are one way to describe software in a two-dimensional format, specifically providing convenient mechanisms to visualize multi-tasking, branching, and function calls. Flowcharts are very useful in the initial design stage of a software system to define complex algorithms. Furthermore, flowcharts can be used in the final documentation stage of a project in order to assist in its use or modification.

Figures throughout this section illustrate the syntax used to draw flowcharts. The oval shapes define **entry** and **exit points**. The main entry point is the starting point of the software. Each function, or subroutine, also has an entry point, which is the place the function starts. If the function has input parameters they are passed in at the entry point. The exit point returns the flow of control back to the place from which the function was called. If the function has return parameters they are returned at the exit point. When the software runs continuously, as is typically the case in an embedded system, there will be no main exit point.

We use rectangles to specify **process blocks**. In a high-level flowchart, a process block might involve many operations, but in a low-level flowchart, the exact operation is defined in the rectangle. The parallelogram will be used to define an **input/output operation**. Some flowchart artists use rectangles for both processes and input/output. Since input/output operations are an important part of embedded systems, we will use the parallelogram format, which will make it easier to identify input/output in our flowcharts. The diamond-shaped objects define a branch point or **decision block**. The rectangle with double lines on the side specifies a **call to a predefined function**. In this book, functions, subroutines and procedures are terms that all refer to a well-defined section of code that performs a specific operation. **Functions** usually return a result parameter, while **procedures** usually do not. Functions and procedures are terms used when describing a high-level language, while subroutines often used when describing assembly language. When a function (or subroutine or procedure) is called, the software execution path jumps to the function, the specific operation is performed, and the execution path returns to the point immediately after the function call. Circles are used as **connectors**.

> **Common error:** In general, it is bad programming style to develop software that requires a lot of connectors when drawing its flowchart.

There are a seemingly unlimited number of tasks one can perform on a computer, and the key to developing great products is to select the correct ones. Just like hiking through the woods, we need to develop guidelines (like maps and trails) to keep us from getting lost. One of the fundamental issues when developing software, regardless whether it is a microcontroller with 1000 lines of assembly code or a large computer system with billions of lines is to maintain a consistent structure. One such framework is called structured programming. A good high-level language will force the programmer to write structured programs. **Structured programs** are built from three basic building blocks: the sequence, the conditional, and the while-loop. At the lowest level, the process block contains simple and well-defined commands. I/O functions are also low-level building blocks. Structured programming involves combining existing blocks into more complex structures, as shown in Figure 1.8.

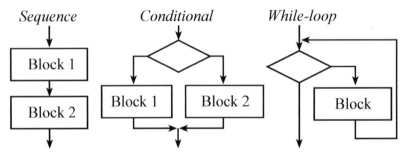

Figure 1.8. Flowchart showing the basic building blocks of structured programming.

Maintenance Tip: Remember to update the flowcharts as modifications are made to the software

Next, we will revisit the pacemaker example in order to illustrate the flowchart syntax. A **thread** is the sequence of actions caused by executing software. The flowchart in Figure 1.9 defines a single-threaded execution because there is one sequence.

Example 1.1 (continued): Use a flowchart to describe an algorithm that a pacemaker might use to regulate and improve heart function.

Solution: This example illustrates a common trait of an embedded system, that is, they perform the same set of tasks over and over forever. The program starts at main when power is applied, and the system behaves like a pacemaker until the battery runs out. Figure 1.9 shows a flowchart for a very simple algorithm. If the heart is beating normally with a rate greater than or equal to 1 beat/sec (60 BPM), then the atrial sensor will detect activity and the first decision will go right. Since this is normal beating, the ventricular activity will occur within the next 200 ms, and the ventricular sensor will also detect activity. In this situation, no output pulses will be issued. If the delay between atrial contraction and ventricular contract were longer than the normal 200 ms, then the pacemaker will activate the ventricles 200 ms after each atrial contraction. If the ventricle is beating faster than 60 BPM without any atrial contractions, then no ventricular stimulations will be issued. If there is no activity from either atrium or the ventricle (or if that rate is slower than 60 BPM), then the ventricles are paced at 60 BPM.

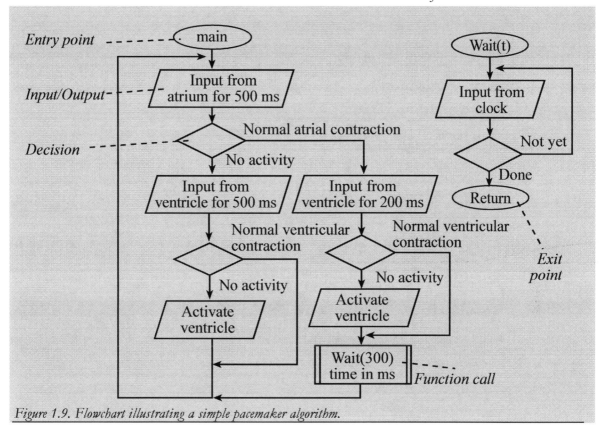

Figure 1.9. Flowchart illustrating a simple pacemaker algorithm.

Checkpoint 1.13: Assume you are given a simple watch that just tells you the time in hours, minutes, and seconds. Let t be an input parameter. Explain how you could use the watch to wait t seconds.

1.3.5. Parallel, distributed, and concurrent programming

Many problems cannot be implemented using the single-threaded execution pattern described in the previous section. **Parallel programming** allows the computer to execute multiple threads at the same time. State-of-the art multi-core processors can execute a separate program in each of its cores. **Fork** and **join** are the fundamental building blocks of parallel programming. After a fork, two or more software threads will be run in parallel. I.e., the threads will run simultaneously on separate processors.

Two or more simultaneous software threads can be combined into one using a join. The flowchart symbols for fork and join are shown in Figure 1.10. Software execution after the join will wait until all threads above the join are complete. As an analogy, if I want to dig a big hole in my back yard, I will invite three friends over and give everyone a shovel. The fork operation changes the situation from me working alone to four of us ready to dig. The four digging tasks are run in parallel. When the overall task is complete, the join operation causes the friends to go away, and I am working alone again. A complex system may employ multiple microcontrollers, each running its own software. We classify this configuration as parallel or **distributed programming**.

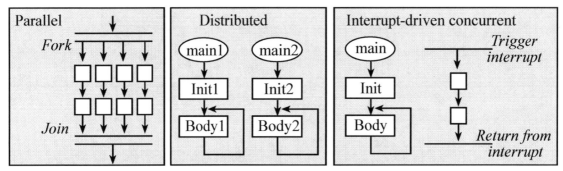

Figure 1.10. Flowchart symbols to describe parallel, distributed, and concurrent programming.

Concurrent programming allows the computer to execute multiple threads, but only one at a time. Interrupts are one mechanism to implement concurrency on real-time systems. **Interrupts** have a hardware trigger and a software action. An interrupt is a parameter-less subroutine call, triggered by a hardware event. The flowchart symbols for interrupts are also shown in Figure 1.10. The trigger is a hardware event signaling it is time to do something. Examples of interrupt triggers we will see in this book include new input data has arrived, output device is idle, and periodic event. The second component of an interrupt-driven system is the software action called an interrupt service routine (ISR). The foreground thread is defined as the execution of the main program, and the background threads are executions of the ISRs.

Consider the analogy of sitting in a comfy chair reading a book. Reading a book is like executing the main program in the foreground. Because there is only one of you, this scenario is analogous to a computer with one processor. You start reading at the beginning of the book and basically read one page at a time in a sequential fashion. You might jump to the back and look something up in the glossary, then jump back to where you were, which is analogous to a function call. Similarly, if you might read the same page a few times, which is analogous to a program loop. Even though you skip around a little, the order of pages you read follows a logical and well-defined sequence. Conversely, if the telephone rings, you place a bookmark in the book, and answer the phone. When you are finished with the phone conversation, you hang up the phone and continue reading in the book where you left off. The ringing phone is analogous to hardware trigger and the phone conversation is like executing the ISR.

Notice in this analogy that there is one person who does multiple tasks by doing one task, halting that task, doing a second task to completion, and returning to the original task. In the computer world we have one processor that does one task running the main program, halts that task, does a second task to completion running the ISR, and then returns to the main program.

Example 1.2 (continued): Use a flowchart to describe an algorithm that a stand-alone smoke detector might use to warn people in the event of a fire.

Solution: This example illustrates a common trait of a low-power embedded system. The system begins with a power on reset, causing it to start at main. The initialization enables the timer interrupts, and then it shuts off the alarm. In a low-power system the microcontroller goes to sleep when there are no tasks to perform. Every 30 seconds the timer interrupt wakens the microcontroller and executes the interrupt service routine. The first task is to read the smoke sensor. If there is no fire, it will flash the LED and return from interrupt. At this point, the main program will put the microcontroller back to sleep. The letters (A-K) in Figure 1.11 specify the software activities in this multithreaded example. Initially it executes A-B-C and goes to sleep. Every 30 seconds, assuming there is no fire, it executes <-D-E-F-G-J-K-J-K-····-J-K-H->C This sequence will execute in about 1 ms, dominated by the time it takes to flash the LED. This is a low-power solution because the microcontroller is powered for about 0.003% of the time, or 1 ms every 30 seconds. We could reduce power even more by sleeping during the 1ms the LED is on.

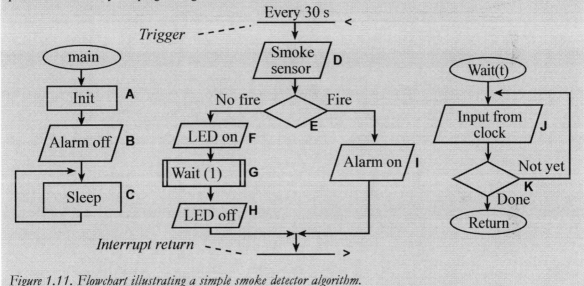

Figure 1.11. Flowchart illustrating a simple smoke detector algorithm.

To illustrate the concept of parallel programming, assume we have a multi-core computer with four processors. Consider the problem of finding the maximum value in a large buffer. First, we divide the buffer into four equal parts. Next, we execute a fork, as shown in the left-most flowchart in Figure 1.10, launching four parallel threads. The four processors run in parallel each finding the maximum of its subset. When all four threads are complete, they perform a join and combine the four results to find the overall maximum. It is important to distinguish parallel programming like this from multithreading implementing concurrent processing with interrupts. Because most microcontrollers have a single processor, this book with focus on concurrent processing with interrupts and distributed processing with a network involving multiple microcontrollers.

1.3.6. Creative discovery using bottom-up design

Figure 1.5 describes top-down design as a cyclic process, beginning with a problem statement and ending up with a solution. With a **bottom-up design** we begin with solutions and build up to a problem statement. Many innovations begin with an idea, "what if...?" In a bottom-up design, one begins with designing, building, and testing low-level components. Figure 1.12 illustrates a two-level process, combining three subcomponents to create the overall product. This hierarchical process could have more levels and/or more components at each level. The low-level designs can occur in parallel. The design of each component is cyclic, iterating through the design-build-test cycle until the performance is acceptable.

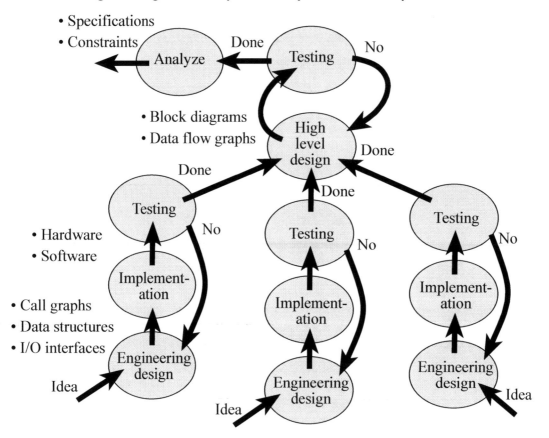

Figure 1.12. System development process illustrating bottom-up design.

Bottom-up design is inefficient because some subsystems are designed, built, and tested, but never used. Furthermore, in a truly creative environment most ideas cannot be successfully converted to operational subsystems. Creative laboratories are filled with finished, half-finished, and failed subcomponents. As the design progresses the components are fit together to make the system more and more complex. Only after the system is completely built and tested does one define its overall specifications.

The bottom-up design process allows creative ideas to drive the products a company develops. It also allows one to quickly test the feasibility of an idea. If one fully understands a problem area and the scope of potential solutions, then a top-down design will arrive at an effective solution most quickly. On the other hand, if one doesn't really understand the problem or the scope of its solutions, a bottom-up approach allows one to start off by learning about the problem.

Observation: A good engineer knows both bottom-up and top-down design methods, choosing the approach most appropriate for the situation at hand.

1.4. Digital Logic and Open Collector

Digital logic has two states, with many enumerations such as high and low, 1 and 0, true and false, on and off. There are four currents of interest, as shown in Figure 1.13, when analyzing if the inputs of the next stage are loading the output. I_{IH} and I_{IL} are the currents required of an input when high and low respectively. Furthermore, I_{OH} and I_{OL} are the maximum currents available at the output when high and low. In order for the output to properly drive all the inputs of the next stage, the maximum available output current must be larger than the sum of all the required input currents for both the high and low conditions.

$$|I_{OL}| \geq \sum |I_{IL}| \qquad \text{and} \qquad |I_{OH}| \geq \sum |I_{IH}|$$

Absolute value operators are put in the above relations because data sheets are inconsistent about specifying positive and negative currents. The arrows in Figure 1.13 define the direction of current regardless of whether the data sheet defines it as a positive or negative current. It is your responsibility to choose parts such that the above inequalities hold.

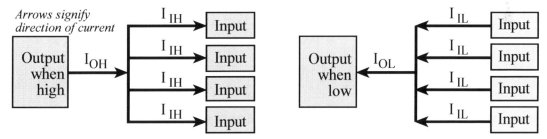

Figure 1.13. Sometimes one output must drive multiple inputs.

Kirchhoff's Current Law (KCL) states the sum of all the currents into one node must be zero. The above inequalities are not a violation of KCL, because the output currents are the available currents and the input currents are the required currents. Once the system is built and running, the actual output current will of course exactly equal the sum of the actual input currents. As a matter of completeness, we include **Kirchhoff's Voltage Law** (KVL), which states the sum of all the voltages in a closed loop must be zero. Table 1.4 shows typical current values for the various digital logic families. The TM4C123 microcontrollers give you two choices of output current for the digital output pins. The TM4C1294 adds a third 12-mA mode.

Family	Example	I_{OH}	I_{OL}	I_{IH}	I_{IL}
Standard TTL	7404	0.4 mA	16 mA	40 µA	1.6 mA
Schottky TTL	74S04	1 mA	20 mA	50 µA	2 mA
Low Power Schottky	74LS04	0.4 mA	4 mA	20 µA	0.4 mA
High Speed CMOS	74HC04	4 mA	4 mA	1 µA	1 µA
Adv High Speed CMOS	74AHC04	4 mA	4 mA	1 µA	1 µA
TM4C 2mA-drive	TM4C123	2 mA	2 mA	2 µA	2 µA
TM4C 4mA-drive	TM4C123	4 mA	4 mA	2 µA	2 µA
TM4C 8mA-drive	TM4C123	8 mA	8 mA	2 µA	2 µA
TM4C 12mA-drive	MSP432E4	12 mA	12 mA	2 µA	2 µA

Table 1.4. The input and output currents of various digital logic families and microcontrollers.

Observation: For TTL devices the logic low currents are much larger than the logic high currents.

When we design circuits using devices all from a single logic family, we can define **fan out** as the maximum number of inputs, one output can drive. For transistor-transistor logic (TTL) logic we can calculate fan out from the input and output currents:

Fan out = minimum((I_{OH}/I_{IH}) , (I_{OL}/I_{IL}))

Conversely, the fan out of high-speed complementary metal-oxide semiconductor (CMOS) devices, which includes most microcontrollers, is determined by **capacitive loading** and not by the currents. Figure 1.14 shows a simple circuit model of a CMOS interface. The ideal voltage of the output device is labeled V_1. For interfaces in close proximity, the resistance R results from the output impedance of the output device, and the capacitance C results from the input capacitance of the input device. However, if the interface requires a cable to connect the two devices, both the resistance and capacitance will be increased by the cable. The voltage labeled V_2 is the effective voltage as seen by the input. If V_2 is below 1.3 V, the TM4C microcontrollers will interpret the signal as low. Conversely, the voltage is above 2.0 V, these microcontrollers will consider it high. The **slew rate** of a signal is the slope of the voltage versus time during the time when the logic level switches between low and high. A similar parameter is the **transition time**, which is the time it takes for an output to switch from one logic level to another. In Figure 1.14, the transition time is defined as the time it takes V_2 to go from 1.3 to 2.0 V. There is a capacitive load for the output and each input. As this capacitance increases the slew rate decreases, which will increase the transition time. Signals with a high slew rate can radiate a lot of noise. So, to reduce noise emissions we sometimes limit the slew rate of the signals.

There are two ways to determine the fan out of CMOS circuits. First, some circuits have a minimum time its input can exist in the transition range. For example, it might specify the signal cannot be above 1.3 and below 2.0 V for more than 20 ns. Clock inputs are often specified this way. A second way is to calculate the time constant τ, which is $R*C$ for this circuit. Let T be the pulse width of the digital signal. If T is large compared to τ, then the CMOS interface functions properly. For circuits that mix devices from one family with another, we must look individually at the input and output currents, voltages and capacitive loads. There is no simple formula.

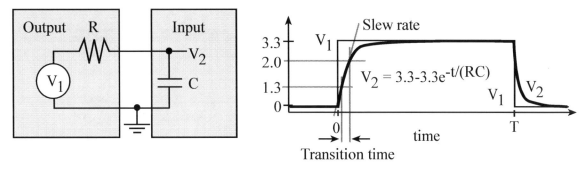

Figure 1.14. Capacitance loading is an important factor when interfacing CMOS devices.

Figure 1.15 compares the input and output voltages for many of the digital logic families. V_{IL} is the voltage below which an input is considered a logic low. Similarly, V_{IH} is the voltage above which an input is considered a logic high. The output voltage depends strongly on current required to drive the inputs of the next stage. V_{OH} is the output voltage when the signal is high. In particular, if the output is a logic high, and the current is less than I_{OH}, then the voltage will be greater than V_{OH}. Similarly, V_{OL} is the output voltage when the signal is low. In particular, if the output is a logic low, and the current is less than I_{OL}, then the voltage will be less than V_{OL}. The digital input pins on the TM4C microcontrollers are **5V-tolerant**, meaning an input high signal can be any voltage from 2.145 to 5.0 V.

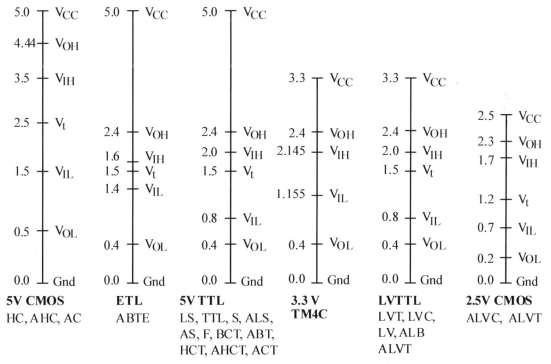

Figure 1.15. Voltage thresholds for various digital logic families.

1. Introduction to Embedded Systems

For the output of one circuit to properly drive the inputs of the next circuit, the output low voltage needs to be low enough, and the output high voltage needs to be high enough.

$$V_{OL} \leq V_{IL} \text{ for all inputs} \quad \text{and} \quad V_{OH} \geq V_{IH} \text{ for all inputs}$$

The maximum output current specification on the TM4C family is 25 mA, which is the current above which will cause damage. However, we can select I_{OH} and I_{OL} to be 2, 4, or 8 mA (also 12 mA on the MSP432E401Y). Normally, we design the system so the output currents are less than I_{OH} and I_{OL}. V_t is the typical threshold voltage, which is the voltage at which the input usually switches between logic low and high. Formally however, an input is considered in the transition region (value indeterminate) for voltages between V_{IL} and V_{IH}. The five parameters that affect our choice of logic families are

- Power supply voltage (e.g., +5V, 3.3V etc.)
- Power supply current (e.g., will the system need to run on batteries?)
- Speed (e.g., clock frequency and propagation delays)
- Output drive, I_{OL}, I_{OH} (e.g., does it need to drive motors or lights?)
- Noise immunity (e.g., electromagnetic field interference)
- Temperature (e.g., electromagnetic field interference)

Checkpoint 1.14: How will the TM4C123 interpret an input pin as the input voltage changes from 0, 1, 2, 3, 4, to 5V? I.e., for each voltage, will it be considered as a logic low, as a logic high or as indeterminate?

Checkpoint 1.15: Considering both voltage and current, can the output of a 74HC04 drive the input of a 74LS04? Assume both are running at 5V.

Checkpoint 1.16: Considering both voltage and current, can the output of a 74LS04 drive the input of a 74HC04? Assume both are running at 5V.

A very important concept used in computer technology is **tristate logic**, which has three output states: high, low, and off. Other names for the off state are HiZ, floating, and tristate. Tristate logic is drawn as a triangle shape with a signal on the top of the triangle. In this Figure 1.16, A is the data input, G is the gate input, and B is the data output. When there is no circle on the gate, it operates in positive logic, meaning if the gate is high, then the output data equals the input data. If the positive-logic gate is low, then the output will float. When there is a circle on the gate, it operates in negative logic, meaning if the gate is low, then the output data equals the input data. If the negative-logic gate is high, then the output will float.

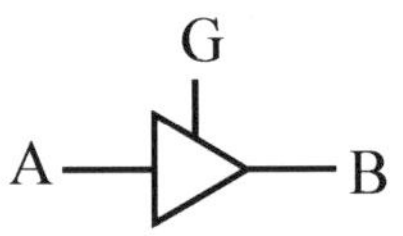

A	G	B
0	0	Z
1	0	Z
0	1	0
1	1	1

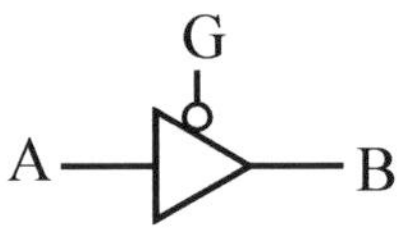

A	G	B
0	0	0
1	0	1
0	1	Z
1	1	Z

Figure 1.16. Digital logic drawing of tristate drivers.

There are a wide range of technologies available for digital logic design. To study these differences consider the 74LVT245, the 74ALVC245, the 74LVC245, the 74ALB245, the 74AC245, the 74AHC245, and the 74LV245.

Each of these chips is an 8-input 8-output bidirectional tristate driver. Table 1.5 lists some of the interfacing parameters for each technology. I_{CC} is the total supply current required to drive the chip. t_{pd} is the propagation delay from input to output. V_{CC} is the supply voltage.

Family Technology	V_{IL} V_{IH}	V_{OL} V_{OH}	I_{OL}	I_{OH}	I_{CC}	t_{pd}
LVT - Low-Voltage BiCMOS	LVTTL	LVTTL	64	-32	190	3.5
ALVC - Advanced Low-Voltage CMOS	LVTTL	LVTTL	24	-24	40	3.0
LVC - Low-Voltage CMOS	LVTTL	LVTTL	24	-24	10	4.0
ALB - Advanced Low-Voltage BiCMOS	LVTTL	LVTTL	25	-25	800	2.0
AC - Advanced CMOS	CMOS	CMOS	12	-12	20	8.5
AHC - Advanced High Speed CMOS	CMOS	CMOS	4	-4	20	11.9
LV - Low-Voltage CMOS	LVTTL	LVTTL	8	-8	20	14
units			mA	mA	µA	ns

Table 1.5. Comparison of the output drive, supply current and speed of various 3.3V logic '245 gates.

Observation: There is an inverse relationship between supply current I_{CC} and propagation delay t_{pd}.

The 74LS04 is a low-power Schottky NOT gate, as shown on the left in Figure 1.17. It is called Schottky logic because the devices are made from Schottky transistors. The output is high when the transistor Q_4 is active, driving the output to V_{cc}. The output is low when the transistor Q_5 is active, driving the output to 0.

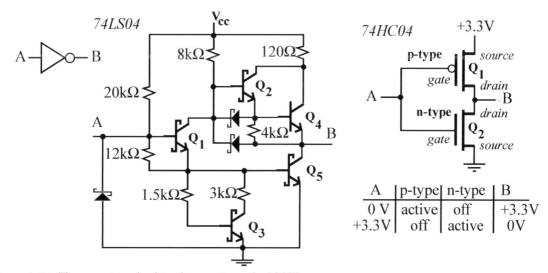

Figure 1.17. Two transistor-level implementations of a NOT gate.

It is obviously necessary to read the data sheet for your microcontroller. However, it is also good practice to review the errata published by the manufacturer about your microcontroller. The errata define situations where the actual chip does not follow specifications in the data sheet. For example, the regular TM4C123 data sheet states the I/O pins are +5V tolerant. However, reading the errata for the LM3S811 version C2 announces that "PB6, PC5, and PC6 are not 5-V tolerant."

The 74HC04 is a high-speed CMOS NOT gate, shown on the right in Figure 1.17. The output is high when the transistor Q_1 is active, driving the output to 3.3V. The output is low when the transistor Q_2 is active, driving the output to 0. Since most microcontrollers are made with high-speed CMOS logic, its outputs behave like the Q_1/Q_2 "push/pull" transistor pair. Output ports are not inverting. I.e., when you write a "1" to an output port, then the output voltage goes high. Similarly, when you write a "0" to an output port, then the output voltage goes low. Analyses of the circuit in Figure 1.17 reveal some of the basic properties of high-speed CMOS logic. First, because of the complementary nature of the P-channel (the one on the top) and N-channel (the one on the bottom) transistors, when the input is constant (continuously high or continuously low), the supply current, I_{cc}, is very low. Second, the gate will require supply current only when the output switches from low to high or from high to low. This observation leads to the design rule that the power required to run a high-speed CMOS system is linearly related to the frequency of its clock, because the frequency of the clock determines the number of transitions per second. Along the same lines, we see that if the voltage on input A exists between V_{IL} and V_{IH} for extended periods of time, then both Q_1 and Q_2 are partially active, causing a short from power to ground. This condition can cause permanent damage to the transistors. Third, since the input A is connected to the gate of the two MOS transistors, the input currents will be very small (≈ 1 μA). In other words, the input impedance (input voltage divide by input current) of the gate is very high. Normally, a high input impedance is a good thing, except if the input is not connected. If the input is not connected then it takes very little input currents to cause the logic level to switch.

> **Common error:** If unused input pins on a CMOS microcontroller are left unconnected, then the input signal may oscillate at high frequencies depending on the EM fields in the environment, wasting power unnecessarily.

> **Observation:** It is a good design practice to connect unused CMOS inputs to ground or connect them to +3.3V.

Now that we understand that CMOS digital logic is built with PNP and NPN transistors, we can revisit the interface requirements for connecting a digital output from one module to a digital input of another module. Figure 1.18 shows the model when the output is high. To make the output high, a PNP transistor in the output module is conducting (Q_1) driving +3.3 V to the output. The high voltage will activate the gate of NPN transistors in the input module (Q_4). The I_{IH} is the current into the input module needed to activate all gates connected to the input. The actual current I will be between 0 and I_{IH}. For a high signal, current flows from +3.3V, across the source-drain of Q_1, into the gate of Q_4, and then to ground. As the actual current I increases, the actual output voltage V will drop. I_{OH} is the maximum output current that guarantees the output voltage will be above V_{OH}. Assuming the actual I is less than I_{OH}, the actual voltage V will be between V_{OH} and +3.3V. If the input voltage is between V_{IH} and +3.3V, the input signal is considered high by the input. For the high signal to be transferred properly, V_{OH} must be larger than V_{IH} and I_{OH} must be larger than I_{IH}.

Figure 1.19 shows the model when the output is low. To make the output low, an NPN transistor in the output module is conducting (Q_2) driving the output to 0V. The low voltage will activate the gate of PNP transistors in the input module (Q_3). The I_{IL} is the current out of the input module needed to activate all gates connected to the input. The actual current I will be between 0 and I_{IL}. For a low signal, current flows from +3.3V in the input module, across the source-gate of Q_3, across the source-drain gate of Q_2, and then to ground. As the actual current I increases, the actual output voltage V will increase. I_{OL} is the maximum output current that guarantees the output voltage will be less than V_{OL}. Assuming the actual I is less than I_{OL}, the actual voltage V will be between 0 and V_{OL}. If the input voltage is between 0 and V_{IL}, the input signal is considered low by the input. For the low signal to be transferred properly, V_{OL} must be less than V_{IL} and I_{OL} must be larger than I_{IL}.

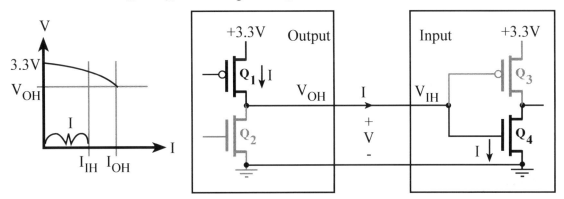

Figure 1.18. Model for the input/output characteristics when the output is high.

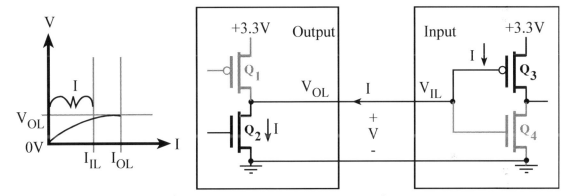

Figure 1.19. Model for the input/output characteristics when the output is low.

Open collector logic has outputs with two states: low and off. The 74LS05 is a low-power Schottky open collector NOT gate, as shown in Figure 1.20. When drawing logic diagrams, we add the 'x' on the output to specify open collector logic.

The 74HC05 is a high-speed CMOS open collector NOT gate is also shown in Figure 1.20. It is called open collector because the collector pin of Q_2 is not connected, or left open. The output is off when there is no active transistor driving the output. In other words, when the input is low, the output floats. This "not driven" condition is called the open collector state. When the input is high, the output will be low, caused by making the transistor Q_2 is active driving the output to 0. Technically, the 74HC05 implements **open drain** rather than open collector, because it is the drain pin of Q_2 that is left open. In this book, we will use the terms open collector and open drain interchangeably to refer to digital logic with two output states (low and off). Because of the multiple uses of open collector, many microcontrollers can implement open collector logic. On TM4C microcontrollers, we can affect this mode by defining an output as open drain.

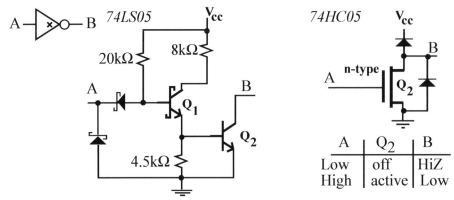

Figure 1.20. Two transistor implementations of an open collector NOT gate.

We can use a bipolar junction transistor (BJT) to source or sink current. For most of the circuits in this book the transistors are used in saturated mode. In general, we will use NPN transistors to sink current to ground. We turn on an NPN transistor by applying a positive V_{be}. This means when the NPN transistor is on, current flows from the collector to the emitter. When the NPN transistor is off, no current flows from the collector to the emitter. Each transistor has an input and output impedance, h_{ie} and h_{oe} respectively. The current gain is h_{fe} or β. The hybrid-pi small signal model for the bipolar NPN transistor is shown in Figure 1.21.

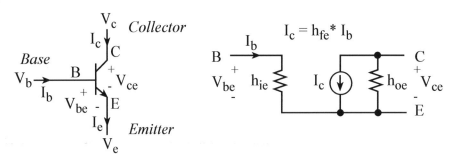

Figure 1.21. NPN transistor model.

There are five basic design rules when using individual bipolar NPN transistors in saturated mode:

1) Normally $V_c > V_e$
2) Current can only flow in the following directions
    from base to emitter (input current)
    from collector to emitter (output current)
    from base to collector (doesn't usually happen, but could if $V_b > V_c$)
3) Each transistor has maximum values for the following terms that should not be exceeded
    I_b I_c V_{ce} and $I_c \cdot V_{ce}$
4) The transistor acts like a current amplifier
    $I_c = h_{fe} \cdot I_b$
5) The transistor will activate if $V_b > V_e + V_{be(SAT)}$
    where $V_{be(SAT)}$ is typically above 0.6V

In general, we will use PNP transistors to source current from a positive voltage supply. We turn on a PNP transistor by applying a positive V_{eb}. This means when the PNP transistor is on, current flows from the emitter to the collector. When the PNP transistor is off, no current flows from the emitter to the collector. Each transistor has an input and output impedance, h_{ie} and h_{oe} respectively. The current gain is h_{fe} or β. The hybrid-pi small signal model for the bipolar PNP transistor is shown in Figure 1.22.

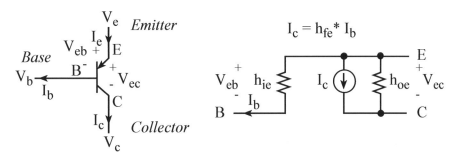

Figure 1.22. PNP transistor model.

There are five basic design rules when using individual bipolar PNP transistors in saturated mode:
1) Normally $V_e > V_c$
2) Current can only flow in the following directions
   from emitter to base (input current)
   from emitter to collector (output current)
   from collector to base (doesn't usually happen, but could if $V_c > V_b$)
3) Each transistor has maximum values for the following terms that should not be exceeded
   I_b I_c V_{ce} and $I_c \cdot V_{ce}$
4) The transistor acts like a current amplifier
   $I_c = h_{fe} \cdot I_b$
5) The transistor will activate if $V_b < V_e - V_{be(SAT)}$
   where $V_{be(SAT)}$ is typically above 0.6V

Performance Tip: A good transistor design is one that the input/output response is independent of h_{fe}. We can design a saturated mode circuit so that I_b is 2 to 5 times as large as needed to supply the necessary I_c.

The Table 1.6 illustrates the wide range of bipolar transistors that we can use.

Type	NPN	PNP	package	$V_{be(SAT)}$	$V_{ce(SAT)}$	h_{fe} min/max	I_c
general purpose	2N3904	2N3906	TO-92	0.85 V	0.2 V	100	10mA
general purpose	PN2222	PN2907	TO-92	1.2 V	0.3 V	100	150mA
general purpose	2N2222	2N2907	TO-18	1.2 V	0.3 V	100/300	500mA
power transistor	TIP29A	TIP30A	TO-220	1.3 V	0.7 V	15/75	1A
power transistor	TIP31A	TIP32A	TO-220	1.8 V	1.2 V	25/50	3A
power transistor	TIP41A	TIP42A	TO-220	2.0 V	1.5 V	15/75	3A
power darlington	TIP120	TIP125	TO-220	2.5 V	2.0 V	1000 min	3A

Table 1.6. Parameters of typical transistors used to source or sink current.

Under most conditions we place a resistor in series with the base of a BJT when using it as a current switch. The value of this resistor is typically 100Ω to 10kΩ. The purpose of this base resistor is to limit the current into the base. The h_{ie} is typically around 60 Ω. If you connect a microcontroller output port directly to the base of a BJT (without the resistor), then when the output is high it will try and generate a current of 3.3V/60Ω = 50 mA, potentially damaging the microcontroller. If there is a 1kΩ resistor between the microcontroller and base of the NPN, then the V_{be} voltage goes to V_{beSAT} (0.7) and (3.3-0.7)/1000Ω or 2.6 mA.

In general, we can use an open collector NOT gate to control the current to a device, such as a relay, a light emitting diode (LED), a solenoid, a small motor, and a small light. The 74HC05 can handle up to 4 mA. The 7405 and 7406 can handle up to 16 and 40 mA respectively, but they must be powered at +5V. For currents up to 150 mA we can use a PN2222 transistor, as shown in Figure 1.23, to create a low-cost but effective solution. When output of the microcontroller is high, the transistor is on, making its output low (V_{OL} or V_{CE}). In this state, a 10 mA current is applied to the diode, and it lights up. But when output of the microcontroller is low, the transistor is off, making its output float, which is neither high nor low. This floating output state causes the LED current to be zero, and the diode is dark. The resistor is selected to set the LED current. Assume the V_{CE} is 0.3V, and the desired LED operating point is 1.9V 10mA. In this case, the correct resistor value is (3.3-1.9-0.3V)/10mA = 110Ω.

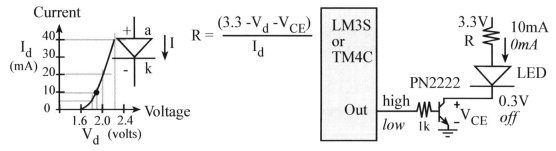

Figure 1.23. Open collector used to interface a light emitting diode.

Checkpoint 1.17: What resistor value would you choose to operate the LED at 2V 20mA?

When needed for digital logic, we can convert an open collector output to a digital signal using a pull-up resistor from the output to V_{CC}. In this way, when the open collector output floats, the signal will be a digital high. How do we select the value of the **pull-up resistor**? In general the smaller the resistor, the larger the I_{OH} it will be able to supply when the output is high. On the other hand, a larger resistor does not waste as much I_{OL} current when the output is low. One way to calculate the value of this pull-up resistor is to first determine the required output high voltage, V_{out}, and output high current, I_{out}. To supply a current of at least I_{out} at a voltage above V_{out}, the resistor must be less than:

$$R \leq (V_{CC} - V_{out})/I_{out}$$

As an example we will calculate the resistor value for the situation where the circuit needs to drive five regular TTL loads, with V_{CC} equal to 5V. We see from Figure 1.15 that V_{out} must be above V_{IH} (2V) in order for the TTL inputs to sense a high logic level. We can add a safety factor and set V_{out} at 3V. In order for the high output to drive all five TTL inputs, I_{out} must be more than five I_{IH}. From Table 1.4, we see that I_{IH} is 40 μA, so I_{out} should be larger than 5 times 40 μA or 0.2mA. For this situation, the resistor must be less than (5-3V)/0.2mA = 10 kΩ.

Another example of open collector logic occurs when interfacing switches to the microcontroller. The circuit on the left of Figure 1.24 shows a mechanical switch with one terminal connected to ground. In this circuit, when the switch is pressed, the voltage *r* is zero. When the switch is not processed, the signal *r* floats.

The circuit on the middle of Figure 1.24 shows the mechanical switch with a 10 kΩ pull-up resistor attached the other side. When the switch is pressed the voltage at *s* still goes to zero, because the resistance of the switch (less than 0.1Ω) is much less than the pull-up resistor. But now when the switch is not pressed the pull-up resistor creates a +3.3V at *s*. This circuit is shown connected to an input pin of the microcontroller. The software, by reading the input port, can determine whether or not the switch is pressed. If the switch is pressed the software will read zero, and if the switch is not pressed the software will read one. This middle circuit is called **negative logic** because the active state, switch is being pressed, has a lower voltage than the inactive state.

The circuit on the right of Figure 1.24 also interfaces a mechanical switch to the microcontroller, but it implements positive logic using a pull-down resistor. The signal *t* will be high if the switch is pressed and low if it is released. This right circuit is called **positive logic** because the active state, switch is being pressed, has a higher voltage than the inactive state.

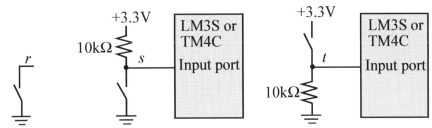

Figure 1.24. Single Pole Single Throw (SPST) Switch interface.

> **Observation:** We can activate pull-up or pull-down resistors on the ports on most microcontrollers, so the interfaces in Figure 1.24 can be made without the external resistor.

Earlier we used a voltage and current argument to determine the value of the pull-up resistor. In this section we present another method one could use to select this resistor. The TM4C microcontrollers have an input current of 2 μA. At 3.3 V, this is the equivalent of an input impedance of about 1 MΩ. A switch has an on-resistance of less than 1 Ω. We want the resistor to be small when compared to 1 MΩ, but large compared to 1 Ω. The 10 kΩ pull-up resistor is 100 times smaller than the input impedance and 10,000 times larger than the switch resistance. For the TM4C, the internal pull-up resistor ranges from 13 to 30 kΩ, and the internal pull-down resistor ranges from 13 to 35 kΩ.

1.5. Digital Representation of Numbers

1.5.1. Fundamentals

Information is stored on the computer in binary form. A binary bit can exist in one of two possible states. In **positive logic**, the presence of a voltage is called the '1', true, asserted, or high state. The absence of a voltage is called the '0', false, not asserted, or low state. Conversely in **negative logic**, the true state has a lower voltage than the false state. Figure 1.25 shows the output of a typical complementary metal oxide semiconductor (CMOS) circuit. The left side shows the condition with a true bit, and the right side shows a false. The output of each digital circuit consists of a p-type transistor "on top of" an n-type transistor. In digital circuits, each transistor is essentially on or off. If the transistor is on, it is equivalent to a short circuit between its two output pins. Conversely, if the transistor is off, it is equivalent to an open circuit between its outputs pins. On TM4C microcontrollers powered with 3.3 V supply, a voltage between 2 and 5 V is considered high, and a voltage between 0 and 1.3 V is considered low. Separating the two regions by 0.7 V allows digital logic to operate reliably at very high speeds. The design of transistor-level digital circuits is beyond the scope of this book. However, it is important to know that digital data exist as binary bits and encoded as high and low voltages.

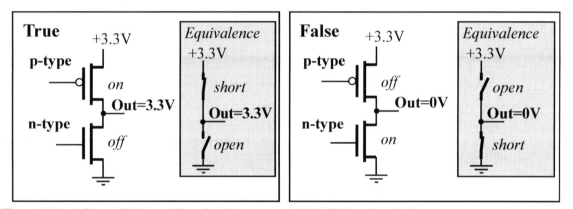

Figure 1.25. A binary bit is true if a voltage is present and false if the voltage is 0.

Numbers are stored on the computer in binary form. In other words, information is encoded as a sequence of 1's and 0's. On most computers, the memory is organized into 8-bit bytes. This means each 8-bit byte stored in memory will have a separate address. **Precision** is the number of distinct or different values. We express precision in alternatives, decimal digits, bytes, or binary bits. **Alternatives** are defined as the total number of possibilities as listed in Table 1.7. Let the operation $[[x]]$ be the greatest integer of x. E.g., $[[2.1]]$ is rounded up to 3. For example, an 8-bit number scheme can represent 256 different numbers, which means 256 alternatives. An 8-bit digital to analog converter (DAC) can generate 256 different analog outputs. An 8-bit analog to digital converter (ADC) can measure 256 different analog inputs.

Binary bits	Bytes	Alternatives
8	1	256
10	2	1024
12	2	4096
14	2	16,384
16	2	65,536
20	3	1,048,576
24	3	16,777,216
30	3	1,073,741,824
32	4	4,294,967,296
64	8	18,446,744,073,709,551,616
n	[[n/8]]	2^n

Table 1.7. Relationship between bits, bytes and alternatives as units of precision.

Observation: A good rule of thumb to remember is $2^{10 \cdot n}$ is approximately $10^{3 \cdot n}$.

Decimal digits are used to specify precision of measurement systems that display results as numerical values, as defined in Table 1.8. A full decimal digit can be any value 0, 1, 2, 3, 4, 5, 6, 7, 8, or 9. A digit that can be either 0 or 1 is defined as a ½ decimal digit. The terminology of a ½ decimal digit did not arise from a mathematical perspective of precision, but rather it arose from the physical width of the LED/LCD module used to display a blank or '1' as compared to the width of a full digit. Similarly, we define a digit that can be + or - also as a half decimal digit, because it has two choices. A digit that can be 0,1,2,3 is defined as a ¾ decimal digit, because it is wider than a ½ digit but narrower than a full digit. We also define a digit that can be -1, -0, +0, or +1 as a ¾ decimal digit, because it also has four choices. We use the expression 4½ decimal digits to mean 20,000 alternatives and the expression 4¾ decimal digits to mean 40,000 alternatives. The use of a ½ decimal digit to mean twice the number of alternatives or one additional binary bit is widely accepted. On the other hand, the use of ¾ decimal digit to mean four times the number of alternatives or two additional binary bits is not as commonly accepted. For example, consider the two ohmmeters. Assume both are set to the 0 to 200 kΩ range. A 3½ digit ohmmeter has a resolution of 0.1 kΩ with measurements ranging from 0.0 to 199.9 kΩ. On the other hand, a 4½ digit ohmmeter has a resolution of 0.01 kΩ with measurements ranging from 0.00 to 199.99 kΩ. Table 1.8 illustrates decimal-digit representation of precision.

Decimal digits	Alternatives
3	1000
3½	2000
3¾	4000
4	10,000
4½	20,000
4¾	40,000
5	100,000
n	10^n

Table 1.8. Definition of decimal digits as a unit of precision.

Checkpoint 1.18: How many binary bits correspond to 2½ decimal digits?

Checkpoint 1.19: How many decimal digits correspond to 10 binary bits?

Checkpoint 1.20: How many binary bits correspond to 6½ decimal digits?

Checkpoint 1.21: About how many decimal digits can be presented in a 64-bit 8-byte number? You can answer this without a calculator, just using the "rule of thumb".

The **hexadecimal** number system uses base 16 as opposed to our regular decimal number system that uses base 10. Hexadecimal is a convenient mechanism for humans to represent binary information, because it is extremely simple for us to convert back and forth between binary and hexadecimal. Hexadecimal number system is often abbreviated as "hex". A nibble is defined as four binary bits, which will be one hexadecimal digit. In mathematics, a subscript of 2 means binary, but in this book we will define binary numbers beginning with %. In assembly language however, we will use hexadecimal format when we need to define binary numbers. The hexadecimal digits are 0, 1, 2, 3, 4, 5, 6, 7, 8, 9, A, B, C, D, E, and F. Some assembly languages use the prefix $ to signify hexadecimal, and in C we use the prefix 0x. To convert from binary to hexadecimal, you simply separate the binary number into groups of four binary bits (starting on the right), then convert each group of four bits into one hexadecimal digit. For example, if you wished to convert 10100111_2, first you would group it into nibbles 1010 0111, then you would convert each group 1010=A and 0111=7, yielding the result of 0xA7. To convert hexadecimal to binary, you simply substitute the 4-bit binary for each hexadecimal digit. For example, if you wished to convert 0xB5D1, you substitute B=1011, 5=0101, D=1101, and 1=0001, yielding the result of 1011010111010001_2.

Checkpoint 1.22: Convert the binary number 111011101011_2 to hexadecimal.

Checkpoint 1.23: Convert the hex number 0x3800 to binary.

Checkpoint 1.24: How many binary bits does it take to represent 0x12345?

A great deal of confusion exists over the abbreviations we use for large numbers. In 1998 the International Electrotechnical Commission (IEC) defined a new set of abbreviations for the powers of 2, as shown in Table 1.9. These new terms are endorsed by the Institute of Electrical and Electronics Engineers (IEEE) and International Committee for Weights and Measures (CIPM) in situations where the use of a binary prefix is appropriate. The confusion arises over the fact that the mainstream computer industry, such as Microsoft, Apple, and Dell, continues to use the old terminology. According to the companies that market to consumers, a 1 GHz is 1,000,000,000 Hz but 1 Gbyte of memory is 1,073,741,824 bytes. The correct terminology is to use the SI-decimal abbreviations to represent powers of 10, and the IEC-binary abbreviations to represent powers of 2. The scientific meaning of 2 kilovolts is 2000 volts, but 2 kibibytes is the proper way to specify 2048 bytes. The term **kibibyte** is a contraction of kilo binary byte and is a unit of information or computer storage, abbreviated KiB.

1 KiB = 2^{10} bytes = 1024 bytes
1 MiB = 2^{20} bytes = 1,048,576 bytes
1 GiB = 2^{30} bytes = 1,073,741,824 bytes

These abbreviations can also be used to specify the number of binary bits. The term **kibibit** is a contraction of kilo binary bit, and is a unit of information or computer storage, abbreviated Kibit.

1 Kibit = 2^{10} bits = 1024 bits
1 Mibit = 2^{20} bits = 1,048,576 bits
1 Gibit = 2^{30} bits = 1,073,741,824 bits

A **mebibyte** (1 MiB is 1,048,576 bytes) is approximately equal to a megabyte (1 MB is 1,000,000 bytes), but mistaking the two has nonetheless led to confusion and even legal disputes. In the engineering community, it is appropriate to use terms that have a clear and unambiguous meaning.

Value	SI Decimal	SI Decimal	Value	IEC Binary	IEC Binary
1000^1	k	kilo-	1024^1	Ki	kibi-
1000^2	M	mega-	1024^2	Mi	mebi-
1000^3	G	giga-	1024^3	Gi	gibi-
1000^4	T	tera-	1024^4	Ti	tebi-
1000^5	P	peta-	1024^5	Pi	pebi-
1000^6	E	exa-	1024^6	Ei	exbi-
1000^7	Z	zetta-	1024^7	Zi	zebi-
1000^8	Y	yotta-	1024^8	Yi	yobi-

Table 1.9. Common abbreviations for large numbers.

1.5.2. 8-bit numbers

A **byte** contains 8 bits as shown in Figure 1.26, where each bit $b_7,...,b_0$ is binary and has the value 1 or 0. We specify b_7 as the most significant bit or MSB, and b_0 as the least significant bit or LSB. In C, the **unsigned char** or **uint8_t** data type creates an unsigned 8-bit number.

b_7	b_6	b_5	b_4	b_3	b_2	b_1	b_0

*Figure 1.26. 8-bit binary format, created using either char or unsigned char (in C99 **int8_t** or **uint8_t**).*

If a byte is used to represent an unsigned number, then the value of the number is

$N = 128 \cdot b_7 + 64 \cdot b_6 + 32 \cdot b_5 + 16 \cdot b_4 + 8 \cdot b_3 + 4 \cdot b_2 + 2 \cdot b_1 + b_0$

Notice that the significance of bit n is 2^n. There are 256 different unsigned 8-bit numbers. The smallest unsigned 8-bit number is 0 and the largest is 255. For example, 10000100_2 is 128+4 or 132.

Checkpoint 1.25: Convert the binary number 01101001_2 to unsigned decimal.

Checkpoint 1.26: Convert the hex number 0x23 to unsigned decimal.

The basis of a number system is a subset from which linear combinations of the basis elements can be used to construct the entire set. The basis represents the "places" in a "place-value" system. For positive integers, the basis is the infinite set {1, 10, 100...} and the "values" can range from 0 to 9. Each positive integer has a unique set of values such that the dot-product of the value-vector times the basis-vector yields that number. For example, 2345 is (..., 2,3,4,5)•(..., 1000,100,10,1), which is 2*1000+3*100+4*10+5. For the unsigned 8-bit number system, the basis is

{1, 2, 4, 8, 16, 32, 64, 128}

The values of a binary number system can only be 0 or 1. Even so, each 8-bit unsigned integer has a unique set of values such that the dot-product of the values times the basis yields that number. For example, 69 is (0,1,0,0,0,1,0,1)•(128,64,32,16,8,4,2,1), which equals 0*128+1*64+0*32+0*16+0*8+1*4+0*2+1*1.

Checkpoint 1.27: Give the representations of decimal 37 in 8-bit binary and hexadecimal.

Checkpoint 1.28: Give the representations of decimal 202 in 8-bit binary and hexadecimal.

One of the first schemes to represent signed numbers was called one's complement. It was called **one's complement** because to negate a number, you complement (logical not) each bit. For example, if 25 equals 00011001 in binary, then –25 is 11100110. An 8-bit one's complement number can vary from 127 to +127. The most significant bit is a sign bit, which is 1 if and only if the number is negative. The difficulty with this format is that there are two zeros +0 is 00000000, and –0 is 11111111. Another problem is that one's complement numbers do not have basis elements. These limitations led to the use of two's complement.

In C, the **char** or **int8_t** data type creates a signed 8-bit number. The **two's complement** number system is the most common approach used to define signed numbers. It was called two's complement because to negate a number, you complement each bit (like one's complement), and then add 1. For example, if 25 equals 00011001 in binary, then -25 is 11100111. If a byte is used to represent a signed two's complement number, then the value is

$$N = -128 \cdot b_7 + 64 \cdot b_6 + 32 \cdot b_5 + 16 \cdot b_4 + 8 \cdot b_3 + 4 \cdot b_2 + 2 \cdot b_1 + b_0$$

There are 256 different signed 8-bit numbers. The smallest signed 8-bit number is -128 and the largest is 127. For example, 10000010_2 equals -128+2 or -126.

Checkpoint 1.29: Are the signed and unsigned decimal representations of the 8-bit hex number 0x35 the same or different?

For the signed 8-bit number system the basis is

{1, 2, 4, 8, 16, 32, 64, -128}

The most significant bit in a two's complement signed number will specify the sign. An error will occur if you use signed operations on unsigned numbers, or use unsigned operations on signed numbers. To improve the clarity of our software, always specify the format of your data (signed versus unsigned) when defining or accessing the data.

Checkpoint 1.30: Give the representations of -31 in 8-bit binary and hexadecimal.

Observation: To take the negative of a two's complement signed number, we first complement (flip) all the bits, then add 1.

Many beginning students confuse a signed number with a negative number. A signed number is one that can be either positive or negative. A negative number is one less than zero. Notice that the same binary pattern of 11111111_2 could represent either 255 or -1. It is very important for the software developer to keep track of the number format. The computer cannot determine whether the 8-bit number is signed or unsigned. You, as the programmer, will determine whether the number is signed or unsigned by the specific assembly instructions you select to operate on the number. Some operations like addition, subtraction, and shift left (multiply by 2) use the same hardware (instructions) for both unsigned and signed operations. On the other hand, multiply, divide, and shift right (divide by 2) require separate hardware (instruction) for unsigned and signed operations.

1.5.3. Character information

We can use bytes to represent characters with the American Standard Code for Information Interchange (ASCII) code. Standard ASCII is actually only 7 bits, but is stored using 8-bit bytes with the most significant byte equal to 0. Some computer systems use the 8th bit of the ASCII code to define additional characters such as graphics and letters in other alphabets. The 7-bit ASCII code definitions are given in the Table 1.10. For example, the letter 'V' is in the 0x50 row and the 6 column. Putting the two together yields hexadecimal 0x56. The NUL character has the value 0 and is used to terminate strings. The '0' character has value 0x30 and represents the zero digit. In C and C99, we use the **char** data type for characters.

	BITS 4 to 6							
		0	1	2	3	4	5	6	7
B	0	NUL	DLE	SP	0	@	P	`	p
I	1	SOH	XON	!	1	A	Q	a	q
T	2	STX	DC2	"	2	B	R	b	r
S	3	ETX	XOFF	#	3	C	S	c	s
	4	EOT	DC4	$	4	D	T	d	t
	5	ENQ	NAK	%	5	E	U	e	u
0	6	ACK	SYN	&	6	F	V	f	v
	7	BEL	ETB	'	7	G	W	g	w
T	8	BS	CAN	(	8	H	X	h	x
O	9	HT	EM	)	9	I	Y	i	y
	A	LF	SUB	*	:	J	Z	j	z
3	B	VT	ESC	+	;	K	[	k	{
	C	FF	FS	,	<	L	\	l	|
	D	CR	GS	-	=	M	]	m	}
	E	SO	RS	.	>	N	^	n	~
	F	SI	US	/	?	O	_	o	DEL

Table 1.10. Standard 7-bit ASCII.

Checkpoint 1.31: How is the character '0' represented in ASCII?

One way to encode a character string is to use null-termination. In this way, the characters of the string are stored one right after the other, and the end of the string is signified by the NUL character (0). For example, the string "Valvano" is encoded as the following eight bytes 0x56, 0x61, 0x6C, 0x76, 0x61, 0x6E, 0x6F, and 0x00.

Checkpoint 1.32: How is "Hello World" encoded as a null-terminated ASCII string?

1.5.4. 16-bit numbers

A **halfword** or **short** contains 16 bits, where each bit $b_{15},...,b_0$ is binary and has the value 1 or 0, as shown in Figure 1.27. When we store 16-bit data into memory it requires two bytes. The memory systems on most computers are byte addressable, which means there is a unique address for each byte. Therefore, there are two possible ways to store in memory the two bytes that constitute the 16-bit data. Data could be stored in either little-endian or big-endian format. **Little endian** means the least significant byte is at the lower address and the most significant byte is at the higher address. **Big endian** means the most significant byte is at the lower address and the least significant byte is at the higher address. Freescale microcontrollers implement the big-endian format. Intel computers implement the little-endian format. Some processors, like the ARM and the PowerPC are **biendian**, because they can be configured to efficiently handle both big and little endian. Bit 15 of the Application Interrupt and Reset Control (APINT) register on the ARM® Cortex™-M specifies little-endian (0) or big-endian (1) data access. The Tiva® microcontrollers (TM4C), however, only use little-endian mode. When communicating data between computers one must know the format used.

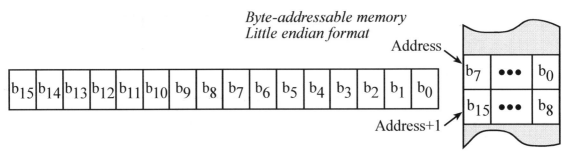

*Figure 1.27. A halfword is a 16-bit binary number. In C99, we use **int16_t** or **uint16_t**.*

If a halfword is used to represent an unsigned number, defined as an **unsigned short** or **uint16_t**, then the value of the number is

$$N = 32768 \cdot b_{15} + 16384 \cdot b_{14} + 8192 \cdot b_{13} + 4096 \cdot b_{12} + 2048 \cdot b_{11} + 1024 \cdot b_{10} + 512 \cdot b_9 + 256 \cdot b_8 + 128 \cdot b_7 + 64 \cdot b_6 + 32 \cdot b_5 + 16 \cdot b_4 + 8 \cdot b_3 + 4 \cdot b_2 + 2 \cdot b_1 + b_0$$

There are 65536 different unsigned 16-bit numbers. The smallest unsigned 16-bit number is 0 and the largest is 65535. For example, 0010000110000100_2 or 0x2184 is 8192+256+128+4 or 8580. For the unsigned 16-bit number system the basis is

{1, 2, 4, 8, 16, 32, 64, 128, 256, 512, 1024, 2048, 4096, 8192, 16384, 32768}

There are also 65536 different signed 16-bit numbers, defined either as **short**, **signed short**, or **int16_t**. The smallest two's complement signed 16-bit number is -32768 and the largest is 32767. For example, 1101000000000100_2 or 0xD004 is 32768+16384+4096+4 or 12284. If a halfword is used to represent a signed two's complement number, then the value is

$$N = -32768 \cdot b_{15} + 16384 \cdot b_{14} + 8192 \cdot b_{13} + 4096 \cdot b_{12} + 2048 \cdot b_{11} + 1024 \cdot b_{10} + 512 \cdot b_9 \\ + 256 \cdot b_8 + 128 \cdot b_7 + 64 \cdot b_6 + 32 \cdot b_5 + 16 \cdot b_4 + 8 \cdot b_3 + 4 \cdot b_2 + 2 \cdot b_1 + b_0$$

An error will occur if you use 16-bit operations on 8-bit numbers, or use 8-bit operations on 16-bit numbers. To improve the clarity of your software, always specify the precision of your data when defining or accessing the data. For the signed 16-bit number system the basis is

{1, 2, 4, 8, 16, 32, 64, 128, 256, 512, 1024, 2048, 4096, 8192, 16384, -32768}

1.5.5. 32-bit numbers

The native number on the ARM is a 32-bit **word**, where each bit $b_{31},...,b_0$ is binary and has the value 1 or 0, as shown in Figure 1.28, which is stored in little-endian format.

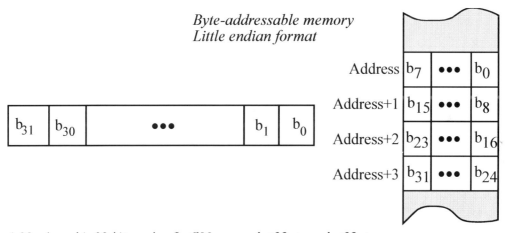

*Figure 1.28. A word is 32-bit number. In C99, we use **int32_t** or **uint32_t**.*

We define **unsigned long** or **uint32_t** to be an unsigned 32-bit number. The value is

$$N = 2^{31} \cdot b_{31} + 2^{30} \cdot b_{30} + ... + 2^2 \cdot b_2 + 2^1 \cdot b_1 + 2^0 b_0$$

There are 4,294,967,296 (2^{32}) different unsigned 32-bit numbers. The smallest unsigned 32-bit number is 0 and the largest is 4,294,967,295 (2^{32}). For the unsigned 32-bit number system the basis is

$$\{2^0, 2^1, 2^2, ..., 2^{30}, 2^{31}\}$$

We define **long, signed long,** or **int32_t** to be a signed 32-bit number. The value is

$$N = -2^{31} \cdot b_{31} + 2^{30} \cdot b_{30} + \ldots + 4 \cdot b_2 + 2 \cdot b_1 + b_0$$

There are 4,294,967,296 (2^{32}) different signed 32-bit numbers. The smallest unsigned 32-bit number is -2,147,483,648 (-2^{31}) and the largest is +2,147,483,647 ($2^{31}-1$). For the signed 32-bit number system the basis is

$$\{2^0, 2^1, 2^2, \ldots, 2^{30}, -2^{31}\}$$

When dealing with 16 or 32-bit numbers we normally would not pick out individual bytes, but rather capture the entire multiple-byte data as one non-divisible piece of information. On the other hand, if each byte in a multiple-byte data structure is individually addressable, then both the big- and little-endian schemes store the data in first to last sequence. For example, assume we wish to store the four ASCII characters "LM3S" as a string. These five bytes, which are 0x4C4D335300, would exist at five locations in memory. The first letter, the ASCII 'L'=0x4C would be stored in first location, regardless of which endian format the computer uses.

The terms "big and little endian" comes from Jonathan Swift's satire Gulliver's Travels. In Swift's book, the little people of Blefuscu believed the correct way to crack an egg is on the big end; hence they were called Big-Endians. In the rival kingdom, the little people of Lilliput were called Little-Endians because they insisted that the only proper way is to break an egg on the little end. The Lilliputians considered the people of Blefuscu as inferiors. The Big- and Little-Endians fought a long and senseless war over which end is best to crack an egg. Lilliput and Blefuscu were satirical references to 18th century Great Britain and France. However, one might argue they also refer to Intel and Motorola during the 1980's.

1.5.6. Fixed-point numbers

We will use **fixed-point** numbers when we wish to express values in our computer that have noninteger values. A fixed-point number contains two parts. The first part is a variable integer, called *I*. The second part of a fixed-point number is a fixed constant, called the **resolution** Δ. The integer may be signed or unsigned. An unsigned fixed-point number is one that has an unsigned variable integer. A signed fixed-point number is one that has a signed variable integer. The **precision** of a number system is the total number of distinguishable values that can be represented. The precision of a fixed-point number is determined by the number of bits used to store the variable integer. On the ARM, we can use 8, 16 or 32 bits for the integer. Extended precision with more the 32 bits can be implemented, but the execution speed will be slower because the calculations will have to be performed using software algorithms rather than with hardware instructions. This integer part is saved in memory and is manipulated by software. These manipulations include but are not limited to add, subtract, multiply, divide, and square root. The resolution is fixed, and cannot be changed during execution of the program. The resolution is not stored in memory. Usually we specify the value of the resolution using software comments to explain our fixed-point algorithm. The value of the fixed-point number is defined as the product of the variable integer and the fixed constant:

Fixed-point value $\equiv I \cdot \Delta$

Observation: If the range of numbers is known and small, then the numbers can be represented in a fixed-point format.

We specify the **range** of a fixed-point number system by giving the smallest and largest possible value. The range depends on both the variable integer and the fixed constant. For example, if the system used a 16-bit unsigned variable, then the integer part can vary from 0 to 65535. Therefore, the range of an unsigned 16-bit fixed-point system is 0 to 65535•Δ. In general, the range of the fixed-point system is

Smallest fixed-point value = $I_{min} \cdot \Delta$, where I_{min} is the smallest integer value
Largest fixed-point value = $I_{max} \cdot \Delta$, where I_{max} is the largest integer value

Checkpoint 1.33: What is the range of a 16-bit signed fixed-point number with Δ = 0.001?

When interacting with a human operator, it is usually convenient to use **decimal fixed-point**. With decimal fixed-point the fixed constant is a power of 10.

Decimal fixed-point value ≡ $I \cdot 10^m$ for some constant integer m

Again, the m is fixed and is not stored in memory. Decimal fixed-point will be easy to display, while binary fixed-point will be easier to use when performing mathematical calculations. The ARM processor is very efficient performing left and right shifts. With **binary fixed-point** the fixed constant is a power of 2. An example is shown in Figure 1.29.

Binary fixed-point value ≡ $I \cdot 2^n$ for some constant integer n

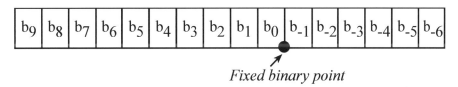

Fixed binary point

Figure 1.29. 16-bit binary fixed-point format with Δ=2^{-6}.

It is good practice to express the fixed-point resolution with units. For example, a decimal fixed-point number with a resolution of 0.001 V is really the same thing as an integer with units of mV. Consider an analog to digital converter (ADC) that converts an analog voltage in the range of 0 to +5 V into a digital number between 0 and 255. This ADC has a precision of 8 bits because it has 256 distinct alternatives. ADC resolution is defined as the smallest difference in input voltage that can be reliably distinguished. Because the 256 alternatives are spread evenly across the 0 to +5V range, we expect the ADC resolution to be about 5V/256 or 0.02V. When we choose a fixed-point number system to represent the voltages we must satisfy two constraints. First, we want the resolution of the number format to be better than the ADC resolution (Δ < 0.02). Second, we want the range of the number system to encompass all of the voltages in the range of the ADC (65535•Δ > 5). It would be appropriate to store voltages as 16-bit unsigned decimal fixed-point numbers with a resolution of 0.01V, 0.001V, or 0.0001V.

Using $\Delta=0.01$V, we store 4.23 V by making the integer part equal to 423. If we wished to use binary fixed-point, then we could choose a resolution anywhere in the range of 2^{-6} to 2^{-13} V. In general, we want to choose the largest resolution that satisfies both constraints, so the integer parts have smaller values. Smaller numbers are less likely to cause overflow during calculations. More discussion of overflow will be presented in the next chapter.

Checkpoint 1.34: Give an approximation of π using the decimal fixed-point with $\Delta = 0.001$.

Checkpoint 1.35: Give an approximation of π using the binary fixed-point with $\Delta = 2^{-8}$.

Microcontrollers in the LM3S family have a 10-bit ADC and a range of 0 to +3 V. Microcontrollers in the TM4C family provide a 12-bit ADC and a range of 0 to +3.3 V. With a 12-bit ADC, the resolution is 3.3V/4096 or about 0.001V. It would be appropriate to store voltages as 16-bit unsigned fixed-point numbers with a resolution of either 10^{-3} or 2^{-10} V. Let V_{in} be the analog voltage in volts and N be the integer ADC output, then the analog to digital conversion is approximately

$N = 4096 * V_{in} / 3.3$

Assume we use a fixed-point resolution of 10^{-3} V. We use this equation to calculate the integer part of a fixed-point number given the ADC result N. The definition of the fixed-point is

$V_{in} = I \cdot 10^{-3}$

Combining the above two equations yields

$I = (3300 * N) / 4096$

It is very important to carefully consider the order of operations when performing multiple integer calculations. There are two mistakes that can happen when we calculate 3300*N/4096. The first error is overflow, and it is easy to detect. **Overflow** occurs when the result of a calculation exceeds the range of the number system. In this example, if the multiply is implemented as 16-bit operation, then 3000*N can overflow the 0 to 65535 range. One solution of the overflow problem is promotion. Promotion is the action of increasing the inputs to a higher precision, performing the calculation at the higher precision, checking for overflow, then demoting the result back to the lower precision. In this example, the 3300, N, and 4096 are all converted to 32-bit unsigned numbers. (3300*N)/4096 is calculated in 32-bit precision. Because we know the range of N is 0 to 4095, we know the calculation of I will yield numbers between 0 and 3300, and therefore it will fit back in a 16-bit variable during demotion. The other error is called drop-out. **Drop-out** occurs during a right shift or a divide, and the consequence is that an intermediate result loses its ability to represent all of the values. To avoid drop-out, it is very important to divide last when performing multiple integer calculations. If we divided first, e.g., $I=3300*(N/4096)$, then the values of I would always be 0. We could have calculated $I=(3300*N+2048)/4096$ to implement rounding to the closest integer. The value 2048 is selected because it is about one half of the denominator. Sometimes we can simplify the numbers in an attempt to prevent overflow. In this cause we could have calculated $I=(825*N+256)/1024$. However, this formulation could still overflow 16-bit math and requires promotion to 32 bits to operate correctly.

When adding or subtracting two fixed-point numbers with the same Δ, we simply add or subtract their integer parts. First, let x, y, and z be three fixed-point numbers with the same Δ. Let $x=I\bullet\Delta$, $y=J\bullet\Delta$, and $z=K\bullet\Delta$. To perform $z=x+y$, we simply calculate $K=I+J$. Similarly, to perform $z=x-y$, we simply calculate $K=I-J$.

When adding or subtracting fixed-point numbers with different fixed parts, then we must first convert the two inputs to the format of the result before adding or subtracting. This is where binary fixed-point is more efficient, because the conversion process involves shifting rather than multiplication/division. Many instructions on the ARM allow a data shift operation to be performed at no added execution time.

For multiplication, we have $z=x\bullet y$. Again, we substitute the definitions of each fixed-point parameter, and solve for the integer part of the result. If all three variables have the same resolution, then $z=x\bullet y$ becomes $K\bullet\Delta = I\bullet\Delta\bullet J\bullet\Delta$ yielding $K = I\bullet J\bullet\Delta$. If the three variables have different resolutions, such as $x=I\bullet 2^n$, $y=J\bullet 2^m$, and $z=K\bullet 2^p$, then $z=x\bullet y$ becomes $K\bullet 2^p = I\bullet 2^n \bullet J\bullet 2^m$ yielding $K = I\bullet J\bullet 2^{n+m-p}$.

For division, we have $z=x/y$. Again, we substitute the definitions of each fixed-point parameter, and solve for the integer part of the result. If all three variables have the same resolution, then $z=x/y$ becomes $K\bullet\Delta = (I\bullet\Delta)/(J\bullet\Delta)$ yielding $K = I/J/\Delta$. If the three variables have different resolutions, such as $x=I\bullet 2^n$, $y=J\bullet 2^m$, and $z=K\bullet 2^p$, then $z=x/y$ becomes $K\bullet 2^p = (I\bullet 2^n)/(J\bullet 2^m)$ yielding $K = (I/J)\bullet 2^{n-m-p}$. Again, it is very important to carefully consider the order of operations when performing multiple integer calculations. We must worry about overflow and drop-out. If $(n-m-p)$ is positive then the left shift $(I\bullet 2^{n-m-p})$ should be performed before the divide $(/J)$. Conversely, if $(n-m-p)$ is negative then the right shift should be performed after the divide $(/J)$.

We can approximate a non-integer constant as the quotient of two integers. For example, the difference between 41/29 and $\sqrt{2}$ is 0.00042. If we need a more accurate representation, we can increase the size of the integers; the difference between 239/169 and $\sqrt{2}$ is only 1.2E-05. Using a binary fixed-point approximation will be faster on the ARM because of the efficiency of the shift operation. For example, approximating $\sqrt{2}$ as 181/128 yields an error of 0.0002. Furthermore, approximating $\sqrt{2}$ as 11585/8192 yields an error of only 2.9E-05.

Observation: For most real numbers in the range of 0.5 to 2, we can find two 3-digit integers I and J such that the difference between the approximation I/J and truth is less than 1E-5.

Checkpoint 1.36: What is the error in approximating $\sqrt{5}$ by 161/72? By 682/305?

We can use fixed-point numbers to perform complex operations using the integer functions of our microcontroller. For example, consider the following digital filter calculation.

$$y = x - 0.0532672\bullet x1 + x2 + 0.0506038\bullet y1 - 0.9025\bullet y2$$

In this case, the variables y, $y1$, $y2$, x, $x1$, and $x2$ are all integers, but the constants will be expressed in binary fixed-point format. The value -0.0532672 will be approximated by $-14\bullet 2^{-8}$. The value 0.0506038 will be approximated by $13\bullet 2^{-8}$. Lastly, the value -0.9025 will be approximated by $-231\bullet 2^{-8}$. The fixed-point implementation of this digital filter is

$$y = x + x2 + (-14\bullet x1 + 13\bullet y1 - 231\bullet y2) >> 8$$

Common Error: Lazy or incompetent programmers use floating-point in many situations where fixed-point would be preferable.

Example 1.5: Implement a function to calculate the surface area of a cylinder using fixed-point calculations. r is radius of the cylinder, which can vary from 0 to 1 cm. The radius is stored as a fixed-point number with resolution 0.001 cm. The software variable containing the integer part of the radius is **n**, which can vary from 0 to 1000. The height of the cylinder is 1 cm. The surface area is approximated by

$$s = 2\pi * (r^2 + r)$$

Solution: The surface area can range from 0 to 12.566 cm² ($2\pi*(1^2 + 1)$). The surface area is stored as a fixed-point number with resolution 0.001 cm². The software variable containing the integer part of the surface area is **m**, which can vary from 0 to 12566. In order to better understand the problem, we make a table of expected results.

r	**n**	s	**m**
0.000	0	0.000	0
0.001	1	0.006	6
0.010	10	0.063	63
0.100	100	0.691	691
1.000	1000	12.566	12566

To solve this problem we use the definition of a fixed-point number. In this case, r is equal to **n**/1000 and s is equal to **m**/1000. We substitution these definitions into the desired equation.

$$s = (6.283)*(\mathbf{r}^2 + \mathbf{r})$$
$$\mathbf{m}/1000 = 6.283*((\mathbf{n}/1000)^2 + (\mathbf{n}/1000))$$
$$\mathbf{m} = 6.283*(\mathbf{n}^2/1000 + \mathbf{n})$$
$$\mathbf{m} = 6283*(\mathbf{n}^2 + 1000*\mathbf{n})/1000000$$
$$\mathbf{m} = (6283*(\mathbf{n} +1000)*\mathbf{n})/1000000$$

If we wish to round the result to the closest integer we can add ½ the divisor before dividing.

$$\mathbf{m} = (6283*(\mathbf{n} +1000)*\mathbf{n}+500000)/1000000$$

One of the problems with this equation is the intermediate result can overflow a 32-bit calculation. One way to remove the overflow is to approximate 2π by 6.28. However, this introduces error. A better way to eliminate overflow is to approximate 2π by 289/46.

$$\mathbf{m} = (289*(\mathbf{n} +1000)*\mathbf{n}+23000)/46000$$

If we set **n** to its largest value, **n** =1000, we calculate the largest value the numerator can be as $(289*(1000 +1000)* 1000 +23000) = 578023000$, which fits in a 30-bit number.

Observation: As the fixed constant is made smaller, the accuracy of the fixed-point representation is improved, but the variable integer part also increases. Unfortunately, larger integers will require more bits for storage and calculations.

Checkpoint 1.37: Using a fixed constant of 10^{-3}, rewrite the digital filter y = x+0.0532672•x1+x2+0.0506038•y1-0.9025•y2 in decimal fixed-point format.

1.5.7. Floating-point numbers

We can use fixed-point when the range of values is small and known. Therefore, we will not need floating-point operations for most embedded system applications because fixed-point is sufficient. Furthermore, if the processor does not have floating-point instructions then a floating-point implementation will run much slower than the corresponding fixed-point implementation. However, it is appropriate to know the definition of floating-point. NASA believes that there are on the order of 10^{21} stars in our Universe. Manipulating large numbers like these is not possible using integer or fixed-point formats. Other limitation with integer or fixed-point numbers is there are some situations where the range of values is not known at the time the software is being designed. In a Physics research project, you might be asked to count the rate at which particles strike a sensor. Since the experiment has never been performed before, you do not know in advance whether there will be 1 per second or 1 trillion per second. The applications with numbers of large or unknown range can be solved with floating-point numbers. **Floating-point** is similar in format to binary fixed-point, except the exponent is allowed to change at run time. Consequently, both the exponent and the mantissa will be stored. Just like with fixed-point numbers we will use binary exponents for internal calculations, and decimal exponents when interfacing with humans. This number system is called floating-point because as the exponent varies the binary point or decimal point moves.

Observation: If the range of numbers is unknown or large, then the numbers must be represented in a floating-point format.

Observation: Floating-point implementations on computers like the Cortex™-M3 that do not have hardware support are extremely long and very slow. So, if you really need floating point, an TM4C with floating point hardware support is highly desirable.

The IEEE Standard for Binary Floating-Point Arithmetic or ANSI/IEEE Std 754-1985 is the most widely-used format for floating-point numbers. The single precision floating point operations on the LM4F/TM4C microcontrollers are compatible with this standard. There are three common IEEE formats: single-precision (32-bit), double-precision (64-bit), and double-extended precision (80-bits). Only the 32-bit short real format is presented here. The floating-point format, f, for the single-precision data type is shown in Figure 1.30. Computers use binary floating-point because it is faster to shift than it is to multiply/divide by 10.

Bit 31	Mantissa sign, $s=0$ for positive, $s=1$ for negative
Bits 30:23	8-bit biased binary exponent $0 \leq e \leq 255$
Bits 22:0	24-bit mantissa, m, expressed as a binary fraction, A binary 1 as the most significant bit is implied. $m = 1.m_1 m_2 m_3 ... m_{23}$

| s | e_7 | e_0 | m_1 | m_{23} |

Figure 1.30. 32-bit single-precision floating-point format.

The value of a single-precision floating-point number is

$$f = (-1)^s \cdot 2^{e-127} \cdot m$$

The range of values that can be represented in the single-precision format is about $\pm 10^{-38}$ to $\pm 10^{+38}$. The 24-bit mantissa yields a precision of about 7 decimal digits. The floating-point value is zero if both e and m are zero. Because of the sign bit, there are two zeros, positive and negative, which behave the same during calculations. To illustrate floating-point, we will calculate the single-precision representation of the number 10. To find the binary representation of a floating-point number, first extract the sign.

$$10 = (-1)^0 \cdot 10$$

Step 2, multiply or divide by two until the mantissa is greater than or equal to 1, but less than 2.

$$10 = (-1)^0 \cdot 2^3 \cdot 1.25$$

Step 3, the exponent e is equal to the number of divide by twos plus 127.

$$10 = (-1)^0 \cdot 2^{130-127} \cdot 1.25$$

Step 4, separate the 1 from the mantissa. Recall that the 1 will not be stored.

$$10 = (-1)^0 \cdot 2^{130-127} \cdot (1+0.25)$$

Step 5, express the mantissa as a binary fixed-point number with a fixed constant of 2-23.

$$10 = (-1)^0 \cdot 2^{130-127} \cdot (1+2097152 \cdot 2^{-23})$$

Step 6, convert the exponent and mantissa components to hexadecimal.

$$10 = (-1)^0 \cdot 2^{\$82-127} \cdot (1+\$200000 \cdot 2^{-23})$$

Step 7, extract s, e, m terms, convert hexadecimal to binary

$$10 = (0,\$82,\$200000) = (0,10000010,01000000000000000000000)$$

Sometimes this conversion does not yield an exact representation, as in the case of 0.1. In particular, the fixed-point representation of 0.6 is only an approximation.

Step 1 $0.1 = (-1)^0 \cdot 0.1$
Step 2 $0.1 = (-1)^0 \cdot 2^{-4} \cdot 1.6$
Step 3 $0.1 = (-1)^0 \cdot 2^{123-127} \cdot 1.6$
Step 4 $0.1 = (-1)^0 \cdot 2^{123-127} \cdot (1+0.6)$
Step 5 $0.1 \approx (-1)^0 \cdot 2^{123-127} \cdot (1+5033165 \cdot 2^{-23})$
Step 6 $0.1 \approx (-1)^0 \cdot 2^{\$7B-127} \cdot (1+\$4CCCCD \cdot 2^{-23})$
Step 7 $0.1 \approx (0,\$7B,\$4CCCCD) = (0,01111011,10011001100110011001101)$

The following example shows the steps in finding the floating-point approximation for π.

Step 1 $\pi = (-1)^0 \cdot \pi$
Step 2 $\pi \approx (-1)^0 \cdot 2^1 \cdot 1.570796327$
Step 3 $\pi \approx (-1)^0 \cdot 2^{128-127} \cdot 1.570796327$
Step 4 $\pi \approx (-1)^0 \cdot 2^{128-127} \cdot (1+0.570796327)$
Step 5 $\pi \approx (-1)^0 \cdot 2^{128-127} \cdot (1+4788187 \cdot 2^{-23})$
Step 6 $\pi \approx (-1)^0 \cdot 2^{\$80-127} \cdot (1+\$490FDB \cdot 2^{-23})$
Step 7 $\pi \approx (0,\$80,\$490FDB) = (0,10000000,10010010000111111011011)$

There are some special cases for floating-point numbers. When e is 255, the number is considered as plus or minus infinity, which probably resulted from an overflow during calculation. When e is 0, the number is considered as **denormalized**. The value of the mantissa of a denormalized number is less than 1. A denormalized short result number has the value,

$$f = (-1)^s \cdot 2^{-126} \cdot m \qquad \text{where } m = 0.m_1 m_2 m_3 ... m_{23}$$

Observation: The floating-point zero is stored in denormalized format.

When two floating-point numbers are added or subtracted, the smaller one is first **unnormalized**. The mantissa of the smaller number is shifted right and its exponent is incremented until the two numbers have the same exponent. Then, the mantissas are added or subtracted. Lastly, the result is normalized. To illustrate the floating-point addition, consider the case of 10+0.1. First, we show the original numbers in floating-point format. The mantissa is shown in binary format.

$$10.0 = (-1)^0 \cdot 2^3 \cdot 1.01000000000000000000000$$
$$+\ 0.1 = (-1)^0 \cdot 2^{-4} \cdot 1.10011001100110011001101$$

Every time the exponent is incremented the mantissa is shifted to the right. Notice that 7 binary digits are lost. The 0.1 number is unnormalized, but now the two numbers have the same exponent. Often the result of the addition or subtraction will need to be normalized. In this case the sum did not need normalization.

$$10.0 = (-1)^0 \cdot 2^3 \cdot 1.01000000000000000000000$$
$$+\ 0.1 = (-1)^0 \cdot 2^3 \cdot 0.00000011001100110011001\ 1001101$$
$$10.1 = (-1)^0 \cdot 2^3 \cdot 1.01000011001100110011001$$

When two floating-point numbers are multiplied, their mantissas are multiplied and their exponents are added. When dividing two floating-point numbers, their mantissas are divided and their exponents are subtracted. After multiplication and division, the result must be normalized. To illustrate the floating-point multiplication, consider the case of 10*0.1. Let $m1$, $m2$ be the values of the two mantissas. Since the range is $1 \leq m1, m2 < 2$, the product $m1*m2$ will vary from $1 \leq m1*m2 < 4$.

$$10.0 = (-1)^0 \cdot 2^3 \cdot 1.01000000000000000000000$$
$$*\ 0.1 = (-1)^0 \cdot 2^{-4} \cdot 1.10011001100110011001101$$
$$1.0 = (-1)^0 \cdot 2^{-1} \cdot 10.0000000000000000000000$$

The result needs to be normalized.

$$1.0 = (-1)^0 \cdot 2^0 \cdot 1.00000000000000000000000$$

Roundoff is the error that occurs as a result of an arithmetic operation. For example, the multiplication of two 32-bit mantissas yields a 64-bit product. The final result is normalized into a normalized floating-point number with a 32-bit mantissa. Roundoff is the error caused by discarding the least significant bits of the product. Roundoff during addition and subtraction can occur in two places. First, an error can result when the smaller number is shifted right. Second, when two n-bit numbers are added the result is n+1 bits, so an error can occur as the n+1 sum is squeezed back into an n-bit result.

Truncation is the error that occurs when a number is converted from one format to another. For example, when an 80-bit floating-point number is converted to 32-bit floating-point format, 40 bits are lost as the 64-bit mantissa is truncated to fit into the 24-bit mantissa. Recall, the number 0.1 could not be exactly represented as a short real floating-point number. This is an example of truncation as the true fraction was truncated to fit into the finite number of bits available.

If the range is known and small and a fixed-point system can be used, then a 32-bit fixed-point number system will have better resolution than a 32-bit floating-point system. For a fixed range of values (i.e., one with a constant exponent), a 32-bit floating-point system has only 23 bits of precision, while a 32-bit fixed-point system has 9 more bits of precision.

Performance Tip: The single precision floating-point programs written in assembly on the TM4C run much faster than equivalent C code because you can write assembly to perform operations in the native floating point assembly instructions.

1.6. Ethics

Because embedded systems are employed in many safety-critical devices, injury or death may result if there are hardware and/or software faults. Table 1.11 lists dictionary definitions of the related terms morals and ethics. A moral person is one who knows right from wrong, but an ethical person does the right thing.

Morals	Ethics
1. of, pertaining to, or concerned with the principles or rules of right conduct or the distinction between right and wrong; ethical: *moral attitudes*.	1. (*used with a singular or plural verb*) a system of moral principles: *the ethics of a culture*.
2. expressing or conveying truths or counsel as to right conduct, as a speaker or a literary work; moralizing: *a moral novel*.	2. the rules of conduct recognized in respect to a particular class of human actions or a particular group, culture, etc.: *medical ethics*; *Christian ethics*.
3. founded on the fundamental principles of right conduct rather than on legalities, enactment, or custom: *moral obligations*.	3. moral principles, as of an individual: *His ethics forbade betrayal of a confidence*.
4. capable of conforming to the rules of right conduct: *a moral being*.	4. (*usually used with a singular verb*) that branch of philosophy dealing with values relating to human conduct, with respect to the rightness and wrongness of certain actions and to the goodness and badness of the motives and ends of such actions.
5. conforming to the rules of right conduct (opposed to immoral): *a moral man*.	
6. virtuous in sexual matters; chaste.	
7. of, pertaining to, or acting on the mind, feelings, will, or character: *moral support*.	
8. resting upon convincing grounds of probability; virtual: *a moral certainty*.	

Table 1.11. Dictionary definitions of morals and ethics http://dictionary.reference.com

Most companies have a specific and detailed code of ethics, similar to the IEEE Code of Ethics presented below. Furthermore, patent and copyright laws provide a legal perspective to what is right and wrong. Nevertheless, many situations present themselves in the grey area. In these cases, you should seek advice from people whose ethics you trust. However, you are ultimately responsible for your own actions.

> *IEEE Code of Ethics*
> *We, the members of the IEEE, in recognition of the importance of our technologies in affecting the quality of life throughout the world, and in accepting a personal obligation to our profession, its members and the communities we serve, do hereby commit ourselves to the highest ethical and professional conduct and agree:*
> *1. to accept responsibility in making decisions consistent with the safety, health, and welfare of the public, and to disclose promptly factors that might endanger the public or the environment;*
> *2. to avoid real or perceived conflicts of interest whenever possible, and to disclose them to affected parties when they do exist;*
> *3. to be honest and realistic in stating claims or estimates based on available data;*
> *4. to reject bribery in all its forms;*
> *5. to improve the understanding of technology; its appropriate application, and potential consequences;*
> *6. to maintain and improve our technical competence and to undertake technological tasks for others only if qualified by training or experience, or after full disclosure of pertinent limitations;*
> *7. to seek, accept, and offer honest criticism of technical work, to acknowledge and correct errors, and to credit properly the contributions of others;*
> *8. to treat fairly all persons regardless of such factors as race, religion, gender, disability, age, or national origin;*
> *9. to avoid injuring others, their property, reputation, or employment by false or malicious action;*
> *10. to assist colleagues and co-workers in their professional development and to support them in following this code of ethics.*

A great volume of software exists in books and on the Internet. How you use this information in your classes is up to your professor. When you become a practicing engineer making products for profit, you will wish to use software written by others. Examples of software in books and on the internet are comprised of two components. The first component is the software code itself, and the second component is the algorithm used to solve the problem. To use the algorithm, you should search to see if it has patent protection. If it is protected, you could purchase or license the technology. If the algorithm is not protected and you wish to use the software code, you should ask permission from the author and give citation to source. If the algorithm is not protected and the author does not grant permission, you can still implement the algorithm by writing your own software. In all cases, you are responsible for testing.

A very difficult situation results when you leave one company and begin work for another. Technical expertise (things you know) and procedures (things you know how to do) that you have learned while working for a company belong to you, not your employer. This is such a huge problem that many employers have a detailed and legal contract employees must sign to be hired. A **non-compete clause** (NCC), also called a **covenant not to compete** (CNC), certifies the employee agrees not to pursue a similar job with any company in competition with the employer. Companies use these agreements to prevent present and former employees from working with their competitors. An example agreement follows:

EMPLOYEE NON-COMPETE AGREEMENT

For good consideration and as an inducement for _____ (Company) to employ _____ (Employee), the undersigned Employee hereby agrees not to directly or indirectly compete with the business of the Company and its successors and assigns during the period of employment and for a period of _____ years following termination of employment and notwithstanding the cause or reason for termination. The term "not compete" as used herein shall mean that the Employee shall not own, manage, operate, consult or to be employee in a business substantially similar to or competitive with the present business of the Company or such other business activity in which the Company may substantially engage during the term of employment. The Employee acknowledges that the Company shall or may in reliance of this agreement provide Employee access to trade secrets, customers and other confidential data and good will. Employee agrees to retain said information as confidential and not to use said information on his or her behalf or disclose same to any third party. This agreement shall be binding upon and inure to the benefit of the parties, their successors, assigns, and personal representatives.

Signed this _____ day of _____

_____Company

_____Employee

1.7. Exercises

1.1 Is RAM volatile or nonvolatile?

1.2 Is flash ROM volatile or nonvolatile?

1.3 For each term give a definition in 16 words or less: microprocessor, microcomputer, and microcontroller.

1.4 For each term give a definition in 16 words or less: bandwidth, real-time, latency.

1.5 For each term give a definition in 16 words or less: volatile, nonvolatile.

1.6 List the four components of a processor and define each in 16 words or less

1.7 For each parameter give a definition in 16 words or less: precision, range, resolution

1.8 What are the differences between CISC and RISC processors?

1.9 Describe structured programming in 16 words or less.

1.10 What are the differences between parallel programming and concurrent programming?

1.11 Define distributed programming in 16 words or less.

1.12 What are the differences between tristate and open collector logic?

1.13 Define open drain logic in 16 words or less?

1.14 Define 5-V tolerant in 16 words or less?

1.15 Considering just current, how many 74S Schottky inputs can one microcontroller output drive running in 8 mA output mode?

1.16 Considering just current, how many 74LS low-power Schottky inputs can one microcontroller output drive running in 2 mA output mode?

1.17 What is the qualitative difference in supply current between the CMOS devices and the non-CMOS devices? What is the explanation for the difference?

1.18 Using the circuit in Figure 1.23, what resistor value operates an LED at 1.8 V and 15 mA?

1.19 Using the circuit in Figure 1.23, what resistor value operates an LED at 1.6 V and 12 mA?

1.20 In 16 words or less describe the differences between positive logic and negative logic.

1.21 For each ADC parameter give a definition in 20 words or less: precision, range, resolution

1.22 How many alternatives does a 12-bit ADC have?

1.23 If a system uses an 11-bit ADC, about how many decimal digits will it have?

1.24 What is the difference between the terms kilobit and kibibit?

1.25 How many alternatives does a 13-bit ADC have?

1.26 If a system uses an 14-bit ADC, about how many decimal digits will it have?

1.27 If a system requires 3½ decimal digits of precision, what is the smallest number of bits the ADC needs to have?

1.28 If a system requires 5 decimal digits of precision, what is the smallest number of bits the ADC needs to have?

1.29 Convert the following decimal numbers to 8-bit unsigned binary: 26, 65, 124, and 202.

1.30 Convert the following decimal numbers to 8-bit signed binary: 23, 61, -122, and -5.

1.31 Convert the following hex numbers to unsigned decimal: 0x2A, 0x69, 0xB3, and 0xDE.

1.32 Convert the 16-bit binary number 0010001001101010_2 to unsigned decimal.

1.33 Convert the 16-bit hex number 0x5678 to unsigned decimal.

1.34 Convert the unsigned decimal number 12345 to 16-bit hexadecimal.

1.35 Convert the unsigned decimal number 20000 to 16-bit binary.

1.36 Convert the 16-bit hex number 0x7654 to signed decimal.

1.37 Convert the 16-bit hex number 0xBCDE to signed decimal.

1.38 Convert the signed decimal number 23456 to 16-bit hexadecimal.

1.39 Convert the signed decimal number –20000 to 16-bit binary.

1.40 Give an approximation of $\sqrt{7}$ using the decimal fixed-point ($\Delta = 0.001$) format.

1.41 Give an approximation of $\sqrt{7}$ using the binary fixed-point ($\Delta = 2^{-8}$) format.

1.42 Give an approximation of $\sqrt{103}$ using the decimal fixed-point ($\Delta = 0.01$) format.

1.43 Give an approximation of $\sqrt{93}$ using the binary fixed-point ($\Delta = 2^{-4}$) format.

1.44 A signed 16-bit binary fixed-point number system has a Δ resolution of 1/256. What is the corresponding value of the number if the integer part stored in memory is 385?

1.45 An unsigned 16-bit decimal fixed-point number system has a Δ resolution of 1/100. What is the corresponding value of the number if the integer part stored in memory is 385?

1.46 Give the short real floating-point representation of $\sqrt{2}$.

1.47 Give the short real floating-point representation of –134.4.

1.48 Give the short real floating-point representation of –0.0123.

D1.49 Draw a flow chart for the embedded system described in Example 1.3.

D1.50 Draw a flow chart for the embedded system in a simple watch that just tells time.

D1.51 Search the internet for a design of a flash ROM cell that uses 2 transistors. Label on the circuit the voltages occurring when the bit is zero, and when the bit is high.

D1.52 Search the internet for a design of a RAM cell that uses 6 transistors. Label on the circuit the voltages occurring when the bit is zero, and the voltages occurring when the bit is high.

D1.53 Design the circuit that interfaces a 1.5V 5mA LED to the microcontroller.

D1.53 Design the circuit that interfaces a 2.5V 1mA LED to the microcontroller.

D1.55 Assume M and N are two integers, each less than 1000. Find the best set of M and N, such that M/N is approximately $\sqrt{6}$. (Like 27/11, but much more accurate).

D1.56 Assume M and N are two integers, each less than 1000. Find the best set of M and N, such that M/N is approximately $\sqrt{7}$. (Like 8/3, but much more accurate).

D1.57 Assume M and N are two integers, each less than 1000. Find the best set of M and N, such that M/N is approximately e. (Like 19/7, but much more accurate).

D1.58 Assume M and N are two integers, each less than 1000. Find the best set of M and N, such that M/N is approximately ln(2). (Like 9/13, but much more accurate).

D1.59 First, rewrite the following digital filter using decimal fixed-point math. Assume the inputs are unsigned 8-bit values (0 to 255). Then, rewrite it so that it can be calculated with integer math using the fact that 0.22222 is about 2/9 and 0.088889 is about 4/45 and 0.8 is 4/5. In both cases, the calculations are to be performed in 16-bit unsigned integer form without overflow. y = 0.22222•x +0.08889•x1 + 0.80000•y1

D1.60 Perform the operation 2+π in short real floating-point format. Determine the difference between what you got and what you should have gotten (2+π). This error has two components: truncation error that results in the approximation π itself and roundoff error that occurs during the addition.

D1.61 Perform the operation 0.2*0.2 in short real floating-point format. Determine the difference between what you got and what you should have gotten (0.04). This error has two components: truncation error that results in the approximation of 0.2 itself and roundoff error that occurs during the multiplication.

D1.62 Perform the operation 1 + 1E9 in short real floating-point format. Determine the difference between what you got and what you should have gotten.

D1.63 Consider the following situation: Suppose that you are a development engineer with responsibility for an embedded system employed in one of your company's major products. You seek to improve the efficiency of the embedded system and, following some research, you discover an algorithm posted on the Web that would provide a vast improvement for your system. The algorithm is written in the same language as that used by your system.
    a) Would it ever be ethical to copy the code that implements the algorithm and incorporate it in your embedded system?
    b) Would it ever be good engineering practice to incorporate the code that implements the algorithm and in your embedded system?

D1.64 Suppose that the Web article containing the code states that it may be copied and used in any manner providing that it is not used in a product for sale. Are there any circumstances that would permit the ethical use of the algorithm?

D1.65 You are a development engineer that has recently left a position with a large corporation to work for a small embedded system company. Your team at the new company is working on a project that would be vastly improved through the use of a new procedure that was developed by your previous company. While you did not participate in the procedure's development, you are aware of all the technical details necessary to effectively employ it. Please answer and explain your response to each of the following questions:
    a) Would it ever be ethical to disclose the procedure to your team at the new company?
    b) Would it ever be good engineering practice to incorporate the procedure in your new team's embedded system?
    c) Assuming that you feel it could be ethical to disclose the procedure, what considerations or circumstances would influence your decision?
    d) Suppose that the considerations/circumstances in (c) above lead you to the conclusion that it would not be ethical to disclose the procedure. What changes to the considerations and circumstances would be necessary to permit you to ethically disclose the procedure to your new team?

1.8. Lab Assignments

Lab 1.1 Your microcontroller development board comes with starter code. Find the example that flashes an LED. Use this example to discover how to perform the following tasks. 1) How do you open a software project? 2) How do you compile the software project? 3) Can the software be run in a simulator? If so, how do you run on the simulator? 4) How do you download object code onto the real board? 5) Within the debugger how do you perform these operations: see the registers, observe RAM, start/stop execution, and set breakpoints? Change the software so the LED blinks twice as slow.

Lab 1.2 Your microcontroller development board comes with starter code. Find the example that outputs to either a serial port or an LCD. Use this example to discover how to perform the following tasks. 1) How do you open a software project? 2) How do you compile the software project? 3) Can the software be run in a simulator? If so, how do you run on the simulator? 4) How do you download object code onto the real board? 5) Within the debugger how do you perform these operations: see the registers, observe RAM, start/stop execution, and set breakpoints? Change the software so the system outputs your name.

Lab 1.3 The system has one LED, two switches, and resistors. In this lab you will not use a microcontroller. Design, implement, and test a circuit that turns on an LED if both switches are pressed. Using different resistors measure five different voltage and current points on the LED operating curve like Figure 1.23. Compare measured data to parameters from the LED data sheet.

Lab 1.4 The system has one LED, two switches, and resistors. In this lab you will not use a microcontroller. Design, implement, and test a circuit that turns on an LED if either switch is pressed. Using different resistors measure five different voltage and current points on the LED operating curve like Figure 1.23. Compare measured data to parameters from the LED data sheet.

2. ARM Cortex-M Processor

Chapter 2 objectives are to:

- Introduce Cortex™-M processor architecture
- Present an overview of the Cortex™-M core assembly language
- Define the memory-mapped I/O structure of the TM4C family
- Describe the parallel ports on the TM4C family
- Present the SysTick timer
- Describe the system clocks
- Present general thoughts about how to choose a microcontroller

In this chapter we present a general description of the ARM Cortex™-M processor. Rather than reproducing the voluminous details that can be found in the data sheets, we will present general concepts and give specific examples illustrating these concepts. After reading this chapter, you should be able to look up and understand detailed specifics in the ARM Cortex™-M Technical Reference Manual. Data sheets can be found on the web sites of either ARM or the companies that make the microcontrollers, like Texas Instruments. Some of these data sheets are also posted on the web site accompanying this book. This web site can be found at **http://users.ece.utexas.edu/~valvano/arm**.

There are two reasons we must learn the assembly language of the computer which we are using. Sometimes, but not often, we wish to optimize our application for maximum execution speed or minimum memory size, and writing pieces of our code in assembly language is one approach to such optimizations. The most important reason, however, is that by observing the assembly code generated by the compiler for our C code we can truly understand what our software is doing. Based on this understanding, we can evaluate, debug, and optimize our system.

Our first input/output interfaces will use the parallel ports, allowing us to exchange digital information with the external world. Specifically, we will learn how to connect switches and LEDs to the microcontroller. The second technique we will learn is to control time. We can select the execution speed of the microcontroller using the phase-lock-loop, and we can perform time delays using the SysTick timer.

Even though we will design systems based specifically on the TM4C family, these solutions can, with little effort, be implemented on other versions of the Cortex™-M family. We will discuss prototyping methods to build embedded systems and present a simple example with binary inputs and outputs.

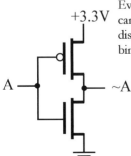

2.1. Cortex™-M Architecture

Figure 2.1 shows a simplified block diagram of a microcontroller based on the ARM® Cortex™-M processor. It is a **Harvard architecture** because it has separate data and instruction buses. The Cortex™-M instruction set combines the high performance typical of a 32-bit processor with high code density typical of 8-bit and 16-bit microcontrollers. Instructions are fetched from flash ROM using the ICode bus. Data are exchanged with memory and I/O via the system bus interface. On the Cortex™-M4 there is a second I/O bus for high-speed devices like USB. There are many sophisticated debugging features utilizing the DCode bus. The nested vectored interrupt controller (NVIC) manages **interrupts**, which are hardware-triggered software functions. Some internal peripherals, like the NVIC communicate directly with the processor via the private peripheral bus (PPB). The tight integration of the processor and interrupt controller provides fast execution of interrupt service routines (ISRs), dramatically reducing the interrupt latency.

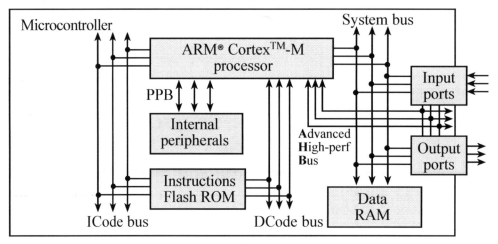

Figure 2.1. Harvard architecture of an ARM® Cortex™-M-based microcontroller.

2.1.1. Registers

The registers are depicted in Figure 2.2. R0 to R12 are general purpose registers and contain either data or addresses. Register R13 (also called the stack pointer, SP) points to the top element of the stack. Actually, there are two stack pointers: the main stack pointer (MSP) and the process stack pointer (PSP). Only one stack pointer is active at a time. In a high-reliability operating system, we could activate the PSP for user software and the MSP for operating system software. This way the user program could crash without disturbing the operating system. Because of the simple and dedicated nature of the embedded systems developed in this book, we will exclusively use the main stack pointer. Register R14 (also called the link register, LR) is used to store the return location for functions. The LR is also used in a special way during exceptions, such as interrupts. Interrupts are covered in Chapter 5. Register R15 (also called the program counter, PC) points to the next instruction to be fetched from memory. The processor fetches an instruction using the PC and then increments the PC by 2 or 4.

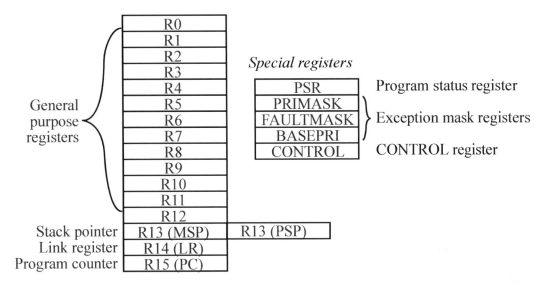

Figure 2.2. Registers on the ARM® Cortex™-M processor.

The ARM Architecture **Procedure Call Standard**, AAPCS, part of the ARM **Application Binary Interface** (ABI), uses registers R0, R1, R2, and R3 to pass input parameters into a C function. Also according to AAPCS we place the return parameter in Register R0.

There are three status registers named Application Program Status Register (APSR), the Interrupt Program Status Register (IPSR), and the Execution Program Status Register (EPSR) as shown in Figure 2.3. These registers can be accessed individually or in combination as the **Program Status Register** (PSR). The N, Z, V, C, and Q bits give information about the result of a previous ALU operation. In general, the **N bit** is set after an arithmetical or logical operation signifying whether or not the result is negative. Similarly, the **Z bit** is set if the result is zero. The **C bit** means carry and is set on an unsigned overflow, and the **V bit** signifies signed overflow. The **Q bit** is the sticky saturation flag, indicating that "saturation" has occurred, and is set by the `SSAT` and `USAT` instructions.

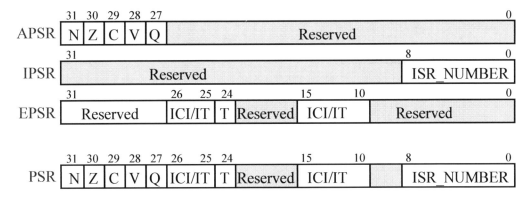

Figure 2.3. The program status register of the ARM® Cortex™-M processor.

The **T bit** will always be 1, indicating the ARM® Cortex™-M is executing Thumb instructions. The ICI/IT bits are used by interrupts and by the IF-THEN instructions. The ISR_NUMBER indicates which interrupt if any the processor is handling. Bit 0 of the special register **PRIMASK** is the interrupt mask bit. If this bit is 1 most interrupts and exceptions are not allowed. If the bit is 0, then interrupts are allowed. Bit 0 of the special register **FAULTMASK** is the fault mask bit. If this bit is 1 all interrupts and faults are not allowed. If the bit is 0, then interrupts and faults are allowed. The nonmaskable interrupt (NMI) is not affected by these mask bits. The **BASEPRI** register defines the priority of the executing software. It prevents interrupts with lower or equal priority but allows higher priority interrupts. For example if **BASEPRI** equals 3, then requests with level 0, 1, and 2 can interrupt, while requests at levels 3 and higher will be postponed. The details of interrupt processing will be presented in Chapter 5.

2.1.2. Memory

Microcontrollers within the same family differ by the amount of memory and by the types of I/O modules. All LM3S and LM4F/TM4C microcontrollers have a Cortex™-M processor. There are hundreds of members in this family; some of them are listed in Table 2.1.

Part number	RAM	Flash	I/O	I/O modules
LM3S811	8	64	32	PWM
LM3S1968	64	256	52	PWM
LM3S2110	16	64	40	PWM, CAN
LM3S8962	64	256	42	PWM, CAN, Ethernet, IEEE1588
LM4F120H5QR	32	256	43	floating point, CAN, DMA, USB
TM4C123GH6PGE	32	256	105	floating point, CAN, DMA, USB, PWM
TM4C123GH6PM	32	256	43	floating point, CAN, DMA, USB, PWM
TM4C123GH6ZRB	32	256	120	floating point, CAN, DMA, USB, PWM
TM4C1294NCPDT	256	1024	90	floating point, CAN, DMA, USB, PWM, Ethernet
MSP432E401Y	256	1024	90	floating point, CAN, DMA, USB, PWM, Ethernet
	KiB	KiB	pins	

Table 2.1. Memory and I/O modules (all have SysTick, RTC, timers, UART, I²C, SSI, and ADC).

The memory map of TM4C123 is illustrated in Figure 2.4. Although specific for the TM4C123, all ARM® Cortex™-M microcontrollers have similar memory maps. In general, Flash ROM begins at address 0x0000.0000, RAM begins at 0x2000.0000, the peripheral I/O space is from 0x4000.0000 to 0x5FFF.FFFF, and I/O modules on the private peripheral bus exist from 0xE000.0000 to 0xE00F.FFFF. In particular, the only differences in the memory map for the various members of the LM3S and LM4F/TM4C families are the ending addresses of the flash and RAM. Having multiple buses means the processor can perform multiple tasks in parallel. The following is some of the tasks that can occur in parallel

ICode bus	Fetch opcode from ROM
DCode bus	Read constant data from ROM
System bus	Read/write data from RAM or I/O, fetch opcode from RAM
PPB	Read/write data from internal peripherals like the NVIC
AHB	Read/write data from high-speed I/O and parallel ports (M4 only)

The ARM® Cortex™-M uses **bit-banding** to allow read/write access to individual bits in RAM and some bits in the I/O space. There are two parameters that define bit-banding: the address and the bit you wish to access. Assume you wish to access bit b of RAM address 0x2000.0000+n, where b is a number 0 to 7. The aliased address for this bit will be

0x2200.0000 + 32*n + 4*b

Reading this address will return a 0 or a 1. Writing a 0 or 1 to this address will perform an atomic read-modify-write modification to the bit.

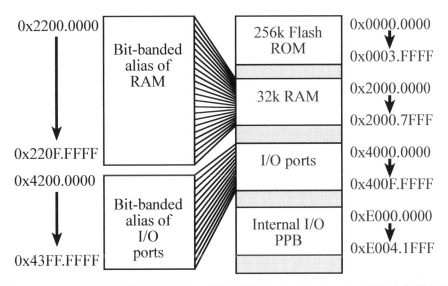

Figure 2.4. Memory map of the TM4C123. The MSP432E4 is similar but with 1024k ROM, 256k RAM.

If we consider 32-bit word-aligned data in RAM, the same bit-banding formula still applies. Let the word address be 0x2000.0000+n. n starts at 0 and increments by 4. In this case, we define b as the bit from 0 to 31. In little-endian format, bit 1 of the byte at 0x2000.0001 is the same as bit 9 of the word at 0x2000.0000. The aliased address for this bit will still be

0x2200.0000 + 32*n + 4*b

Examples of bit-banded addressing are listed in Table 2.2. Writing a 1 to location 0x2200.0018 will set bit 6 of RAM location 0x2000.0000. Reading location 0x2200.0024 will return a 0 or 1 depending on the value of bit 1 of RAM location 0x2000.0001.

Checkpoint 2.1: What address do you use to access bit 5 of the byte at 0x2000.1003?

Checkpoint 2.2: What address do you use to access bit 20 of the word at 0x2000.1000?

The other bit-banding region is the I/O space from 0x4000.0000 through 0x400F.FFFF. In this region, let the I/O address be 0x4000.0000+n, and let b represent the bit 0 to 7. The aliased address for this bit will be

0x4200.0000 + 32*n + 4*b

RAM address	Offset n	Bit b	Bit-banded alias
0x2000.0000	0	0	0x2200.0000
0x2000.0000	0	1	0x2200.0004
0x2000.0000	0	2	0x2200.0008
0x2000.0000	0	3	0x2200.000C
0x2000.0000	0	4	0x2200.0010
0x2000.0000	0	5	0x2200.0014
0x2000.0000	0	6	0x2200.0018
0x2000.0000	0	7	0x2200.001C
0x2000.0001	1	0	0x2200.0020
0x2000.0001	1	1	0x2200.0024

Table 2.2. Examples of bit-banded addressing.

Checkpoint 2.3: What address do you use to access bit 2 of the byte at 0x4000.0003?

2.1.3. Stack

The **stack** is a last-in-first-out temporary storage. To create a stack, a block of RAM is allocated for this temporary storage. On the ARM® Cortex™-M, the stack always operates on 32-bit data. The stack pointer (SP) points to the 32-bit data on the top of the stack. The stack grows downwards in memory as we push data on to it so, although we refer to the most recent item as the "top of the stack" it is actually the item stored at the lowest address! To push data on the stack, the stack pointer is first decremented by 4, and then the 32-bit information is stored at the address specified by SP. To pop data from the stack, the 32-bit information pointed to by SP is first retrieved, and then the stack pointer is incremented by 4. SP points to the last item pushed, which will also be the next item to be popped. The processor allows for two stacks, the main stack and the process stack, with two independent copies of the stack pointer. The boxes in Figure 2.5 represent 32-bit storage elements in RAM. The grey boxes in the figure refer to actual data stored on the stack, and the white boxes refer to locations in memory that do not contain stack data. This figure illustrates how the stack is used to push the contents of Registers R0, R1, and R2 in that order. Assume Register R0 initially contains the value 1, R1 contains 2 and R2 contains 3. The drawing on the left shows the initial stack. The software executes these six instructions

```
PUSH {R0}
PUSH {R1}
PUSH {R2}
POP  {R3}
POP  {R4}
POP  {R5}
```

The instruction **PUSH {R0}** saves the value of R0 on the stack. It first decrements SP by 4, and then it stores the 32-bit contents of R0 into the memory location pointed to by SP. The four bytes are stored little endian. The right-most drawing shows the stack after the push occurs three times. The stack contains the numbers 1 2 and 3, with 3 on top.

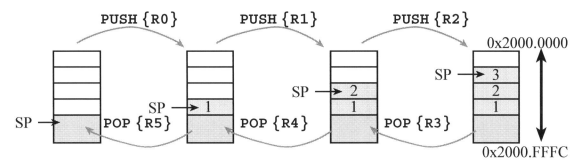

Figure 2.5. Stack picture showing three numbers first being pushed, then three numbers being popped.

The instruction **POP {R3}** retrieves data from the stack. It first moves the value from memory pointed to by SP into R3, and then it increments SP by 4. After the pop occurs three times the stack reverts to its original state and registers R3, R4 and R5 contain 3 2 1 respectively. We define the 32-bit word pointed to by SP as the **top** entry of the stack. If it exists, we define the 32-bit data immediately below the top, at SP+4, as **next** to top. Proper use of the stack requires following these important rules

1. Functions should have an equal number of pushes and pops
2. Stack accesses (push or pop) should not be performed outside the allocated area
3. Stack reads and writes should not be performed within the free area
4. Stack push should first decrement SP, then store the data
5. Stack pop should first read the data, and then increment SP

Functions that violate rule number 1 will probably crash when incorrect data are popped off at a later time. Violations of rule number 2 can be caused by a stack underflow or overflow. Overflow occurs when the number of elements became larger than the allocated space. Stack underflow is caused when there are more pops than pushes, and is always the result of a software bug. A stack overflow can be caused by two reasons. If the software mistakenly pushes more than it pops, then the stack pointer will eventually overflow its bounds. Even when there is exactly one pop for each push, a stack overflow can occur if the stack is not allocated large enough. The processor will generate a **bus fault** when the software tries read from or write to an address that doesn't exist. If valid RAM exists below the stack then pushing to an overflowed stack will corrupt data in this memory.

First, we will consider the situation where the allocated stack area is placed at the beginning of RAM. For example, assume we allocate 4096 bytes for the stack from 0x2000.0000 to 0x2000.0FFF, see the left side of Figure 2.6. The SP is initialized to 0x2000.1000, and the stack is considered empty. If the SP becomes less than 0x2000.0000 a **stack overflow** has occurred. The stack overflow will cause a bus fault because there is nothing at address 0x1FFF.FFFC. If the software tries to read from or write to any location greater than or equal to 0x2000.1000 then a **stack underflow** has occurred. At this point the stack and global variables may exist at overlapping addresses. Stack underflow is a very difficult bug to recognize, because the first consequence will be unexplained changes to data stored in global variables.

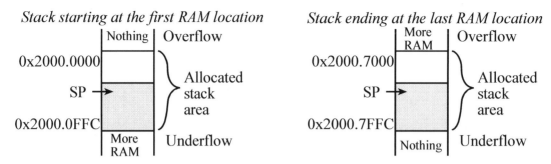

Figure 2.6. Drawings showing two possible ways to allocate the stack area in RAM.

Next, we will consider the situation where the allocated stack area is placed at the end of RAM. The TM4C123 has 32 KiB of RAM from 0x2000.0000 to 0x2000.7FFF. So in this case we allocate the 4096 bytes for the stack from 0x2000.7000 to 0x2000.7FFF, shown on the right side of Figure 2.6. The SP is initialized to 0x2000.8000, and the stack is considered empty. If the SP becomes less than 0x2000.7000 a stack overflow has occurred. The stack overflow will not cause a bus fault because there is memory at address 0x2000.6FFC. Stack overflow in this case is a very difficult bug to recognize, because the first consequence will be unexplained changes to data stored below the stack region. If the software tries to read from or write to any location greater than or equal to 0x2000.8000 then a stack underflow has occurred. In this case, stack underflow will cause a bus fault.

Executing an interrupt service routine will automatically push information on the stack. Since interrupts are triggered by hardware events, exactly when they occur is not under software control. Therefore, violations of rules 3, 4, and 5 will cause erratic behavior when operating with interrupts. Rules 4 and 5 are followed automatically by the **PUSH** and **POP** instructions.

2.1.4. Operating modes

The ARM® Cortex™-M has two privilege levels called privileged and unprivileged. Bit 0 of the **CONTROL** register is the **thread mode privilege level** (TPL). If TPL is 1 the processor level is privileged. If the bit is 0, then processor level is unprivileged. Running at the unprivileged level prevents access to various features, including the system timer and the interrupt controller. Bit 1 of the CONTROL register is the active stack pointer selection (ASPSEL). If ASPSEL is 1, the processor uses the PSP for its stack pointer. If ASPSEL is 0, the MSP is used. When designing a high-reliability operating system, we will run the user code at an unprivileged level using the PSP and the OS code at the privileged level using the MSP.

The processor knows whether it is running in the foreground (i.e., the main program) or in the background (i.e., an interrupt service routine). ARM defines the foreground as **thread mode**, and the background as **handler mode**. Switching from thread mode to handler mode occurs when an interrupt is triggered. The processor begins in thread mode, signified by ISR_NUMBER=0. Whenever it is servicing an interrupt it switches to handler mode, signified by setting ISR_NUMBER to specify which interrupt is being processed. All interrupt service routines run using the MSP. At the end of the interrupt service routine the processor is switched back to thread mode, and the main program continues from where it left off.

2.1.5. Reset

A reset occurs immediately after power is applied and can also occur by pushing the reset button available on most boards. After a reset, the processor is in thread mode, running at a privileged level, and using the MSP stack pointer. The 32-bit value at flash ROM location 0 is loaded into the SP. All stack accesses are word aligned. Thus, the least significant two bits of SP must be 0. A reset also loads the 32-bit value at location 4 into the PC. This value is called the reset vector. All instructions are halfword aligned. Thus, the least significant bit of PC must be 0. However, the assembler will set the least significant bit in the reset vector, so the processor will properly initialize the thumb bit (T) in the PSR. On the ARM® Cortex™-M, the T bit should always be set to 1. On reset, the processor initializes the LR to 0xFFFFFFFF.

2.2. Texas Instruments TM4C I/O pins

Table 2.1 listed the memory configuration for some of the Texas Instruments microcontrollers. In this section, we present the I/O pin configurations for the TM4C123 and TM4C1294/MSP432E401Y microcontrollers. The regular function of a pin is to perform parallel I/O, described later in Section 2.4. Most pins, however, have an alternative function. For example, port pins PA1 and PA0 can be either regular parallel port pins, or an asynchronous serial port called universal asynchronous receiver/transmitter (UART).

Joint Test Action Group (**JTAG**), standardized as the IEEE 1149.1, is a standard test access port used to program and debug the microcontroller board. Each microcontroller uses four or five port pins for the JTAG interface. Even though it is possible to use the four/five JTAG pins as general I/O, debugging most microcontroller boards will be more stable if these pins are left dedicated to the JTAG debugger. The following lists I/O devices found on microcontrollers.

- **UART**   Universal asynchronous receiver/transmitter
- **SSI**   Synchronous serial interface
- **I²C**   Inter-integrated circuit
- **I²S**   Inter-IC Sound, Integrated Interchip Sound (not on TM4C)
- **Timer**   Periodic interrupts, input capture, and output compare
- **PWM**   Pulse width modulation
- **ADC**   Analog to digital converter, measurement analog signals
- **Analog Comparator** Comparing two analog signals
- **QEI**   Quadrature encoder interface
- **USB**   Universal serial bus
- **Ethernet**   High speed network (TM4C1294/MSP432E401Y)
- **CAN**   Controller area network

The **UART** can be used for serial communication between computers. It is asynchronous and allows for simultaneous communication in both directions. In this book we will use a UART channel to connect to wifi-enabled devices. The **SSI** is alternately called serial peripheral interface (SPI). It is used to interface medium-speed I/O devices. In this book, we will use it to interface a graphics display, a secure digital card (SDC), and a digital to analog converter

(DAC). I^2C is a simple I/O bus that we will use to interface low speed peripheral devices. The inter-IC sound, or integrated interchip sound I^2S protocol is used to communicate sound information between audio devices. Input capture and output compare will be used to create periodic interrupts, and take measurements period, pulse width, phase and frequency. **PWM** outputs will be used to apply variable power to motor interfaces. In a typical motor controller, input capture measures rotational speed and PWM controls power. A PWM output can also be used to create a DAC. The **ADC** will be used to measure the amplitude of analog signals, and will be important in data acquisition systems. The analog comparator takes two analog inputs and produces a digital output depending on which analog input is greater. The **QEI** can be used to interface a brushless DC motor. **USB** is a high-speed serial communication channel. The **Ethernet** port can be used to bridge the microcontroller to the Internet or a local area network. The **CAN** creates a high-speed communication channel between microcontrollers and is commonly found in automotive and other distributed control applications.

2.2.1. TI TM4C123 LaunchPad I/O pins

Figure 2.7 draws the I/O port structure for the TM4C123GH6PM. This microcontroller is used on the EK-TM4C123GXL LaunchPad. Pins on the TM4C family can be assigned to as many as eight different I/O functions. Pins can be configured for digital I/O, analog input, timer I/O, or serial I/O. For example, PD3 can be digital, analog, SSI, I^2C, PWM, or timer. There are two buses used for I/O. The digital I/O ports are connected to both the advanced peripheral bus and the advanced high-performance bus. Because of the multiple buses, the microcontroller can simultaneously perform I/O bus cycles with instruction fetches from flash ROM. The TM4C123GH6PM has eight UART ports, four SSI ports, four I2C ports, two 12-bit ADCs, twelve timers, a CAN port, a USB interface, and 16 PWM outputs. There are 43 I/O lines. There are twelve ADC inputs; each ADC can convert up to 1 million samples per second. Table 2.3 lists the regular and alternate names of the port pins.

Each pin has one configuration bit in the **AMSEL** register. We set this bit to connect the port pin to the ADC or analog comparator. For digital functions, each pin also has four bits in the **PCTL** register, which we set to specify the alternative function for that pin (0 means regular I/O port). Table 2.3 shows the 4-bit **PCTL** configuration used to connect each pin to its alternate function. For example, column "7" means set 4-bit field in **PCTL** to 0111_2.

Pins PC3 – PC0 were left off Table 2.4 because these four pins are reserved for the JTAG debugger and should not be used for regular I/O. Notice, most alternate function modules (e.g., U0Rx) only exist on one pin (PA0). While other functions could be mapped to two or three pins (e.g., CAN0Rx could be mapped to one of the following: PB4, PE4, or PF0).

For example, if we wished to use SSI2 on pins PB7–4, we would set bits 7–4 in the **DEN** register (enable digital), clear bits 7–4 in the **AMSEL** register (disable analog), write a 0010,0010,0010,0010 to bits 31–16 in the **PCTL** register (enable SSI2 functionality), and set bits 7–4 in the **AFSEL** register (enable alternate function). If we wished to sample an analog signal on PD3, we would set bit 3 in the alternate function select register **AFSEL**, clear bit 3 in the digital enable register **DEN** (disable digital), set bit 3 in the analog mode select register **AMSEL** (enable analog), and activate one of the ADCs to sample channel 4. Additional examples will be presented throughout the book.

The Texas Instruments LaunchPad evaluation board (Figure 2.8) is a low-cost development board available as part number EK-TM4C123GXL from www.ti.com and from regular electronic distributors like Digikey, Mouser, Newark, Arrow, and Avnet. The kit provides an integrated In-Circuit Debug Interface (ICDI), which allows programming and debugging of the onboard TM4C123 microcontroller. One USB cable is used by the debugger (ICDI), and the other USB allows the user to develop USB applications (device). The user can select board power to come from either the debugger (ICDI) or the USB device (device) by setting the Power selection switch.

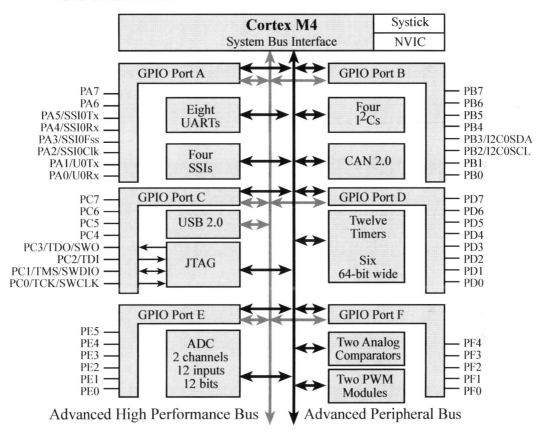

Figure 2.7. I/O port pins for the TM4C123GH6PM microcontrollers. The TM4C123 USB supports device, host, and on-the-go (OTG) modes.

Pins PA1 – PA0 create a serial port, which is linked through the debugger cable to the PC. The serial link is a physical UART as seen by the TM4C and mapped to a virtual COM port on the PC. The USB device interface uses PD4 and PD5. The JTAG debugger requires pins PC3 – PC0. **The LaunchPad connects PB6 to PD0, and PB7 to PD1. If you wish to use both PB6 and PD0 you will need to remove the R9 resistor. Similarly, to use both PB7 and PD1 remove the R10 resistor.** The USB connector on the side of the TM4C123 LaunchPad has five wires because it supports device, host, and OTG modes.

IO	Ain	0	1	2	3	4	5	6	7	8	9	14
PA0		Port	U0Rx							*CAN1Rx*		
PA1		Port	U0Tx							*CAN1Tx*		
PA2		Port		SSI0Clk								
PA3		Port		SSI0Fss								
PA4		Port		SSI0Rx								
PA5		Port		SSI0Tx								
PA6		Port			I₂C1SCL		M1PWM2					
PA7		Port			I₂C1SDA		M1PWM3					
PB0	USB0ID	Port	U1Rx						T2CCP0			
PB1	USB0VBUS	Port	U1Tx						T2CCP1			
PB2		Port			I₂C0SCL				T3CCP0			
PB3		Port			I₂C0SDA				T3CCP1			
PB4	Ain10	Port		SSI2Clk		M0PWM2			T1CCP0	CAN0Rx		
PB5	Ain11	Port		SSI2Fss		M0PWM3			T1CCP1	CAN0Tx		
PB6		Port		SSI2Rx		M0PWM0			T0CCP0			
PB7		Port		SSI2Tx		M0PWM1			T0CCP1			
PC4	C1-	Port	U4Rx	U1Rx		M0PWM6		IDX1	WT0CCP0	U1RTS		
PC5	C1+	Port	U4Tx	U1Tx		M0PWM7		PhA1	WT0CCP1	U1CTS		
PC6	C0+	Port	U3Rx					PhB1	WT1CCP0	USB0epen		
PC7	C0-	Port	U3Tx						WT1CCP1	USB0pflt		
PD0	Ain7	Port	SSI3Clk	SSI1Clk	I₂C3SCL	M0PWM6	M1PWM0		WT2CCP0			
PD1	Ain6	Port	SSI3Fss	SSI1Fss	I₂C3SDA	M0PWM7	M1PWM1		WT2CCP1			
PD2	Ain5	Port	SSI3Rx	SSI1Rx		M0Fault0			WT3CCP0	USB0epen		
PD3	Ain4	Port	SSI3Tx	SSI1Tx				IDX0	WT3CCP1	USB0pflt		
PD4	USB0DM	Port	U6Rx						WT4CCP0			
PD5	USB0DP	Port	U6Tx						WT4CCP1			
PD6		Port	U2Rx			M0Fault0		PhA0	WT5CCP0			
PD7		Port	U2Tx					PhB0	WT5CCP1	NMI		
PE0	Ain3	Port	U7Rx									
PE1	Ain2	Port	U7Tx									
PE2	Ain1	Port										
PE3	Ain0	Port										
PE4	Ain9	Port	U5Rx		I₂C2SCL	M0PWM4	M1PWM2			CAN0Rx		
PE5	Ain8	Port	U5Tx		I₂C2SDA	M0PWM5	M1PWM3			CAN0Tx		
PF0		Port	U1RTS	SSI1Rx	CAN0Rx		M1PWM4	PhA0	T0CCP0	NMI	C0o	
PF1		Port	U1CTS	SSI1Tx			M1PWM5	PhB0	T0CCP1		C1o	TRD1
PF2		Port		SSI1Clk		M0Fault0	M1PWM6		T1CCP0			TRD0
PF3		Port		SSI1Fss	CAN0Tx		M1PWM7		T1CCP1			TRCLK
PF4		Port					M1Fault0	IDX0	T2CCP0	USB0epen		

Table 2.4. PMCx bits in the GPIO_PORTx_PCTL_R register on the TM4C specify alternate functions.
PB1, PB0, PD4 and *PD5* are hardwired to the USB device. *PA0* and *PA1* are hardwired to the serial port.
PWM is not available on LM4F120.

Each 32-bit `GPIO_PORTx_PCTL_R` register defines the alternate function for the eight pins of that port, 4 bits for each pin. For example, if we wished to specify PA5-2 as SSI0, we would set Port A PCTL bits 23-16 to 0x2222 like this:

```
GPIO_PORTA_PCTL_R = (GPIO_PORTA_PCTL_R&0xFF0000FF)+0x00222200;
```

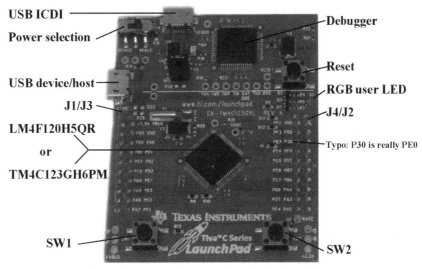

Figure 2.8. Texas Instruments LaunchPad based on the TM4C123GH6PM.

The Texas Instruments LaunchPad evaluation board has two switches and one 3-color LED, as shown in Figure 2.9. The switches are negative logic and will require activation of the internal pull-up resistors. In particular, you will set bits 0 and 4 in `GPIO_PORTF_PUR_R` register. The LED interfaces on PF3 – PF1 are positive logic. To use the LED, make the PF3 – PF1 pins an output. To activate the red color, output a one to PF1. The blue color is on PF2, and the green color is controlled by PF3. The 0-Ω resistors (R1, R2, R11, R12, R13, R25, and R29) connect the corresponding pin to hardware circuits on the LaunchPad board. LaunchPads come with R25 and R29 removed.

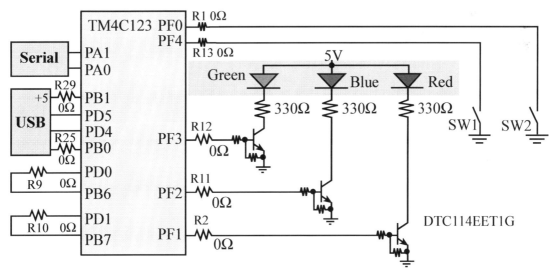

Figure 2.9. Switch and LED interfaces on the Texas Instruments LaunchPad Evaluation Board. The zero ohm resistors can be removed so the corresponding pin can be used without connection to the external circuits. We suggest you remove R9 and R10.

The LaunchPad has four 10-pin connectors, labeled as J1 J2 J3 J4 in Figures 2.8 and 2.10, to which you can attach your external signals. The top side of these connectors has male pins and the bottom side has female sockets. The intent is to stack boards together to make a layered system, see Figure 2.10. Texas Instruments also supplies Booster Packs, which are pre-made external devices that will plug into this 40-pin connector. The Booster Packs for the MSP430 LaunchPad are compatible (one simply plugs these 20-pin connectors into the outer two rows) with this board. The inner 10-pin headers (connectors J3 and J4) are not intended to be compatible with other TI LaunchPads. J3 and J4 apply only to Cortex-M4 Booster Packs.

There are two methods to connect external circuits to the LaunchPad. One method is to purchase a male to female jumper cable (e.g., item number 826 at www.adafruit.com). You could create low-cost male to female jumper wires by soldering a solid wire into a female sockets (e.g., Hirose DF11-2428SCA). A second method is to use solid 22-gauge or 24-gauge wire and connect one end of the solid wire into the bottom or female side of the LaunchPad.

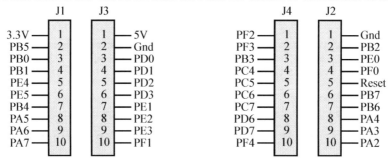

Figure 2.10. Interface connectors on the Texas Instruments TM4C123 LaunchPad Evaluation Board.

2.2.2. TI MSP432E401Y/TM4C1294 LaunchPad I/O pins

Figure 2.11 shows the 90 I/O pins available on the MSP432E401Y/TM4C1294NCPDT. These are identical microcontrollers, however each has its own LaunchPad. In this book, wherever we mention MSP432E401Y it applies equally to the TM4C1294NCPDT. Pins on the TM4C family can be assigned to as many as seven different I/O functions, see Table 2.5. Pins can be configured for digital I/O, analog input, timer I/O, or serial I/O. For example PA0 can be digital I/O, serial input, I2C clock, Timer I/O, or CAN receiver. There are two buses used for I/O. Unlike the TM4C123, the digital I/O ports are only connected to the advanced high-performance bus. The microcontroller can perform I/O bus cycles simultaneous with instruction fetches from flash ROM. The MSP432E401Y has eight UART ports, four SSI ports, ten I2C ports, two 12-bit ADCs, eight timers, two CAN ports, a USB interface, 8 PWM outputs, and an Ethernet port. Of the 90 I/O lines, twenty pins can be used for analog inputs to the ADC. The ADC can convert up to 1M samples per second. Table 2.5 lists the regular and alternate functions of the port pins.

Each 32-bit `GPIO_PORTx_PCTL_R` register defines the alternate function for the eight pins of that port, 4 bits for each pin. For example, if we wished to specify PD5–PD4 as UART2, we would set Port D PCTL bits 23-16 to 0x11 like this:

```
GPIO_PORTD_PCTL_R = (GPIO_PORTD_PCTL_R&0xFF00FFFF)+0x00110000;
```

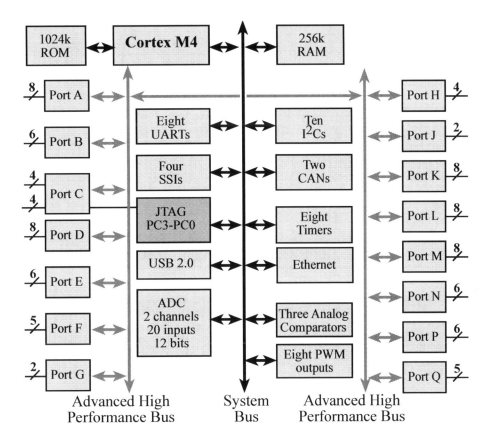

Figure 2.11. I/O port pins for the MSP432E401Y microcontroller.

Figure 2.12 shows the pin locations of the two Booster Pack connectors. There are three methods to connect external circuits to the Connected LaunchPad. One method uses male to female jumper cable (e.g., item number 826 at www.adafruit.com) or solder a solid wire into a female socket (e.g., Hirose DF11-2428SCA) creating a male-to-female jumper wire. In this method, you connect the female socket to the top of the LaunchPad and the male pin into a solderless breadboard. The second method uses male-to-male wires interfacing to the bottom of the LaunchPad. The third method uses two 49-pin right-angle headers so the entire LaunchPad can be plugged into a breadboard. You will need one each of Samtec parts TSW-149-09-L-S-RE and TSW-149-08-L-S-RA. This configuration is shown in Figure 2.13, and directions can be found at **http://users.ece.utexas.edu/~valvano/arm/TM4C1294soldering.pdf**

The Connected LaunchPad has two switches and four LEDs. Switch SW1 is connected to pin PJ0, and SW2 is connected to PJ1. These two switches are negative logic and require enabling the internal pull up (**PUR**). A reset switch will reset the microcontroller and your software will start when you release the switch. Positive logic LEDs D1, D2, D3, and D4 are connected to PN1, PN0, PF4, and PF0 respectively. A power LED indicates that 3.3 volt power is present on the board. R19 is a 0 Ω resistor connecting PA3 and PQ2. Similarly, R20 is a 0 Ω resistor connecting PA2 and PQ3. You need to remove R19 if you plan to use both PA3 and PQ2. You need to remove R20 if you plan to use both PA2 and PQ3. See Figures 2.13 and 2.14.

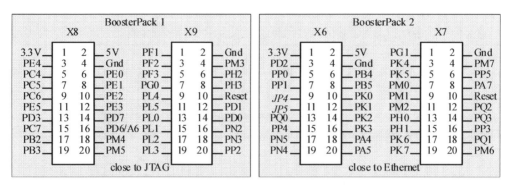

Figure 2.12. Interface connector for LaunchPad. On the MSP-EXP432E401Y X8 pin 16 is PA6. On the EK-TM4C1294-XL X8 pin 16 is PA6.

Jumper JP1 has six pins creating three rows of two. Exactly one jumper should be connected in the JP1 block, which selects the power source. The top position is for BoosterPack power. The middle position draws power from the USB connector, labeled OTG, on the left side of the board near the Ethernet jack. We recommend placing the JP1 jump in the bottom position so power is drawn from the ICDI (Debug) USB connection. Under normal conditions, you should place jumpers in both J2 and J3. Jumpers J2 and J3 facilitate measuring current to the microcontroller. We recommend you place JP4 and JP5 in the "UART" position so PA1 and PA0 are connected to the PC as a virtual COM port. Your code runs on the 128-pin MSP432E401Y microcontroller. There is a second TM4C microcontroller on the board, which acts as the JTAG debugger for your MSP432E401Y. You connect the Debug USB to a PC in order to download and debug software on the board. The other USB is for user applications.

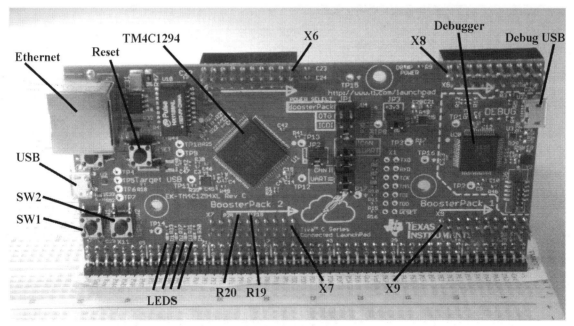

Figure 2.13. MSP-EXP432E401Y / EK-TM4C1294-XL Connected LaunchPad.

Pin	Analog	1	2	3	5	6	7	11	13	14	15
PA0	-	U0Rx	I2C9SCL	T0CCP0	-	-	CAN0Rx	-	-	-	-
PA1	-	U0Tx	I2C9SDA	T0CCP1	-	-	CAN0Tx	-	-	-	-
PA2	-	U4Rx	I2C8SCL	T1CCP0	-	-	-	-	-	-	SSI0Clk
PA3	-	U4Tx	I2C8SDA	T1CCP1	-	-	-	-	-	-	SSI0Fss
PA4	-	U3Rx	I2C7SCL	T2CCP0	-	-	-	-	-	-	SSI0XDAT0
PA5	-	U3Tx	I2C7SDA	T2CCP1	-	-	-	-	-	-	SSI0XDAT1
PA6	-	U2Rx	I2C6SCL	T3CCP0	USB0EPEN	-	-	-	SSI0XDAT2	-	EPI0S8
PA7	-	U2Tx	I2C6SDA	T3CCP1	USB0PFLT	-	-	USB0EPEN	SSI0XDAT3	-	EPI0S9
PB0	USB0ID	U1Rx	I2C5SCL	T4CCP0	-	-	CAN1Rx	-	-	-	-
PB1	USB0VBUS	U1Tx	I2C5SDA	T4CCP1	-	-	CAN1Tx	-	-	-	-
PB2	-	-	I2C0SCL	T5CCP0	-	-	-	-	-	USB0STP	EPI0S27
PB3	-	-	I2C0SDA	T5CCP1	-	-	-	-	-	USB0CLK	EPI0S28
PB4	AIN10	U0CTS	I2C5SCL	-	-	-	-	-	-	-	SSI1Fss
PB5	AIN11	U0RTS	I2C5SDA	-	-	-	-	-	-	-	SSI1Clk
PC4	C1-	U7Rx	-	-	-	-	-	-	-	-	EPI0S7
PC5	C1+	U7Tx	-	-	-	-	RTCCLK	-	-	-	EPI0S6
PC6	C0+	U5Rx	-	-	-	-	-	-	-	-	EPI0S5
PC7	C0-	U5Tx	-	-	-	-	-	-	-	-	EPI0S4
PD0	AIN15	-	I2C7SCL	T0CCP0	C0o	-	-	-	-	-	SSI2XDAT1
PD1	AIN14	-	I2C7SDA	T0CCP1	C1o	-	-	-	-	-	SSI2XDAT0
PD2	AIN13	-	I2C8SCL	T1CCP0	C2o	-	-	-	-	-	SSI2Fss
PD3	AIN12	-	I2C8SDA	T1CCP1	-	-	-	-	-	-	SSI2Clk
PD4	AIN7	U2Rx	-	T3CCP0	-	-	-	-	-	-	SSI1XDAT2
PD5	AIN6	U2Tx	-	T3CCP1	-	-	-	-	-	-	SSI1XDAT3
PD6	AIN5	U2RTS	-	T4CCP0	USB0EPEN	-	-	-	-	-	SSI2XDAT3
PD7	AIN4	U2CTS	-	T4CCP1	USB0PFLT	-	-	-	-	-	SSI2XDAT2
PE0	AIN3	U1RTS	-	-	-	-	-	-	-	-	-
PE1	AIN2	U1DSR	-	-	-	-	-	-	-	-	-
PE2	AIN1	U1DCD	-	-	-	-	-	-	-	-	-
PE3	AIN0	U1DTR	-	-	-	-	-	-	-	-	-
PE4	AIN9	U1RI	-	-	-	-	-	-	-	-	SSI1XDAT0
PE5	AIN8	-	-	-	-	-	-	-	-	-	SSI1XDAT1
PF0	-	-	-	-	EN0LED0	M0PWM0	-	-	-	SSI3XDAT1	TRD2
PF1	-	-	-	-	EN0LED2	M0PWM1	-	-	-	SSI3XDAT0	TRD1
PF2	-	-	-	-	-	M0PWM2	-	-	-	SSI3Fss	TRD0
PF3	-	-	-	-	-	M0PWM3	-	-	-	SSI3Clk	TRCLK
PF4	-	-	-	-	EN0LED1	M0FAULT0	-	-	-	SSI3XDAT2	TRD3
PG0	-	-	I2C1SCL	-	EN0PPS	M0PWM4	-	-	-	-	EPI0S11
PG1	-	-	I2C1SDA	-	-	M0PWM5	-	-	-	-	EPI0S10
PH0	-	U0RTS	-	-	-	-	-	-	-	-	EPI0S0
PH1	-	U0CTS	-	-	-	-	-	-	-	-	EPI0S1
PH2	-	U0DCD	-	-	-	-	-	-	-	-	EPI0S2
PH3	-	U0DSR	-	-	-	-	-	-	-	-	EPI0S3
PJ0	-	U3Rx	-	-	EN0PPS	-	-	-	-	-	-
PJ1	-	U3Tx	-	-	-	-	-	-	-	-	-
PK0	AIN16	U4Rx	-	-	-	-	-	-	-	-	EPI0S0
PK1	AIN17	U4Tx	-	-	-	-	-	-	-	-	EPI0S1
PK2	AIN18	U4RTS	-	-	-	-	-	-	-	-	EPI0S2
PK3	AIN19	U4CTS	-	-	-	-	-	-	-	-	EPI0S3
PK4	-	-	I2C3SCL	-	EN0LED0	M0PWM6	-	-	-	-	EPI0S32
PK5	-	-	I2C3SDA	-	EN0LED2	M0PWM7	-	-	-	-	EPI0S31
PK6	-	-	I2C4SCL	-	EN0LED1	M0FAULT1	-	-	-	-	EPI0S25
PK7	-	U0RI	I2C4SDA	-	RTCCLK	M0FAULT2	-	-	-	-	EPI0S24
PL0	-	-	I2C2SDA	-	-	M0FAULT3	-	-	-	USB0D0	EPI0S16
PL1	-	-	I2C2SCL	-	-	PhA0	-	-	-	USB0D1	EPI0S17
PL2	-	-	-	-	C0o	PhB0	-	-	-	USB0D2	EPI0S18
PL3	-	-	-	-	C1o	IDX0	-	-	-	USB0D3	EPI0S19
PL4	-	-	-	T0CCP0	-	-	-	-	-	USB0D4	EPI0S26

Pin	Analog	1	2	3	5	6	7	11	13	14	15
PL5	-	-	-	T0CCP1	-	-	-	-	-	USB0D5	EPI0S33
PL6	*USB0DP*	-	-	T1CCP0	-	-	-	-	-	-	-
PL7	*USB0DM*	-	-	T1CCP1	-	-	-	-	-	-	-
PM0	-	-	-	T2CCP0	-	-	-	-	-	-	EPI0S15
PM1	-	-	-	T2CCP1	-	-	-	-	-	-	EPI0S14
PM2	-	-	-	T3CCP0	-	-	-	-	-	-	EPI0S13
PM3	-	-	-	T3CCP1	-	-	-	-	-	-	EPI0S12
PM4	TMPR3	U0CTS	-	T4CCP0	-	-	-	-	-	-	-
PM5	TMPR2	U0DCD	-	T4CCP1	-	-	-	-	-	-	-
PM6	TMPR1	U0DSR	-	T5CCP0	-	-	-	-	-	-	-
PM7	TMPR0	U0RI	-	T5CCP1	-	-	-	-	-	-	-
PN0	-	U1RTS	-	-	-	-	-	-	-	-	-
PN1	-	U1CTS	-	-	-	-	-	-	-	-	-
PN2	-	U1DCD	U2RTS	-	-	-	-	-	-	-	EPI0S29
PN3	-	U1DSR	U2CTS	-	-	-	-	-	-	-	EPI0S30
PN4	-	U1DTR	U3RTS	I2C2SDA	-	-	-	-	-	-	EPI0S34
PN5	-	U1RI	U3CTS	I2C2SCL	-	-	-	-	-	-	EPI0S35
PP0	C2+	U6Rx	-	-	-	-	-	-	-	-	SSI3XDAT2
PP1	C2-	U6Tx	-	-	-	-	-	-	-	-	SSI3XDAT3
PP2	-	U0DTR	-	-	-	-	-	-	-	USB0NXT	EPI0S29
PP3	-	U1CTS	U0DCD	-	-	-	RTCCLK	-	-	USB0DIR	EPI0S30
PP4	-	U3RTS	U0DSR	-	-	-	-	-	-	USB0D7	-
PP5	-	U3CTS	I2C2SCL	-	-	-	-	-	-	USB0D6	-
PQ0	-	-	-	-	-	-	-	-	-	SSI3Clk	EPI0S20
PQ1	-	-	-	-	-	-	-	-	-	SSI3Fss	EPI0S21
PQ2	-	-	-	-	-	-	-	-	-	SSI3XDAT0	EPI0S22
PQ3	-	-	-	-	-	-	-	-	-	SSI3XDAT1	EPI0S23
PQ4	-	U1Rx	-	-	-	-	DIVSCLK	-	-	-	-

Table 2.5. PMCx bits in the GPIO_PORTx_PCTL_R register on the MSP432E401Y specify alternate functions. PD7 can be NMI by setting PCTL bits 31-28 to 8. *PL6* and *PL7* are hardwired to the USB.

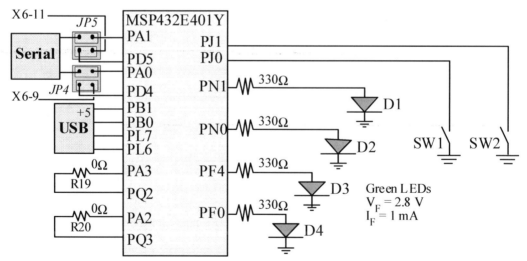

Figure 2.14. Switch and LED interfaces on the MSP432E401Y/TM4C1294 LaunchPad Evaluation Board. The zero ohm resistors can be removed so all the pins can be used.

Each pin has one configuration bit in the **AMSEL** register. We set this bit to connect the port pin to the ADC or analog comparator. For digital functions, each pin also has four bits in the **PCTL** register, which we set to specify the alternative function for that pin (0 means regular I/O port). Table 2.5 shows the 4-bit **PCTL** configuration used to connect each pin to its alternate function. For example, column "3" means set 4-bit field in **PCTL** to 0011.

Pins PC3 – PC0 were left off Table 2.5 because these four pins are reserved for the JTAG debugger and should not be used for regular I/O. Notice, some alternate function modules (e.g., U0Rx) only exist on one pin (PA0). While other functions could be mapped to two or three pins. For example, T0CCP0 could be mapped to one of the following: PA0, PD0, or PL4.

The PCTL bits in Table 2.5 can be tricky to understand. For example, if we wished to use UART6 on pins PP0 and PP1, we would set bits 1,0 in the **DEN** register (enable digital), clear bits 1,0 in the **AMSEL** register (disable analog), write a 0001,0001 to bits 7–0 in the **PCTL** register (enable UART6 functionality), and set bits 1,0 in the **AFSEL** register (enable alternate function). If we wished to sample an analog signal on PD0, we would set bit 0 in the alternate function select register **AFSEL**, clear bit 0 in the digital enable register **DEN** (disable digital), set bit 0 in the analog mode select register **AMSEL** (enable analog), and activate one of the ADCs to sample channel 15. Additional examples will be presented throughout the book.

Jumpers JP4 and JP5 select whether the serial port on UART0 (PA1 – PA0) or on UART2 (PD5 – 4) is linked through the debugger cable to the PC. The serial link is a physical UART as seen by the MSP432E401Y and is mapped to a virtual COM port on the PC. The USB device interface uses PL6 and PL7. The JTAG debugger requires pins PC3 – PC0.

To use the negative logic switches, make the pins digital inputs, and activate the internal pull-up resistors. In particular, you will activate the Port J clock, clear bits 0 and 1 in `GPIO_PORTJ_DIR_R` register, set bits 0 and 1 in `GPIO_PORTJ_DEN_R` register, and set bits 0 and 1 in `GPIO_PORTJ_PUR_R` register. The LED interfaces are positive logic. To use the LEDs, make the PN1, PN0, PF4, and PF0 pins an output. You will activate the Port N clock, set bits 0 and 1 in `GPIO_PORTN_DIR_R` register, and set bits 0 and 1 in `GPIO_PORTN_DEN_R` register. You will activate the Port F clock, set bits 0 and 4 in `GPIO_PORTF_DIR_R` register, and set bits 0 and 4 in `GPIO_PORTF_DEN_R` register.

2.3. ARM® Cortex™-M Assembly Language

This section focuses on the ARM® Cortex™-M assembly language. There are many ARM® processors, and this book focuses on Cortex™-M microcontrollers, which executes Thumb® instructions extended with Thumb-2 technology. This section does not present all the Thumb instructions. Rather, we present a few basic instructions in order to understand how the processor works. For further details, please refer to Volume 1 (Embedded Systems: Introduction to ARM® Cortex™-M Microcontrollers), and to the ARM® Cortex™-M Technical Reference Manual.

2.3.1. Syntax

Assembly language instructions have four fields separated by spaces or tabs. The **label field** is optional and starts in the first column and is used to identify the position in memory of the current instruction. You must choose a unique name for each label. The **opcode field** specifies the processor command to execute. The **operand field** specifies where to find the data to execute the instruction. Thumb instructions have 0, 1, 2, 3, or 4 operands, separated by commas. The **comment field** is also optional and is ignored by the assembler, but allows you to describe the software making it easier to understand. You can add optional spaces between operands in the operand field. However, a semicolon must separate the operand and comment fields. Good programmers add comments to explain the software.

```
Label  Opcode  Operands      Comment
Func   MOV     R0, #100      ; this sets R0 to 100
       BX      LR            ; this is a function return
```

When describing assembly instructions we will use the following list of symbols

Ra Rd Rm Rn Rt and **Rt2** represent registers
#imm12 represents a 12-bit constant, 0 to 4095
#imm16 represents a 16-bit constant, 0 to 65535
operand2 represents the flexible second operand as described in Section 2.2.2
{cond} represents an optional logical condition as listed in Table 2.6
{type} encloses an optional data type as listed in Table 2.7
{S} is an optional specification that this instruction sets the condition code bits
Rm {, shift} specifies an optional shift on **Rm** as described in Section 2.2.2
Rn {, #offset} specifies an optional offset to **Rn** as described in Section 2.2.2

Suffix	Flags	Meaning
EQ	Z = 1	Equal
NE	Z = 0	Not equal
CS or HS	C = 1	Higher or same, unsigned ≥
CC or LO	C = 0	Lower, unsigned <
MI	N = 1	Negative
PL	N = 0	Positive or zero
VS	V = 1	Overflow
VC	V = 0	No overflow
HI	C = 1 and Z = 0	Higher, unsigned >
LS	C = 0 or Z = 1	Lower or same, unsigned ≤
GE	N = V	Greater than or equal, signed ≥
LT	N ≠ V	Less than, signed <
GT	Z = 0 and N = V	Greater than, signed >
LE	Z = 1 or N ≠ V	Less than or equal, signed ≤
AL	Can have any value	Always. This is the default when no suffix is specified.

Table 2.6. Condition code suffixes used to optionally execution instruction.

For example, the general description of the addition instruction
ADD{cond} {Rd,} Rn, #imm12

could refer to any of the following examples

```
ADD    R0, #1            ; R0=R0+1
ADD    R0,R1,#10         ; R0=R1+10
ADDGE  R5,#100           ; if N==V, then R5=R5+100
ADDEQ  R12,R1,#100       ; if Z=1, then R12=R1+100
```

All object code is halfword-aligned. This means instructions can be 16 or 32 bits wide, and the program counter bit 0 will always be 0. The stack must remain word aligned, meaning the bottom two bits of the SP will always remain 0.

2.3.2. Addressing modes and operands

A fundamental issue in program development is the differentiation between data and address. It is in assembly language programming in general and addressing modes in specific that this differentiation becomes clear. When we put the number 1000 into register R0, whether this is data or address depends on how the 1000 is used. To run efficiently we try to keep frequently accessed data in registers. However, we need to access memory to fetch parameters or save results. The **addressing mode** is the format the instruction uses to specify the memory location to read or write data. All instructions begin by fetching the machine instruction (op code and operand) pointed to by the PC. Some instructions operate completely within the processor and require no memory data fetches. For example, the **ADD R1,R2** instruction performs R1+R2 and stores the sum back into R1. If the data is found in the instruction itself, like **MOV R0,#1**, the instruction uses **immediate addressing** mode. A register that contains the address or location of data is called a **pointer** or **index** register. **Indexed addressing** mode uses a register pointer to access memory. The addressing mode that uses the PC as the pointer is called **PC-relative addressing** mode. It is used for branching, for calling functions, and accessing constant data stored in ROM. The addressing mode is called PC-relative because the machine code contains the address difference between where the program is now and the address to which the program will access. There are many more addressing modes, but for now, these few addressing modes, as illustrated below, are enough to get us started. The **LDR** instruction will read a 32-bit word from memory and place the data in a register. With PC-relative addressing, the assembler automatically calculates the correct PC offset.

```
Func   PUSH   {R1,R2,LR}     ; save registers and return address
       MOV    R2,#100        ; R2=100, immediate addressing
       LDR    R1,=Count      ; R1 points to variable Count, using PC-relative
       LDR    R0,[R1]        ; R0= value of variable Count
       LDR    R0,[R1,#4]     ; R0= word pointed to by R1+4
       LDR    R0,[R1,#4]!    ; first R1=R1+4, then R0= word pointed to by R1
       LDR    R0,[R1],#4     ; R0= word pointed to by R1, then R1=R1+4
       LDR    R0,[R1,R2]     ; R0= word pointed to by R1+R2
       LDR    R0,[R1,R2, LSL #2] ; R0= word pointed to by R1+4*R2
       BL     Subroutine     ; call Subroutine, using PC-relative addressing
       POP    {R1,R2,PC}     ; restore registers and return
```

Checkpoint 2.4: What is the addressing mode used for?

Checkpoint 2.5: Assume R3 equals 0x2000.0000 at the time **LDR R2, [R3,#8]** is executed. What address will be accessed? If R3 is changed, to what value will R3 become?

Checkpoint 2.6: Assume R3 equals 0x2000.0000 at the time **LDR R2, [R3],#8** is executed. What address will be accessed? If R3 is changed, to what value will R3 become?

The operations caused by the first two **LDR** instructions are illustrated in Figure 2.15. Assume a 32-bit variable **Count** is located in data space at RAM address 0x2000.0000. First, **LDR R1,=Count** makes R1 equal to 0x2000.0000. I.e., R1 points to **Count**. The assembler places a constant 0x2000.0000 in code space and translates the **=Count** into the correct PC-relative access to the constant (e.g., **LDR R1, [PC,#28]**). Second, the **LDR R0, [R1]** instruction will dereference this pointer, bringing the contents at location 0x2000.0000 into R0. Since **Count** is located at 0x2000.0000, this instruction will read the value of the variable into R0.

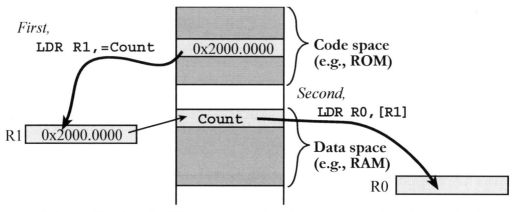

Figure 2.15. Indexed addressing using R1 as a register pointer to access memory. Data is moved into R0. Code space is where we place programs and data space is where we place variables.

Many general data processing instructions have a flexible second operand. This is shown as **Operand2** in the descriptions of the syntax of each instruction. **Operand2** can be a constant or a register with optional shift. We specify an **Operand2** constant in the form **#constant**:

```
ADD Rd, Rn, #constant    ;Rd = Rn+constant
```

where **constant** can be (**X** and **Y** are hexadecimal digits):

- Constant produced by shifting an 8-bit value left by any number of bits
- Constant of the form **0x00XY00XY**
- Constant of the form **0xXY00XY00**
- Constant of the form **0xXYXYXYXY**

We can also specify an **Operand2** register in the form **Rm {,shift}**. For example:

```
ADD Rd, Rn, Rm {,shift}   ;Rd = Rn+Rm
```

where **Rm** is the register holding the data for the second operand, and **shift** is an optional shift to be applied to **Rm**. **shift** can be one of:

ASR #n	Arithmetic shift right **n** bits, $1 \leq n \leq 32$.
LSL #n	Logical shift left **n** bits, $1 \leq n \leq 31$.
LSR #n	Logical shift right **n** bits, $1 \leq n \leq 32$.
ROR #n	Rotate right **n** bits, $1 \leq n \leq 31$.
RRX	Rotate right one bit, with extend.

If we omit the shift, or specify **LSL #0**, the instruction uses the value in **Rm**. If we specify a shift, the shift is applied to the value in **Rm**, and the resulting 32-bit value is used by the instruction. However, the contents in the register **Rm** remain unchanged. For example,

```
ADD R0,R1,LSL #4    ; R0 = R0 + R1*16 (R1 unchanged)
ADD R0,R1,ASR #4    ; signed R0 = R0 + R1/16 (R1 unchanged)
```

An **aligned access** is an operation where a word-aligned address is used for a word, dual word, or multiple word access, or where a halfword-aligned address is used for a halfword access. Byte accesses are always aligned. The **T** specifies the instruction is unprivileged. The Cortex™-M processor supports unaligned access only for the following instructions:

- **LDR, LDRT**   Load 32-bit word
- **LDRH, LDRHT**   Load 16-bit unsigned halfword
- **LDRSH, LDRSHT**   Load 16-bit signed halfword
- **STR, STRT**   Store 32-bit word
- **STRH, STRHT**   Store 16-bit halfword

All other read and write memory operations generate a usage fault exception if they perform an unaligned access, and therefore their accesses must be address aligned.

Common Error: Since not every instruction supports every addressing mode, it would be a mistake to use an addressing mode not available for that instruction.

2.3.3. Memory access instructions

This section presents mechanisms to read from and write to memory. As illustrated in Figure 2.15, to access memory we first establish a pointer to the object, then use indexed addressing. Usually code space is in ROM, but it is possible to assign code space to RAM. Data space is where we place variables. There are four types of memory objects, and typically we use a specific register to access it.

Memory object type	Register	Example operand
Constants in code space	PC	**=Constant**
Local variables on the stack	SP	**[SP,#0x04]**
Global variables in RAM	R0–R12	**[R0]**
I/O ports	R0–R12	**[R0]**

We use the **LDR** instruction to load data from memory into a register. There is a special form of **LDR** which instructs the assembler to load a constant or address into a register. This is a "pseudo-instruction" and the assembler will output suitable instructions to generate the specified value in the register. This form for **LDR** is

```
LDR{cond} Rd, =number
LDR{cond} Rd, =label
```

where **{cond}** is an optional condition (see Table 2.6), **Rd** is the destination register, and **label** is a label anywhere in memory. Figure 2.15 illustrates how to create a pointer to a variable in RAM. A similar approach can be used to access I/O ports. On the TM4C123, Port A exists at address 0x4000.43FC. After executing the first **LDR** instruction, R5 equals 0x4000.43FC, which is a pointer to Port A, and after executing the second **LDR** instruction, R6 contains the value at Port A.

```
Input   LDR R5,=0x400043FC  ;R5=0x400043FC, R5  = &PortA
        LDR R6,[R5]         ;Input from PortA into R6
;       ...
        BX  LR
```

The assembler translated the above assembly into this equivalent

```
Input   LDR R5,[PC,#16]     ;PC+16 is the address of the DCD
        LDR R6,[R5]
;       ...
        BX  LR
        DCD 0x400043FC
```

We use the **LDR** instruction to load data from RAM to a register and the **STR** instruction to store data from a register to RAM. In real life, when we *move* a box to the basement, *push* a broom across the floor, *load* bullets into a gun, *store* spoons in a drawer, *pop* a candy into your mouth, or *transfer* employees to a new location, there is a physical object and the action changes the location of that object. Assembly language uses these same verbs, but the action will be different. In most cases, it creates a copy of the data and places the copy at the new location. In other words, since the original data still exists in the previous location, there are now two copies of the information. The exception to this memory-access-creates-two-copies-rule is a stack pop. When we pop data from the stack, it no longer exists on the stack leaving us just one copy. For example in Figure 2.15, the instruction **LDR R0, [R1]** loads the contents of the variable **Count** into R0. At this point, there are two copies of the data, the original in RAM and the copy in R0. If we next add 1 to R0, the two copies have different values. When we learn about interrupts in Chapter 5, we will take special care to handle shared information stored in global RAM, making sure we access the proper copy.

When accessing memory data, the type of data can be 8, 16, 32, or 64 bits wide. For 8-bit and 16-bit accesses the type can also be signed or unsigned. To specify the data type we add an optional modifier, as listed in Table 2.7. When we load an 8-bit or 16-bit unsigned value into a register, the most significant bits are filled with 0, called **zero pad**.

When we load an 8-bit or 16-bit signed value into a register, the sign bit of the value is filled into the most significant bits, called **sign extension**. This way, if we load an 8-bit -10 (0xF6) into a 32-bit register, we get the 32-bit -10 (0xFFFF.FFF6). When we store an 8-bit or 16-bit value, only the least significant bits are used.

{type}	Data type	Meaning	
	32-bit word	0 to 4,294,967,295	or -2,147,483,648 to +2,147,483,647
B	Unsigned 8-bit byte	0 to 255,	Zero pad to 32 bits on load
SB	Signed 8-bit byte	-128 to +127,	Sign extend to 32 bits on load
H	Unsigned 16-bit halfword	0 to 65535,	Zero pad to 32 bits on load
SH	Signed 16-bit halfword	-32768 to +32768,	Sign extend to 32 bits on load
D	64-bit data		Uses two registers

Table 2.7. Optional modifier to specify data type when accessing memory.

Most of the addressing modes listed in the previous section can be used with load and store. The following lists the general form for some of the load and store instructions

```
LDR{type}{cond} Rd, [Rn]         ; load memory at [Rn] to Rd
STR{type}{cond} Rt, [Rn]         ; store Rt to memory at [Rn]
LDR{type}{cond} Rd, [Rn, #n]     ; load memory at [Rn+n] to Rd
STR{type}{cond} Rt, [Rn, #n]     ; store Rt to memory [Rn+n]
LDR{type}{cond} Rd, [Rn,Rm,LSL #n] ; load memory at [Rn+Rm*2ⁿ] to Rd
STR{type}{cond} Rt, [Rn,Rm,LSL #n] ; store Rt to memory [Rn+Rm*2ⁿ]
```

Program 2.1 sets each element of an array to the index. The **AREA DATA** directive specifies the following lines are placed in data space (typically RAM). The **Data SPACE 40** allocates ten uninitialized words. The **AREA CODE** directive specifies the following lines are placed in code space (typically ROM). The |.text| connects this program to the C code generated by the compiler. **ALIGN=2** will force the machine code to be halfword-aligned as required. The local variable **i** contains the array index. In assembly, the index **i** is kept in register R0. The **LDR** instruction establishes R1 as a pointer to the beginning of the array, or the **base** address. Since each array element is 32 bits, the address of the ith element of the array is **base**+4*i. The logical shift left by 2 implements the multiply by 4. In particular, the addressing mode **[R1,R0,LSL #2]** creates an effective address of R1+4*R0, with neither R1 nor R0 being changed by the instruction.

	AREA	DATA	
Data	SPACE	40 ; 32-bit data, length=10	
	AREA	|.text|, CODE, READONLY, ALIGN=2	
Set	MOVS	R0,#0x00 ; index i=0	
	LDR	R1,=Data ; R1 = &Data	
loop	STR	R0,[R1,R0,LSL #2]	
	ADDS	R0,R0,#1 ; i=i+1	
	CMP	R0,#10	
	BLT	loop ; repeat if i<10	
	BX	LR	

```
// C language implementation
uint32_t Data[10];

void Set(void){
int i;
  for(i=0; i<10; i++){
    Data[i] = i;
  }
}
```

Program 2.1. Assembly and C versions that initialize a global array of ten elements.

Checkpoint 2.7: Explain how to change Program 2.1 if the array were ten 16-bit numbers?

2.3.4. Logical operations

Software uses logical and shift operations to combine information, to extract information and to test information. A **unary operation** produces its result given a single input parameter. Examples of unary operations include negate, complement, increment, and decrement. In discrete digital logic, the **complement** operation is called a NOT gate, previously shown in Figure 1.17, see also Table 2.8.

A	~A
0	1
1	0

Table 2.8. Logical complement.

A **binary operation** produces a single result given two inputs. The **logical and** (&) operation yields a true result if both input parameters are true. We can use the **and** operation to extract, or **mask**, individual bits from a value. We can also use it to clear bits. The **logical or** (|) operation yields a true result if either input parameter is true. We can use the **or** operation to set bits. The **exclusive or** (^) operation yields a true result if exactly one input parameter is true. We can use the **exclusive or** to toggle or flip bits. The logical operators are summarized in Table 2.9. The logical instructions on the ARM Cortex^M-M take two inputs, one from a register and the other from the flexible second operand. These operations are performed in a bit-wise fashion on two 32-bit parameters yielding a 32-bit result. The result is stored into the destination register. For example, the calculation $r=m\&n$ means each bit is calculated separately, $r_{31}=m_{31}\&n_{31}$, $r_{30}=m_{30}\&n_{30}$, ..., $r_0=m_0\&n_0$. In C, when we write `r=m&n; r=m|n; r=m^n;` the logical operation occurs in a bit-wise fashion as described by Table 2.9. However, when we write `r=m&&n; r=m||n;`, the logical operation occurs in a word-wise fashion. For example, `r=m&&n;` means `r` will become zero if either `m` is zero or `n` is zero. Conversely, `r` will become 1 if both `m` is nonzero and `n` is nonzero.

A Rn	B Operand2	A&B AND	A|B ORR	A^B EOR	A&(~B) BIC	A|(~B) ORN
0	0	0	0	0	0	1
0	1	0	1	1	0	0
1	0	0	1	1	1	1
1	1	1	1	0	0	1

Table 2.9. Logical operations performed by the Cortex-M.

All instructions place the result into the destination register **Rd**. If **Rd** is omitted, the result is placed into **Rn**, which is the register holding the first operand. If the optional **S** suffix is specified, the N and Z condition code bits are updated on the result of the operation. Let **B** be the 32-bit value generated by the flexible second operand, **Operand2**. Some flexible second operands may affect the C bit. These logical instructions will leave the V bit unchanged.

```
AND{S}{cond} {Rd,} Rn, Operand2      ;Rd=Rn&B
ORR{S}{cond} {Rd,} Rn, Operand2      ;Rd=Rn|B
EOR{S}{cond} {Rd,} Rn, Operand2      ;Rd=Rn^B
BIC{S}{cond} {Rd,} Rn, Operand2      ;Rd=Rn&(~B)
ORN{S}{cond} {Rd,} Rn, Operand2      ;Rd=Rn|(~B)
```

The output of an open collector gate, drawn with the 'x', has two states low (0V) and HiZ (floating.) Consider the operation of the transistor-level circuit for the 74HC05. If *A* is high (+3.3V), the transistor is active, and the output is low (0V). If *A* is low (0V), the transistor is off, and the output is neither high nor low. In general, we can use an **open collector** NOT gate to control the current to a device, such as a relay, an LED, a solenoid, or a small motor. The 74HC05, the 7405, and the 7406 are all open collector NOT gates. 74HC04 is high speed CMOS and can only sink up to 4 mA when its output is low. Since the 7405 and 7406 are transistor-transistor-logic (TTL) they can sink more current. In particular, the 7405 has a maximum output low current (I_{OL}) of 16 mA, whereas the 7406 has a maximum I_{OL} of 40 mA.

Checkpoint 2.8: Which way does current flow in the open collector output?

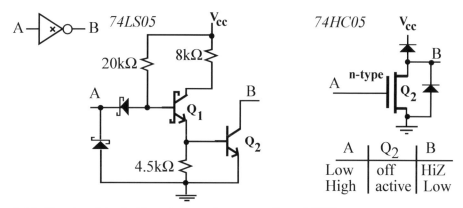

Figure 2.16. Two transistor implementations of an open collector NOT gate.

Digital storage devices are essential components used to make registers and memory. The simplest storage device is the set-reset flip flop. One way to build one is shown on the left side of Figure 2.17. If the inputs are $S^*=0$ and $R^*=1$, then the Q output will be 1. Conversely, if the inputs are $S^*=1$ and $R^*=0$, then the Q output will be 0. Normally, we leave both the S^* and R^* inputs high. We make the signal S^* go low, then back high to set the flip-flip, making $Q=1$. Conversely, we make the signal R^* go low, then back high to reset the flip-flip, making $Q=0$. If both S^* and R^* are 1, the value on Q will be remembered or stored. This flip flop enters an unpredictable mode with S^* and R^* are simultaneously low.

The gated D flip flop is also shown in Figure 2.17. The front-end circuits take a data input, D, and a control signal, W, and produce the S^* and R^* commands for the set-reset flip flop. For example, if W=0, then the flip flop is in its quiescent state, remembering the value on Q that was previously written. However, if W=1, then the data input is stored into the flip flop. In particular, if D=1 and W=1, then $S^*=0$ and $R^*=1$, making Q=1. Furthermore, if D=0 and W=1, then $S^*=1$ and $R^*=0$, making Q=0. So, to use the gated flip flop, we first put the data on the D input, next we make W go high, then we make W go low. This causes the data value to be stored at Q. After W goes low, the data does not need to exist at the D input anymore. If the D input changes while W is high, then the Q output will change correspondingly. However, the last value on the D input is remembered or latched when the W falls, as shown in Table 2.10.

The D flip-flop, shown on the right of Figure 2.17, can also be used to store information. D flip-flips are the basic building block of RAM and registers on the computer. To save information, we first place the digital value we wish to remember on the *D* input, and then give a rising edge to the **clock** input. After the rising edge of the **clock**, the value is available at the *Q* output, and the *D* input is free to change. The operation of the clocked D flip flop is defined on the right side of Table 2.10. The 74HC374 is an 8-bit D flip-flop, such that all 8 bits are stored on the rising edge of a single clock. The 74HC374 is similar in structure and operation to a register, which is high speed memory inside the processor. If the gate (*G*) input on the 74HC374 is high, its outputs will be HiZ (floating), and if the gate is low, the outputs will be high or low depending on the stored values on the flip flop.

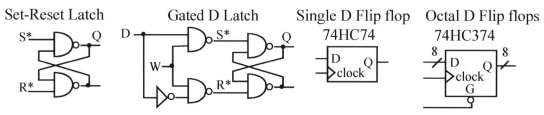

Figure 2.17. Digital storage elements.

D	W	Q	D	clock	Q
0	0	Q_{old}	0	0	Q_{old}
1	0	Q_{old}	0	1	Q_{old}
0	1	0	1	0	Q_{old}
1	1	1	1	1	Q_{old}
0	↓	0	0	↑	0
1	↓	1	1	↑	1

Table 2.10. D flip-flop operation. Q_{old} is the value of the **D** input at the time of the active edge of on **W** or clock.

The tristate driver, shown in Figure 2.18, can be used dynamically control signals within the computer. The tristate driver is an essential component from which computers are built. To active the driver, we make its gate (*G**) low. When the driver is active, its output (*Y*) equals its input (*A*). To deactivate the driver, we make its *G** high. When the driver is not active, its output *Y* floats independent of *A*. We saw this floating state with the open collector logic, and it is also called HiZ or high impedance. The HiZ output means the output is neither driven high nor low. The operation of a tristate driver is defined in Table 2.11. The 74HC244 is an 8-bit tristate driver, such that all 8 bits are active or not active controlled by a single gate. The 74HC374 8-bit D flip-flop includes tristate drivers on its outputs. Normally, we can't connect to digital outputs together. The tristate driver provides a way to connect multiple outputs to the same signal, as long as at most one of the gates is active at a time.

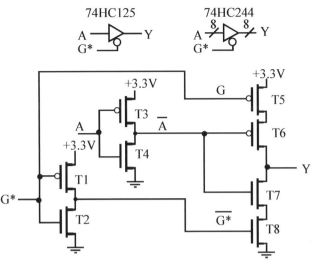

Figure 2.18. A 1-bit and an 8-bit tristate driver (G* is in negative logic).

Table 2.11 describes how a tristate driver in Figure 2.18 works. Transistors T1 and T2 create the logical complement of G^*. Similarly, transistors T3 and T4 create the complement of A. An input of $G^*=0$ causes the driver to be active. In this case, both T5 and T8 will be on. With T5 and T8 on, the circuit behaves like a cascade of two NOT gates, so the output Y equals the input A. However, if the input $G^*=1$, both T5 and T8 will be off. Since T5 is in series with the +3.3V, and T8 in series with the ground, the output Y will be neither high nor low. I.e., it will float.

A	G*	T1	T2	T3	T4	T5	T6	T7	T8	Y
0	0	on	off	on	off	on	off	on	on	0
1	0	on	off	off	on	on	on	off	on	1
0	1	off	on	on	off	off	off	on	off	HiZ
1	1	off	on	off	on	off	on	off	off	HiZ

Table 2.11. Tristate driver operation. HiZ is the floating state, such that the output is not high or low.

2.3.5. Shift operations

Like programming in C, the assembly shift is a binary operation. In C, the $<<$ and $>>$ operators take two inputs and yield one output, e.g., the right shift is $R = M>>N$. The **logical shift right** (LSR) is similar to an unsigned divide by 2^n, where n is the number of bits shifted, as shown in Figure 2.19. A zero is shifted into the most significant position, and the carry flag will hold the bit shifted out. The right shift operations do not round. In general, the LSR discards bits shifted out, and the UDIV truncates towards 0. Thus, when using UDIV to divide unsigned numbers by a power of 2, UDIV and LSR yield identical results. The **arithmetic shift right** (ASR) is similar to a signed divide by 2^n. Notice that the sign bit is preserved, and the carry flag will hold the bit shifted out. This right shift operation also does not round. In general, the ASR discards bits shifted out, and the SDIV truncates towards 0. The **logical shift left** (LSL) operation works for both unsigned and signed multiply by 2^n. A zero is shifted into the least significant position, and the carry bit will contain the bit that was shifted out. The two **rotate** operations can be used to create multiple-word shift functions. There is no rotate left instruction, because a rotate left 10 bits is the same as rotate right 22 bits.

All instructions place the result into the destination register **Rd**. **Rm** is the register holding the value to be shifted. The number of bits to shift is either in register **Rs**, or specified as a constant **n**. If the optional **S** suffix is specified, the N and Z condition code bits are updated on the result of the operation. The C bit is the carry out after the shift as shown in Figure 2.19. These shift instructions will leave the V bit unchanged.

```
LSR{S}{cond} Rd, Rm, Rs     ; logical shift right Rd=Rm>>Rs  (unsigned)
LSR{S}{cond} Rd, Rm, #n     ; logical shift right Rd=Rm>>n   (unsigned)
ASR{S}{cond} Rd, Rm, Rs     ; arithmetic shift right Rd=Rm>>Rs (signed)
ASR{S}{cond} Rd, Rm, #n     ; arithmetic shift right Rd=Rm>>n (signed)
LSL{S}{cond} Rd, Rm, Rs     ; shift left Rd=Rm<<Rs (signed and unsigned)
LSL{S}{cond} Rd, Rm, #n     ; shift left Rd=Rm<<n  (signed and unsigned)
ROR{S}{cond} Rd, Rm, Rs     ; rotate right
ROR{S}{cond} Rd, Rm, #n     ; rotate right
RXX{S}{cond} Rd, Rm         ; rotate right with extension
```

Observation: Use logic shift operations on unsigned numbers and use arithmetic shift operations on signed numbers.

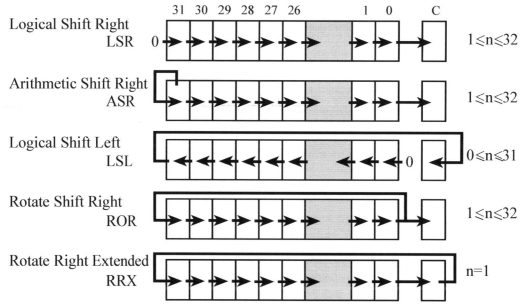

Figure 2.19. Shift operations.

2.3.6. Arithmetic operations

When software executes arithmetic instructions, the operations are performed by digital hardware inside the processor. Even though the design of such logic is complex, we will present a brief introduction, in order to provide a little insight as to how the computer performs arithmetic. It is important to remember that arithmetic operations (addition, subtraction, multiplication, and division) have constraints when performed with finite precision on a processor. An overflow error occurs when the result of an arithmetic operation cannot fit into the finite precision of the register into which the result is to be stored.

For example, consider an 8-bit unsigned number system, where the numbers can range from 0 to 255. If we add two numbers together the result can range from 0 to 510, which is a 9-bit unsigned number. These numbers are similar to the numbers 1–12 on a clock, as drawn in Figure 2.20. If it is 11 o'clock and we wait 3 hours, it becomes 2 o'clock. Shown in the middle of Figure 2.20, if we add 64 to 224, the result becomes 32. In most cases, we would consider this an error. An unsigned overflow occurs during addition when we cross the 255–0 barrier (carry set on overflow). If we subtract two 8-bit unsigned numbers the result can range from -255 to +255, which is a 9-bit signed number. Subtraction moves in a counter-clockwise direction on the number wheel. As shown on the right side of Figure 2.20, if we subtract 64 from 32 (32-64), we get the incorrect result of 224. An unsigned overflow occurs during subtraction if we cross the 255–0 barrier in the other direction (carry clear on overflow). After a subtraction on the Cortex-M the carry is clear if an error occurred, and the carry is set if no error occurred and the answer is correct.

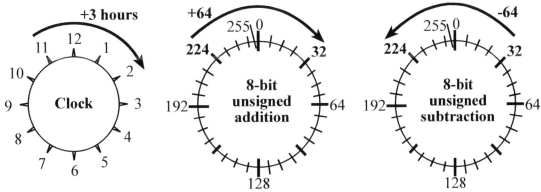

Figure 2.20. The carry bit is set on addition when crossing the 255–0 boundary. The carry bit is cleared on subtraction when crossing the 255–0 boundary.

Similarly, when two 32-bit numbers are added or subtracted, the result may not fit back into a 32-bit register. The same addition and subtraction hardware (instructions) can be used to operate on either unsigned or signed numbers. Although we use the same instructions, we must use separate overflow detection for signed and unsigned operations.

Checkpoint 2.9: How many bits does it take to store the result of two unsigned 32-bit numbers added together?

Checkpoint 2.10: How many bits does it take to store the result of two signed 32-bit numbers added together?

Checkpoint 2.11: Where is the barrier (discontinuity) on a signed 8-bit number wheel?

Let M be the 32-bit value specified by the `#imm12` constant or generated by the flexible second operand, `Operand2`. When `Rd` is absent, the result is placed back in `Rn`. The compare instructions `CMP` and `CMN` do not save the result of the subtraction, but always set the condition code. The compare instructions are used to create conditional execution, such as if-then, for loops, and while loops. The compiler may use `RSB` or `CMN` to optimize execution speed.

```
ADD{S}{cond} {Rd,} Rn, Operand2       ;Rd = Rn + M
ADD{S}{cond} {Rd,} Rn, #imm12         ;Rd = Rn + M
SUB{S}{cond} {Rd,} Rn, Operand2       ;Rd = Rn - M
SUB{S}{cond} {Rd,} Rn, #imm12         ;Rd = Rn - M
RSB{S}{cond} {Rd,} Rn, Operand2       ;Rd = M - Rn
RSB{S}{cond} {Rd,} Rn, #imm12         ;Rd = M - Rn
CMP{cond} Rn, Operand2                ;Rn - M
CMN{cond} Rn, Operand2                ;Rn - (-M)
```

If the optional `S` suffix is present, addition and subtraction set the condition code bits as shown in Table 2.12. The addition and subtraction instructions work for both signed and unsigned values. As designers, we must know in advance whether we have signed or unsigned numbers. The computer cannot tell from the binary which type it is, so it sets both C and V. Our job as programmers is to look at the C bit if the values are unsigned and look at the V bit if the values are signed.

Bit	Name	Meaning after addition or subtraction
N	negative	Result is negative
Z	zero	Result is zero
V	overflow	Signed overflow
C	carry	Unsigned overflow

Table 2.12. Condition code bits contain the status of the previous arithmetic operation.

If the two inputs to an addition operation are considered as unsigned, then the C bit (carry) will be set if the result does not fit. In other words, after an unsigned addition, the C bit is set if the answer is wrong. If the two inputs to a subtraction operation are considered as unsigned, then the C bit (carry) will be clear if the result does not fit. If the two inputs to an addition or subtraction operation are considered as signed, then the V bit (overflow) will be set if the result does not fit. In other words, after a signed addition, the V bit is set if the answer is wrong. If the result is unsigned, the N=1 means the result is greater than or equal to 2^{31}. Conversely, if the result is signed, the N=1 means the result is negative. Assuming the optional `S` suffix is present, condition code bits are set after the addition $R=X+M$, where X is initial register value and R is the final register value.

N: result is negative $N = R_{31}$

Z: result is zero $Z = \overline{R_{31}} \ \& \ \overline{R_{30}} \ \& \cdots \& \ \overline{R_0}$

V: signed overflow $V = X_{31} \ \& \ M_{31} \ \& \ \overline{R_{31}} \ | \ \overline{X_{31}} \ \& \ \overline{M_{31}} \ \& \ R_{31}$

C: unsigned overflow $C = X_{31} \ \& \ M_{31} \ | \ M_{31} \ \& \ \overline{R_{31}} \ | \ \overline{R_{31}} \ \& \ X_{31}$

If the optional **s** suffix is present, condition code bits are set after the subtraction $R=X-M$, where X is initial register value and R is the final register value. If the C bit is clear after an unsigned subtraction ($R=X-M$), then the result is incorrect because an unsigned overflow occurred.

N: result is negative $\qquad N = R_{31}$

Z: result is zero $\qquad Z = \overline{R_{31}} \ \& \ \overline{R_{30}} \ \& \cdots \& \ \overline{R_0}$

V: signed overflow $\qquad V = X_{31} \ \& \ M_{31} \ \& \ \overline{R_{31}} \ \ | \ \ \overline{X_{31}} \ \& \ \overline{M_{31}} \ \& \ R_{31}$

C: unsigned overflow $\qquad C = \overline{X_{31} \ \& \ \overline{M_{31}} \ \ | \ \ M_{31} \ \& \ R_{31} \ \ | \ \ R_{31} \ \& \ \overline{X_{31}}}$

We begin the design of an adder circuit with a simple subcircuit called a binary full adder, as shown in Figure 2.21. There are two binary data inputs A, B and a carry input, C_{in}. There is one data output S_{out}, and one carry output, C_{out}. As shown in Table 2.13, C_{in} A, and B are three independent binary inputs each of which could be 0 or 1. These three inputs are added together (the sum could be 0, 1, 2, or 3) and the result is encoded in the two-bit binary result with C_{out} as the most significant bit and S_{out} as the least significant bit. C_{out} is true if the sum is 2 or 3, and S_{out} is true if the sum is 1 or 3.

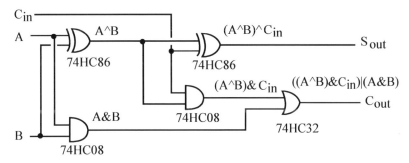

Figure 2.21. A binary full adder.

A	B	C_{in}	$A+B+C_{in}$	C_{out}	S_{out}
0	0	0	0	0	0
0	0	1	1	0	1
0	1	0	1	0	1
0	1	1	2	1	0
1	0	0	1	0	1
1	0	1	2	1	0
1	1	0	2	1	0
1	1	1	3	1	1

Table 2.13. Input/output response of a binary full adder.

We build 32-bit adder by concatenating 32 binary full adders together. The carry into the 32-bit adder is zero, and the carry out will be saved in the carry bit.

Checkpoint 2.12: How many bits does it take to store the result of two unsigned 32-bit numbers multiplied together?

Checkpoint 2.13: How many bits does it take to store the result of two signed 32-bit numbers multiplied together?

Multiply (**MUL**), multiply with accumulate (**MLA**), and multiply with subtract (**MLS**) use 32-bit operands, and producing a 32-bit result. These three multiply instructions only save the bottom 32 bits of the result. They can be used for either signed or unsigned numbers, but no overflow flags are generated. If the **Rd** register is omitted, the **Rn** register is the destination. If the **S** suffix is added to **MUL**, then the Z and N bits are set according to the result. The division instructions do not set condition code flags, and will round towards zero if the division does not evenly divide into an integer quotient.

```
MUL{S}{cond} {Rd,} Rn, Rm       ; Rd = Rn * Rm
MLA{cond} Rd, Rn, Rm, Ra        ; Rd = Ra + Rn*Rm
MLS{cond} Rd, Rn, Rm, Ra        ; Rd = Ra - Rn*Rm
UDIV{cond} {Rd,} Rn, Rm         ; Rd = Rn/Rm       unsigned
SDIV{cond} {Rd,} Rn, Rm         ; Rd = Rn/Rm       signed
```

The following four multiply instructions use 32-bit operands and produce a 64-bit result. The two registers **RdLo** and **RdHi** contain the least significant and most significant parts respectively of the 64-bit result, signified as **Rd**. These multiply instructions do not set condition code flags

```
UMULL{cond} RdLo, RdHi, Rn, Rm  ; Rd = Rn * Rm
SMULL{cond} RdLo, RdHi, Rn, Rm  ; Rd = Rn*Rm
UMLAL{cond} RdLo, RdHi, Rn, Rm  ; Rd = Rd + Rn * Rm
SMLAL{cond} RdLo, RdHi, Rn, Rm  ; Rd = Rd + Rn*Rm
```

Checkpoint 2.14: Can the 32 by 32 bit multiply instructions UMULL or SMULL overflow?

2.3.7. Functions and control flow

Normally the computer executes one instruction after another in a linear fashion. In particular, the next instruction to execute is found immediately following the current instruction. We use branch instructions to deviate from this straight line path. Table 2.6 lists the conditional execution available on the ARM® Cortex™-M. In this section, we will use the conditional branch instruction to implement if-then, while-loop and for-loop control structures.

```
B{cond}   label      ; branch to label
BX{cond}  Rm         ; branch indirect to location specified by Rm
BL{cond}  label      ; branch to subroutine at label
BLX{cond} Rm         ; branch to subroutine indirect specified by Rm
```

Subroutines, **procedures**, and **functions** are programs that can be called to perform specific tasks. They are important conceptual tools because they allow us to develop modular software. The programming languages Pascal, Fortran, and Ada distinguish between functions, which return values, and procedures, which do not. On the other hand, the programming languages C, C++, Java, and Lisp do not make this distinction and treat functions and procedures as synonymous. Object-oriented programming languages use the term **method** to describe programs that are part of objects; it is also used in conjunction with type classes. In assembly language, we use the term subroutine for all subprograms whether or not they return a value. Modular programming allows us to build complex systems using simple components. In this section we present a short introduction on the syntax for defining subroutines. We define a subroutine by giving it a name in the label field, followed by instructions, which when executed, perform the desired effect. The last instruction in a subroutine will be `BX LR`, which we use to return from the subroutine. In Program 2.2, we define the subroutine named `Change`, which adds 25 to the variable `Num`. The flowchart for this example is drawn in Figure 2.22. In assembly language, we will use the `BL` instruction to call this subroutine. At run time, the `BL` instruction will save the return address in the LR register. The return address is the location of the instruction immediately after the `BL` instruction. At the end of the subroutine, the `BX LR` instruction will get the return address from the LR register, returning the program to the place from which the subroutine was called. More precisely, it returns to the instruction immediately after the instruction that performed the subroutine call. The comments specify the order of execution. The while-loop causes instructions 4–10 to be repeated over and over.

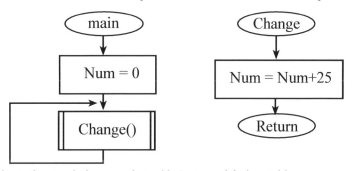

Figure 2.22. A flowchart of a simple function that adds 25 to a global variable.

```
Change LDR    R1,=Num      ; 5)  R1  = &Num
       LDR    R0,[R1]      ; 6)  R0  = Num
       ADD    R0,R0,#25    ; 7)  R0  = Num+25
       STR    R0,[R1]      ; 8)  Num = Num+25
       BX     LR           ; 9)  return
main   LDR    R1,=Num      ; 1)  R1  = &Num
       MOV    R0,#0        ; 2)  R0  = 0
       STR    R0,[R1]      ; 3)  Num = 0
loop   BL     Change       ; 4)  function call
       B      loop         ; 10) repeat
```

```
uint32_t Num;
void Change(void){
  Num = Num+25;
}
void main(void){
  Num = 0;
  while(1){
    Change();
  }
}
```

Program 2.2. Assembly and C versions that define a simple function.

In C, input parameters, if any, are passed in R0–R3. The output parameter, if needed, is returned in R0. Recall that all object code is halfword aligned, meaning bit 0 of the PC is always clear. When the `BL` instruction is executed, bits 31–1 of register LR are loaded with the address of the instruction after the `BL`, and bit 0 is set to one. When the `BX LR` instruction is executed, bits 31–1 of register LR are put back into the PC, and bit 0 of LR goes into the T bit. On the ARM® Cortex™-M, the T bit should always be 1, meaning the processor is always in the Thumb state. Normally, the proper value of the T bit is assigned automatically.

Decision making is an important aspect of software programming. Two values are compared and certain blocks of program are executed or skipped depending on the results of the comparison. In assembly language it is important to know the precision (e.g., 16-bit, 32-bit) and the format of the two values (e.g., unsigned, signed). It takes three steps to perform a comparison. We begin by reading the first value into a register. The second step is to compare the first value with the second value. We can use either a subtract instruction (**subs**) or a compare instruction (**cmp**). These instructions set the condition code bits. The last step is a conditional branch. The available conditions are listed in Table 2.6. The branch will occur if the condition is true.

Program 2.3 illustrates an if-then structure involving testing for unsigned greater than or equal to. It will increment **Num** if it is less than 25600. Since the variable is unsigned, we use an unsigned conditional. Furthermore, we want to execute the increment if **Num** is less than 25600, so we perform the opposite conditional branch (greater than or equal to) to skip over.

```
Change LDR   R1,=Num      ; R1 = &Num         uint32_t Num;
       LDR   R0,[R1]      ; R0 = Num          void Change(void){
       CMP   R0,#25600                          if(Num < 25600){
       BHS   skip                                 Num = Num+1;
       ADD   R0,R0,#1     ; R0 = Num+1         }
       STR   R0,[R1]      ; Num = Num+1       }
skip   BX    LR           ; return
```

Program 2.3. Assembly and C software showing an if-then control structure.

Program 2.4 illustrates an if-then-else structure involving signed numbers. It will increment **Num** if it is less than 100, otherwise it will set it to -100. Since the variable is signed, we use an signed conditional. Again, we want to execute the increment if **Num** is less than 100, so we perform the opposite conditional branch (greater than or equal to) to skip over.

```
Change LDR   R1,=Num      ; R1 = &Num         int32_t Num;
       LDR   R0,[R1]      ; R0 = Num          void Change(void){
       CMP   R0,#100                            if(Num < 100){
       BGE   else                                 Num = Num+1;
       ADD   R0,R0,#1     ; R0 = Num+1         }
       B     skip                              else{
else   MOV   R0,#-100     ; -100                 Num = -100;
skip   STR   R0,[R1]      ; update Num        }
       BX    LR           ; return            }
```

Program 2.4. Assembly and C software showing an if-then-else control structure.

Checkpoint 2.15: Why does Program 2.3 use **BHS** and Program 2.4 use **BGE**?

If-then-else control structures are commonly found in computer software. If the **BHS** in Program 2.3 or the **BGE** in Program 2.4 were to branch, the instruction pipeline would have to be flushed and refilled. In order to optimize execution speed for short if-then and if-then-else control structures, the ARM® Cortex™-M employs conditional execution. The conditional execution begins with the **IT** instruction, which specifies the number of instructions in the control structure (1 to 4) and the conditional for the first instruction. The syntax is

IT{x{y{z}}} cond

where **x y** and **z** specify the existence of the optional second, third, or fourth conditional instruction respectively. We can specify **x y** and **z** as **T** for execute if true or **E** for else. The **cond** field choices are listed in Table 2.6. The conditional suffixes for the 1 to 4 following instruction must match the conditional field of the **IT** instruction. In particular, the conditional for the true instructions exactly match the conditional for the **IT** instruction. Furthermore, the else instructions must have the logical complement conditional. If the condition is true the instruction is executed. If the condition is false, the instruction is fetched, but not executed. For example, Program 2.3 could have been written as follows. The two T's in **ITT** means there are two true instructions.

```
Change  LDR    R1,=Num       ; R1 = &Num
        LDR    R0,[R1]       ; R0 = Num
        CMP    R0,#25600
        ITT    LO
        ADDLO  R0,R0,#1      ; if(R0<25600) R0 = Num+1
        STRLO  R0,[R1]       ; if(R0<25600) Num = Num+1
        BX     LR            ; return
```

Program 2.4 could have been written as follows. The one T and one E in **ITE** means there is one true and one else instruction.

```
Change  LDR    R1,=Num       ; R1 = &Num
        LDR    R0,[R1]       ; R0 = Num
        CMP    R0,#100
        ITE    LT
        ADDLT  R0,R0,#1      ; if(R0< 100) R0 = Num+1
        MOVGE  R0,#-100      ; if(R0>=100) R0 = -100
        STR    R0,[R1]       ; update Num
        BX     LR            ; return
```

The following assembly converts one hex digit (0–15) in R0 to ASCII in R1. The one T and one E in **ITE** means there is one true and one else instruction.

```
        CMP    R0,#9         ; Convert R0 (0 to 15) into ASCII
        ITE    GT            ; Next 2 are conditional
        ADDGT  R1,R0,#55     ; Convert 0xA -> 'A'
        ADDLE  R1,R0,#48     ; Convert 0x0 -> '0'
```

By themselves, the conditional branch instructions do not require a preceding **IT** instruction. However, a conditional branch can be used as the last instruction of an **IT** block. There are a lot of restrictions on IT. For more details, refer to the programming reference manual.

2.3.8. Stack usage

Figure 2.5 shows the **push** and **pop** instructions can be used to store temporary information on the stack. If a subroutine modifies a register, it is a matter of programmer style as to whether or not it should save and restore the register. According to AAPCS a subroutine can freely change R0–R3 and R12, but save and restore any other register it changes. In particular, if one subroutine calls another subroutine, then it must save and restore the LR. AAPCS also requires pushing and popping multiples of 8 bytes, which means an even number of registers. In the following example, assume the function modifies register R0, R4, R7, R8 and calls another function. The programming style dictates registers R4 R7 R8 and LR be saved. Notice the return address is pushed on the stack as LR, but popped off into PC. When multiple registers are pushed or popped, the data exist in memory with the lowest numbered register using the lowest memory address. In other words, the registers in the { } can be specified in any order. Of course remember to balance the stack by having the same number of pops as pushes.

```
Func    PUSH  {R4,R7,R8,LR}   ; save registers as needed
        ; body of the function
        POP   {R4,R7,R8,PC}   ; restore registers and return
```

The ARM processor has a lot of registers, and we appropriately should use them for temporary information such as function parameters and local variables. However, when there are a lot of parameters or local variables we can place them on the stack. Program 2.5 is similar to Program 2.1, except the **data** buffer is now local, and placed on the stack. The **SUB** instruction allocates 10 words on the stack. Figure 2.23 shows the stack before and after the allocation. The SP points to the first location of **data**. The local variable **i** is held in R0. The flexible second operand for the STR instruction uses SP as the base pointer, and R0*4 as the offset. The **ADD** instruction deallocates the local variable, balancing the stack.

Set	SUB	sp,sp,#0x28	;allocate	`// C language`
	MOVS	r0,#0x00	;i=0	
	B	test		`void Set(void){`
loop	STR	r0,[sp,r0,LSL #2]		`uint32_t data[10];`
	ADDS	r0,r0,#1	;i++	`int i;`
test	CMP	r0,#0x0A		`  for(i=0; i<10; i++){`
	BLT	loop		`    data[i] = i;`
	ADD	sp,sp,#0x28	;deallocate	`  }`
	BX	LR		`}`

Program 2.5. Assembly and C versions that initialize a local array of ten elements.

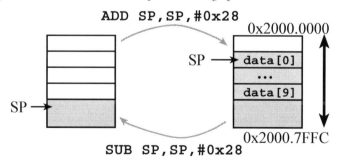

Figure 2.23. A stack picture showing a local array of ten elements (40 bytes).

2.3.9. Keil assembler directives

We use assembler directives to assist and control the assembly process. The following directives change the way the code is assembled.

`AREA CODE`	; places code in code space (flash ROM)
`AREA DATA`	; places objects in data space (RAM)
`THUMB`	; uses Thumb instructions
`ALIGN`	; skips 0 to 3 bytes to make next word aligned
`END`	; end of file

The following directives can add variables and constants.

`DCB expr{,expr}`	; places 8-bit byte(s) into memory
`DCW expr{,expr}`	; places 16-bit halfword(s) into memory
`DCD expr{,expr}`	; places 32-bit word(s) into memory
`SPACE size`	; reserves size bytes, uninitialized

The **EQU** directive gives a symbolic name to a numeric constant, a register-relative value or a program-relative value. * is a synonym for **EQU**. We will use it to define I/O port addresses. For example, these four definitions will be used to initialize and operate Port D.

```
GPIO_PORTD_DATA_R  equ  0x400073FC
GPIO_PORTD_DIR_R   equ  0x40007400
GPIO_PORTD_DEN_R   equ  0x4000751C
SYSCTL_RCGCGPIO_R  equ  0x400FE608
```

2.4. Parallel I/O ports

2.4.1. Basic concepts of input and output ports

The simplest I/O port on a microcontroller is the parallel port. A parallel I/O port is a simple mechanism that allows the software to interact with external devices. It is called parallel because multiple signals can be accessed all at once. An **input port**, which allows the software to read external digital signals, is read only. That means a read cycle access from the port address returns the values existing on the inputs at that time. In particular, the tristate driver (triangle shaped circuit in Figure 2.24) will drive the input signals onto the data bus during a read cycle from the port address. A write cycle access to an input port usually produces no effect. The digital values existing on the input pins are copied into the microcontroller when the software executes a read from the port address. There are no input-only ports on TM4C microcontrollers. TM4C microcontrollers have 5V-tolerant inputs, meaning an input high signal can be any voltage from 2.0 to 5.0 V. On the STMicroelectronics STM32F10xx, some inputs are 5-V tolerant and others are not. On the MSP432 none of the pins are 5-V tolerant.

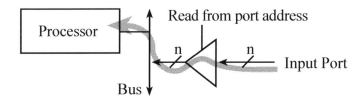

Figure 2.24. A read only input port allows the software to sense external digital signals.

Checkpoint 2.16: What happens if the software writes from an input port like Figure 2.24?

Common Error: Many program errors can be traced to confusion between I/O ports and regular memory. For example, you cannot write to an input port.

While an input device usually just involves the software reading the port, an output port can participate in both the read and write cycles very much like a regular memory. Figure 2.25 describes a **readable output port**. A write cycle to the port address will affect the values on the output pins. In particular, the microcontroller places information on the data bus and that information is clocked into the D flip flops. Since it is a readable output, a read cycle access from the port address returns the current values existing on the port pins. There are no output-only ports on TM4C microcontrollers.

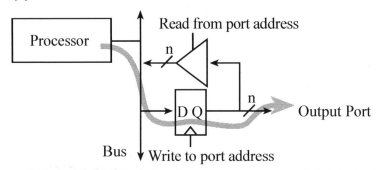

Figure 2.25. A readable output port allows the software to generate external digital signals.

Checkpoint 2.17: What happens if the software reads from an output port like Figure 2.25?

To make the microcontroller more marketable, most ports can be software-specified to be either inputs or outputs. Microcontrollers use the concept of a **direction register** to determine whether a pin is an input (direction register bit is 0) or an output (direction register bit is 1), as shown in Figure 2.26. We define an initialization ritual as a program executed during start up that initializes hardware and software. If the ritual software makes direction bit zero, the port behaves like a simple input, and if it makes the direction bit one, it becomes a readable output port. Each digital port pin has a direction bit. This means some pins on a port may be inputs while others are outputs. The digital port pins on most microcontrollers are bidirectional, operating similar to Figure 2.26.

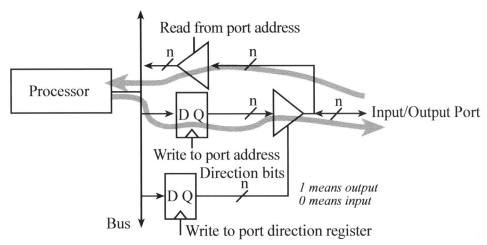
Figure 2.26. A bidirectional port can be configured as a read-only input port or a readable output port.

2.4.2. I/O Programming and the direction register

On most embedded microcontrollers, the I/O ports are memory mapped. This means the software accesses an input/output port simply by reading from or writing to the appropriate address. To make our software more readable we include symbolic definitions for the I/O ports. We set the direction register (e.g., **GPIO_PORTD_DIR_R**) to specify which pins are input and which are output. By default, the alternate function register is zero, specifying the corresponding bits are regular port pins (e.g., **GPIO_PORTD_AFSEL_R**). We will set bits in the alternative function register when we wish to activate the functions listed in Tables 2.4 and 2.5. Typically, we write to the direction and alternate function registers once during the initialization phase. We use the data register (e.g., **GPIO_PORTD_DATA_R**) to perform input/output on the port. Conversely, we read and write the data register multiple times to perform input and output respectively during the running phase. Table 2.14 shows the parallel port registers for the TM4C123. The other Texas Instruments microcontrollers are similar. The CR, AMSEL, PCTL, and LOCK registers exist only on LM4F/TM4C. For the LM3S software, simply remove accesses to these four registers. The only differences among various members of the Texas Instruments microcontroller family are the number of ports and available pins in each port.

For example, the MSP432E401Y has fifteen digital I/O ports A (8 bits), B (6 bits), C (8 bits), D (8 bits), E (6 bits), F (5 bits), G (2 bits), H (4 bits), J (2 bits), K (8 bits), L (8 bits), M (8 bits), N(6 bits), P (6 bits), and Q (5 bits). Furthermore, the MSP432E401Y has different addresses for ports. Refer to the file **msp432e401y.h** or to the data sheet for more the specific addresses of its I/O ports. We use the **PUR** register to activate an internal pull-up resistor, and we use the **PDR** register to activate an internal pull-down resistor.

Address	7	6	5	4	3	2	1	0	Name
0x400FE608	-	-	GPIOF	GPIOE	GPIOD	GPIOC	GPIOB	GPIOA	SYSCTL_RCGCGPIO_R
0x400FEA08	-	-	GPIOF	GPIOE	GPIOD	GPIOC	GPIOB	GPIOA	SYSCTL_PRGPIO_R
0x400043FC	DATA	DATA	DATA	DATA	DATA	DATA	DATA	DATA	GPIO_PORTA_DATA_R
0x40004400	DIR	DIR	DIR	DIR	DIR	DIR	DIR	DIR	GPIO_PORTA_DIR_R
0x40004420	SEL	SEL	SEL	SEL	SEL	SEL	SEL	SEL	GPIO_PORTA_AFSEL_R
0x40004510	PUE	PUE	PUE	PUE	PUE	PUE	PUE	PUE	GPIO_PORTA_PUR_R
0x4000451C	DEN	DEN	DEN	DEN	DEN	DEN	DEN	DEN	GPIO_PORTA_DEN_R
0x40004524	1	1	1	1	1	1	1	1	GPIO_PORTA_CR_R
0x40004528	0	0	0	0	0	0	0	0	GPIO_PORTA_AMSEL_R
0x400053FC	DATA	DATA	DATA	DATA	DATA	DATA	DATA	DATA	GPIO_PORTB_DATA_R
0x40005400	DIR	DIR	DIR	DIR	DIR	DIR	DIR	DIR	GPIO_PORTB_DIR_R
0x40005420	SEL	SEL	SEL	SEL	SEL	SEL	SEL	SEL	GPIO_PORTB_AFSEL_R
0x40005510	PUE	PUE	PUE	PUE	PUE	PUE	PUE	PUE	GPIO_PORTB_PUR_R
0x4000551C	DEN	DEN	DEN	DEN	DEN	DEN	DEN	DEN	GPIO_PORTB_DEN_R
0x40005524	1	1	1	1	1	1	1	1	GPIO_PORTB_CR_R
0x40005528	0	0	AMSEL	AMSEL	0	0	0	0	GPIO_PORTB_AMSEL_R
0x400063FC	DATA	DATA	DATA	DATA	JTAG	JTAG	JTAG	JTAG	GPIO_PORTC_DATA_R
0x40006400	DIR	DIR	DIR	DIR	JTAG	JTAG	JTAG	JTAG	GPIO_PORTC_DIR_R
0x40006420	SEL	SEL	SEL	SEL	JTAG	JTAG	JTAG	JTAG	GPIO_PORTC_AFSEL_R
0x40006510	PUE	PUE	PUE	PUE	JTAG	JTAG	JTAG	JTAG	GPIO_PORTC_PUR_R
0x4000651C	DEN	DEN	DEN	DEN	JTAG	JTAG	JTAG	JTAG	GPIO_PORTC_DEN_R
0x40006524	1	1	1	1	JTAG	JTAG	JTAG	JTAG	GPIO_PORTC_CR_R
0x40006528	AMSEL	AMSEL	AMSEL	AMSEL	JTAG	JTAG	JTAG	JTAG	GPIO_PORTC_AMSEL_R
0x400073FC	DATA	DATA	DATA	DATA	DATA	DATA	DATA	DATA	GPIO_PORTD_DATA_R
0x40007400	DIR	DIR	DIR	DIR	DIR	DIR	DIR	DIR	GPIO_PORTD_DIR_R
0x40007420	SEL	SEL	SEL	SEL	SEL	SEL	SEL	SEL	GPIO_PORTD_AFSEL_R
0x40007510	PUE	PUE	PUE	PUE	PUE	PUE	PUE	PUE	GPIO_PORTD_PUR_R
0x4000751C	DEN	DEN	DEN	DEN	DEN	DEN	DEN	DEN	GPIO_PORTD_DEN_R
0x40007524	CR	1	1	1	1	1	1	1	GPIO_PORTD_CR_R
0x40007528	0	0	AMSEL	AMSEL	AMSEL	AMSEL	AMSEL	AMSEL	GPIO_PORTD_AMSEL_R
0x400243FC	-	-	DATA	DATA	DATA	DATA	DATA	DATA	GPIO_PORTE_DATA_R
0x40024400	-	-	DIR	DIR	DIR	DIR	DIR	DIR	GPIO_PORTE_DIR_R
0x40024420	-	-	SEL	SEL	SEL	SEL	SEL	SEL	GPIO_PORTE_AFSEL_R
0x40024510	-	-	PUE	PUE	PUE	PUE	PUE	PUE	GPIO_PORTE_PUR_R
0x4002451C	-	-	DEN	DEN	DEN	DEN	DEN	DEN	GPIO_PORTE_DEN_R
0x40024524	-	-	1	1	1	1	1	1	GPIO_PORTE_CR_R
0x40024528	-	-	AMSEL	AMSEL	AMSEL	AMSEL	AMSEL	AMSEL	GPIO_PORTE_AMSEL_R
0x400253FC	-	-	-	DATA	DATA	DATA	DATA	DATA	GPIO_PORTF_DATA_R
0x40025400	-	-	-	DIR	DIR	DIR	DIR	DIR	GPIO_PORTF_DIR_R
0x40025420	-	-	-	SEL	SEL	SEL	SEL	SEL	GPIO_PORTF_AFSEL_R
0x40025510	-	-	-	PUE	PUE	PUE	PUE	PUE	GPIO_PORTF_PUR_R
0x4002551C	-	-	-	DEN	DEN	DEN	DEN	DEN	GPIO_PORTF_DEN_R
0x40025524	-	-	-	1	1	1	1	CR	GPIO_PORTF_CR_R
0x40025528	-	-	-	0	0	0	0	0	GPIO_PORTF_AMSEL_R

Address	31-28	27-24	23-20	19-16	15-12	11-8	7-4	3-0	Name
0x4000452C	PMC7	PMC6	PMC5	PMC4	PMC3	PMC2	PMC1	PMC0	GPIO_PORTA_PCTL_R
0x4000552C	PMC7	PMC6	PMC5	PMC4	PMC3	PMC2	PMC1	PMC0	GPIO_PORTB_PCTL_R
0x4000652C	PMC7	PMC6	PMC5	PMC4	0x1	0x1	0x1	0x1	GPIO_PORTC_PCTL_R
0x4000752C	PMC7	PMC6	PMC5	PMC4	PMC3	PMC2	PMC1	PMC0	GPIO_PORTD_PCTL_R
0x4002452C	----	----	PMC5	PMC4	PMC3	PMC2	PMC1	PMC0	GPIO_PORTE_PCTL_R
0x4002552C	----	----	----	PMC4	PMC3	PMC2	PMC1	PMC0	GPIO_PORTF_PCTL_R
0x40006520	LOCK (write 0x4C4F434B to unlock, other locks) (reads 1 if locked, 0 if unlocked)								GPIO_PORTC_LOCK_R
0x40007520	LOCK (write 0x4C4F434B to unlock, other locks) (reads 1 if locked, 0 if unlocked)								GPIO_PORTD_LOCK_R
0x40025520	LOCK (write 0x4C4F434B to unlock, other locks) (reads 1 if locked, 0 if unlocked)								GPIO_PORTF_LOCK_R

Table 2.14. Some TM4C123 parallel ports. Each register is 32 bits wide. For PCTL bits, see Tables 2.4 and 2.5. JTAG means do not use these pins and do not change any of these bits.

To initialize a TM4C I/O port for general use we perform seven steps. Because a hardware reset will assume digital GPIO, steps 3, 4 and 6 could be skipped. First, we activate the clock for the port. We need to add a short delay between activating the clock and accessing the port registers. Second, we unlock the pin if needed. PC3-0 are locked to the debugger. Only PD7 and PF0 on the TM4C123 need to be unlocked. On the MSP432E401Y only PD7 needs to be unlocked. All the other bits on the two microcontrollers are always unlocked. Third, we disable the analog function. Fourth, we select GPIO mode in PCTL. Fifth, we set its direction register. The direction register specifies bit for bit whether the corresponding pins are input (0) or output (1). Sixth, we disable the alternate function. Seventh, we enable the digital port.

Common Error: You will get a bus fault if you access a port without enabling its clock. Also, you have to wait about 5 bus cycles after enabling the clock, before you access the registers.

In this first example we will make PD7-4 input, and we will make PD3-0 output, as shown in Program 2.6. To use a port we first must activate its clock in the **SYSCTL_RCGCGPIO_R** register. The second step is to unlock the port (if needed), by writing a special value to the **LOCK** register, followed by setting bits in the **CR** register. The third step is to disable the analog functionality (reset will configure to digital), by clearing bits in the **AMSEL** register. The fourth step is to select GPIO functionality (reset will configure to GPIO), by clearing bits in the **PCTL** register, as described in Tables 2.4 and 2.5. The fifth step is to specify whether the pin is an input or an output by clearing or setting bits in the **DIR** register. Because we are using the pins as regular digital I/O, the sixth step is to clear the corresponding bits in the **AFSEL** register. The last step is to enable the corresponding I/O pins by writing ones to the **DEN** register. To run this example on the Texas Instruments LaunchPad, we also set bits in the **PUR** register for the two switch inputs (Figure 2.9) to have an internal pull-up resistor.

When the software reads from location 0x400073FC the bottom 8 bits are returned with the current values on Port D. The top 24 bits are returned zero. As shown in Figure 2.26, the input pins show the current digital state, and the output pins show the value last written to the port. The function **PortD_Input** will read from the four input pins and return a value, 0x00 to 0x0F, depending on the current status of the inputs. The function **PortD_Output** will write new values to the four output pins.

```c
#define GPIO_PORTD_DATA_R    (*((volatile uint32_t *)0x400073FC))
#define GPIO_PORTD_DIR_R     (*((volatile uint32_t *)0x40007400))
#define GPIO_PORTD_AFSEL_R   (*((volatile uint32_t *)0x40007420))
#define GPIO_PORTD_DEN_R     (*((volatile uint32_t *)0x4000751C))
#define GPIO_PORTD_LOCK_R    (*((volatile uint32_t *)0x40007520))
#define GPIO_PORTD_AMSEL_R   (*((volatile uint32_t *)0x40007528))
#define GPIO_PORTD_PCTL_R    (*((volatile uint32_t *)0x4000752C))
#define SYSCTL_RCGCGPIO_R    (*((volatile uint32_t *)0x400FE608))
void PortD_Init(void){
  SYSCTL_RCGCGPIO_R |= 0x08;           // 1) activate clock for Port D
  while((SYSCTL_PRGPIO_R&0x08) == 0){};// ready?
  GPIO_PORTD_LOCK_R = 0x4C4F434B;      // 2) unlock GPIO Port D
  GPIO_PORTD_CR_R = 0xFF;              // allow changes to PD7
  GPIO_PORTD_AMSEL_R = 0x00;           // 3) disable analog on PD
  GPIO_PORTD_PCTL_R = 0x00000000;      // 4) PCTL GPIO on PD7-0
  GPIO_PORTD_DIR_R = 0x0F;             // 5) PD7-4 in, PD3-0 out
  GPIO_PORTD_AFSEL_R = 0x00;           // 6) disable alt funct on PD7-0
```

```
  GPIO_PORTD_DEN_R = 0xFF;              // 7) enable digital I/O on PD7-0
}
uint32_t PortD_Input(void){
  return (GPIO_PORTD_DATA_R>>4); // read PD7-PD4 inputs
}
void PortD_Output(uint32_t data){
  GPIO_PORTD_DATA_R = data;       // write PD3-PD0 outputs
}
```
Program 2.6. A set of functions using PD7–PD4 as inputs and PD3–PD0 as outputs.

Checkpoint 2.18: Does the entire port need to be defined as input or output, or can some pins be input while others are output?

In Program 2.6 the assumption was the software module had access to all of Port D. In other words, this software owned all eight pins of Port D. In most cases, a software module needs access to only some of the port pins. If two or more software modules access the same port, a conflict will occur if one module changes modes or output values set by another module. It is good software design to write **friendly** software, which only affects the individual pins as needed. Friendly software does not change the other bits in a shared register. Conversely, **unfriendly** software modifies more bits of a register than it needs to. The difficulty of unfriendly code is each module will run properly when tested by itself, but weird bugs result when two or more modules are combined.

Consider the problem that a software module need to output to just Port D bit 1. After enabling the clock for Port D, we use read-modify-write software to initialize just pin 1. Remember only PD7 and PF0 require unlocking on the TM4C123, and only PD7 requires unlocking on the MSP432E401Y, so this code does not need to unlock.

```
SYSCTL_RCGCGPIO_R |= 0x08;            // 1) activate clock for Port D
while((SYSCTL_PRGPIO_R&0x08) == 0){};// ready?
GPIO_PORTD_DIR_R |= 0x02;             // PD1 is an output
GPIO_PORTD_AFSEL_R &= ~0x02;          // regular port function
GPIO_PORTD_AMSEL_R &= ~0x02;          // disable analog on PD1
GPIO_PORTD_PCTL_R &= ~0x000000F0;     // PCTL GPIO on PD1
GPIO_PORTD_DEN_R |= 0x02;             // PD1 is enabled as a digital port
```

There is no conflict if two or more modules enable the clock for Port D. There are two ways on the Cortex™-M to access individual port bits. The first method is to use read-modify-write software to change just pin 1. A read-or-write sequence can be used to set one or more bits.

```
GPIO_PORTD_DATA_R |= 0x02;            // make PD1 high
```

A read-and-write sequence can be used to clear one or more bits.

```
GPIO_PORTD_DATA_R &= ~0x02;           // make PD1 low
```

The second method uses the **bit-specific addressing**. The TM4C family implements a more flexible way to access port pins than the bit-banding described earlier in the chapter. This bit-specific addressing doesn't work for all the I/O registers, just the parallel port data registers. The TM4C mechanism allows collective access to 1 to 8 bits in a data port. We define eight address offset constants in Table 2.15. Basically, if we are interested in bit b, the constant is

$4*2^b$. There are 256 possible bit combinations we might be interested in accessing, from all of them to none of them. Each possible bit combination has a separate address for accessing that combination. For each bit we are interested in, we add up the corresponding constants from Table 2.15 and then add that sum to the base address for the port. The base addresses for the data ports can be found in GPIO chapter of the microcontroller data sheet. For example, assume we are interested in Port A bits 1, 2, and 3. The base address for Port A is 0x4000.4000, and the constants are 0x0020, 0x0010 and 0x008. The sum of 0x4000.4000+0x0020+0x0010+0x008 is the address 0x4000.4038. If we read from 0x4000.4038 only bits 1, 2, and 3 will be returned. If we write to this address only bits 1, 2, and 3 will be modified.

If we wish to access bit	Constant
7	0x0200
6	0x0100
5	0x0080
4	0x0040
3	0x0020
2	0x0010
1	0x0008
0	0x0004

Table 2.15. Address offsets used to specify individual data port bits.

The base address for Port D is 0x4000.7000. If we want to read and write all 8 bits of this port, the constants will add up to 0x03FC. Notice that the sum of the base address and the constants yields the 0x4000.73FC address used in Program 2.6. In other words, read and write operations to `GPIO_PORTD_DATA_R` will access all 8 bits of Port D. If we are interested in just bit 1 of Port D, we add 0x0008 to 0x4000.7000, and we can define this in C as

```
#define PD1     (*((volatile uint32_t *)0x40007008))
```

Now, a simple write operation can be used to set PD1. The following code is friendly because it does not modify the other 7 bits of Port D.

```
PD1 = 0x02;       // make PD1 high
```

A simple write sequence will clear PD1. The following code is also friendly.

```
PD1 = 0x00;       // make PD1 low
```

A read from `PD1` will return 0x01 or 0x00 depending on whether the pin is high or low, respectively. The following code is also friendly.

```
PD1 = PD1^0x01;   // toggle PD1
```

Checkpoint 2.19: According to Tables 2.4 and 2.5, what happens to Port D bit 5 if we set bit 5 in its alternative function register and PCTL = 0001?

Checkpoint 2.20: What happens if we write to location 0x4000.7000?

Checkpoint 2.21: Specify a #define that allows us to access bits 7 and 2 of Port D. Use this #define to make both bits 7 and 2 of Port D high.

Checkpoint 2.22: Specify a #define that allows us to access bits 6, 5, 0 of Port B. Use this #define to make bits 6, 5 and 0 of Port B high.

Example 2.1: The goal is develop a means for the microcontroller to turn on and turn off an AC-powered appliance. The interface will use a solid state relay (SSR) having a control portion equivalent to an LED with parameters of 2V and 10 mA. Include appropriate functions.

Solution: Since we need to interface an LED, we use an open collector NOT gate just like Figure 1.23. We choose an electronic circuit that has an output current larger than the 10 mA needed by the SSR. Since the maximum I_{CE} of the PN2222 is 150 mA, it can sink the 10 mA required by the SSR. A 7405 or 7406 could also have been used, but they require a +5V supply. The resistor is selected to control the current to the diode. Using the LED design equation, $R = (3.3-V_d-V_{CE})/I_d = (3.3-2-0.3\text{V})/0.01\text{A} = 100\ \Omega$. There is a standard value 5% resistor at 100 Ω. The specification $V_{CE}=0.3\text{V}$ is a maximum. If V_{CE} is actually between 0.1 and 0.3V, then 10 to 12 mA will flow, and the relay will still activate properly. When the input to the PN2222 is high (**p**=3.3V), the output is low (**q**=0.3V), see Figure 2.27. In this state, a 10 mA current is applied to the diode, and relay switch activates. This causes 120 VAC power to be delivered to the appliance. But, when the input is low (**p**=0), the output floats (**q**=HiZ, which is neither high nor low). This floating output state causes the LED current to be zero, and the relay switch opens. In this case, no AC power is delivered to the appliance.

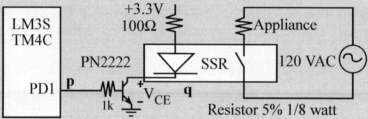

Figure 2.27. Solid state relay interface using a PN2222 NPN transistor.

The initialization will set bit 1 of the direction register to make PD1 an output, see Program 2.7. This function should be called once at the start of the system. After initialization, the on and off functions can be called to control the appliance.

```
#define PD1 (*((volatile uint32_t *)0x40007008))
void SSR_Init(void){ // reset clears AFSEL, AMSEL, PCTL
  SYSCTL_RCGCGPIO_R |= 0x08;          // 1) activate clock for Port D
  while((SYSCTL_PRGPIO_R&0x08) == 0){};// ready?
  GPIO_PORTD_DIR_R |= 0x02;           // PD1 is an output
  GPIO_PORTD_DEN_R |= 0x02;           // PD1 is enabled as a digital port
}
void SSR_Off(void){
  PD1 = 0x00;        // turn off the appliance
}
void SSR_On(void){
  PD1 = 0x02;        // turn on the appliance
}
```
Program 2.7. A set of functions using PD1 as an output (SSR_xxx).

Example 2.2: Interface a push button switch to the microcontroller and write software functions that initialize and read the switch.

Solution: The first step is to draw a hardware circuit connecting the switch to an input port of the microcontroller. We will use positive logic interface because we want the digital signal to be high if and only if the switch is pressed, as shown in Figure 2.28. Similar to Figure 1.24, **PB1** contains a signal that is high or low depending on the position of the switch. If the switch is not pressed, the 10 kΩ resistor creates a 0 V signal on the port pin, and virtually no current flows through the resistor (I_{IL} is 2 µA). If the switch is pressed, a 3.3 V signal is on the port pin and 0.33 mA flows through the 10 kΩ resistor. Some switches bounce, which means there will be multiple open/closed cycles when the switch is changed. This simple solution can be used if the switch doesn't bounce or if the bouncing doesn't matter. The software solution requires two functions. The initialization function is called once when the system starts. Whenever the software wishes to know the switch status, it calls the input function. When the computer reads Port B it gets all 8 bits of the input port. The following C code will set a variable to true (nonzero) if and only if the switch is pressed.

```
Pressed = GPIO_PORTB_DATA_R&0x02;   // true if the switch is pressed
```

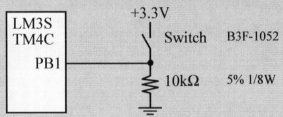

Figure 2.28. Positive logic interface of a switch to a microcontroller input.

The initialization in Program 2.8 activates the clock, clears the direction register bit for PB1, and enables Port B bit 1 as a digital port. The definition of **PB1** in Program 2.8 uses bit-specific addressing so the software just sees bit 1. Calling the module **Switch_** rather than **PortB_** separates what it does (input from a switch) from how it works (input from PB1). This abstraction makes it easier to understand, and easier to change.

```
#define PB1 (*((volatile uint32_t *)0x40005008))
void Switch_Init(void){ // reset clears AFSEL, AMSEL, PCTL
  SYSCTL_RCGCGPIO_R |= 0x02;        // activate clock for Port B
  while((SYSCTL_PRGPIO_R&0x02) == 0){};// ready?
  GPIO_PORTB_DIR_R &= ~0x02;        // PB1 is an input
  GPIO_PORTB_DEN_R |= 0x02;         // PB1 enabled as a digital port
}
uint32_t Switch_Input(void){
  return PB1;      // 0x02 if pressed, 0x00 if not pressed
}
```
Program 2.8. A set of functions that interface an input switch to PB1 (Switch_xxx).

Maintenance Tip: Using abstract function names like **SSR_** and **Switch_** make the software easier to understand, which in turn makes them easier to debug, easier to maintain, and facilitates reuse of the code.

Some problems are so unique that they require the engineer to invent completely original solutions. Most of the time, however, the engineer can solve even complex problems by building the system from components that already exist. Creativity will still be required in selecting the proper components, making small changes in their behavior (tweaking), arranging them in an effective and efficient manner, and then verifying the system satisfies both the requirements and constraints. When young engineers begin their first job, they are sometimes surprised to see that education does not stop with college graduation, but rather is a life-long activity. In fact, it is the educational goal of all engineers to continue to learn both processes (rules about how to solve problems) and products (hardware and software components). As the engineer becomes more experienced, he or she has a larger toolbox from which processes and components can be selected.

The hardest step for most new engineers is the first one: where to begin? We begin by analyzing the problem to create a set of specifications and constraints in the form of a requirements document. Next, we look for components, in the form of previously debugged solutions, which are similar to our needs. Often during the design process, additional questions or concerns arise. We at that point consult with our clients to clarify the problem. Next we rewrite the requirements document and get it reapproved by the clients.

It is often difficult to distinguish whether a parameter is a specification or a constraint. In actuality, when designing a system it often doesn't matter into which category a parameter falls, because the system must satisfy all specifications and constraints. Nevertheless, when documenting the device it is better to categorize parameters properly. Specifications generally define in a quantitative manner the overall system objectives as given to us by our customers.

Constraints, on the other hand, generally define the boundary space within which we must search for a solution to the problem. If we must use a particular component, it is often considered a constraint. In this book, we constrain most designs to include a TM4C microcontroller. Constraints also are often defined as an inequality, such as the cost must be less than $50, or the battery must last for at least one week. Specifications on the other hand are often defined as a quantitative number, and the system satisfies the requirement if the system operates within a specified tolerance of that parameter. Tolerance can be defined as a percentage error or as a range with minimum and maximum values.

The high-level design uses data flow graphs. We then combine the pieces and debug the system. As the pieces are combined we can draw a call graph to organize the parts. If new components are designed, we can use flowcharts to develop new algorithms. The more we can simulate the system, the more design possibilities we can evaluate, and the quicker we can make changes. Debugging involves both making sure it works, together with satisfying all requirements and constraints.

Observation: Defining realistic tolerances on specifications will have a profound effect on cost.

Checkpoint 2.23: What are the effects of specifying a tighter tolerance (e.g., 1% when the problem asked for 5%)?

Checkpoint 2.24: What are the effects of specifying a looser tolerance (e.g., 10% when the problem asked for 5%)?

2.5. Phase-Lock-Loop

Normally, the execution speed of a microcontroller is determined by an external crystal. Both LaunchPad boards have a 16 MHz crystal. Most microcontrollers have a phase-lock-loop (PLL) that allows the software to adjust the execution speed of the computer. Typically, the choice of frequency involves the tradeoff between software execution speed and electrical power. In other words, slowing down the bus clock will require less power to operate and generate less heat. Speeding up the bus clock obviously allows for more calculations per second at the cost of requiring more power to operate and generating more heat.

The default bus speed for the TM4C internal oscillator is 16 MHz ±1%. The internal oscillator is significantly less precise than the crystal, but it requires less power and does not need an external crystal. This means for most applications we will activate the main oscillator and the PLL so we can have a stable bus clock.

There are two ways to activate the PLL. We could call a library function, or we could access the clock registers directly. In general, using library functions creates a better design because the solution will be more stable (less bugs) and will be more portable (easier to switch microcontrollers). However, the objective of the book is to present microcontroller fundamentals. Showing the direct access does illustrate some concepts of the PLL. First, we can include the Stellaris/Tiva library and call the **SysCtlClockSet** function to change the speed. This function is defined in the **sysctl.c** file. The library function activates the PLL because of the **SYSCTL_USE_PLL** parameter. The main oscillator is the one with the external crystal attached. The last parameter specifies the frequency of the attached crystal. Assume we wish to run an TM4C with a 16 MHz crystal at 80 MHz. The divide by 2.5 creates a bus frequency of 80 MHz, implemented as 400 MHz divided by 5.

```
SysCtlClockSet( SYSCTL_SYSDIV_2_5 | SYSCTL_USE_PLL |
                SYSCTL_OSC_MAIN | SYSCTL_XTAL_16MHZ);
```

To make our code more portable, it is a good idea to use library functions whenever possible. However, we will present an explicit example illustrating how the PLL works. An external crystal is attached to the TM4C microcontroller, as shown in Figure 2.29. The PLLs on the other Texas Instruments microcontrollers operate in the same basic manner. Table 2.16 shows the clock registers used to define what speed the processor operates. The output of the main oscillator (Main Osc) is a clock at the same frequency as the crystal. By setting the OSCSRC bits to 0, the multiplexer control will select the main oscillator as the clock source.

For example, the main oscillator for the TM4C on the evaluation board will be 16 MHz. This means the reference clock (Ref Clk) input to the phase/frequency detector will be 16 MHz. For a 16 MHz crystal, we set the XTAL bits to 10101 (see Table 2.16). In this way, a 400 MHz output of the voltage controlled oscillator (VCO) will yield a 16 MHz clock at the other input of the phase/frequency detector. If the 400 MHz clock is too slow, the **up** signal will add charge, increasing the input to the VCO, leading to an increase in the 400 MHz frequency. If the 400 MHz clock is too fast, **down** signal to the charge pump will subtract charge, decreasing the input to the VCO, leading to a decrease in the 400 MHz frequency. The feedback loop in the PLL will drive the output to a stable 400 MHz frequency.

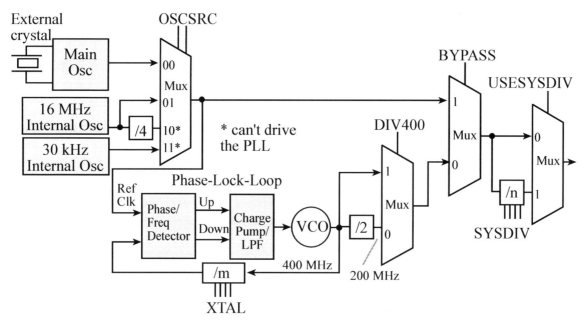

Figure 2.29. Block diagram of the main clock tree on the TM4C including the PLL (others are similar).

XTAL	Crystal Freq (MHz)	XTAL	Crystal Freq (MHz)	XTAL	Crystal Freq (MHz)
0x4	3.579545 MHz	0xC	6.144 MHz	0x14	14.31818 MHz
0x5	3.6864 MHz	0xD	7.3728 MHz	0x15	16.0 MHz
0x6	4 MHz	0xE	8 MHz	0x16	16.384 MHz
0x7	4.096 MHz	0xF	8.192 MHz	0x17	18.0 MHz
0x8	4.9152 MHz	0x10	10.0 MHz	0x18	20.0 MHz
0x9	5 MHz	0x11	12.0 MHz	0x19	24.0 MHz
0xA	5.12 MHz	0x12	12.288 MHz	0x1A	25.0 MHz
0xB	6 MHz (reset value)	0x13	13.56 MHz	others	reserved

Address	26-23	22	13	11	10-6	5-4	Name
$400FE060	SYSDIV	USESYSDIV	PWRDN	BYPASS	XTAL	OSCSRC	SYSCTL_RCC_R
$400FE050					PLLRIS		SYSCTL_RIS_R

	31	30	28-22	13	11	6-4	
$400FE070	USERCC2	DIV400	SYSDIV2	PWRDN2	BYPASS2	OSCSRC2	SYSCTL_RCC2_R

Table 2.16. Main clock registers (other values of XTAL are reserved).

Program 2.9 shows a program to activate a microcontroller with a 16 MHz main oscillator to run at 80 MHz. 0) Use RCC2 because it provides for more options. 1) The first step is set BYPASS2 (bit 11). At this point the PLL is bypassed and there is no system clock divider. 2) The second step is to specify the crystal frequency in the four XTAL bits using the code in Table 2.16. The OSCSRC2 bits are cleared to select the main oscillator as the oscillator clock source. 3) The third step is to clear PWRDN2 (bit 13) to activate the PLL. 4) The fourth step is to configure and enable the clock divider using the 7-bit SYSDIV2 field. If the 7-bit SYSDIV2 is **n**, then the clock will be divided by **n**+1. To get the desired 80 MHz from the 400 MHz PLL, we need to divide by 5. So, we place a 4 into the SYSDIV2 field. 5) The fifth step is to wait for the PLL to stabilize by waiting for PLLRIS (bit 6) in the **SYSCTL_RIS_R** to become high. 6) The last step is to connect the PLL by clearing the BYPASS2 bit. To modify this program to operate on other microcontrollers, you will need to change XTAL and the SYSDIV2.

```
void PLL_Init(void){
  SYSCTL_RCC2_R |=  0x80000000;  // 0) Use RCC2
  SYSCTL_RCC2_R |=  0x00000800;  // 1) bypass PLL while initializing
  SYSCTL_RCC_R = (SYSCTL_RCC_R &~0x000007C0)+0x00000540;   //2) 16 MHz
  SYSCTL_RCC2_R &= ~0x00000070;  // configure for main oscillator source
  SYSCTL_RCC2_R &= ~0x00002000;  // 3) activate PLL by clearing PWRDN
  SYSCTL_RCC2_R |=  0x40000000;  // 4) use 400 MHz PLL
  SYSCTL_RCC2_R = (SYSCTL_RCC2_R&~ 0x1FC00000)+(4<<22);   // 80 MHz
  while((SYSCTL_RIS_R&0x00000040)==0){};   // 5) wait for the PLL to lock
  SYSCTL_RCC2_R &= ~0x00000800; // 6) enable PLL by clearing BYPASS
}
```
Program 2.9. Activate the TM4C with a 16 MHz crystal to run at 80 MHz (PLL_xxx).

Checkpoint 2.25: How would you change Program 2.9 if your TM4C microcontroller had an 8 MHz crystal and you wish to run at 50 MHz?

We can make a first order estimate of the relationship between work done in the software and electrical power required to run the system. There are two factors involved in the performance of software. We define software work as the desired actions performed by executing software:

*Software work = algorithm * speed* (in instructions/sec)

In other words, if we want to improve software performance we can write better software or increase the rate at which the computer executes instructions. Recall that the compiler converts our C software into Cortex M machine code, so the efficiency of the compiler will also affect this relationship. Furthermore, most compilers have optimization settings that allow you to make your software run faster at the expense of using more memory. On the Cortex M, most instructions execute in 1 or 2 bus cycles. See section 3.3 in **CortexM4_TRM_r0p1.pdf** for more details. In CMOS logic, most of the electrical power required to run the system occurs in making signals change, that is, when a digital signal rises from 0 to 1, or falls from 1 to 0. Therefore we see a linear relationship between bus frequency and electrical power. Let *m* be the slope of this linear relationship

*Power = m * f_{Bus}*

Some of the factors that affect the slope m are operating voltage and fundamental behavior of how the CMOS transistors are designed. If we approximate the Cortex M processor as being able to execute one instruction every two bus cycles, we can combine the above two equations to see the speed-power tradeoff.

*Software work = algorithm * ½ f_{Bus} = algorithm * ½ Power/m*

Observation: To save power, we slow down the bus frequency removing as much of the wasted bus cycles while still performing all of the required tasks.

2.6. SysTick Timer

SysTick is a simple counter that we can use to create time delays and generate periodic interrupts. It exists on all Cortex™-M microcontrollers, so using SysTick means the system will be easy to port to other microcontrollers. Table 2.17 shows some of the register definitions for SysTick. **CURRENT** is a 24-bit down counter that runs at the bus clock frequency.

Address	31-24	23-17	16	15-3	2	1	0	Name
$E000E010	0	0	COUNT	0	CLK_SRC	INTEN	ENABLE	NVIC_ST_CTRL_R
$E000E014	0	24-bit RELOAD value					NVIC_ST_RELOAD_R
$E000E018	0	24-bit CURRENT value of SysTick counter					NVIC_ST_CURRENT_R

Table 2.17. SysTick registers.

There are four steps to initialize the SysTick timer. First, we clear the **ENABLE** bit to turn off SysTick during initialization. Second, we set the **RELOAD** register. Third, we write to the **NVIC_ST_CURRENT_R** value to clear the counter. Lastly, we write the desired mode to the control register, **NVIC_ST_CTRL_R**. We set the **CLK_SRC** bit specifying the core clock will be used. We must set **CLK_SRC**=1, because **CLK_SRC**=0 mode is not implemented on the LM3S family. LM4F/TM4C microcontrollers do support **CLK_SRC**=0 internal oscillator mode. In Chapter 5, we will set **INTEN** to enable interrupts, but in this first example we clear **INTEN** so interrupts will not be requested. We need to set the **ENABLE** bit so the counter will run. When the **CURRENT** value counts down from 1 to 0, the **COUNT** flag is set. On the next clock, the **CURRENT** is loaded with the **RELOAD** value. In this way, the SysTick counter (**CURRENT**) is continuously decrementing. If the **RELOAD** value is n, then the SysTick counter operates at modulo $n+1$ (...n, $n-1$, $n-2$ ... 1, 0, n, $n-1$, ...). In other words, it rolls over every $n+1$ counts. The **COUNT** flag could be configured to trigger an interrupt. However, in this first example interrupts will not be generated. We set **RELOAD** to 0x00FFFFFF for a general counter. For a delay timer or a periodic interrupt the value (**RELOAD**+1)*busperiod will determine the delay time or interrupt period. If we activate the PLL to run the microcontroller at 80 MHz, then the SysTick counter decrements every 12.5 ns. In general, if the period of the core bus clock is t, then the **COUNT** flag will be set every $(n+1)t$. Reading the **NVIC_ST_CTRL_R** control register will return the **COUNT** flag in bit 16, and then clear the flag. Also, writing any value to the **NVIC_ST_CURRENT_R** register will reset the counter to zero and clear the **COUNT** flag.

Program 2.10 uses the SysTick timer to implement a time delay. For example, the user calls **SysTick_Wait10ms(123);** and the function returns 1.23 seconds later. In the function **SysTick_Wait()**, the **NVIC_ST_RELOAD_R** value is set to specify the delay. Writing to **CURRENT** clears the **COUNT** flag and reloads the counter. When the counter goes from 1 to 0, the flag **COUNT** is set.

The accuracy of SysTick depends on the accuracy of the clock. We use the PLL to derive a bus clock based on the 16 MHz crystal, the time measured or generated using SysTick will be very accurate. More specifically, the accuracy of the NX5032GA crystal on the LaunchPad board is ±50 parts per million (PPM), which translates to 0.005%, which is about ±5 seconds per day. One could spend more money on the crystal and improve the accuracy by a factor of 10. Not only are crystals accurate, they are stable. The NX5032GA crystal will vary only ±150 PPM as temperature varies from -40 to +150 °C. Crystals are more stable than they are accurate, typically varying by less than 5 PPM per year.

```
#define NVIC_ST_CTRL_R          (*((volatile uint32_t *)0xE000E010))
#define NVIC_ST_RELOAD_R        (*((volatile uint32_t *)0xE000E014))
#define NVIC_ST_CURRENT_R       (*((volatile uint32_t *)0xE000E018))
void SysTick_Init(void){
  NVIC_ST_CTRL_R = 0;                    // 1) disable SysTick during setup
  NVIC_ST_RELOAD_R = 0x00FFFFFF;         // 2) maximum reload value
  NVIC_ST_CURRENT_R = 0;                 // 3) any write to current clears it
  NVIC_ST_CTRL_R = 0x00000005;           // 4) enable SysTick with core clock
}
void SysTick_Wait(uint32_t delay){ // delay is in 12.5ns units
  NVIC_ST_RELOAD_R = delay-1;  // number of counts to wait
  NVIC_ST_CURRENT_R = 0;       // any value written to CURRENT clears
  while((NVIC_ST_CTRL_R&0x00010000)==0){ // wait for COUNT flag
  }
}
void SysTick_Wait10ms(uint32_t delay){ // delay is in 10ms units
  uint32_t i;
  for(i=0; i<delay; i++){
    SysTick_Wait(800000);   // 800000*12.5ns equals 10ms
  }
}
```

Program 2.10. Timer functions that implement a time delay (SysTick_xxx).

Checkpoint 2.26: How would you change `SysTick_Wait10ms` in Program 2.10 if your microcontroller were running at 50 MHz?

Example 2.3: Design an embedded system that spins a stepper motor by outputting a 0101, 0110, 1010, 1001 binary repeating pattern, separated by 100 ms between outputs.

Solution: This system will need four output pins. Since the problem didn't specify the voltage, current, or torque of the stepper motor, we could either put off these decisions until the engineering design stage in order to simplify the design or minimize cost, or we could go back to the clients and ask for additional requirements. In this case, the clients have a 12 V, ½ A stepper. We decide to use a L293 stepper driver not because it is best, but because it is simple, meets specifications, and we have used it before. Due to the nature of this book, we will constrain all designs to include a TM4C microcontroller. Table 2.18 summarizes the specifications and constraints. We will use standard 5% resistors to minimize cost.

Specifications	Constraints
Repeating pattern of 5, 6, 10, 9	TM4C123-based
100 ms delay between outputs	Minimize cost
12 V, ½ A stepper motor	Use L293 interface chip

Table 2.18. Specifications and constraints of the embedded system.

Tolerance for this system says it is acceptable if the delay is greater than 95 ms and less than 105 ms. Similarly, it will be acceptable as long as the motor voltage is between 10 V to 14 V for up to ½ A. The data flow graph in Figure 2.30 shows information as it flows from the controller software to the four outputs. The data flow graph will be important during the subsequent design phases because the hardware blocks can be considered as a preliminary hardware block diagram of the system. The call graph, also shown in Figure 2.30, illustrates this is master/slave configuration where the controller software will manipulate the four outputs.

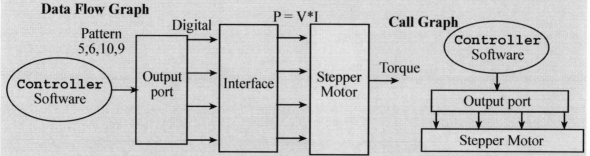

Figure 2.30. *Data flow graph and call graph of the stepper motor system.*

The output voltage of the microcontroller is 0 or 3.3V. From Table 1.4 we see TM4C microcontrollers can source or sink up to 8 mA, thus we will need an interface to deliver the 12 V at ½ A. The details of the stepper motor and interface will be presented later in Section 4.7.2. This circuit is shown in Figure 2.31. Notice the similarity in structure between the data flow graph and the electrical circuit.

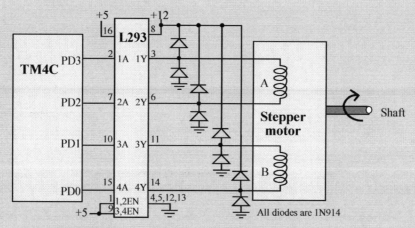

Figure 2.31. *Hardware circuit for the stepper motor output system.*

Pseudo-code is similar to high-level languages, but without a rigid syntax. This means we utilize whatever syntax we like. Flowcharts are good when the software involves complex algorithms with many decisions points causing control paths. On the other hand, pseudo-code may be better when the software is more sequential and involves complex mathematical calculations.

The software design of this system also involves using examples presented earlier with some minor tweaking. The only data required in this problem is the 5–6–10–9 sequence. Later in Chapter 3, we will consider solutions to this type of problem using data structures, but in this first example, we will take a simple approach, not using a data structure. Figure 2.32 illustrates a software design process using flowcharts. We start with general approach on the left. Flowchart 1 shows the software will initialize the output port, and perform the output sequence.

As we design the software system, we fill in the details. This design process is called successive refinement. It is also classified as top-down, because we begin with high-level issues, and end at the low-level. In Flowchart 2, we set the direction register, and then output the sequence 5-6-10-9. It is at this stage we figured out how to create the repeating sequence. Flowchart 3 fills in the remaining details. Because we perform a similar operation each time we output, we will use a helper function called **step**.

Many software developers use pseudo-code rather than flowcharts, because the pseudo-code itself can be embedded into the software as comments. Program 2.11 shows the C implementation for this system. Notice the similarity in structure between Flowchart 3 and this code. The **STEPPER** definition implements friendly access to pins PD3 – PD0.

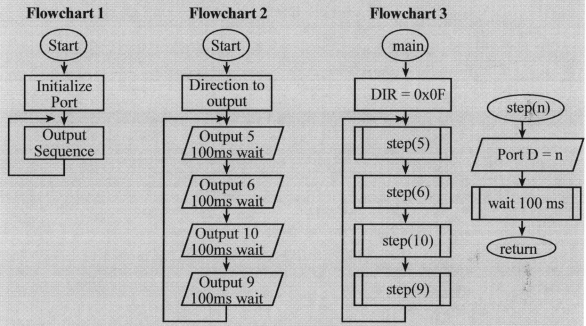

Figure 2.32. Software design for the stepper motor system using flowcharts.

In order to test the system we need to build a prototype. One option is simulation. A second option is to use a development system like the ones shown in Figures 2.8 and 2.13. In this approach, you build the external circuits on a protoboard and use the debugger to download and test the software. A third approach is typically used after a successful evaluation with one of the previous methods. In this approach, we design a printed circuit board (PCB) including both the external circuits and the microcontroller itself.

Portability is a measure of how easy it is to convert software that runs on one machine to run on another machine. Notice the abstractions in this example make it more portable.

Before we connect the motor, we need to test the software to verify the values and times are correct. We use the software debugger to single step our program, which correctly outputs the 1010, 1001, 0101, 0110 binary repeating pattern. During this single stepping the port outputs go high and low in the proper pattern. Using a voltmeter on the circuit we observe a 0.25V signal on the output of the microcontroller when low, and a 3.2 V voltage when the output is high. We test the system at full speed and observe the four outputs on a logic analyzer, collecting data presented as Figure 2.33. As we will see later in Section 4.7.2, the time delay between outputs will determine the speed of the motor rotation.

```c
#define STEPPER  (*((volatile uint32_t *)0x4000703C))
static void step(uint32_t n){
  STEPPER = n;            // output to stepper causing it to step once
  SysTick_Wait10ms(10);   // program 2.10
}
int main(void){ // reset clears AFSEL, PCTL, AMSEL
  SysTick_Init();
  SYSCTL_RCGCGPIO_R |= 0x08;                  // activate clock for Port D
  while((SYSCTL_PRGPIO_R&0x08) == 0){};// ready?
  GPIO_PORTD_DIR_R |= 0x0F;   // PD3-0 is an output
  GPIO_PORTD_DEN_R |= 0x0F;   // PD3-0 enabled as a digital port
  while(1){
    step(5);   // motor is 0101
    step(6);   // motor is 0110
    step(10);  // motor is 1010
    step(9);   // motor is 1001
  }
}
```
Program 2.11. C software for the stepper motor system (GPIO_xxx).

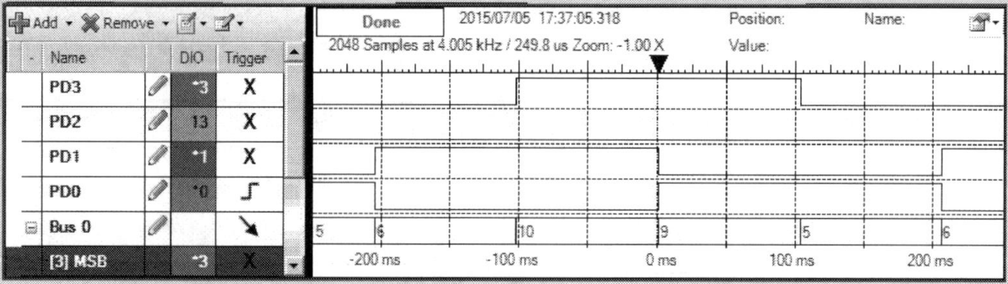

Figure 2.33. Logic analyzer waveforms collected during the testing the stepper motor system.

2.7. Choosing a Microcontroller

I chose to focus this book on the TM4C family of microcontrollers, because it has a rich set of features needed to teach the fundamentals required for both today's and tomorrow's embedded systems. Sometimes, the computer engineer is faced with the task of selecting the microcontroller for the project. When faced with this decision some engineers will only consider those devices for which they have hardware and software experience. Fortunately, this blind approach often still yields an effective and efficient product, because many microcontrollers overlap in their cost and performance. In other words, if a familiar microcontroller can implement the desired functions for the project, then it is often efficient to bypass that more perfect piece of hardware in favor of a faster development time. On the other hand, sometimes we wish to evaluate all potential candidates. It may be cost-effective to hire or train the engineering personnel so that they are proficient in a wide spectrum of potential microcontroller devices. There are many factors to consider when selecting a microcontroller.

A first group of factors deals with cost, maintenance and production:

- Labor costs includes training, development, and testing
- Material costs includes parts and supplies
- Manufacturing costs depend on the number and complexity of the components
- Maintenance costs involve revisions to fix bugs and perform upgrades
- Second source availability

A second group of factors deals with memory and the processor:

- ROM size must be big enough to hold instructions and fixed data for the software
- RAM size must be big enough to hold locals, parameters, and global variables
- EEPROM to hold nonvolatile fixed constants that are field-configurable
- Processor must be capable of performing all calculations in real time
- 8-, 16-, or 32-bit data size should match most of the data to be processed
- Numerical operations like multiply, divide, saturation, floating point
- Special functions like multiply/accumulate, fuzzy logic, complex numbers
- Availability of high-level language cross-compilers, simulators, and debuggers

A third group of factors deals with input and output:

- I/O bandwidth determines the input/output rate
- Parallel ports for the input/output digital signals
- Serial ports to interface with other computers or I/O devices
- Timer functions to generate signals, measure frequency, and measure period
- Pulse width modulation for the output signals in many control applications
- ADC that is used to convert analog inputs to digital numbers
- DAC that is used to convert digital numbers to analog outputs
- Special I/O functions such as CAN, Ethernet, wifi, Bluetooth, and USB

A fourth group of factors deals with system level design:

- Package size and environmental issues affect many embedded systems
- Power requirements because many systems will be battery operated

When considering speed it is best to compare time to execute a benchmark program similar to your specific application, rather than just comparing bus frequency. One of the difficulties is that the microcontroller selection depends on the speed and size of the software, but the software cannot be written without the computer. Given this uncertainty, it is best to select a family of devices with a range of execution speeds and memory configurations. In this way a prototype system with large amounts of memory and peripherals can be purchased for software and hardware development, and once the design is in its final stages, the specific version of the computer can be selected now knowing the memory and speed requirements for the project. In conclusion, while this book focuses on the ARM® Cortex™-M microcontrollers, it is expected that once the study of this book is completed, the reader will be equipped with the knowledge to select the proper microcontroller and complete the software design.

2.8. Exercises

2.1 What is special about Register 13? Register 14? Register 15?

2.2 In 20 words or less describe the differences between von Neumann and Harvard architectures.

2.3 What happens when you load a value into Register 15 with bit 0 set?

2.4 Write C code that sets bit 31 of memory location 0x2000.1234 using bit-banding.

2.5 Write C code that clears bit 16 of memory location 0x2000.8000 using bit-banding.

2.6 Write C code that sets bit 1 of memory location 0x4000.5400 using bit-banding. What effect does this operation have?

2.7 Write C code that clears bit 2 of memory location 0x4000.7400 using bit-banding. What effect does this operation have?

2.8 How much RAM and ROM are in TM4C123? What are the specific address ranges of these memory components?

2.9 How much RAM and ROM are in LM3S1968? What are the specific address ranges of these memory components?

2.10 How much RAM and ROM are in LM3S8962? What are the specific address ranges of these memory components?

2.11 What are the bits in the Program Status Register (PSR) of ARM® Cortex™-M?

2.12 What happens if you execute these four assembly instructions?
```
PUSH {R1}
PUSH {R2}
POP  {R1}
POP  {R2}
```

2.13 Write assembly code that pushes registers R1 R3 and R5 onto the stack.

2.14 How do you initialize the stack?

2.15 How do you specify where to begin execution after a reset?

2.16 What does word-aligned mean?

2.17 When does the LR have to be pushed on the stack?

2.18 Does the associative principle hold for signed integer multiply and divide? Assume `Out1 Out2 A B C` are all the same precision (e.g., 32 bits). In particular do these two C calculations always achieve identical outputs? If not, give an example.
```
Out1 = (A*B)/C;
Out2 = A*(B/C);
```

2.19 Does the associative principle hold for signed integer addition and subtraction? Assume `Out3 Out4 A B C` are all the same precision (e.g., 32 bits). In particular do these two C calculations always achieve identical outputs? If not, give an example.
```
Out3 = (A+B)-C;
Out4 = A+(B-C);
```

2.20 What are parallel ports are available on the TM4C123?

2.21 What are parallel ports are available on the LM3S1968?

2.22 What are parallel ports are available on the LM3S8962?

2.23 What is a direction register? Why does the microcontroller have direction registers?

2.24 What is the alternative function register?

D2.25 Write software that initializes TM4C Port A, so pins 7,5,3,1 are output and the rest are input.

D2.26 Write software that initializes TM4C Port A, so pins 5,4 are output and the rest are input.

D2.27 Write software that initializes TM4C Port A, so pins 5, 4, and 3 are output. Make the initialization friendly. Design an output function that takes a 3-bit parameter (0 to 7) and writes the value to these three pins. Use bit-specific addressing for the output.

D2.28 Write software that initializes TM4C Port E, so pin 1 is an output. Make the initialization friendly. Design an output function that takes a 1-bit parameter (0 or 1) and writes the value to this pin. Use bit-specific addressing for the output.

D2.29 Redesign the SSR interface shown in Figure 2.27 using a +5V source. In particular, recalculate the required resistor value if we were to change the +3.3V to +5V.

D2.30 Interface four LEDs to an output port. Each LED operates at 1.9 V and 1 mA. In particular, calculate the required resistor values needed for these LEDs.

D2.31 Rewrite the software in Program 2.11 so the pattern changes every 1 sec.

D2.32 Design a switch interface that it is negative logic. I.e., the input is low if the switch is pressed and high if the switch is not pressed.

2.9. Lab Assignments

The labs in this book involve the following steps:

Part a) During the analysis phase of the project determine additional specifications and constraints. In particular, discover which microcontroller you are to use, whether you are to develop in assembly language or in C, and whether the project is to be simulated then built, just built or just simulated. For example, inputs can be created with switches and outputs can be generated with LEDs. The UART can be interfaced to a PC, and a communication program like PuTTY can be used to interact with the system.

Part b) Design, build, and test the hardware interfaces. Use a computer-aided-drawing (CAD) program to draw the hardware circuits. Label all pins, chips, and resistor values. In this chapter, there will be one switch for each input and one LED for each output. Connect the switch interfaces to microcontroller input pins, and connect the LED interfaces to microcontroller output pins. Pressing the switch will signify a high input logic value. You should activate the LED to signify a high output logic value.

Part c) Design, implement and test the software that initializes the I/O ports and performs the specified function. Often a main program is used to demonstrate the system.

Lab 2.1 The overall objective is to create a NOT gate. The system has one digital input and one digital output, such that the output is the logical complement of the input. Implement the design such that the complement function occurs in the software of the microcontroller.

Lab 2.2 The overall objective is to create a 3-input AND gate. The system has three digital inputs and one digital output, such that the output is the logical and of the three inputs. Implement the design such that the AND function occurs in the software of the microcontroller.

Lab 2.3 The overall objective is to create a 2-input EXCLUSIVE OR gate. The system has two digital inputs and one digital output, such that the output is the logical exclusive or of the two inputs. Implement the design such that the EXCLUSIVE OR function occurs in the software.

Lab 2.4 The overall objective is to create a 3-input voting logic. The system has three digital inputs and one digital output, such that the output is high if and only if two or more inputs are high. This means the output will be low if two for more inputs are low. Implement the design such that the voting function occurs in the software of the microcontroller.

Lab 2.5 The overall objective is to a variable frequency oscillator. The system has two digital inputs and two digital outputs. If input1 is true the digital output1 oscillates at 262 Hz. If the input1 is false the output1 remains low. If input2 is true the digital output2 oscillates at 392 Hz. If the input2 is false the output2 remains low. If you connect each output to a 10 kΩ resistor as shown in the figure, then you can hear the tones as middle C and middle G.

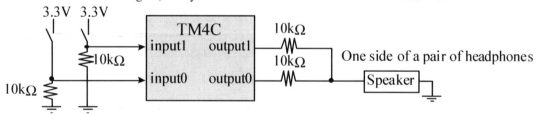

3. Software Design

Chapter 3 objectives are to:

- Present the software design process
- Describe a software coding style
- Define modules, board support package, and device drivers
- Present a design method using finite state machines
- Define the concept of threads
- Implement FIFO queues
- Present a simple memory manager as an introduction to the heap
- Introduce the art of debugging

The ultimate success of an embedded system project depends both on its software and hardware. Computer scientists pride themselves in their ability to develop quality software. Similarly electrical engineers are well-trained in the processes to design both digital and analog electronics. Manufacturers, in an attempt to get designers to use their products, provide application notes for their hardware devices. The main objective of this book is to combine effective design processes together with practical software techniques in order to develop quality embedded systems. As the size and complexity of the software increase, software development changes from simple "coding" to "software engineering". Naturally, as the system complexity increases so do the engineering skills required to design such systems. These software skills presented in this chapter include modular design, layered architecture, abstraction, and verification. Even if real-time embedded systems are on the small end of the size scale, never the less, these systems can be quite complex. Therefore, the above mentioned skills are essential for developing embedded systems. This chapter on software development is placed early in the book because writing good software is an art that must be developed and cannot be added on at the end of a project. Good software combined with average hardware will always outperform average software on good hardware. In this chapter we will outline various techniques for developing quality software and then apply these techniques throughout the remainder of the book.

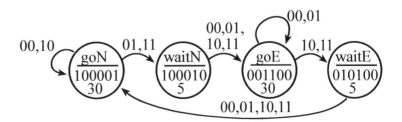

3.1. Attitude

Good engineers employ well-defined design processes when developing complex systems. When we work within a structured framework, it is easier to prove our system works (verification) and to modify our system in the future (maintenance). As our software systems become more complex, it becomes increasingly important to employ well-defined software design processes. Throughout this book, a very detailed set of software development rules will be presented. This book focuses on real-time embedded systems written in C. At first, it may seem radical to force such a rigid structure to software. We might wonder if creativity will be sacrificed in the process. True creativity is more about effective solutions to important problems and not about being sloppy and inconsistent. Because software maintenance is a critical task, the time spent organizing, documenting, and testing during the initial development stages will reap huge dividends throughout the life of the software project.

Observation: The easiest way to debug is to write software without any bugs.

We define **clients** as people who will use our software. Sometimes, the client is the end-user who uses the embedded system. Other times, we develop hardware/software components that plug into a larger system. In this case, the client develops software that will call our functions. We define **coworkers** as engineers who will maintain our software. We must make it easy for a coworker to debug, use, and extend our software.

Writing quality software has a lot to do with attitude. We should be embarrassed to ask our coworkers to make changes to our poorly written software. Since so much software development effort involves maintenance, we should create software modules that are easy to change. In other words, we should expect each piece of our code will be read by another engineer in the future, whose job it will be to make changes to our code. We might be tempted to quit a software project once the system is running, but this short time we might save by not organizing, documenting, and testing will be lost many times over in the future when it is time to update the code.

As project managers, we must reward good behavior and punish bad behavior. A company, in an effort to improve the quality of their software products, implemented the following policies. "The employees in the customer relations department receive a bonus for every software bug that they can identify. These bugs are reported to the software developers, who in turn receive a bonus for every bug they fix."

Checkpoint 3.1: Why did the above policy fail horribly?

We should demand of ourselves that we deliver bug-free software to our clients. Again, we should be embarrassed when our clients report bugs in our code. We should be ashamed when other programmers find bugs in our code. There are four steps we can take to facilitate this important aspect of software design.

Test it now. When we find a bug, fix it immediately. The longer we put off fixing a mistake the more complicated the system becomes, making it harder to find. Remember that bugs do not go away automatically, but we can make the system so complex that the bugs will manifest themselves in a mysterious and obscure fashion. For the same reason, we should completely test each module individually, before combining them into a larger system. We should not add new features before we are convinced the existing features are bug-free. In this way, we start with a working system, add features, and then debug this system until it is working again.

This incremental approach makes it easier to track progress. It allows us to undo bad decisions, because we can always revert back to a previous working system. Adding new features before the old ones are debugged is very risky. With this sloppy approach, we could easily reach the project deadline with 100% of the features implemented, but have a system that doesn't run. In addition, once a bug is introduced, the longer we wait to remove it, the harder it will be to correct. This is particularly true when the bugs interact with each other. Conversely, with the incremental approach, when the project schedule slips, we can deliver a working system at the deadline that supports some of the features.

Maintenance Tip: Go from working system to working system.

Plan for testing. How to test should be considered at the beginning, middle, and end of a project. In particular, testing should be included as part of the initial design. Our testing and the client's usage go hand in hand. In particular, how we test the software module will help the client understand the context and limitations of how our software is to be used. It often makes sense to explain the testing procedures to the client as an effort to communicate the features and limitations of the module. Furthermore, a clear understanding of how the client wishes to use our software is critical for both the software design and its testing. For example, after seeing how you tested the module, the client may respond, "That's nice, but what I really want it to do is …". If this happens, it makes sense to rewrite the requirements document to reflect this new understanding of the client's expectation.

Maintenance Tip: It is better to have some parts of the system that run with 100% reliability than to have the entire system with bugs.

Get help. Use whatever features are available for organization and debugging. Pay attention to warnings, because they often point to misunderstandings about data or functions. Misunderstanding of assumptions can cause bugs when the software is upgraded, or reused in a different context than originally conceived. Remember that computer time is a lot cheaper than programmer time. It is a mistake to debug an embedded system simply by observing its inputs and outputs. We need to use both software and hardware debugging tools to visualize internal parameters within the system.

Maintenance Tip: It is better to have a system that runs slowly than to have one that doesn't run at all.

Divide and conquer. In the early days of microcomputer systems, software size could be measured in hundreds of lines of source code or thousands of bytes of object code. These early systems, due to their small size, were inherently simple. The explosion of hardware technology (both in speed and size) has led to a similar increase in the size of software systems. The only hope for success in a large software system will be to break it into simple modules. In most cases, the complexity of the problem itself cannot be avoided. E.g., there is just no simple way to get to the moon. Nevertheless, a complex system can be created out of simple components. A real creative effort is required to orchestrate simple building blocks into larger modules, which themselves are grouped. We use our creativity to break a complex problem into simple components, rather than developing complex solutions to simple problems.

Observation: There are two ways of constructing a software design: one way is to make it so simple that there are obviously no deficiencies and the other way is make it so complicated that there are no obvious deficiencies. C.A.R. Hoare, "The Emperor's Old Clothes," CACM Feb. 1981.

3.2. Quality Programming

Software development is similar to other engineering tasks. We can choose to follow well-defined procedures during the development and evaluation phases, or we can meander in a haphazard way and produce code that is hard to test and harder to change. The ultimate goal of the system is to satisfy the stated objectives such as accuracy, stability, and input/output relationships. Nevertheless it is appropriate to separately evaluate the individual components of the system. Therefore in this section, we will evaluate the quality of our software. There are two categories of performance criteria with which we evaluate the "goodness" of our software. Quantitative criteria include dynamic efficiency (speed of execution), static efficiency (ROM and RAM program size), and accuracy of the results. Qualitative criteria center on ease of software maintenance. Another qualitative way to evaluate software is ease of understanding. If your software is easy to understand then it will be:

- Easy to debug, including both finding and fixing mistakes
- Easy to verify, meaning we can prove it is correct
- Easy to maintain, meaning we can add new features

Common error: Programmers who sacrifice clarity in favor of execution speed often develop software that runs fast but is error-prone and difficult to change.

Golden Rule of Software Development: Write software for others as you wish they would write for you.

3.2.1. Quantitative Performance Measurements

In order to evaluate our software quality, we need performance measures. The simplest approaches to this issue are quantitative measurements. **Dynamic efficiency** is a measure of how fast the program executes. It is measured in seconds or processor bus cycles. Because of the complexity of the Cortex™-M, it will be hard to estimate execution speed by observing the assembly language generated by the compiler. Rather, we will employ methods to experimentally measure execution speed. **Static efficiency** is the number of memory bytes required. Since most embedded computer systems have both RAM and ROM, we specify memory requirement in global variables, stack space, fixed constants, and program object code. The global variables plus maximum stack size must fit into the available RAM. Similarly, the fixed constants plus program size must fit into the available ROM. We can judge our software system according to whether or not it satisfies given constraints, like software development costs, memory available, and time table. Many of the system specifications are quantitative, and hence the extend to which the system meets specifications is an appropriate measure of quality.

3.2.2. Qualitative Performance Measurements

Qualitative performance measurements include those parameters to which we cannot assign a direct numerical value. Often in life the most important questions are the easiest to ask, but the hardest to answer. Such is the case with software quality. So therefore we ask the following qualitative questions. Can we prove our software works? Is our software easy to understand? Is our software easy to change? Since there is no single approach to writing quality software, I can only hope to present some techniques that you may wish to integrate into your own software style. In fact, we will devote most this chapter to the important issue of developing quality software. In particular, we will study self-documented code, abstraction, modularity, and layered software. These parameters indeed play a profound effect on the bottom-line financial success of our projects. Although quite real, because there is often not an immediate and direct relationship between software quality and profit, we may be tempted to dismiss its importance.

> **Observation:** Most people get better with practice. So if you wish to become a better programmer, I suggest you write great quantities of software.

To get a benchmark on how good a programmer you are, I challenge you to two tests. In the first test, find a major piece of software that you have written over 12 months ago, and then see if you can still understand it enough to make minor changes in its behavior. The second test is to exchange with a peer a major piece of software that you have both recently written (but not written together), then in the same manner, if you can make minor changes to each other's software.

> **Observation:** You can tell if you are a good programmer if 1) you can understand your own code 12 months later, and 2) others can make changes to your code.

3.3. Software Style Guidelines

One of the recurring themes of this software style section is consistency. Maintaining a consistent style will help us locate and understand the different components of our software, as well as prevent us from forgetting to include a component or worse including it twice.

3.3.1. Organization of a code file

The following regions should occur in this order in every code file (e.g., **file.c**).

Opening comments. The first line of every file should contain the file name. This is because some printers do not automatically print the name of the file. Remember that these opening comments will be duplicated in the corresponding header file (e.g., **file.h**) and are intended to be read by the client, the one who will use these programs. If major portions of this software are copied from copyrighted sources, then we must satisfy the copyright requirements of those sources. The rest of the opening comments should include

- The overall purpose of the software module
- The names of the programmers
- The creation (optional) and last update dates
- The hardware/software configuration required to use the module
- Copyright information

Including .h files. Next, we will place the `#include` statements that add the necessary header files. Adding other code files, if necessary, will occur at the end of the file, but here at the top of the file we include just the header files. Normally the order doesn't matter, so we will list the include files in a hierarchical fashion starting with the lowest level and ending at the highest high. If the order of these statements is important, then write a comment describing both what the proper order is and why the order is important. Putting them together at the top will help us draw a call graph, which will show us how our modules are connected. In particular, if we consider each code file to be a separate module, then the list of `#include` statements specifies which other modules can be called from this module. Of course one header file is allowed to include other header files. However, we should avoid having one header file include other header files. This restriction makes the organizational structure of the software system easier to observe. Be careful to include only those files that are absolutely necessary. Adding unnecessary include statements will make our system seem more complex than it actually is.

extern references. After including the header files, we can declare any external variables or functions. External references will be resolved by the linker, when various modules are linked together to create a single executable application. Placing them together at the top of the file will help us see how this software system fits together (i.e., is linked to) other systems.

#define statements. After external references, we should place the `#define` macros. These macros can define operations or constants. Since these definitions are located in the code file (e.g., **file.c**), they will be private. This means they are available within this file only. If the client does not need to use or change the macro operation or constant, then it should be made private by placing it here in the code file. Conversely, if we wish to create public macros, then we place them in the header file for this module.

struct union enum statements. After the define statements, we should create the necessary data structures using `struct union` and `enum`. Again, since these definitions are located in the code file (e.g., **file.c**), they will be private.

Global variables and constants. After the structure definitions, we should include the global variables and constants. There are two aspects of data that are important. First, we can specify where the data is allocated. If it is a variable that needs to exist permanently, we will place it in RAM as a global variable. If it is a constant that needs to exist permanently, we will place it in ROM using `const`. If the data is needed temporarily, we can define it as a local. The compiler will allocate locals in registers or on the stack in whichever way is most efficient.

```
int32_t PublicGlobal;                    // accessible by any module
static int32_t PrivateGlobal;            // accessible in this file only
const int32_t Constant=1234567;          // in ROM
void function(void){
  static int32_t veryPrivateGlobal;      // accessible by this function only
  int32_t privateLocal;                  // accessible by this function only
}
```

We define a **global** variable as one with permanent allocation. In the above examples, `PublicGlobal PrivateGlobal` and `veryPrivateGlobal` are global. `Constant` will be defined in ROM, and cannot be changed. We define a *local* variable as one with temporary allocation. The variable `privateLocal` is local and may exist on the stack or in a register.

The second aspect of the data is its scope. **Scope** specifies which software can access the data. **Public** variables can be accessed by any software. **Private** variables have restricted scope, which can be limited to the one file, the one function, or even to one { } program block. In general, we wish to minimize the scope of our data. Minimizing scope reduces complexity and simplifies testing. If we specify the global with `static`, then it will be private and can only be accessed by programs in this file. If we do not specify the global with `static` then it will be public, and can be accessed any program. For example, the `PublicGlobal` variable can be defined in other files using `extern` and the linker will resolve the reference. However, the `PrivateGlobal` cannot be accessed from software outside of the one file in which this variable is defined. Again, we classify `PrivateGlobal` as private because its scope is restricted. We put all the globals together before any function definitions to symbolize the fact that any function in this file has access to these globals. If we have a permanent variable that is only accessed by one function, then it should be defined inside the function with `static`. For example, the variable `veryPrivateGlobal` is permanently allocated in RAM, but can only be accessed by the function.

Maintenance Tip: Reduce complexity in our system by restricting direct access to our data. E.g., make global variables static if possible.

Prototypes of private functions. After the globals, we should add any necessary prototypes. Just like global variables, we can restrict access to private functions by defining them as static. Prototypes for the public functions will be included in the corresponding header file. In general, we will arrange the code implementations in a top-down fashion. Although not necessary, we will include the parameter names with the prototypes. Descriptive parameter names will help document the usage of the function. For example, which of the following prototypes is easier to understand?

```
static void plot(int16_t, int16_t);
static void plot(int16_t time, int16_t pressure);
```

Maintenance Tip: Reduce complexity in our system by restricting the software that can call a function E.g., make functions static if possible.

Implementations of the functions. The heart of the implementation file will be, of course, the implementations. Again, private functions should be defined as static. The functions should be sequenced in a logical manner. The most typical sequence is top-down, meaning we begin with the highest level and finish with the lowest level. Another appropriate sequence mirrors the manner in which the functions will be used. For example, start with the initialization functions, followed by the operations, and end with the shutdown functions. For example:

- Open
- Input
- Output
- Close

Including .c files. If the compiler does not support projects, then we would end the file with **#include** statements that add the necessary code files. Since most compilers support projects, we should use its organizational features and avoid including code files. The project simplifies the management of large software systems by providing organizational structure to the software system. Again, if we use projects, then including code files will be unnecessary, and hence should be avoided.

Employ run-time testing. If our compiler supports **assert()** functions, use them liberally. In particular, place them at the beginning of functions to test the validity of the input parameters. Place them after calculations to test the validity of the results. Place them inside loops to verify indices and pointers are valid. There is a secondary benefit to using **assert()**. The **assert()** statements provide inherent documentation of the assumptions.

3.3.2. Organization of a header file

Once again, maintaining a consistent style facilitates understanding and helps to avoid errors of omission. Definitions made in the header file will be public, i.e., accessible by all modules. As stated earlier, it is better to make global variables private rather than placing them in the header file. Similarly, we should avoid placing actual code in a header file.

There are two types of header files. The first type of header file has no corresponding code file. In other words, there is a **file.h**, but no **file.c**. In this type of header, we can list global constants and helper macros. Examples of global constants are data types (see **integer.h**), I/O port addresses (see **tm4c123ge6pm.h**), and calibration coefficients. Debugging macros could be grouped together and placed in a **debug.h** file. We will not consider software in these types of header files as belonging to a particular module.

The second type of header file does have a corresponding code file. The two files, e.g., **file.h**, and **file.c**, form a software module. In this type of header, we define the prototypes for the public functions of the module. The **file.h** contains the policies (behavior or what it does) and the **file.c** file contains the mechanisms (functions or how it works.) The following regions should occur in this order in every header file.

Opening comments. The first line of every file should contain the file name. This is because some printers do not automatically print the name of the file. Remember that these opening comments should be duplicated in the corresponding code file (e.g., **file.c**) and are intended to be read by the client, the one who will use these programs. We should repeat copyright information as appropriate. The rest of the opening comments should include

- The overall purpose of the software module
- The names of the programmers
- The creation (optional) and last update dates
- The hardware/software configuration required to use the module
- Copyright information

Including .h files. Nested includes in the header file should be avoided. As stated earlier, nested includes obscure the manner in which the modules are interconnected.

#define statements. Public constants and macros are next. Special care is required to determine if a definition should be made private or public. One approach to this question is to begin with everything defined as private, and then shift definitions into the public category only when deemed necessary for the client to access in order to use the module. If the parameter relates to what the module does or how to use the module, then it should probably be public. On the other hand, if it relates to how it works or how it is implemented, it should probably be private.

struct union enum statements. The definitions of public structures allow the client software to create data structures specific for this module.

Global variables and constants. If at all possible, public global variables should be avoided. Public constants follow the same rules as public definitions. If the client must have access to a constant to use the module, then it could be placed in the header file.

Prototypes of public functions. The prototypes for the public functions are last. Just like the implementation file, we will arrange the code implementations in a top-down fashion. Comments should be directed to the client, and these comments should clarify what the function does and how the function can be used. Examples of how to use the module could be included in the comments.

Often we wish to place definitions in the header file that must be included only once. If multiple files include the same header file, the compiler will include the definitions multiple times. Some definitions, such as function prototypes, can be defined then redefined. However, a common approach to header files uses **#ifndef** conditional compilation. If the object is not defined, then the compiler will include everything from the **#ifndef** until the matching **#endif**. Inside of course, we define that object so that the header file is skipped on subsequent attempts to include it. Each header file must have a unique object. One way to guarantee uniqueness is to use the name of the header file itself in the object name.

```
#ifndef __File_H__
#define __File_H__
struct Position{
   int bValid;     // true if point is valid
   int16_t x;      // in cm
   int16_t y;      // in cm
};
typedef struct Position PositionType;
#endif
```

3.3.3. Formatting

The rules set out in this subsection are not necessary for the program to compile or to run. Rather the intent of the rules are to make the software easier to understand, easier to debug, and easier to change. Just like beginning an exercise program, these rules may be hard to follow at first, but the discipline will pay dividends in the future.

Make the software easy to read. I strongly object to hardcopy printouts of computer programs during the development phase of a project. At this time, there are frequent updates made by multiple members of the software development team. Because a hardcopy printout will be quickly obsolete, we should develop and debug software by observing it on the computer screen. In order to eliminate horizontal scrolling, no line of code should be wider than the size of the editor screen. If we do make hard copy printouts of the software at the end of a project, this rule will result in a printout that is easy to read.

Indentation should be set at 2 spaces. When transporting code from one computer to another, the tab settings may be different. So, tabs that look good on one computer may look ugly on another. For this reason, we should avoid tabs and use just spaces. Local variable definitions can go on the same line as the function definition, or in the first column on the next line.

Be consistent about where we put spaces. Similar to English punctuation, there should be no space before a comma or a semicolon, but there should be at least one space or a carriage return after a comma or a semicolon. There should be no space before or after open or close parentheses. Assignment and comparison operations should have a single space before and after the operation. One exception to the single space rule is if there are multiple assignment statements, we can line up the operators and values. For example

```
voltage   =   1;
pressure |=   100;
status   &=  ~0x02;
```

Be consistent about where we put braces {}. Misplaced braces cause both syntax and semantic errors, so it is critical to maintain a consistent style. Place the opening brace at the end of the line that opens the scope of the multi-step statement. The only code that can go on the same line after an opening brace is a local variable declaration or a comment. Placing the open brace near the end of the line provides a visual clue that a new code block has started. Place the closing brace on a separate line to give a vertical separation showing the end of the multi-step statement. The horizontal placement of the close brace gives a visual clue that the following code is in a different block. For example

```
void main(void){ int i, j, k;
  j = 1;
  if(sub0(j)){
    for(i = 0; i < 6; i++){
      sub1(i);
    }
    k = sub2(i, j);
  }
  else{
    k = sub3();
  }
}
```

Use braces after all **if**, **else**, **for**, **do**, **while**, **case**, and **switch** commands, even if the block is a single command. This forces us to consider the scope of the block making it easier to read and easier to change. For example, assume we start with the following code.

```
if(flag)
    n = 0;
```

Now, we add a second statement that we want to execute also if the flag is true. The following error might occur if we just add the new statement.

```
if(flag)
    n = 0;
    c = 0;
```

If all of our blocks are enclosed with braces, we would have started with the following.

```
if(flag){
    n = 0;
}
```

Now, when we add a second statement, we get the correct software.

```
if(flag){
    n = 0;
    c = 0;
}
```

3.3.4. Code Structure

Make the presentation easy to read. We define presentation as the look and feel of our software as displayed on the screen. If at all possible, the size of our functions should be small enough so the majority of a "single idea" fits on a single computer screen. We must consider the presentation as a two-dimensional object. Consequently, we can reduce the 2-D area of our functions by encapsulating components and defining them as private functions, or by combining multiple statements on a single line. In the horizontal dimension, we are allowed to group multiple statements on a single line only if the collection makes sense. We should list multiple statements on a single line, if we can draw a circle around the statements and assign a simple collective explanation to the code.

> **Observation:** Most professional programmers do not create hard copy printouts of the software. Rather, software is viewed on the computer screen, and developers use a code repository like Git or SVN to store and share their software.

Another consideration related to listing multiple statements on the same line is debugging. The compiler often places debugging information on each line of code. Breakpoints in some systems can only be placed at the beginning of a line. Consider the following three presentations. Since the compiler generates exactly the same code in each case, the computer execution will be identical. Therefore, we will focus on the differences in style. The first example has a horrific style.

```
void testFilter(int32_t start, int32_t stop, int32_t step){ int32_t x,y;
    initFilter();UART_OutString("x(n) y(n)"); UART_OutChar(CR);
    for(x=start;x<=stop; x=x+step){ y=filter(x); UART_OutUDec(x);
UART_OutChar(SP); UART_OutUDec(y); UART_OutChar(CR);} }
```

The second example places each statement on a separate line. Although written in an adequate style, it is unnecessarily vertical.

```
void testFilter(int32_t start, int32_t stop, int32_t step){
int32_t x;
int32_t y;
  initFilter();
  UART_OutString("x(n)  y(n)");
  UART_OutChar(CR);
  for(x = start; x <= stop; x = x+step){
    y = filter(x);
    UART_OutUDec(x);
    UART_OutChar(SP);
    UART_OutUDec(y);
    UART_OutChar(CR);
  }
}
```

The following implementation groups the two variable definitions together because the collection can be considered as a single object. The variables are related to each other. Obviously, **x** and **y** are the same type (32-bit signed), but in a physical sense, they would have the same units. For example, if **x** represents a signal in mV, then **y** is also a signal in mV. Similarly, the UART output sequences cause simple well-defined operations.

```
void testFilter(int32_t start, int32_t stop, int32_t step){ int32_t x, y;
  initFilter();
  UART_OutString("x(n)  y(n)"); UART_OutChar(CR);
  for(x = start; x <= stop; x = x+step){
    y = filter(x);
    UART_OutUDec(x); UART_OutChar(SP); UART_OutUDec(y); UART_OutChar(CR);
  }
}
```

The "make the presentation easy to read" guideline sometimes comes in conflict with the "be consistent where we place braces" guideline. For example, the following example is obviously easy to read, but violates the placement of brace rule.

```
for(i = 0; i < 6; i++) dataBuf[i] = 0;
```

When in doubt, we will always be consistent where we place the braces. The correct style is also easy to read.

```
for(i = 0; i < 6; i++){
  dataBuf[i] = 0;
}
```

Employ modular programming techniques. Complex functions should be broken into simple components, so that the details of the lower-level operations are hidden from the overall algorithms at the higher levels. An interesting question arises: *Should a subfunction be defined if it will only be called from a single place?* The answer to this question, in fact the answer to all questions about software quality, is yes if it makes the software easier to understand, easier to debug, and easier to change.

Minimize scope. In general, we hide the implementation of our software from its usage. The scope of a variable should be consistent with how the variable is used. In a military sense, we ask the question, "Which software has the need to know?" Global variables should be used only when the lifetime of the data is permanent, or when data needs to be passed from one thread to another. Otherwise, we should use local variables. When one module calls another, we should pass data using the normal parameter-passing mechanisms. As mentioned earlier, we consider I/O ports in a manner similar to global variables. There is no syntactic mechanism to prevent a module from accessing an I/O port, since the ports are at fixed and known absolute addresses. Processors used to build general purpose computers have a complex hardware system to prevent unauthorized software from accessing I/O ports, but the details are beyond the scope of this book. In most embedded systems, however, we must rely on the does-access rather than the can-access method when dealing with I/O devices. In other words, we must have the discipline to restrict I/O port access only in the module that is designed to access it. For similar reasons, we should consider each interrupt vector address separately, grouping it with the corresponding I/O module, even though there will be one file containing all the vectors.

Use types. Using a **typedef** will clarify the format of a variable. It is another example of the separation of mechanism and policy. New data types and structures will begin with an upper case letter. The **typedef** allows us to hide the representation of the object and use an abstract concept instead. For example

```
typedef int16_t Temperature;
void main(void){ Temperature lowT, highT;
}
```

This allows us to change the representation of temperature without having to find all the temperature variables in our software. Not every data type requires a **typedef**. We will use types for those objects of fundamental importance to our software, and for those objects for which a change in implementation is anticipated. As always, the goal is to clarify. If it doesn't make it easier to understand, easier to debug, or easier to change, don't do it.

Prototype all functions. Public functions obviously require a prototype in the header file. In the implementation file, we will organize the software in a top-down hierarchical fashion. Since the highest level functions go first, prototypes for the lower-level private functions will be required. Grouping the low-level prototypes at the top provides a summary overview of the software in this module. Include both the type and name of the input parameters. Specify the function as void even if it has no parameters. These prototypes are easy to understand:

```
void start(int32_t period, void(*functionPt)(void));
int16_t divide(int16_t dividend, int16_t divisor);
```

These prototypes are harder to understand:

```
start(int32_t, (*)());
int16_t divide(int16_t, int16_t);
```

Declare data and parameters as const whenever possible. Declaring an object as const has two advantages. The compiler can produce more efficient code when dealing with parameters that don't change. The second advantage is to catch software bugs, i.e., situations where the program incorrectly attempts to modify data that it should not modify.

goto statements are not allowed. Debugging is hard enough without adding the complexity generated when using `goto`. A corollary to this rule is when developing assembly language software, we should restrict the branching operations to the simple structures allowed in C.

++ and -- should not appear in complex statements. These operations should only appear as commands by themselves. Again, the compiler will generate the same code, so the issue is readability. The statement

```
*(--pt) = buffer[n++];
```

should have been written as

```
--pt;
*(pt) = buffer[n];
n++;
```

If it makes sense to group, then put them on the same line. The following code is allowed

```
buffer[n] = 0; n++;
```

Be a parenthesis zealot. When mixing arithmetic, logical, and conditional operations, explicitly specify the order of operations. Do not rely on the order of precedence. As always, the major style issue is clarity. Even if the following code were actually to perform the intended operation (which in fact it does not), it would be poor style.

```
if( x + 1 & 0x0F == y | 0x04)
```

The programmer assigned to modify it in the future will have a better chance if we had written

```
if(((x + 1) & 0x0F) == (y | 0x04))
```

Use enum instead of #define or const. The use of enum allows for consistency checking during compilation, and provides for easy to read software. A good optimizing compiler will create exactly the same object code for the following four implementations of the same operation. So once again, we focus on style. In the first implementation we needed comments to explain the operations. In the second implementation no comments are needed because of the two #define statements.

```
// implementation 1
int Mode;   // 0 means error
void function1(void){
  Mode = 1; // no error
}
void function2(void){
  if(Mode == 0){ // error?
    UART_OutString("error");
  }
}
```

```
// implementation 2
#define NOERROR 1
#define ERROR 0
int Mode;
void function1(void){
  Mode = NOERROR;
}
void function2(void){
  if(Mode == ERROR){
    UART_OutString("error");
  }
}
```

In the third implementation, shown below on the left, the compiler performs a type-match, making sure **Mode**, **NOERROR**, and **ERROR** are the same type. Consider a fourth implementation that uses enumeration to provide a check of both type and value. We can explicitly set the values of the enumerated types if needed.

```
// implementation 3
const int NOERROR = 1;
const int ERROR = 0;
int Mode;
void function1(void){
  Mode = NOERROR;
}
void function2(void){
  if(Mode == ERROR){
    UART_OutString("error");
  }
}
```

```
// implementation 4
enum Mode_state{ ERROR,
                 NOERROR};
enum Mode_state Mode;
void function1(void){
  Mode = NOERROR;
}
void function2(void){
  if(Mode == ERROR){
    UART_OutString("error");
  }
}
```

#define statements, if used properly, can clarify our software and make our software easy to change. It is proper to use size in all places that refer to the size of the data array.

```
#define SIZE 10
int16_t Data[SIZE];
void initialize(void){ int16_t j;
  for(j = 0; j < SIZE; j++)
    Data[j] = 0;
}
```

Don't use bit-shift for arithmetic operations. Computer architectures and compilers used to be so limited that it made sense to perform multiply/divide by 2 using a shift operation. For example, when multiplying a number by 4, we might be tempted to write **data<<2**. This is wrong; if the operation is multiply, we should write **data*4**. Compiler optimization has developed to the point where the compiler can choose to implement **data*4** as either a shift or multiply depending on the instruction set of the computer. To make code easy to understand, we will use * for multiply, / for divide, and << >> for shift.

3.3.5. Naming convention

Choosing names for variables and functions involves creative thought, and it is intimately connected to how we feel about ourselves as programmers. Of the policies presented in this section, naming conventions may be the hardest habit for us to change. The difficulty is that there are many conventions that satisfy the "easy to understand" objective. Good names reduce the need for documentation. Poor names promote confusion, ambiguity, and mistakes. Poor names can occur because code has been copied from a different situation and inserted into our system without proper integration (i.e., changing the names to be consistent with the new situation.) They can also occur in the cluttered mind of a second-rate programmer, who hurries to deliver software before it is finished.

Names should have meaning. If we observe a name out of the context of the place at which it was defined, the meaning of the object should be obvious. The object **TxFifo** is clearly a transmit first in first out circular queue. The function **UART_OutString** will output a string to the serial port.

Avoid ambiguities. Don't use variable names in our system that are vague or have more than one meaning. For example, it is vague to use **temp**, because there are many possibilities for temporary data, in fact, it might even mean temperature. Don't use two names that look similar, but have different meanings.

Give hints about the type. We can further clarify the meaning of a variable by including phrases in the variable name that specify its type. For example, **dataPt**, **timePt**, and **putPt** are pointers. Similarly, **voltageBuf**, **timeBuf**, and **pressureBuf** are data buffers. Other good phrases include **Flag Mode U16 L Index Cnt**, which refer to Boolean flag, system state, unsigned 16-bit, signed 32-bit, index into an array, and a counter respectively.

Use the same name to refer to the same type of object. For example, everywhere we need a local variable to store an ASCII character we could use the name **letter**. Another common example is to use the names **i**, **j**, and **k** for indices into arrays. The names **V1** and **R1** might refer to a voltage and a resistance. The exact correspondence is not part of the policies presented in this section, just the fact that a correspondence should exist. Once another programmer learns which names we use for which object types, understanding our code becomes easier.

Use a prefix to identify public objects. In this style policy, an underline character will separate the module name from the function name. As an exception to this rule, we can use the underline to delimit words in all upper-case name (e.g., **#define MIN_PRESSURE 10**). Functions that can be accessed outside the scope of a module will begin with a prefix specifying the module to which it belongs. It is poor style to create public variables, but if they need to exist, they too would begin with the module prefix. The prefix matches the file name containing the object. For example, if we see a function call, **UART_OutString("Hello world");** we know this public function belongs to the UART module, where the policies are defined in **UART.h** and the implementation in **UART.c**. Notice the similarity between this syntax (e.g., **UART_Init()**) and the corresponding syntax we would use if programming the module as a class in object-oriented language like C++ or Java (e.g., **UART.Init()**). Using this convention, we can easily distinguish public and private objects.

Use upper and lower case to specify the allocation of an object. We will define I/O ports and constants using no lower-case letters, like typing with caps-lock on. In other words, names without lower-case letters refer to objects with fixed values. **TRUE**, **FALSE**, and **NULL** are good examples of fixed-valued objects. As mentioned earlier, constant names formed from multiple words will use an underline character to delimit the individual words. E.g., **MAX_VOLTAGE**, **UPPER_BOUND**, and **FIFO_SIZE**. Permanently-allocated globals will begin with a capital letter, but include some lower-case letters. Local variables will begin with a lower-case letter, and may or may not include upper case letters. Since all functions are permanently allocated, we can start function names with either an upper-case or lower-case letter. Using this convention, we can distinguish constants, globals and locals. An object's properties (public/private, local/global, constant/variable) are always perfectly clear at the place where the object is defined. The importance of the naming policy is to extend that clarity also to the places where the object is used.

Use capitalization to delimit words. Names that contain multiple words should be defined using a capital letter to signify the first letter of the word. Creating a single name output of multiple words by capitalizing the middle words and squeezing out the spaces is called **CamelCase**. Recall that the case of the first letter specifies whether is the local or global. Some programmers use the underline as a word-delimiter, but except for constants, we will reserve underline to separate the module name from the variable name. Table 3.1 presents examples of the naming convention used in this book.

Type	Examples
Constants	`CR SAFE_TO_RUN PORTA STACK_SIZE START_OF_RAM`
Local variables	`maxTemperature lastCharTyped errorCnt`
Private global variable	`MaxTemperature LastCharTyped ErrorCnt`
Public global variable	`DAC_MaxVoltage Key_LastCharTyped Network_ErrorCnt`
Private function	`ClearTime wrapPointer InChar`
Public function	`Timer_ClearTime RxFifo_Put Key_InChar`

Table 3.1. Examples of names. Use underline to define the module name. Use uppercase for constants. Use CamelCase for variables and functions.

Checkpoint 3.2: How can you tell if a function is private or public?

Checkpoint 3.3: How can you tell if a variable is local or global?

3.3.6. Comments

Discussion about comments was left for last, because they are the least important aspect involved in writing quality software. It is much better to write well-organized software with simple interfaces having operations so easy to understand that comments are not necessary. The goal of this section is to present ideas concerning software documentation in general, and writing comments in particular. Because maintenance is the most important phase of software development, documentation should assist software maintenance. In many situations the software is not static, but continuously undergoing changes. Because of this liquidity, I believe that flowchart and software manuals are not good mechanisms for documenting programs because it is difficult to keep these types of documentation up to date when modifications are made. Therefore, the term documentation in this book refers almost exclusively to comments that are included in the software itself.

The beginning of every file should include the file name, purpose, hardware connections, programmer, date, and copyright. For example, we could write:

```
// filename  adtest.c
// Test of TM4C123 ADC
// 1 Hz sampling on PD3 and output to the serial port
// Last modified 8/6/17 by Jonathan W. Valvano
// Copyright 2017 by Jonathan W. Valvano
//    You may use, edit, run or distribute this file
//    as long as the above copyright notice remains
```

The beginning of every function should include a line delimiting the start of the function, purpose, input parameters, output parameters, and special conditions that apply. The comments at the beginning of the function explain the policies (e.g., how to use the function.) These comments, which are similar to the comments for the prototypes in the header file, are intended to be read by the client. For example, we could explain a function this way:

```
//-------------------UART_InUDec---------------------
// InUDec accepts ASCII input in unsigned decimal
//      and converts to a 32-bit unsigned number
//      valid range is 0 to 4294967295
// Input: none
// Output: 32-bit unsigned number
// If you enter a number above 2^32-1, it will truncate
// Backspace will remove last digit typed
```

Comments can be added to a variable or constant definition to clarify the usage. In particular, comments can specify the units of the variable or constant. For complicated situations, we can use additional lines and include examples. E.g.,

```
int16_t V1;        // voltage at node 1 in mV,
                   // range -5000 mV to +5000 mV
uint16_t Fs;       // sampling rate in Hz
int FoundFlag;     // 0 if keyword not yet found,
                   // 1 if found
uint16_t Mode;     // determines system action,
// as one of the following three cases
#define IDLE 0
#define COLLECT 1
#define TRANSMIT 2
```

Comments can be used to describe complex algorithms. These types of comments are intended to be read by our coworkers. The purpose of these comments is to assist in changing the code in the future, or applying this code into a similar but slightly different application. Comments that restate the function provide no additional information, and actually make the code harder to read. Examples of bad comments include:

```
time++;       // add one to time
mode = 0;     // set mode to zero
```

Good comments explain why the operation is performed, and what it means:

```
time++;       // maintain elapsed time in msec
mode = 0;     // switch to idle mode because no data
```

We can add spaces so the comment fields line up. As stated earlier, we avoid tabs because they often do not translate from one system to another. In this way, the software is on the left and the comments can be read on the right.

Maintenance Tip: If it is not written down, it doesn't exist.

As software developers, our goal is to produce code that not only solves our current problem but can also serve as the basis of our future solutions. In order to reuse software we must leave our code in a condition such that future programmers (including ourselves) can easily understand its purpose, constraints, and implementation. Documentation is not something tacked onto software after it is done, but rather it is a discipline built into it at each stage of the development. Writing comments as we develop the software forces us to think about what the software is doing and more importantly why we are doing it. Therefore, we should carefully develop a programming style that provides appropriate comments. I feel a comment that tells us why we perform certain functions is more informative than comments that tell us what the functions are.

Common error: A comment that simply restates the operation does not add to the overall understanding.

Common error: Putting a comment on every line of software often hides the important information.

Good comments assist us now while we are debugging and will assist us later when we are modifying the software, adding new features, or using the code in a different context. When a variable is defined, we should add comments to explain how the variable is used. If the variable has units then it is appropriate to include them in the comments. It may be relevant to specify the minimum and maximum values. A typical value and what it means often will clarify the usage of the variable. For example:

```
int16_t SetPoint;
// The desired temperature for the control system
// 16-bit signed temperature with resolution of 0.5C,
// The range is -55C to +125C
// A value of 25 means 12.5C,
// A value of -25 means -12.5C
```

When a constant is used, we could add comments to explain what the constant means. If the number has units then it is appropriate to include them in the comments. For example:

```
V = 999;   // 999mV is the maximum voltage
Err = 1;   // error code of 1 means out of range
```

There are two types of readers of our comments. Our client is someone who will use our software incorporating it into a larger system. Client comments focus on the policies of the software. What are the possible valid inputs? What are the resulting outputs? What are the error conditions? Just like a variable, it may be relevant to specify the minimum and maximum values for the input/output parameters. Typical input/output values and what they mean often will clarify the usage of the function. Often we give entire software examples showing how the functions could be used.

The second type of comments is directed to the programmer responsible for debugging and software maintenance (coworker). Coworker comments focus on the mechanisms of the software. These comments explain how the function works.

Generally we separate these comments from the ones intended for the user of the function. This separation is the first of many examples in this book of the concept "separation of policies from mechanisms". The policy is what the function does, and the mechanism is how it works. Specifically, we place this second type of comments within the body of the function. If we are developing in C, then these comments should be included in the *.c file along with the function implementation.

Self-documenting code is software written in a simple and obvious way, such that its purpose and function are self-apparent. Descriptive names for variables, constants, and functions will go a long way to clarify their usage. To write wonderful code like this, we first must formulate the problem by organizing it into clear well-defined subproblems. How we break a complex problem into small parts goes a long way toward making the software self-documenting. The concepts of abstraction, modularity, and layered software, all presented later in this chapter, address this important issue of software organization.

Observation: The purpose of a comment is to assist in debugging and maintenance.

We should use careful indenting and descriptive names for variables, functions, labels, and I/O ports. Liberal use of **#define** provide explanation of software function without cost of execution speed or memory requirements. A disciplined approach to programming is to develop patterns of writing that you consistently follow. Software developers are unlike short story writers. When writing software it is a good design practice to use the same function outline over and over again.

Observation: It is better to write clear and simple software that is easy to understand without comments than to write complex software that requires a lot of extra explanation to understand.

3.4. Modular Software

In this section we introduce the concept of modular programming and demonstrate that it is an effective way to organize our software projects. There are three reasons for forming modules. First, functional abstraction allows us to reuse a software module from multiple locations. Second, complexity abstraction allows us to divide a highly complex system into smaller less complicated components. The third reason is portability. If we create modules for the I/O devices then we can isolate the rest of the system from the hardware details. Portability will be enhanced when we create a device driver or board support package.

3.4.1. Variables

Variables are an important component of software design, and there are many factors to consider when creating variables. Some of the obvious considerations are the allocation, size, and format of the data. However, an important factor involving modular software is **scope**. The scope of a variable defines which software modules can access the data. Variables with a restricted access are classified as **private**, and variables shared between multiple modules are

public. We can restrict the scope to a single file, a single function, or even a single program block within a matching pair of braces, {}. In general, when we limit the scope of our variables a system is easier to design (because the modules are smaller and simpler), easier to change (because code can be reused), and easier to verify (because interactions between modules are well-defined). However, since modules are not completely independent we need a mechanism to transfer information from one to another. The **allocation** of a variable specifies where or how it exists. Because their contents are allowed to change, all variables must be allocated in registers or in RAM, but not in ROM. Constants can and should be allocated in ROM. **Global** variables contain information that is permanent and are usually assigned a fixed location in RAM. On the other hand, **local** variables contain temporary information and are stored in a register or allocated on the stack. In summary, there are three types of variables: public globals (shared permanent), private globals (unshared permanent), and private locals (unshared temporary). We will learn later in Section 3.7 how to create temporary variables on the heap, which can be public or private as needed. Because there is no appropriate way to create a public local variable, we usually refer to private local variables simply as local variables, and the fact that they are private is understood.

A **local** variable has temporary allocation because we create local variables on the stack or in registers. Because the stack and registers are unique to each function, this information cannot be shared with other software modules. Therefore, under most situations, we can further classify these variables as private. Local variables are allocated, used, and then deallocated, in this specific order. For speed reasons, we wish to assign local variables to a register. When we assign local variable to a register, we can do so in a formal manner. There will be a certain line in the assembly software at which the register begins to contain the variable (allocation), followed by lines where the register contains the information (access or usage), and a certain line in the software after which the register no longer contains the information (deallocation). In C, we define local variables after an opening brace.

```
void MyFunction(void){ uint16_t i;       // i is a local
  for(i = 0; i < 10; i++){ uint32_t j; // j is a local
    j = i+100;
    UART_OutUDec(j);
  }
}
```

The information stored in a local variable is not permanent. This means if we store a value into a local variable during one execution of the module, the next time that module is executed the previous value is not available. Examples include loop counters and temporary sums. We use a local variable to store data that are temporary in nature. We can implement a local variable using the stack or registers. Some reasons why we choose local variables over global variables:

- Dynamic allocation/release allows for reuse of RAM
- Limited scope of access (making it private) provides for data protection
   Only the program that created the local variable can access it
- Since an interrupt will save registers and create its own stack frame
   Works correctly if called from multiple concurrent threads (reentrant)
- Since absolute addressing is not used, the code is relocatable

A **global** variable is allocated at a permanent and fixed location in RAM. A public global variable contains information that is shared by more than one program module. We must use global variables to pass data between the main program (i.e., foreground thread) and an ISR (i.e., background thread). If a function called from the foreground belongs to the same module as the ISR, then a global variable used to pass data between the function and the ISR is classified as a private global (assuming software outside the module does not directly access the data). Global variables are allocated at compile time and never deallocated. Allocation of a global variable means the compiler assigns the variable a fixed location in RAM. The information they store is permanent. Examples include time of day, date, calibration tables, user name, temperature, FIFO queues, and message boards. When dealing with complex data structures, pointers to the data structures are shared. In general, it is a poor design practice to employ public global variables. On the other hand, private global variables are necessary to store information that is permanent in nature. In C, we define global variables outside of the function.

```
int32_t Count=0;   // Count is a global variable
void MyFunction(void){
   Count++;       // number of times function was called
}
```

Checkpoint 3.4: How do you create a local variable in C?

Sometimes we store temporary information in global variables out of laziness. This practice is to be discouraged because it wastes memory and may cause the module to work incorrectly if called from multiple concurrent threads (non-reentrant). Non-reentrant programs can produce very sneaky bugs, since they might only crash in rare situations when the same code called from different threads when the first thread is in a particular critical section. Such a bug is difficult to reproduce and diagnose. In general, it is good design to limit the scope of a variable as much as possible.

Checkpoint 3.5: How do you create a global variable in C?

In C, a **static** local has permanent allocation, which means it maintains its value from one call to the next. It is still local in scope, meaning it is only accessible from within the function. I.e., modifying a local variable with **static** changes its allocation (it is now permanent), but doesn't change its scope (it is still private). In the following example, **count** contains the number of times **MyFunction** is called. The initialization of a static local occurs just once, during startup.

```
void MyFunction(void){ static int32_t count=0;
   count++;       // number of times function was called
}
```

In C, we create a private global variable using the **static** modifier. Modifying a global variable with **static** does not change its allocation (it is still permanent), but does reduce its scope. Regular globals can be accessed from any function in the system (public), whereas a **static** global can only be accessed by functions within the same file. **Static** globals are private to that particular file. Functions can be static also, meaning they can be called only from other functions in the file. E.g.,

```
static int16_t myPrivateGlobalVariable; // this file only
void static MyPrivateFunction(void){
}
```

In C, a **const** global is read-only. It is allocated in the ROM. Constants, of course, must be initialized at compile time. E.g.,

```
const int16_t Slope=21;
const uint8_t SinTable[8]={0,50,98,142,180,212,236,250};
```

Checkpoint 3.6: How does the **static** modifier affect locals, globals, and functions in C?

Checkpoint 3.7: How does the **const** modifier affect a global variable in C?

Common error: If you leave off the **const** modifier in the **SinTable** example, the table will be allocated twice, once in ROM containing the initial values, and once in RAM containing data to be used at run time. Upon startup, the system copies the ROM-version into the RAM-version.

Maintenance Tip: It is good practice to specify the units of a variable (e.g., volts, cm etc.).

Common error: In C, global variables are initialized to zero by default, but local variables are not initialized.

3.4.2. Dividing tasks into subtasks

The key to completing any complex task is to break it down into manageable subtasks. **Modular programming** is a style of software development that divides the software problem into distinct and independent modules. The parts are as small as possible, yet relatively independent. Complex systems designed in a modular fashion are easier to debug because each module can be tested separately. Industry experts estimate that 50 to 90% of software development cost is spent in **maintenance**. All five aspects of software maintenance are simplified by organizing the software system into modules.

- Correcting mistakes
- Adding new features
- Optimizing for execution speed or program size
- Porting to new computers or operating systems
- Reconfiguring the software to solve similar related programs

The approach is particularly useful when a task is large enough to require several programmers.

A **program module** is a self-contained software task with clear entry and exit points. We make the distinction between module and a C language function. A module can be a collection of functions that in its entirety performs a well-defined set of tasks. The collection of serial port I/O functions presented later in section 3.4.4 can be considered one module. A collection of 32-bit math operations is another example of a module. The main program and other high-level functions may constitute a module, but usually a set of functions that perform a well-defined task can also be written as modules. Modular programming involves both the specification of the individual modules and the connection scheme by which the modules are connected together to form the software system. While the module may be called from many locations throughout the system, there should be well-defined **entry points**.

The overall goal of modular programming is to enhance clarity. The smaller the task, the easier it will be to understand. **Coupling** is defined as the influence one module's behavior has on another module. In order to make modules more independent we strive to minimize coupling. Obvious and appropriate examples of coupling are the input/output parameters explicitly passed from one module to another. On the other hand, information stored in shared global variables can be quite difficult to track. In a similar way shared accesses to I/O ports can also introduce unnecessary complexity. Global variables cause coupling between modules that complicate the debugging process because now the modules may not be able to be separately tested. On the other hand, we must use global variables to pass information into and out of an interrupt service routine, and from one call to an interrupt service routine to the next call.

Another problem specific to embedded systems is the need for fast execution. For this reason the ARM architecture has enough registers so that some can be used to store local variables. Allocating local variables in registers produces shorter and faster code as compared to globals and stack-based locals. When passing information through global variables is required, it is better to use a well-defined abstract technique like a mailbox or first-in-first-out (FIFO) queue.

Assign a logically complete task to each module. The module is logically complete when it can be separated from the rest of the system and placed into another application. The **interfaces** are extremely important. The interfaces determine the policies of our modules. In other words, the interfaces define the operations of our software system. The interfaces also represent the coupling between modules. In general we wish to minimize the amount of information passing between the modules yet maximize the number of modules. Of the following three objectives when dividing a software project into subtasks, it is really only the first one that matters.

- Make the software project easier to understand
- Increase the number of modules
- Decrease the interdependency (minimize coupling)

We can develop and connect modules in a hierarchical manner. Construct new modules by combining existing modules. In a hierarchical system the modules are organized into a tree-structured call graph. In the call graph, an arrow points from the calling routine to the module it calls. The I/O ports are organized into groups (e.g., all the serial port I/O registers are in one group). The call graph allows us to see the organization of the project. To make simpler call graphs on large projects we can combine multiple related functions into a single module. The main program is at the top and the I/O ports are at the bottom. In a hierarchical system the modules are organized both in a horizontal fashion (grouped together by function) and in a vertical fashion (overall policies decisions at the top and implementation details at the bottom). Since one of the advantages of breaking a large software project into subtasks is concurrent development, it makes sense to consider concurrency when dividing the tasks. In other words, the modules should be partitioned in such a way that multiple programmers can develop the subtasks as independently as possible. On the other hand careful and constant supervision is required as modules are connected together and tested.

Observation: If module A calls module B, and module B calls module A, then you have created a special situation that must account for these mutual calls.

There are two approaches to **hierarchical programming**. The top-down approach starts with a general overview, like an outline of a paper, and builds refinement into subsequent layers. A top-down programmer was once quoted as saying,

> *"Write no software until every detail is specified"*

It provides a better global approach to the problem. Managers like **top-down** because it gives them tighter control over their workers. The top-down approach work well when an existing operational system is being upgraded or rewritten. On the other hand the **bottom-up** approach starts with the smallest detail, builds up the system "one brick at a time." The bottom-up approach provides a realistic appreciation of the problem because we often cannot appreciate the difficulty or the simplicity of a problem until we have tried it. It allows programmers to start immediately coding, and gives programmers more input into the design. For example, a low-level programmer may be able to point out features that are not possible and suggest other features that are even better. Some software projects are flawed from their conception. With bottom-up design, the obvious flaws surface early in the development cycle.

I believe bottom-up is better when designing a complex system and specifications are open-ended. On the other hand, top-down is better when you have a very clear understanding of the problem specifications and the constraints of your computer system. The best software I have ever produced was actually written twice. The first pass was programmed bottom up and served only to provide a clear understanding of the problem, clarification of the features I wanted, and the limitations of my hardware. I literally threw all the source code in the trash, and programmed the second pass in a top-down manner.

Arthur C. Clarke's Third Law: Any sufficiently advanced technology is indistinguishable from magic.

J. Porter Clark's Law: Sufficiently advanced incompetence is indistinguishable from malice.

One of the biggest mistakes beginning programmers make is the inappropriate usage of I/O calls (e.g., screen output and keyboard input). An explanation for their foolish behavior is that they haven't had the experience yet of trying to reuse software they have written for one project in another project. Software portability is diminished when it is littered with user input/output. To reuse software with user I/O in another situation, you will almost certainly have to remove the input/output statements. In general, we avoid interactive I/O at the lowest levels of the hierarchy, rather return data and flags and let the higher level program do the interactive I/O. Often we add keyboard input and screen output calls when testing our software. It is important to remove the I/O that not directly necessary as part of the module function. This allows you to reuse these functions in situations where screen output is not available or appropriate. Obviously screen output is allowed if that is the purpose of the routine.

Common Error: Performing unnecessary I/O in a subroutine makes it harder to reuse at a later time.

From a formal perspective, I/O devices are considered as global. This is because I/O devices reside permanently at fixed addresses. From a syntactic viewpoint any module has access to any I/O device. In order to reduce the complexity of the system we will restrict the number of modules that actually do access the I/O device. It will be important to clarify which modules have access to I/O devices and when they are allowed to access it. When more than one module accesses an I/O device, then it is important to develop ways to arbitrate (which module goes first if two or more want to access simultaneously) or synchronize (make a second module wait until the first is finished.) These arbitration issues will be presented in Chapters 4 and 5.

Information hiding is similar to minimizing coupling. It is better to separate the mechanisms of software from its policies. We should separate what the function does (the relationship between its inputs and outputs) from how it does it. It is good to hide certain inner workings of a module, and simply interface with the other modules through the well-defined input/output parameters. For example we could implement a FIFO by maintaining the current number of elements in a global variable, `Count`. A good module will hide how `Count` is implemented from its users. If the user wants to know how many elements are in the FIFO, it calls a `TxFifo_Size()` routine that returns the value of `Count`. A badly written module will not hide `Count` from its users. The user simply accesses the global variable `Count`. If we update the FIFO routines, making them faster or better, we might have to update all the programs that access `Count` too. The object-oriented programming environments provide well-defined mechanisms to support information hiding. This separation of policies from mechanisms can be seen also in layered software.

The *Keep It Simple Stupid* approach tries to generalize the problem so that it fits an abstract model. Unfortunately, the person who defines the software specifications may not understand the implications and alternatives. Sometimes we can restate the problem to allow for a simpler (and possibly more powerful) solution. As a software developer, we always ask ourselves these questions:

> "How important is this feature?"
> "What alternative ways could this system be structured?"
> "How can I redefine the problem to make it simpler?"

We can classify the coupling between modules as highly coupled, loosely coupled, or uncoupled. A highly-coupled system is not desirable, because there is a great deal of interdependence between modules. A loosely-coupled system is optimal, because there is some dependence but interconnections are weak. An uncoupled system, one with no interconnections at all, is typically inappropriate in an embedded system, because all components should be acting towards a common objective. There are three ways in which modules can be coupled. The natural way in which modules are coupled is where one module calls or invokes a function in a second module. This type of coupling, called **invocation coupling**, can be visualized with a call graph. A second way modules can be coupled is by data transfer. If information flows from one module to another, we classify this as **bandwidth coupling**. Bandwidth, which is the information transfer rate, is a quantitative measure of coupling. Bandwidth coupling can be visualized with a data flow graph. The third type of coupling, called **control coupling**, occurs when actions in one module affect the control path within another module. For example, if Module A sets a global flag and Module B uses the global flag to decide its execution path. Control coupling is hard to visualize and hard to debug. Therefore, it is a poor design to employ this type of coupling.

Another way to categorize coupling is to examine how information is passed or shared between modules. We will list the mechanisms from poor to excellent. It is extremely poor design to allow Module A directly modify data or flags within Module B. Similarly poor design is to organize important data into a common shared global space and allow modules to read and write these data. It is ok to allow Module A to call Module B and pass it a control flag. This control flag will in turn affect the execution within Module B. It is good design to have one module pass data to another module. A **stamp** is defined as structured data passed from one module to another. **Primitive data** passed between modules is unstructured.

Coupling is a way to describe how modules connect with each other, but it is also important to analyze how various components within one module interact with each other. **Cohesion** is the degree of interrelatedness of internal parts within the module. In general, we wish to maximize cohesion. A cohesive module has all components of the module are directed towards and essential for the same task. It is also important to analyze how components are related as we design modules. **Coincidental cohesion** occurs when components of the module are unrelated, resulting poor design. Examples of coincidental cohesion would be a collection of frequently used routines, a collection of routines written by a single programmer, or a collection of routines written during a certain time interval.

Logical cohesion is a grouping of components into a single module (because they perform similar functions). An example of logical cohesion is a collection of serial output, LCD output, and printer output routines into one module (because all routines perform output). Organizing modules in this fashion is a poor design and results in modules that are hard to reuse.

Temporal cohesion combines components if they are connected in time sequence. If we are baking bread, we activate the yeast in warm water in one bowl, and then we combine the flour, sugar, and spices in another bowl. These two steps are connected only in a sense that we first do one, and then we do another when making bread. If we were making cookies, we would need the flour, sugar, and spices but not the yeast. We want to mix and match existing modules to create new designs, as such, we expect the sequence of module execution to change.

Another poor design, called **procedural cohesion**, groups functions together in order to ensure mandatory ordering. For example, an embedded system might have an input port, an output port, and a timer module. In order to work properly, all three must be initialized. It would be hard to reuse code if we placed all three initialization routines into one module.

We next present appropriate reasons to group components into one module. **Communicational cohesion** exists when components operate on the same data. An example of communicational cohesion would be a collection of routines that filter and extract features from the data.

Sequential cohesion occurs when components are grouped into one module, because the output from one component is the input to another component. Sequential cohesion is a natural consequence of minimizing bandwidth between modules. An example of sequential cohesion is a fuzzy logic controller. This controller has five stages: crisp input, fuzzification, rules, defuzzification, and crisp output. The output of each stage is the input to the next stage. The input bandwidth to the controller and the output bandwidth from the controller can be quite low, but the amount of information transferred between stages can be thousands of times larger.

The best kind of cohesion is **functional cohesion**, where all components combine to implement a single objective, and each component has a necessary contribution to the objective. I/O device drivers, which are a collection of routines for a single I/O device, exhibit functional cohesion.

Another way to classify good and bad modularity is to observe fan in and fan out behavior. In a data flow graph, the tail of an arrow defines a data output, and the head of an arrow defines a data input. The **fan in** of a module is the number of other modules that have direct control on that particular module. Fan in can be visualized by counting the number of arrowheads that terminate on the module in the data flow graph, shown previously in Figure 1.9. The **fan out** of a module is number of other modules directly controlled by this module. Fan out can be visualized by counting the number of tails of arrows that originate on the module in the data flow graph. In general, a system with high fan out is poorly designed, because that one module may constitute a bottleneck or a critical safety path. In other words, the module with high fan out is probably doing too much, performing the tasks that should be distributed to other modules. High fan in is not necessarily a poor design, depending on the application.

3.4.3. Device Drivers and Board Support Package

As the size and complexity of our software systems increase, we learn to anticipate the changes that our software must undergo in the future. In particular, we can expect to redesign our system to run on new and more powerful hardware platforms. A similar expectation is that better algorithms may become available. The objective of this section is to use a layered software approach to facilitate these types of changes.

We can use the call graph like the one drawn in Figure 3.1 to visualize software layers. The arrows point from the calling function to the function it calls. Figure 3.1 shows only one module at each layer, but a complex system might have multiple modules at each layer. A function in a layer can call a function within the same module, or it can call a public function in a module of the same or lower layer. Some layered systems restrict the calls only to modules in the most adjacent layer below it. If we place all the functions that access the I/O hardware in the bottom most layer, we can call this layer a **hardware abstraction layer** (HAL). This bottom-most layer can also be called a **board support package** (BSP) if I/O devices are referenced in an abstract manner. Each middle layer of modules only calls lower level modules, but not modules at a higher level. Usually the top layer consists of the main program. In a multi-threaded environment there can be multiple main programs at the top-most level, but for now assume there is only one main program.

An example of a layered system is Transmission Control Protocol/Internet Protocol (TCP/IP), which consists of at least four distinct layers: application (http, telnet, SMTP, FTP), transport (UDP, TCP), internet (IP, ICMP, IGMP), and network layers (Ethernet).

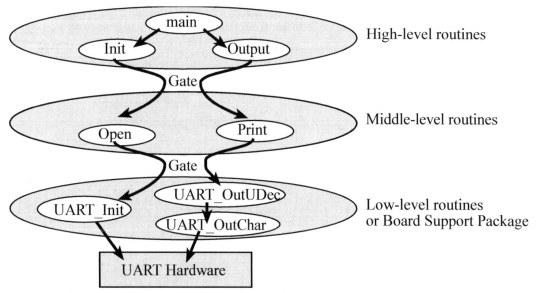

Figure 3.1. A layered approach to interfacing a printer. The bottom layer is the BSP.

To develop a **layered software system** we begin with a modular system. The main advantage of layered software is the ability to separate the modules into groups or layers such that one layer may be replaced without affecting the other layers. For example, you could change which microcontroller you are using, by modifying the low level without any changes to the other levels. Figure 3.1 depicts a layered implementation of a printer interface. In a similar way, you could replace the printer with a solid state disk by replacing just the middle and lower layers. If we were to employ buffering and/or data compression to enhance communication bandwidth, then these algorithms would be added to the middle level. A layered system should allow you to change the implementation of one layer without requiring redesign of the other layers.

A **gate** is used to connect one layer to the next. Another name for this gate is **application program interface** or API. The gates provide a mechanism to link the layers. Because the size of the software on an embedded system is small, it is possible and appropriate to implement a layered system using standard function calls by simply compiling and linking all software together. We will see in the next section that the gate can be implemented by creating a header file with prototypes to public functions. The following rules apply to layered software systems:

1. A module may make a simple call to other modules in the same layer.
2. A module may make a call to a lower level module only by using the gate.
3. A module may not directly access any function or variable in another layer without going through the gate.
4. A module may not call a higher level routine.
5. A module may not modify the vector address of another level's handler(s).
6. (optional) A module may not call farther down than the immediately adjacent lower level.
7. (optional) All I/O hardware access is grouped in the lowest level.
8. (optional) All user interface I/O is grouped in the highest level unless it is the purpose of the module itself to do such I/O.

The purpose of rule 6 is to allow modifications at the low layer to not affect operation at the highest layer. On the other hand, for efficiency reasons you may wish to allow module calls further down than the immediately adjacent lower layer. To get the full advantage of layered software, it is critical to design functionally complete interfaces between the layers. The interface should support all current functions as well as provide for future expansions.

A **device driver** consists of the software routines that provide the functionality of an I/O device. A device driver usually does not hide what type of I/O module it is. E.g., in the next section, we consider a device driver for a serial port. A board support package is similar to a device driver, except that there is more of an attempt to hide what the I/O device actually is. A board support package provides a higher level of abstraction than a regular device driver. The driver consists of the interface routines that the operating system or software developer's program calls to perform I/O operations as well as the low-level routines that configure the I/O device and perform the actual input/output. The issue of the separation of policies from mechanisms is very important in device driver design. The policies of a driver include the list of functions and the overall expected results. In particular, the policies can be summarized by the interface routines that the OS or software developer can call to access the device. The mechanisms of the device driver include the specific hardware and low-level software that actually perform the I/O. As an example, consider the wide variety of mass storage devices that are available. Floppy disk, RAM disks, integrated device electronics (IDE) hard drive, Serial

Advanced Technology Attachment (SATA) hard drive, flash EEPROM drive, and even a network can be used to save and recall data files. A simple mass storage system might have the following C-level interface functions, as explained in the following prototypes (in each case the functions return 0 if successful and an error code if the operation fails:

```
int eFile_Init(void);              // initialize file system
int eFile_Create(char name[]);     // create new file, make it empty
int eFile_WOpen(char name[]);      // open a file for writing
int eFile_Write(int8_t data);      // stream data into open file
int eFile_WClose(void);            // close the file for writing
int eFile_ROpen(char name[]);      // open a file for reading
int eFile_ReadNext(int8_t *pt);    // stream data out of open file
int eFile_RClose(void);            // close the file for reading
int eFile_Delete(char name[]);     // remove this file
```

Building a hardware abstraction layer (HAL) is the same idea as separation of policies from mechanisms. A diagram of this layered concept was shown in Figure 3.1. In the above file example, a HAL or BSP would treat all the potential mass storage devices through the same software interface. Another example of this abstraction is the way some computers treat pictures on the video screen and pictures printed on the printer. With the abstraction layer, the software developer's program draws lines and colors by passing the data in a standard format to the device driver, and the OS redirects the information to the video graphics board or color LaserWriter as appropriate. This layered approach allows one to mix and match hardware and software components but does suffer some overhead and inefficiency.

Low-level **device drivers** normally exist in the Basic Input/Output System (BIOS) ROM and have direct access to the hardware. They provide the interface between the hardware and the rest of the software. Good low-level device drivers allow:

1. New hardware to be installed;
2. New algorithms to be implemented
   a. Synchronization with busy wait, interrupts, or DMA
   b. Error detection and recovery methods
   c. Enhancements like automatic data compression
3. Higher level features to be built on top of the low level
   a. OS features like blocking semaphores
   b. Additional features like function keys

and still maintain the same software interface. In larger systems like the personal computer (PC), the low-level I/O software is compiled and burned in ROM separate from the code that will call it, it makes sense to implement the device drivers as software interrupts (sometimes called traps) and specify the calling sequence language-independent. We define the "client programmer" as the software developer that will use the device driver. In embedded systems like we use, it is appropriate to provide **device.h** and **device.c** files that the client programmer can compile with their application. In a commercial setting, you may be able to deliver to the client only the **device.h** together with the object file. **Linking** is the process of resolving addresses to code and programs that have been complied separately. In this way, the routines can be called from any program without requiring complicated linking. In other words, when the device driver is implemented with a software interrupt, the linking occurs at run time through the vector address of the software interrupt. In our embedded system however, the linking will be static occurring at the time of compilation.

3.4.4. Serial Port Driver

The concept of a device driver can be illustrated with the following design of a serial port device driver. In this section, the contents of the header file (**UART.h**) will be presented, and the implementations will be developed in the next chapter. The device driver software is grouped into four categories. Protected items can only be directly accessed by the device driver itself, and public items can be accessed by other modules.

1. Data structures: global (private) The first component of a device driver includes private global data structures. To be private global means only programs within the driver itself may directly access these variables. If the user of the device driver (e.g., a client) needs to read or write to these variables, then the driver will include public functions that allow appropriate read/write functions. One example of a private global variable might be an **OpenFlag**, which is true if the serial port has been properly initialized. **static** used in this way makes the variable private to the file, but does have permanent allocation. The implementation developed in Chapter 4 will have no private global variables, but the UART implementation developed in Chapter 5 will include a private FIFO queue.

```
int static OpenFlag = 0;   // true if driver has been initialized
```

2. Initialization routines (public, called by the client once in the beginning) The second component of a device driver includes the public functions used to initialize the device. To be public means the user of this driver can call these functions directly. A prototype to public functions will be included in the header file (**UART.h**). The names of public functions will begin with **UART_**. The purpose of this function is to initialize the UART hardware.

```
//------------UART_Init------------
// Initialize Serial port UART
// Input: none
// Output: none
void UART_Init(void);
```

3. Regular I/O calls (public, called by client to perform I/O) The third component of a device driver consists of the public functions used to perform input/output with the device. Because these functions are public, prototypes will be included in the header file (**UART.h**). The input functions are grouped, followed by the output functions.

```
//------------UART_InChar------------
// Wait for new serial port input
// Input: none
// Output: ASCII code for key typed
char UART_InChar(void);

//------------UART_InString------------
// Wait for a sequence of serial port input
// Input: maxSize is the maximum number of characters to look for
// Output: Null-terminated string in buffer
void UART_InString(char *buffer, uint16_t maxSize);
```

```
//------------UART_InUDec------------
// InUDec accepts ASCII input in unsigned decimal format
// and converts to a 32-bit unsigned number (0 to 4294967295)
// Input: none
// Output: 32-bit unsigned number
uint32_t UART_InUDec(void);

//------------UART_OutChar------------
// Output 8-bit to serial port
// Input: letter is an 8-bit ASCII character to be transferred
// Output: none
void UART_OutChar(char letter);
//------------UART_OutString------------
// Output String (NULL termination)
// Input: pointer to a NULL-terminated string to be transferred
// Output: none
void UART_OutString(char *buffer);

//------------UART_OutUDec------------
// Output a 32-bit number in unsigned decimal format
// Input: 32-bit number to be transferred
// Output: none
// Variable format 1-10 digits with no space before or after
void UART_OutUDec(uint32_t number);
```

4. Support software (private) The last component of a device driver consists of private functions. Because these functions are private, prototypes will not be included in the header file (**UART.h**). We place helper functions and interrupt service routines in the category.

Notice that this UART example implements a layered approach, similar to Figure 3.1. The low-level functions provide the mechanisms and are protected (hidden) from the client programmer. The high-level functions provide the policies and are accessible (public) to the client. When the device driver software is separated into **UART.h** and **UART.c** files, you need to pay careful attention as to how many details you place in the **UART.h** file. A good device driver separates the policy (overall operation, how it is called, what it returns, what it does, etc.) from the implementation (access to hardware, how it works, etc.) In general, you place the policies in the **UART.h** file (to be read by the client) and the implementations in the **UART.c** file (to be read by you and your coworkers). Think of it this way: if you were to write commercial software that you wished to sell for profit and you delivered the **UART.h** file and its compiled object file, how little information could you place in the **UART.h** file and still have the software system be fully functional. In summary, the policies will be public, and the implementations will be private.

Observation: A layered approach to I/O programming makes it easier for you to upgrade to newer technology.

Observation: A layered approach to I/O programming allows you to do concurrent development.

3.4.5. Abstract Output Device Driver

In the UART driver shown in the previous section, the routines clearly involve a UART. Another approach to I/O is to provide a high-level abstraction in such a way that the I/O device itself is hidden from the user. There are multiple projects on the book's web site that implement this abstraction. The overall purpose of these examples is to provide an output stream for the standard **printf** feature to which most C programmers are accustomed. For the TM4C123 and MSP432E401Y LaunchPad boards, we can send output to the PC using UART0, to a ST7735 color graphics LCD, or to a low-cost Nokia 5110 graphics LCD. The implementations for the LM3S Stellaris® kits use the organic light emitting diode (OLED) display. Even though all these displays are quite different, they all behave in a similar fashion.

In C, we can specify the output stream used by **printf** by writing a **fputc** function. The **fputc** function is a private and implemented inside the driver. It sends characters to the display and manages the cursor, tab, line feed and carriage return functionalities. The user controls the display by calling the following five public functions.

```
void Output_Init(void);    // Initializes the display interface.
void Output_Clear(void);   // Clears the display
void Output_Off(void);     // Turns off the display
void Output_On(void);      // Turns on the display
void Output_Color(uint32_t newColor); // Set color of future output
```

The user performs output by calling **printf**. This abstraction clearly separates what it does (output information) from how it works (sends pixel data to the display over UART, SSI, or I²C). In these examples all output is sent to the display; however, we could modify the **fputc** function and redirect the output stream to other devices such as the USB, Ethernet, or disk. For the LM3S boards, this example is called **OLED_xxx**. For the TM4C boards, this example can be found as **ST7735_xxx**, **Printf_Nokia5110_xxx**, and **Printf_UART_xxx**.

3.5. Finite State Machines

Software abstraction is when we define a complex problem with a set of basic abstract principles. If we can construct our software system using these building blocks, then we have a better understanding of the problem. This is because we can separate what we are doing from the details of how we are getting it done. This separation also makes it easier to optimize. It provides for a proof of correct function and simplifies both extensions and customization. A good example of abstraction is the **Finite State Machine** (FSM) implementations. The abstract principles of FSM development are the inputs, outputs, states, and state transitions. If we can take a complex problem and map it into a FSM model, then we can solve it with a simple FSM software tools. Our FSM software implementation will be easy to understand, debug, and modify. Other examples of software abstraction include Proportional Integral Derivative digital controllers, fuzzy logic digital controllers, neural networks, and linear systems of differential equations (e.g., PSPICE.) In each case, the problem is mapped into well-defined model with a set of abstract yet powerful rules. Then, the software solution is a matter of implementing the rules of the model.

Linked lists are lists or nodes where one or more of the entries is a (link) to other nodes of similar structure. We can have statically-allocated fixed-size linked lists that are defined at assemble or compile time and exist throughout the life of the software. On the other hand, we implement dynamically-allocated variable-size linked lists that are constructed at run time and can grow and shrink in size. We will use a data structure similar to a linked list called a **linked structure** to build a finite state machine controller. Linked structures are very flexible and provide a mechanism to implement abstractions.

A well-defined model or framework is used to solve our problem (implemented with a linked structure). The three advantages of abstraction are 1) it can be faster to develop because a lot of the building blocks preexist; 2) it is easier to debug (prove correct) because it separates conceptual issues from implementation; and 3) it is easier to change.

An important factor when implementing finite state machines using linked structures is that there should be a clear and one-to-one mapping between the finite state machine and the linked structure. I.e., there should be one structure for each state.

We will present two types of finite state machines. The **Moore FSM** has an output that depends only on the state, and the next state depends on both the input and current state. We will use a Moore implementation if there is an association between a state and an output. There can be multiple states with the same output, but the output defines in part what it means to be in that state. For example, in a traffic light controller, the state of green light on the North road (red light on the East road) is caused by outputting a specific pattern to the traffic light.

On the other hand, the **Mealy FSM** has an output that depends on both the input and the state, and the next state also depends on input and current state. We will use a Mealy implementation if the output causes the state to change. In this situation, we do not need a specific output to be in that state; rather the outputs are required to cause the state transition. For example, to make a robot stand up, we perform a series of outputs causing the state to change from sitting to standing. Although we can rewrite any Mealy machine as a Moore machine and vice versa, it is better to implement the format that is more natural for the particular problem. In this way the state graph will be easier to understand.

Checkpoint 3.8: What are the differences between a Mealy and Moore finite state machine?

One of the common features in many finite state machines is a time delay. We will learn very elaborate mechanisms to handle time in Chapters 5-6, but in this section we will use the SysTick delay functions presented in Program 2.10.

Example 3.1. Design a line-tracking robot that has two drive wheels and two line sensors using a FSM. The goal is to drive the robot along a line placed in the center of the road. The robot has two drive wheels and a third free turning balance wheel. Figure 3.2 shows that PF1 drives the left wheel and PF2 drives the right wheel. If both motors are on (PF2–1 = 11), the robot goes straight. If just the left motor is on (PF2–1 = 01), the robot will turn right. If just the right motor is on (PF2–1 = 10), the robot will turn left. The line sensors are under the robot and can detect whether or not they see the line. The two sensors are connected to Port F, such that:

  PF4,PF0 equal to 0,0 means we are lost, way off to the right or way off to the left.
  PF4,PF0 equal to 0,1 means we are off just a little bit to the right.
  PF4,PF0 equal to 1,0 means we are off just a little bit to the left.
  PF4,PF0 equal to 1,1 means we are on line.

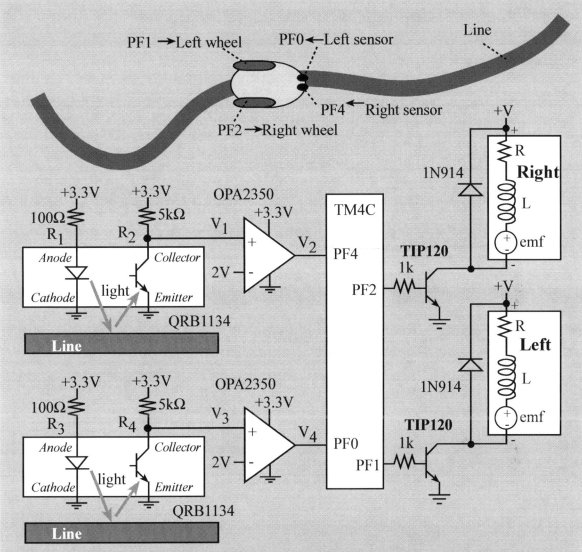

Figure 3.2. Robot with two drive wheels and two line sensors (see sections 6.5 and 8.2 for circuit details).

Solution: The focus of this example is the FSM, but the QRB1134 sensor interface will be described later in example 6.1, and the DC motor interface will be described in Section 6.5. The first step in designing a FSM is to create some states. The outputs of a Moore FSM are only a function of the current state. A Moore implementation was chosen because we define our states by what believe to be true and we will have one action (output) that depends on the state. Each state is given a symbolic name where the state name either describes "what we know" or "what we are doing". We could have differentiated between a little off to the left and way off to the left, but this solution creates a simple solution with 3 states.

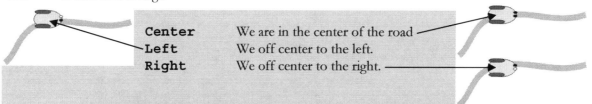

```
Center    We are in the center of the road
Left      We off center to the left.
Right     We off center to the right.
```

The finite state machine implements this line-tracking algorithm. Each state has a 2-bit output value, and four next state pointers. The strategy will be to:

Go straight if we are on the line.
Turn right if we are off to the left.
Turn left if we are off to the right.

Finally, we implement the heuristics by defining the state transitions, as illustrated in Figure 3.3 and Table 3.2. If we used to be in the left state and completely lose the line (input 00), then we know we are left of the line. Similarly if we used to be in the right state and completely lose the line (input 00), then we know we are right of the line. However, we used to be on the center of the line and then completely lose the line (input 00), we do not know we are right or left of the line. The machine will guess we went right of the line. In this implementation we put a constant delay of 10ms in each state. We put the time to wait into the machine as a parameter for each state to provide for clarity of how it works and simplify possible changes in the future. If we are off to the right (input 01), then it will oscillate between **Center** and **Right** states, making a slow turn left. If we are off to the left (input 10), then it will oscillate between **Center** and **Left** states, making a slow turn right.

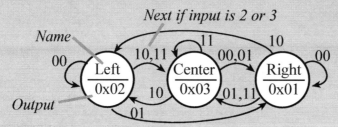

Figure 3.3. Graphical form of a Moore FSM that implements a line tracking robot.

State	Motor	Delay	Input 00	01	10	11
Center	1,1	1	Right	Right	Left	Center
Left	1,0	1	Left	Right	Center	Center
Right	0,1	1	Right	Center	Left	Center

Table 3.2. Tabular form of a Moore FSM that implements a line tracking robot.

The first step in designing the software is to decide on the sequence of operations.

1) Initialize timer and directions registers
2) Specify initial state
3) Perform FSM controller
    a) Output to DC motors, which depends on the state
    b) Delay, which depends on the state
    c) Input from line sensors
    d) Change states, which depends on the state and the input

The second step is to define the FSM graph using a data structure. Program 3.1 shows a table implementation of the Moore FSM. This implementation uses a table data structure, where each state is an entry in the table, and state transitions are defined as indices into this table. The four **Next** parameters define the input-dependent state transitions. The wait times are defined in the software as fixed-point decimal numbers with units of 0.01s. The label **Center** is more descriptive than the state number 0. Notice the 1-1 correspondence between the tabular form in Table 3.2 and the software specification of **fsm[3]**. This 1-1 correspondence makes it possible to prove the software exactly executes the FSM as described in the table.

```c
struct State {
  uint32_t Out;        // 2-bit output
  uint32_t Delay;      // time in 10ms
  uint8_t Next[4];};
typedef const struct State STyp;
#define Center 0
#define Left   1
#define Right  2
STyp fsm[3] = {
  {0x03, 1, { Right, Right,  Left,   Center }},  // Center of line
  {0x02, 1, { Left,  Right,  Center, Center }},  // Left of line
  {0x01, 1, { Right, Center, Left,   Center }}   // Right of line
};
#define PF21 (*((volatile unsigned long *)0x40025018))
#define PF4  (*((volatile unsigned long *)0x40025040))
#define PF0  (*((volatile unsigned long *)0x40025004))
int main(void){ uint32_t S;  // index to the current state
uint32_t input, output;      // state I/O
  Robot_Init();              // Initialize Port F, SysTick
  S = Center;                // initial state
  while(1){
    output = fsm[S].Out;     // set output from FSM
    PF21 = output<<1;        // do output to two motors
    SysTick_Wait10ms(fsm[S].Delay); // wait
    input = PF0+(PF4>>3);    // read sensors
    S = fsm[S].Next[input];  // next depends on input and state
  }
}
void Robot_Init(void){
  SYSCTL_RCGCGPIO_R |= 0x20;         // 1) activate clock for Port F
  SysTick_Init();                    // initialize SysTick (program 2.10)
  GPIO_PORTF_LOCK_R = 0x4C4F434B;    // 2) unlock GPIO Port F
  GPIO_PORTF_CR_R = 0x1F;            // allow changes to PF4-0
  GPIO_PORTF_AMSEL_R = 0x00;         // 3) disable analog on PF
  GPIO_PORTF_PCTL_R = 0x00000000;    // 4) PCTL GPIO on PF4-0
  GPIO_PORTF_DIR_R = 0x0E;           // 5) PF4,PF0 in, PF3-1 out
  GPIO_PORTF_AFSEL_R = 0x00;         // 6) disable alt funct on PF7-0
  GPIO_PORTF_DEN_R = 0x1F;           // 7) enable digital I/O on PF4-0
}
```
Program 3.1. Table implementation of a Moore FSM (TableLineTracker_xxx).

Program 3.2 uses a linked structure, where each state is a node, and state transitions are defined as pointers to other nodes. Again, notice the 1-1 correspondence between Table 3.2 and the software specification of `fsm[3]`.

```c
struct State {
  uint32_t Out;        // 2-bit output
  uint32_t Delay;      // time in 10ms
  const struct State *Next[4];};
typedef const struct State STyp;
#define Center &fsm[0]
#define Left   &fsm[1]
#define Right  &fsm[2]
STyp fsm[3] = {
  {0x03, 1, { Right, Right,  Left,   Center }},  // Center of line
  {0x02, 1, { Left,  Right,  Center, Center }},  // Left of line
  {0x01, 1, { Right, Center, Left,   Center }}   // Right of line
};
int main(void){ STyp *pt;    // state pointer
uint32_t input, output;      // state I/O
  Robot_Init();              // Initialize Port F, SysTick
  pt = Center;               // initial state
  while(1){
    output = pt->Out;                  // set output from FSM
    PF21 = output<<1;                  // do output to two motors
    SysTick_Wait10ms(pt->Delay);       // wait
    input = PF0+(PF4>>3);              // read sensors
    pt = pt->Next[input];              // next depends on input and state
  }
}
```
Program 3.2. Pointer implementation of a Moore FSM (PointerLineTracker_xxx).

You can find a traffic light controller in Volume 1 and on the book web site as **TableTrafficLight_xxx** and **PointerTrafficLight_xxx**, where xxx refers to the specific microcontroller on which the example was tested.

Observation: The table implementation requires less memory space for the FSM data structure, but the pointer implementation will run faster.

Some microcontrollers have ROM that is one-time programmed at the factory. These ROMs cannot be erased and rewritten. On microcontrollers that have both ROM and EEPROM we can place the FSM data structure in EEPROM and the program in ROM. This allows us to make minor modifications to the finite state machine (add/delete states, change input/output values) by changing the linked structure in EEPROM without modifying the program in ROM. In this way small modifications to the finite state machine can be made by reprogramming the EEPROM without having to produce new microcontroller chips.

The purpose of a board support package is to hide as much of the I/O details as possible. We implement a BSP when we expect the high-level system will be deployed onto many low-level platforms. The solution in Program 3.3 can be quickly adapted to any TM4C using any port and any contiguous set of bits simply by changing the `#define` statements.

```c
#define BSP_InPort         GPIO_PORTB_DATA_R
#define BSP_InPort_DIR     GPIO_PORTB_DIR_R
#define BSP_InPort_DEN     GPIO_PORTB_DEN_R
#define BSP_OutPort        GPIO_PORTD_DATA_R
#define BSP_OutPort_DIR    GPIO_PORTD_DIR_R
#define BSP_OutPort_DEN    GPIO_PORTD_DEN_R
#define BSP_GPIO_EN        SYSCTL_RCGCGPIO_R
#define BSP_InPort_Mask    0x00000008   // bit mask for Port D
#define BSP_In_M           0x00000003   // bit mask for pins 1,0
#define BSP_In_Shift       0x00000000   // shift value for Input pins
#define BSP_OutPort_Mask   0x00000002   // bit mask for Port B
#define BSP_Out_M          0x0000003F   // bit mask for pins 5-0
#define BSP_Out_Shift      0x00000000   // shift value for Output pins
struct State {
  uint32_t Out;
  uint32_t Time;
  const struct State *Next[4];};
typedef const struct State STyp;
#define goN   &FSM[0]
#define waitN &FSM[1]
#define goE   &FSM[2]
#define waitE &FSM[3]
STyp FSM[4]={
 {0x21,3000,{goN,waitN,goN,waitN}},
 {0x22, 500,{goE,goE,goE,goE}},
 {0x0C,3000,{goE,goE,waitE,waitE}},
 {0x14, 500,{goN,goN,goN,goN}}};
int main(void){ STyp *pt;   // state pointer
  uint32_t input;      // activate clocks on input and output ports
  BSP_GPIO_EN |= BSP_InPort_Mask|BSP_OutPort_Mask;
  SysTick_Init();           // initialize SysTick timer, program 2.10
  BSP_InPort_DIR &= ~ BSP_In_M; // make InPort pins inputs
  BSP_InPort_DEN |= BSP_In_M;   // enable digital I/O on InPort
  BSP_OutPort_DIR |= BSP_Out_M; // make OutPort pins out
  BSP_OutPort_DEN |= BSP_Out_M; // enable digital I/O on OutPort
  pt = goN;
  while(1){
    BSP_OutPort = (BSP_OutPort&(~BSP_Out_M))|((pt->Out)>>BSP_Out_Shift);
    SysTick_Wait10ms(pt->Time);
    input = (BSP_InPort&BSP_In_M)>>BSP_In_Shift; //00,01,10,11
    pt = pt->Next[input];
  }
}
```

Program 3.3. Enhanced C implementation of a Traffic Light FSM (input on PB1-0, output PD5-0).

Checkpoint 3.9: Change Program 3.3 to place the input on PA5-4 and the output on PB6-1.

The FSM approach makes it easy to change. To change the wait time for a state, we simply change the value in the data structure. To add more states, we simply increase the size of the **fsm[]** structure and define the **Out**, **Time**, and **Next** fields for these new states.

To add more output signals, we simply increase the precision of the **Out** field. To add more input lines, we increase the size of the next field. If there are n input bits, then the size of the next field will be 2^n. For example, if there are four input lines, then there are 16 possible combinations, where each input possibility requires a **Next** value specifying where to go if this combination occurs.

Example 3.2. The goal is to design a finite state machine robot controller, as illustrated in Figure 3.4. Because the outputs cause the robot to change states, we will use a Mealy implementation. The outputs of a Mealy FSM depend on both the input and the current state. This robot has mood sensors that are interfaced to Port B. The robot has four possible conditions:

00	**OK**,	the robot is feeling fine
01	**Tired**,	the robot energy levels are low
10	**Curious**,	the robot senses activity around it
11	**Anxious**,	the robot senses danger

There are four actions this robot can perform, which are triggered by pulsing (make high, then make low) one of the four signals interfaced to Port D.

PD3	**SitDown**,	assuming it is standing, it will perform moves to sit down
PD2	**StandUp**,	assuming it is sitting, it will perform moves to stand up
PD1	**LieDown**,	assuming it is sitting, it will perform moves to lie down
PD0	**SitUp**,	assuming it is sleeping, it will perform moves to sit up

Solution: For this design we can list heuristics describing how the robot is to operate:
If the robot is OK, it will stay in its current state.
If the robot's energy levels are low, it will go to sleep.
If the robot senses activity around it, it will awaken from sleep.
If the robot senses danger, it will stand up.

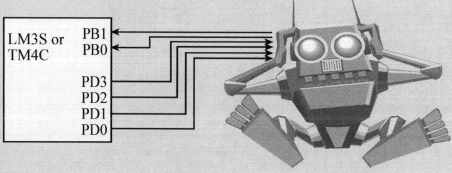

Figure 3.4. Robot interface.

These rules are converted into a finite state machine graph, as shown in Figure 3.5. Each arrow specifies both an input and an output. For example, the "**Tired/SitDown**" arrow from **Stand** to **Sit** states means if we are in the **Stand** state and the input is **Tired**, then we will output the **SitDown** command and go to the **Sit** state. Mealy machines can have time delays, but this example just didn't have time delays.

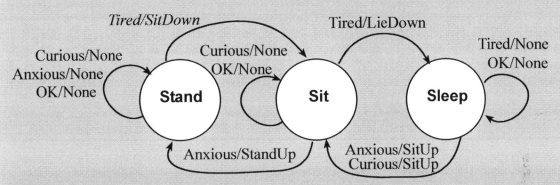

Figure 3.5. Mealy FSM for a robot controller.

The first step in designing the software is to decide on the sequence of operations.
1) Initialize directions registers
2) Specify initial state
3) Perform FSM controller
    a) Input from sensors
    b) Output to the robot, which depends on the state and the input
    c) Change states, which depends on the state and the input

The second step is to define the FSM graph using a linked data structure. Two possible implementations of the Mealy FSM are presented. The implementation in Program 3.4 defines the outputs as simple numbers, where each pulse is defined as the bit mask required to cause that action. The four **Next** parameters define the input-dependent state transitions.

```
struct State{
  uint32_t Out[4];              // outputs
  const struct State *Next[4];  // next
};
typedef const struct State StateType;
#define Stand &FSM[0]
#define Sit   &FSM[1]
#define Sleep &FSM[2]
#define None     0x00
#define SitDown  0x08   // pulse on PD3
#define StandUp  0x04   // pulse on PD2
#define LieDown  0x02   // pulse on PD1
#define SitUp    0x01   // pulse on PD0
```

```c
StateType FSM[3]={
{{None,SitDown,None,None},   //Standing
  {Stand,Sit,Stand,Stand}},
{{None,LieDown,None,StandUp},//Sitting
  {Sit,Sleep,Sit,Stand }},
{{None,None,SitUp,SitUp},    //Sleeping
  {Sleep,Sleep,Sit,Sit}}
};
int main(void){ StateType *pt;    // current state
  uint32_t input;
  SYSCTL_RCGCGPIO_R |= 0x0000000A;  // clock on Ports B and D
  pt = Stand;                 // initial state
  GPIO_PORTB_DIR_R &= ~0x03; // make PB1-0 input from mood sensor
  GPIO_PORTB_AMSEL_R &= ~0x03;       // disable analog on PB
  GPIO_PORTB_PCTL_R &= ~0x000000FF; // PCTL GPIO on PB1-0
  GPIO_PORTB_DEN_R |= 0x03;          // enable digital I/O on PB1-0
  GPIO_PORTD_DIR_R |= 0x0F;          // make PD3-0 output to robot
  GPIO_PORTD_AMSEL_R &= ~0x0F;       // disable analog on PD
  GPIO_PORTD_PCTL_R &= ~0x0000FFFF; // PCTL GPIO on PD3-0
  GPIO_PORTD_DEN_R |= 0x0F;          // enable digital I/O on PD3-0
  while(1){
    input = GPIO_PORTB_DATA_R&0x03;      // input=0-3
    GPIO_PORTD_DATA_R |= pt->Out[Input]; // pulse
    GPIO_PORTD_DATA_R &= ~0x0F;
    pt = pt->Next[Input];   // next state
  }
}
```
Program 3.4. Outputs defined as numbers for a Mealy Finite State Machine.

Program 3.5 uses functions to affect the output. Although the functions in this solution perform simple output, this implementation could be used when the output operations are complex. Again proper memory allocation is required if we wish to implement a stand-alone or embedded system. The **const** qualifier is used to place the FSM data structure in flash ROM. Bit-specific outputs are implemented on Port D.

```c
struct State{
  void *CmdPt[4];               // outputs are function pointers
  const struct State *Next[4]; // next
};
typedef const struct State StateType;
#define Stand &FSM[0]
#define Sit   &FSM[1]
#define Sleep &FSM[2]
void None(void){};
#define GPIO_PORTD0              (*((volatile uint32_t *)0x40007004))
#define GPIO_PORTD1              (*((volatile uint32_t *)0x40007008))
#define GPIO_PORTD2              (*((volatile uint32_t *)0x40007010))
#define GPIO_PORTD3              (*((volatile uint32_t *)0x40007020))
```

```c
void SitDown(void){
  GPIO_PORTD3 = 0x08;
  GPIO_PORTD3 = 0x00; // pulse on PD3
}
void StandUp(void){
  GPIO_PORTD2 = 0x04;
  GPIO_PORTD2 = 0x00; // pulse on PD2
}
void LieDown(void){
  GPIO_PORTD1 = 0x02;
  GPIO_PORTD1 = 0x00; // pulse on PD1
}
void SitUp(void) {
  GPIO_PORTD0 = 0x01;
  GPIO_PORTD0 = 0x00; // pulse on PD0
}
StateType FSM[3]={
 {{(void*)&None,(void*)&SitDown,(void*)&None,(void*)&None},  //Standing
  {Stand,Sit,Stand,Stand}},
 {{(void*)&None,(void*)LieDown,(void*)&None,(void*)&StandUp},//Sitting
  {Sit,Sleep,Sit,Stand }},
 {{(void*)&None,(void*)&None,(void*)&SitUp,(void*)&SitUp},   //Sleeping
  {Sleep,Sleep,Sit,Sit}}
};
int main(void){ StateType *pt;  // current state
  uint32_t input;
  SYSCTL_RCGCGPIO_R  |= 0x0000000A;  // clock on Ports B and D
  pt = Stand;                   // initial state
  GPIO_PORTB_DIR_R &= ~0x03; // make PB1-0 input from mood sensor
  GPIO_PORTB_AMSEL_R &= ~0x03;       // disable analog on PB
  GPIO_PORTB_PCTL_R &= ~0x000000FF;  // PCTL GPIO on PB1-0
  GPIO_PORTB_DEN_R  |= 0x03;         // enable digital I/O on PB1-0
  GPIO_PORTD_DIR_R  |= 0x0F;         // make PD3-0 output to robot
  GPIO_PORTD_AMSEL_R &= ~0x0F;       // disable analog on PD
  GPIO_PORTD_PCTL_R &= ~0x0000FFFF;  // PCTL GPIO on PD3-0
  GPIO_PORTD_DEN_R  |= 0x0F;         // enable digital I/O on PD3-0
  while(1){
    input = GPIO_PORTB_DATA_R&0x03;       // input=0-3
    ((void(*)(void))pt->CmdPt[Input])();  // function
    pt = pt->Next[input];                 // next state
  }
}
```
Program 3.5. Outputs defined as functions for a Mealy Finite State Machine.

Observation: In order to make the FSM respond quicker, we could implement a time delay function that returns immediately if an alarm condition occurs. If no alarm exists, it waits the specified delay. Similarly, we could return from the time delay on a change in input.

Checkpoint 3.10: What happens if the robot is sleeping then becomes anxious?

3.6. Threads

Software (e.g., program, code, module, procedure, function, subroutine etc.) is a list of instructions for the computer to execute. A thread on the other hand is defined as the path of action of software as it executes. The expression "thread" comes from the analogy shown in Figure 3.6. This simple program prints the 8-bit numbers 000 001 002 ... If we connect the statements of our executing program with a continuous line (the thread) we can visualize the dynamic behavior of our software.

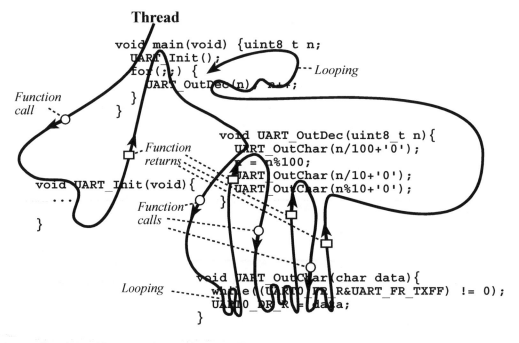

Figure 3.6. Illustration of the definition of a thread. Circle means function call. Rectangle means function return.

The execution of the main program is called the foreground thread. In most embedded applications, the foreground thread executes a loop that never ends. We will learn later, that this thread can be broken (execution suspended, then restarted) by interrupts and direct memory access.

With interrupts we can create multiple threads. Some threads will be created statically, meaning they exist throughout the life of the software, while others will be created and destroyed dynamically. There will usually be one foreground thread running the main program like the above example. In addition to this foreground thread, each interrupt source has its own background thread, which is started whenever the interrupt is requested. Figure 3.7 shows a software system with one foreground thread and two background threads. The "Key" thread is invoked whenever a key is touched on the keyboard and the "Time" thread is invoked every 1ms in a periodic fashion.

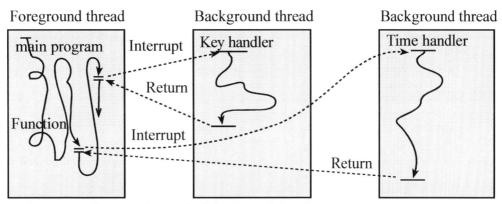

Figure 3.7. Interrupts allow us to have multiple background threads.

Because there is but one processor, the currently running thread must be suspended in order to execute another thread. In the above figure, the suspension of the main program is illustrated by the two breaks in the foreground thread. When a key is touched, the main program is suspended, and a **Keyhandler** thread is created with an "empty" stack and uninitialized registers. When the **Keyhandler** is done it executes a return from interrupt to relinquish control back to the main program. The original stack and registers of the main program will be restored to the state before the interrupt. In a similar way, when the 1 ms timer occurs, the main program is suspended again, and a **Timehandler** thread is created with its own "empty" stack and uninitialized registers. We can think of each thread as having its own registers and its own stack area. In Chapter 5, we will discuss in detail this approach to multithreaded programming. In a real-time operating system (RTOS) there is a preemptive thread scheduler that allows our software to have multiple foreground threads and multiple stacks. The focus of Volume 3 will be the design and analysis of real-time operating systems.

Parallel programming allows the computer to execute multiple threads at the same time. State-of-the-art multi-core processors can execute a separate program in each of its cores. Fork and join are the fundamental building blocks of parallel programming. After a **fork**, two or more software threads will be run in parallel, i.e., the threads will run simultaneously on separate processors. Two or more simultaneous software threads can be combined into one using a **join**. The flowchart symbols for fork and join are shown in Figure 3.8.

Software execution after the join will wait until all threads above the join are complete. As an analogy of parallel execution, when a farmer wants to build a barn, he invites his three neighbors over and gives everyone a hammer. The fork operation changes the situation from the farmer working alone to four people ready to build. The four people now work in parallel to accomplish the single goal of building the barn. When the overall task is complete, the join operation causes the neighbors to go home, and the farmer is working alone again.

Parallel programming is a difficult concept for many software developers, because we have been classically trained to think of computer execution as a single time-linear thread. However, there are numerous real-world scenarios from which to learn the art of parallel programming. A manager of a business, an Army general, and air traffic control are obvious examples that employ parallel operations. From these illustrations we observe hierarchical decision making, delegation of responsibilities, and having an elaborate system of checks and balances. All of these concepts translate into complex software systems involving parallel execution.

To implement parallel execution we need a computer that can execute more than one instruction at a time. A **multi-core processor** has two or more independent central processing units, called **cores**. The cores shared some memory but also have some private storage. The cores can fetch and execute instructions at the same time, thus increasing overall performance. The fork operation in Figure 3.8 will activate three cores launching software to be executed on those three new cores. The join operation will wait until all four branches have completed, and deactivate three of the cores.

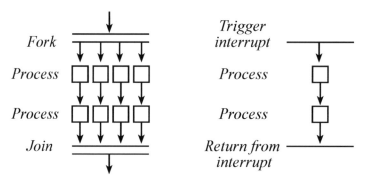

Figure 3.8. Flowchart symbols to describe parallel and concurrent programming.

Concurrent programming allows the computer to execute multiple threads, but only one runs at a time. Interrupts are one mechanism to implement concurrency on real-time systems. Interrupts have a hardware trigger and a software action. An interrupt is a parameter-less subroutine call, triggered by a hardware event. The flowchart symbols for interrupts are also shown in Figure 3.8. The trigger is a hardware event signaling it is time to do something. Examples of interrupt triggers we will see in this book include new input data has arrived, output device is idle, and periodic event. The second component of an interrupt-driven system is the software action called an **interrupt service routine** (ISR). The **foreground** thread is defined as the execution of the main program, and the **background** threads are executions of the ISRs. Consider the analogy of sitting in a comfy chair reading a book. Reading a book is like executing the main program in the foreground. You start reading at the beginning of the book and basically read one page at time in a sequential fashion. You might jump to the back and look something up in the glossary, then jump back to where you where, which is analogous to a function call. Similarly, you might read the same page a few times, which is analogous to a program loop. Even though you skip around a little, the order of pages you read follows a logical and well-defined sequence. Conversely, if the telephone rings, you place a bookmark in the book and answer the phone. When you are finished with the phone conversation, you hang up the phone and continue reading in the book where you left off. The ringing phone is analogous to hardware trigger and the phone conversation is like executing the ISR.

A program segment is **reentrant** if it can be concurrently executed by two (or more) threads. In Figure 3.7 we can conceive of the situation where the main program starts executing a function, is interrupted, and the background thread calls that same function. In order for two threads to share a function, the function must be reentrant. To implement reentrant software, place local variables on the stack, and avoid storing into I/O devices and global memory variables. The issue of reentrancy will be covered in detail later in Chapter 5.

3.7. First In First Out Queue

3.7.1. Classical definition of a FIFO

The first in first out circular queue (**FIFO**) is quite useful for implementing a buffered I/O interface (Figure 3.9). It can be used for both buffered input and buffered output. The order preserving data structure temporarily saves data created by the source (producer) before it is processed by the sink (consumer). The class of FIFOs studied in this section will be statically allocated global structures. Because they are global variables, it means they will exist permanently and can be carefully shared by more than one program. The advantage of using a FIFO structure for a data flow problem is that we can decouple the producer and consumer threads. Without the FIFO we would have to produce 1 piece of data, then process it, produce another piece of data, then process it. With the FIFO, the producer thread can continue to produce data without having to wait for the consumer to finish processing the previous data. This decoupling can significantly improve system performance.

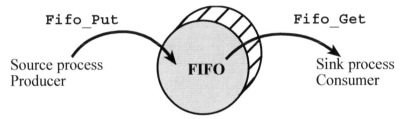

Figure 3.9. The FIFO is used to buffer data between the producer and consumer.

You have probably already experienced the convenience of FIFOs. For example, a FIFO is used while streaming audio from the Internet. As sound data are received from the Internet they are put (calls **Fifo_Put**) in a FIFO. When the sound board needs data it calls **Fifo_Get**. As long as the FIFO never comes full or empty, the sound is played in a continuous manner. A FIFO is also used when you ask the computer to print a file. Rather than waiting for the actual printing to occur character by character, the print command will put the data in a FIFO. Whenever the printer is free, it will get data from the FIFO. The advantage of the FIFO is it allows you to continue to use your computer while the printing occurs in the background. To implement this magic of background printing we will need interrupts. There are many producer/consumer applications. In Table 3.3 the processes on the left are producers that create or input data, while the processes on the right are consumers which process or output data.

Source/Producer	*Sink/Consumer*
Keyboard input	Program that interprets
Program with data	Printer output
Program sends message	Program receives message
Microphone and ADC	Program that saves sound data
Program that has sound data	DAC and speaker

Table 3.3. Producer consumer examples.

The producer puts data into the FIFO. The `Fifo_Put` operation does not discard information already in the FIFO. If the FIFO is full and the user calls `Fifo_Put`, the `Fifo_Put` routine will return a full error signifying the last (newest) data was not properly saved. The sink process removes data from the FIFO. The `Fifo_Get` routine will modify the FIFO. After a get, the particular information returned from the get routine is no longer saved on the FIFO. If the FIFO is empty and the user tries to get, the `Fifo_Get` routine will return an empty error signifying no data could be retrieved. The FIFO is order preserving, such that the information is returned by repeated calls of `Fifo_Get` in the same order as the data was saved by repeated calls of `Fifo_Put`.

There are many ways to implement a statically-allocated FIFO. We can use either a pointer or an index to access the data in the FIFO. We can use either two pointers (or two indices) or two pointers (or two indices) and a counter. The counter specifies how many entries are currently stored in the FIFO. There are even hardware implementations of FIFO queues. We begin with the two-pointer implementation. It is a little harder to implement, but it does have some advantages over the other implementations.

3.7.2. Two-pointer FIFO implementation

The two-pointer implementation has, of course, two pointers. If we were to have infinite memory, a FIFO implementation is easy (Figure 3.10). `GetPt` points to the data that will be removed by the next call to `Fifo_Get`, and `PutPt` points to the empty space where the data will stored by the next call to `Fifo_Put`, see Program 3.6.

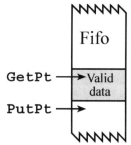

Figure 3.10. The FIFO implementation with infinite memory.

```
int8_t static volatile *PutPt;   // put next
int8_t static volatile *GetPt;   // get next
int Fifo_Put(int8_t data){       // call by value
  *PutPt = data;    // Put
  PutPt++;          // next
  return(1);}       // true if success
int Fifo_Get(int8_t *datapt){
  *datapt = *GetPt; // return by reference
  GetPt++;          // next
  return(1);}       // true if success
```

Program 3.6. Code fragments showing the basic idea of a FIFO.

There are four modifications that are required to the above subroutines. If the FIFO is full when **Fifo_Put** is called then the function should return a full error. Similarly, if the FIFO is empty when **Fifo_Get** is called, then the function should return an empty error. **PutPt** must be wrapped back up to the top when it reaches the bottom (Figure 3.11).

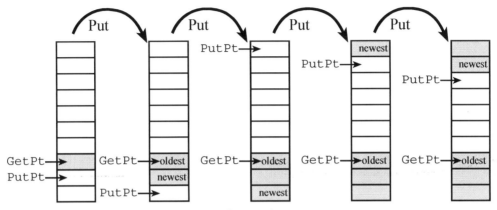

*Figure 3.11. The FIFO **Fifo_Put** operation showing the pointer wrap.*

The **GetPt** must also be wrapped back up to the top when it reaches the bottom (Figure 3.12).

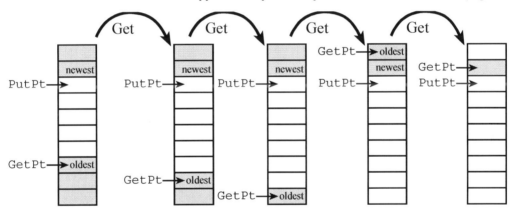

*Figure 3.12. The FIFO **Fifo_Get** operation showing the pointer wrap.*

There are two mechanisms to determine whether the FIFO is empty or full. A simple method is to implement a counter containing the number of bytes currently stored in the FIFO. **Fifo_Get** would decrement the counter and **Fifo_Put** would increment the counter. We will not implement a counter because incrementing and decrementing a counter causes a race condition, meaning the counter could become incorrect when shared in a multithreaded environment. Race conditions and critical sections will be presented in Chapter 5.

The second method is to prevent the FIFO from being completely full. The implementation of this FIFO module is shown in Program 3.7. You can find all the FIFOs of this section on the book web site as **FIFO_xxx**, where xxx refers to the specific microcontroller on which the example was tested.

```
#define FIFOSIZE 10      // can be any size
#define FIFOSUCCESS 1
#define FIFOFAIL    0
typedef int8_t DataType;
DataType volatile *PutPt;  // put next
DataType volatile *GetPt;  // get next
DataType static Fifo[FIFOSIZE];
// initialize FIFO
void Fifo_Init(void){
  PutPt = GetPt = &Fifo[0]; // Empty
}
// add element to FIFO
int Fifo_Put(DataType data){
  DataType volatile *nextPutPt;
  nextPutPt = PutPt+1;
  if(nextPutPt == &Fifo[FIFOSIZE]){
    nextPutPt = &Fifo[0];  // wrap
  }
  if(nextPutPt == GetPt){
    return(FIFOFAIL);      // Failed, FIFO full
  }
  else{
    *(PutPt) = data;       // Put
    PutPt = nextPutPt;     // Success, update
    return(FIFOSUCCESS);
  }
}
// remove element from FIFO
int Fifo_Get(DataType *datapt){
  if(PutPt == GetPt ){
    return(FIFOFAIL);      // Empty if PutPt=GetPt
  }
  *datapt = *(GetPt++);
  if(GetPt == &Fifo[FIFOSIZE]){
     GetPt = &Fifo[0];     // wrap
  }
  return(FIFOSUCCESS);
}
```

Program 3.7. Two-pointer implementation of a FIFO (FIFO_xxx).

For example, if the FIFO had 10 bytes allocated, then the **Fifo_Put** subroutine would allow a maximum of 9 bytes to be stored. If there were already 9 bytes in the FIFO and another **Fifo_Put** were called, then the FIFO would not be modified and a full error would be returned. See Figure 3.13. In this way if **PutPt** equals **GetPt** at the beginning of **Fifo_Get**, then the FIFO is empty. Similarly, if **PutPt+1** equals **GetPt** at the beginning of **Fifo_Put**, then the FIFO is full. Be careful to wrap the **PutPt+1** before comparing it to **Fifo_Get**. This method does not require the length to be stored or calculated.

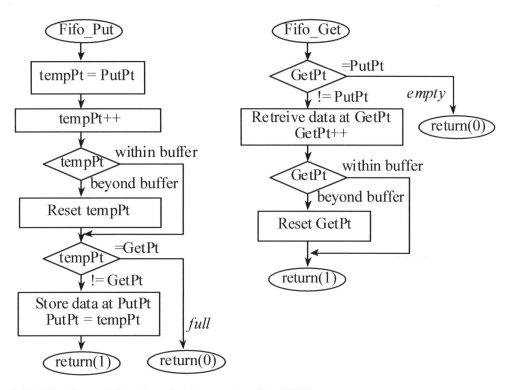

Figure 3.13. Flowcharts of the pointer implementation of the FIFO queue.

To check for FIFO full, the following **Fifo_Put** routine attempts to put using a temporary **PutPt**. If putting makes the FIFO look empty, then the temporary **PutPt** is discarded and the routine is exited without saving the data. This is why a FIFO with 10 allocated bytes can only hold 9 data points. If putting doesn't make the FIFO look empty, then the temporary **PutPt** is stored into the actual **PutPt** saving the data as desired.

To check for FIFO empty, the **Fifo_Get** routine in Program 3.7 simply checks to see if **GetPt** equals **PutPt**. If they match at the start of the routine, **then Fifo_Get** returns with the "empty" condition signified.

Since **Fifo_Put** and **Fifo_Get** have read modify write accesses to global variables they are themselves not reentrant. Similarly **Fifo_Init** has a multiple step write access to global variables. Therefore **Fifo_Init** is not reentrant.

One advantage of this pointer implementation is that if you have a single thread that calls the **Fifo_Get** (e.g., the main program) and a single thread that calls the **Fifo_Put** (e.g., the serial port receive interrupt handler), then this **Fifo_Put** function can interrupt this **Fifo_Get** function without loss of data. So in this particular situation, interrupts would not have to be disabled. It would also operate properly if there were a single interrupt thread calling **Fifo_Get** (e.g., the serial port transmit interrupt handler) and a single thread calling **Fifo_Put** (e.g., the main program.) On the other hand, if the situation is more general, and multiple threads could call **Fifo_Put** or multiple threads could call **Fifo_Get**, then the interrupts would have to be temporarily disabled.

3.7.3. Two index FIFO implementation

The other method to implement a FIFO is to use indices rather than pointers. This FIFO has the restriction that the size must be a power of 2. If **FIFOSIZE** is 16 and the logic **PutI&(FIFOSIZE-1)** returns the bottom four bits of the put index. Similarly, the logic **GetI&(FIFOSIZE-1)** returns the bottom four bits of the get index. Using the bottom bits of the index removes the necessary to check for out of bounds and wrapping. For an implementation see the project FIFO_xxx.

3.7.4. Little's Theorem

In this section we introduce some general theory about queues. Let N be the average number of data packets in the queue plus the one data packet currently being processed by the consumer. Basically, N is the average number of packets in the system. Let λ be the average arrival rate in packets per second (pps). Let R be the average response time of a packet, which includes the time waiting in the queue plus the time for the consumer to process the packet. **Little's Theorem** states

$$N = \lambda R$$

As long as the system is stable, this result is not influenced by the probability distribution of the producer, the probability distribution of the consumer or the service order. Let S be the mean service time for a packet. Thus, $C=1/S$ is defined as the **system capacity** (pps). Stable in this context means the packet arrival rate is less than the system capacity ($\lambda<C$). This means, in most cases, the queue length can be chosen so the queue never fills, and no data are lost. In this case, the arrival rate λ is also the output rate T, or throughput of the system. We can use Little's Theorem to estimate average response time,

$$R = N/T$$

In general, we want T to be high and R to be low. To handle these two conflicting goals, we develop the concept of a power metric for the queue. We can define **utilization factor** as the throughput divided by the capacity, which is a normalized throughput,

$$U = T/C$$

U defines the loading of the queue, because it is. We can define **normalized mean response time**, R/S. We next define **power metric** P as utilization factor divided by normalized mean response time,

$$P = U/(R/S) = (T*S)/(R/S)$$

Substituting Little's Theorem ($R=N/T$), we can write

$$P = U^2/N$$

The goal of the operating system is to maximize P.

3.7.5. FIFO build macros

When we need multiple FIFOs in our system, we could switch over to C++ and define the FIFO as a class, and then instantiate multiple objects to create the FIFOs. A second approach would be to use a text editor, open the source code containing Program 3.7, copy/paste it, and then change names so the functions are unique. A third approach is shown in Programs 3.8 and 3.9, which defines macros allowing us to create as many FIFOs as we need.

```
// macro to create a pointer FIFO
#define AddPointerFifo(NAME,SIZE,TYPE,SUCCESS,FAIL) \
TYPE volatile *NAME ## PutPt;            \
TYPE volatile *NAME ## GetPt;            \
TYPE static NAME ## Fifo [SIZE];         \
void NAME ## Fifo_Init(void){            \
  NAME ## PutPt = NAME ## GetPt = &NAME ## Fifo[0]; \
}                                        \
int NAME ## Fifo_Put (TYPE data){        \
  TYPE volatile *nextPutPt;              \
  nextPutPt = NAME ## PutPt + 1;         \
  if(nextPutPt == &NAME ## Fifo[SIZE]){  \
    nextPutPt = &NAME ## Fifo[0];        \
  }                                      \
  if(nextPutPt == NAME ## GetPt ){       \
    return(FAIL);                        \
  }                                      \
  else{                                  \
    *( NAME ## PutPt ) = data;           \
    NAME ## PutPt = nextPutPt;           \
    return(SUCCESS);                     \
  }                                      \
}                                        \
int NAME ## Fifo_Get (TYPE *datapt){     \
  if( NAME ## PutPt == NAME ## GetPt ){  \
    return(FAIL);                        \
  }                                      \
  *datapt = *( NAME ## GetPt ## ++);     \
  if( NAME ## GetPt == &NAME ## Fifo[SIZE]){ \
    NAME ## GetPt = &NAME ## Fifo[0];    \
  }                                      \
  return(SUCCESS);                       \
}
```

Program 3.8. Two-pointer macro implementation of a FIFO (FIFO_xxx).

To create a 20-element FIFO storing unsigned 16-bit numbers that returns 1 on success and 0 on failure we invoke

```
AddPointerFifo(Rx, 20, uint16_t, 1, 0)
```

creating the three functions **RxFifo_Init()**, **RxFifo_Get()**, and **RxFifo_Put()**.

Program 3.9 is a macro allowing us to create two-index FIFOs.

```c
// macro to create an index FIFO
#define AddIndexFifo(NAME,SIZE,TYPE,SUCCESS,FAIL) \
uint32_t volatile NAME ## PutI;      \
uint32_t volatile NAME ## GetI;      \
TYPE static NAME ## Fifo [SIZE];         \
void NAME ## Fifo_Init(void){           \
  NAME ## PutI = NAME ## GetI = 0;      \
}                                        \
int NAME ## Fifo_Put (TYPE data){       \
  if(( NAME ## PutI - NAME ## GetI ) & ~(SIZE-1)){ \
    return(FAIL);       \
  }                     \
  NAME ## Fifo[ NAME ## PutI &(SIZE-1)] = data; \
  NAME ## PutI ## ++;  \
  return(SUCCESS);       \
}                         \
int NAME ## Fifo_Get (TYPE *datapt){  \
  if( NAME ## PutI == NAME ## GetI ){ \
    return(FAIL);       \
  }                     \
  *datapt = NAME ## Fifo[ NAME ## GetI &(SIZE-1)]; \
  NAME ## GetI ## ++;  \
  return(SUCCESS);       \
}                         \
uint16_t NAME ## Fifo_Size (void){  \
 return ((uint16_t)( NAME ## PutI - NAME ## GetI ));  \
}
```

Program 3.9. Macro implementation of a two-index FIFO. The size must be a power of two (FIFO_xxx).

To create a 32-element FIFO storing signed 32-bit numbers that returns 0 on success and 1 on failure we invoke

AddIndexFifo(Tx, 32, int32_t, 0, 1)

creating the three functions **TxFifo_Init()**, **TxFifo_Get()**, and **TxFifo_Put()**.

Checkpoint 3.11: Show C code to create three FIFOs called CAN1 CAN2 and CAN3. Each FIFO stores 8-bit bytes and must be able to store up to 99 elements.

Checkpoint 3.12: Show C code to create two FIFOs called F1 and F2. Each FIFO stores 16-bit halfwords and must be able to store up to 256 elements.

3.8. Memory Management and the Heap

So far, we have seen two types of allocation: permanent allocation in global variables and temporary allocation in local variables. When we allocate local variables in registers or on the stack these variables must be private to the function and cannot be shared with other functions. Furthermore, each time the function is invoked new local variables are created, and data from previous instantiations are not available. This behavior is usually exactly what we want to happen with local variables. However, we can use the **heap** (or **memory manager**) to have temporary allocation in a way that is much more flexible. In particular, we will be able to explicitly define when data are allocated and when they are deallocated with the only restriction being we first allocate, next we use, and then we deallocate. Furthermore, we can control the scope of the data in a flexible manner.

The use of the heap involves two system functions: `malloc` and `free`. When we wish to allocate space we call `malloc` and specify how many bytes we need. `malloc` will return a pointer to the new object, which we must store in a pointer variable. If the heap has no more space, `malloc` will return a 0, which means null pointer. The heap implements temporary allocation, so when we are done with the data, we return it to the heap by calling `free`. Consider the following simple example with three functions.

```
int32_t *Pt;
void Begin(void){
  Pt = (*int32_t)malloc(4*20); // allocate 20 words
}
void Use(void){ int i;
  for(i = 0; i < 20; i++)
    Pt[i] = i; // put data into array
}
void End(void){
   free(Pt);
}
```

The pointer `Pt` is permanently allocated. The left side of Figure 3.14 shows that initially, even though the pointer exists, it does not point to anything. More specifically, the compiler will initialize it to 0; this 0 is defined as a **null** pointer, meaning it is not valid. When `malloc` is called the pointer is now valid and points to a 20-word array. The array is inside the heap and `Pt` points to it. Any time after `malloc` is called and before `free` is called the array exists and can be accessed via the pointer `Pt`. After you call free, the pointer has the same value as before. However, the array itself does not exist. I.e., these 80 bytes do not belong to your program anymore. In particular, after you call free the heap is allowed to allocate these bytes to some other program. Weird and crazy errors will occur if you attempt to dereference the pointer before the array is allocated, or after it is released.

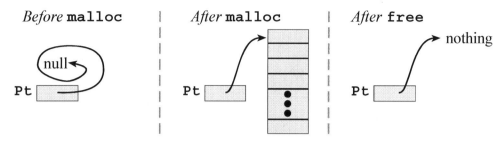

Figure 3.14. The heap is used to dynamically allocate memory.

This array exists and the pointer is valid from when you call **malloc** up until the time you call **free**. In C, the heap does not manage the pointers to allocated block; your program must. If you call **malloc** ten times in a row, you must keep track of the ten pointers you received. The scope of this array is determined by the scope of the pointer, **Pt**. If **Pt** is public, then the array is public. If static were to be added to the definition of **Pt**, then the scope of the array is restricted to software within this file. In the following example, the scope of the array is restricted to the one function. Within one execution of the function, the array is allocated, used, and then deallocated, just like a local variable.

```
void Function(void){ int i;
int32_t *pt;
  pt = (*int32_t)malloc(4*20); // allocate 20 words
  for(i = 0; i < 20; i++)
    pt[i] = i; // put data into array
  free(pt);
}
```

A **memory leak** occurs if software uses the heap to allocate space but forgets to deallocate the space when it is finished. The following is an example of a memory leak. Each time the function is called, a block of memory is allocated. The pointer to the block is stored in a local variable. When the function returns, the pointer no longer exists. This means the allocated block in the heap exists, but the program has no pointer to it. In other words, each time this function returns 80 bytes from the heap are permanently lost.

```
void LeakyFunction(void){ int i;
int32_t *pt;
  pt = (*int32_t)malloc(4*20); // allocate 20 words
  for(i = 0; i < 20; i++)
    pt[i] = i; // put data into array
}
```

In general, the heap manager allows the program to allocate a variable block size, but in this section we will develop a simplified heap manager handles just fixed size blocks. In this example, the block size is specified by the constant **SIZE**. The initialization will create a linked list of all the free blocks (Figure 3.15).

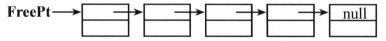

Figure 3.15. The initial state of the heap has all of the free blocks linked in a list.

Program 3.11a shows the global structures for the heap. These entries are defined in RAM. **SIZE** is the number of 8-bit bytes in each block. All blocks allocated and released with this memory manager will be of this fixed size. **NUM** is the number of blocks to be managed. **FreePt** points to the first free block.

```
#define SIZE 80
#define NUM 5
#define NULL 0    // empty pointer
int8_t *FreePt;
int8_t Heap[SIZE*NUM];
```

Program 3.11a. Private global structures for the fixed-block memory manager.

Initialization must be performed before the heap can be used. Program 3.11b shows the software that partitions the heap into blocks and links them together. **FreePt** points to a linear linked list of free blocks. Initially these free blocks are contiguous and in order, but as the manager is used the positions and order of the free blocks can vary. It will be the pointers that will thread the free blocks together.

```
void Heap_Init(void){
int8_t *pt;
  FreePt = &Heap[0];
  for(pt=&Heap[0];
      pt!=&Heap[SIZE*(NUM-1)];
      pt=pt+SIZE){
    *(int32_t *)pt =(int32_t)(pt+SIZE);
  }
  *(int32_t*)pt = NULL;
}
```

Program 3.11b. Functions to initialize the heap.

To allocate a block to manager just removes one block from the free list. Program 3.11c shows the allocate and release functions. The **Heap_Allocate** function will fail and return a null pointer when the heap becomes empty. The **Heap_Release** returns a block to the free list. This system does not check to verify a released block actually was previously allocated.

```
void *Heap_Allocate(void){
int8_t *pt;
  pt = FreePt;
  if (pt != NULL){
    FreePt = (char*) *(char**)pt;
  }
  return(pt);
}
void Heap_Release(void *pt){
int8_t *oldFreePt;
  oldFreePt = FreePt;
  FreePt = (char*)pt;
  *(int32_t *)pt = (int32_t)oldFreePt;
}
```

Program 3.11c. Functions to allocate and release memory blocks (HeapFixedBlock_xxx).

Checkpoint 3.13: There are 5 blocks in this simple heap. How could the memory manager determine if block I (where $0 \leq I \leq 4$) is allocated or free?

Checkpoint 3.14: Using this memory manager, write a malloc and free functions such that the size is restricted to a maximum of 100 bytes. I.e., you may assume the user never asks for more than 100 bytes at a time.

3.9. Introduction to Debugging

Every programmer is faced with the need to debug and verify the correctness of their software. In this section we will study hardware level probes like the oscilloscope, logic analyzer, JTAG, and in-circuit-emulator (ICE); software level tools like simulators, monitors, and profilers; and manual tools like inspection and print statements.

3.9.1. Debugging Tools

Microcontroller-related problems often require the use of specialized equipment to debug the system hardware and software. Useful hardware tools include a logic probe, an oscilloscope, a logic analyzer, and a JTAG debugger. A **logic probe** is a handheld device with an LED or buzzer. You place the probe on your digital circuit and LED/buzzer will indicate whether the signal is high or low. An **oscilloscope**, or scope, graphically displays information about an electronic circuit, where the voltage amplitude versus time is displayed. A scope has one or two channels, with many ways to trigger or capture data. A scope is particularly useful when interfacing analog signals using an ADC or DAC. The PicoScope 2104 (from http://www.picotech.com/) is a low-cost but effective tool for debugging microcontroller circuits. A **logic analyzer** is essentially a multiple channel digital storage scope with many ways to trigger. As shown in Figure 3.16, we can connect the logic analyzer to digital signals that are part of the system, or we can connect the logic analyzer channels to unused microcontroller pins and add software to toggle those pins at strategic times/places. As a troubleshooting aid, it allows the experimenter to observe numerous digital signals at various points in time and thus make decisions based upon such observations. One problem with logic analyzers is the massive amount of information that it generates. To use an analyzer effectively one must learn proper triggering mechanisms to capture data at appropriate times eliminating the need to sift through volumes of output. The logic analyzer figures in this book were collected with a logic analyzer Digilent (from http://www.digilentinc.com/). The Analog Discovery combines a logic analyzer with an oscilloscope, creating an extremely effective debugging tool.

Maintenance Tip: First, find the things that will break you. Second, break them.

Common error: Sometimes the original system operates properly, and the debugging code has mistakes.

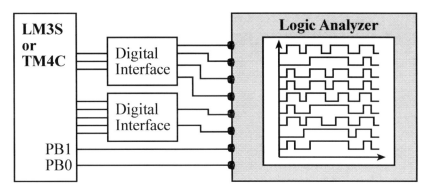

Figure 3.16. A logic analyzer and example output. PB1 and PB0 are extra pins just used for debugging.

Figure 3.17 shows a logic analyzer output, where signals SSI are outputs to the LCD, and UART is transmission between two microcontroller. However PF3 and PF1 are debugging outputs to measuring timing relationships between software execution and digital I/O. The rising edge of PF1 is used to trigger the data collection.

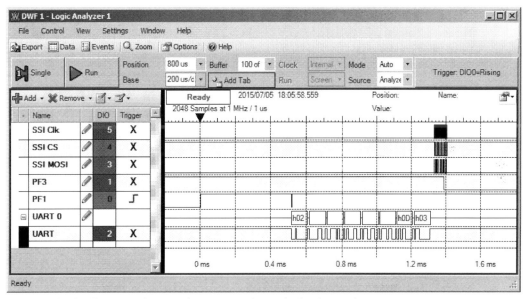

Figure 3.17. Analog Discovery logic analyzer output (www.digilentinc.com).

An emulator is a hardware debugging tool that recreates the input/output signals of the processor chip. To use an emulator, we remove the processor chip and insert the emulator cable into the chip socket. In most cases, the emulator/computer system operates at full speed. The emulator allows the programmer to observe and modify internal registers of the processor. Emulators are often integrated into a personal computer, so that its editor, hard drive, and printer are available for the debugging process.

The only disadvantage of the in-circuit emulator is its cost. To provide some of the benefits of this high-priced debugging equipment, many microcontrollers use a JTAG debugger. The JTAG hardware exists both on the microcontroller chip itself and as an external interface to a personal computer. Although not as flexible as an ICE, the JTAG can provide the ability to observe software execution in real-time, the ability to set breakpoints, the ability to stop the computer, and the ability to read and write registers, I/O ports and memory.

3.9.2. Debugging Theory

Debugging is an essential component of embedded system design. We need to consider debugging during all phases of the design cycle. It is important to develop a structure or method when verifying system performance. This section will present a number of tools we can use when debugging. Terms such as program testing, diagnostics, performance debugging, functional debugging, tracing, profiling, instrumentation, visualization, optimization, verification, performance measurement, and execution measurement have specialized meanings, but they are also used interchangeably, and they often describe overlapping functions. For example, the terms profiling, tracing, performance measurement, or execution measurement may be used to describe the process of examining a program from a time viewpoint. But, tracing is also a term that may be used to describe the process of monitoring a program state or history for functional errors, or to describe the process of stepping through a program with a debugger. Usage of these terms among researchers and users vary.

Black-box testing is simply observing the inputs and outputs without looking inside. Black-box testing has an important place in debugging a module for its functionality. On the other hand, **white-box testing** allows you to control and observe the internal workings of a system. A common mistake made by new engineers is to just perform black box testing. Effective debugging uses both. One must always start with black-box testing by subjecting a hardware or software module to appropriate test-cases. Once we document the failed test-cases, we can use them to aid us in effectively performing the task of white-box testing.

We define a **debugging instrument** as software code that is added to the program for the purpose of debugging. A print statement is a common example of an instrument. Using the editor, we add print statements to our code that either verify proper operation or display run-time errors. A key to writing good debugging instruments is to provide for a mechanism to reliably and efficiently remove all them when the debugging is done. Consider the following mechanisms as you develop your own unique debugging style.

• Place all print statements in a unique column (e.g., first column.), so that the only code that exists in this column will be debugging instruments.

• Define all debugging instruments as functions that all have a specific pattern in their names (e.g., begin with `Debug_`). In this way, the find/replace mechanism of the editor can be used to find all the calls to the instruments.

• Define the instruments so that they test a run time global flag. When this flag is turned off, the instruments perform no function. Notice that this method leaves a permanent copy of the debugging code in the final system, causing it to suffer a runtime overhead, but the debugging code can be activated dynamically without recompiling. Many commercial software applications utilize this method because it simplifies "on-site" customer support.

• Use conditional compilation (or conditional assembly) to turn on and off the instruments when the software is compiled. When the compiler supports this feature, it can provide both performance and effectiveness.

Some compilers support a configuration mode that can be set to debug or release. In debug mode, debugging instruments are added. In release mode, the instruments are removed.

Checkpoint 3.15: Consider the difference between a runtime flag that activates a debugging command versus a compile-time flag. In both cases it is easy to activate/deactivate the debugging statements. List one factor for which each method is superior to the other.

Checkpoint 3.16: What is the advantage of leaving debugging instruments in a final delivered product?

Nonintrusiveness is the characteristic or quality of a debugger that allows the software/hardware system to operate normally as if the debugger did not exist. **Intrusiveness** is used as a measure of the degree of perturbation caused in program performance by the debugging instrument itself. Let t be the time required to execute the instrument, and let Δt be the average time in between executions of the instrument. One quantitative measure of intrusiveness is $t/\Delta t$, which is the fraction of available processor time used by the debugger. For example, a print statement added to your source code may be very intrusive because it might significantly affect the real-time interaction of the hardware and software. Observing signals that already exist as part of the system with an oscilloscope or logic analyzer is nonintrusive. A debugging instrument is classified as **minimally intrusive** if it has a negligible effect on the system being debugged. In a real microcontroller system, breakpoints and single-stepping are also intrusive, because the real hardware continues to change while the software has stopped. When a program interacts with real-time events, the performance can be significantly altered when using intrusive debugging tools. To be effective we must employ nonintrusive or minimally intrusive methods.

Checkpoint 3.17: What does it mean for a debugging instrument to be minimally intrusive? Give both a general answer and a specific criterion.

Although, a wide variety of program monitoring and debugging tools are available today, in practice it is found that an overwhelming majority of users either still prefer or rely mainly upon "rough and ready" manual methods for locating and correcting program errors. These methods include desk-checking, dumps, and print statements, with print statements being one of the most popular manual methods. Manual methods are useful because they are readily available, and they are relatively simple to use. But, the usefulness of manual methods is limited: they tend to be highly intrusive, and they do not provide adequate control over repeatability, event selection, or event isolation. A real-time system, where software execution timing is critical, usually cannot be debugged with simple print statements, because the print statement itself will require too much time to execute.

The first step of debugging is to **stabilize** the system. In the debugging context, we stabilize the problem by creating a test routine that fixes (or stabilizes) all the inputs. In this way, we can reproduce the exact inputs over and over again. Once stabilized, if we modify the program, we are sure that the change in our outputs is a function of the modification we made in our software and not due to a change in the input parameters.

Acceleration means we will speed up the testing process. When we are testing one module we can increase how fast the functions are called in an attempt to expose possible faults. Furthermore, since we can control the test environment, we will **vary** the test conditions over a wide range of possible conditions. **Stress testing** means we run the system beyond the requirements to see at what point it breaks down.

When a system has a small number of possible inputs (e.g., less than a million), it makes sense to test them all. When the number of possible inputs is large we need to choose a set of inputs. **Coverage** defines the subset of possible inputs selected for testing. A **corner case** is defined as a situation at the boundary where multiple inputs are at their maximum or minimum, like the corner of a 3-D cube. At the corner small changes in input may cause lots of internal and external changes. In particular, we need to test the cases we think might be difficult (e.g., the clock output increments one second from 11:59:59 PM December 31, 1999.) There are many ways to decide on the coverage. We can select values:

- Near the extremes and in the middle
- Most typical of how our clients will properly use the system
- Most typical of how our clients will improperly use the system
- That differ by one
- You know your system will find difficult
- Using a random number generator

To stabilize the system we define a fixed set of inputs to test, run the system on these inputs, and record the outputs. Debugging is a process of finding patterns in the differences between recorded behavior and expected results. The advantage of modular programming is that we can perform **modular debugging**. We make a list of modules that might be causing the bug. We can then create new test routines to stabilize these modules and debug them one at a time. Unfortunately, sometimes all the modules seem to work, but the combination of modules does not. In this case we study the interfaces between the modules, looking for intended and unintended (e.g., unfriendly code) interactions.

The emergence of concurrent systems (e.g., distributed networks of microcontrollers), optimizing architectures (e.g., pipelines, cache, branch prediction, out of order execution, conditional execution, and multi-core processors), and the increasing need for security and reliably place further demands on debuggers. The complexities introduced by the interaction of multiple events or time dependent processes are much more difficult to debug than errors associated with sequential programs. The behavior of non-real-time sequential programs is reproducible: for a given set of inputs their outputs remain the same. In the case of concurrent or real-time programs this does not hold true. Control over repeatability, event selection, and event isolation is even more important for concurrent or real-time environments.

Sometimes, the meaning and scope of the term debugging itself is not clear. We hold the view that the goal of debugging is to maintain and improve software, and the role of a debugger is to support this endeavor. We define the debugging process as testing, stabilizing, localizing, and correcting errors. And in our opinion, although testing, stabilizing, and localizing errors are important and essential to debugging, they are auxiliary processes: the primary goal of debugging is to remedy faults and verify the system is operating within specifications.

3.9.3. Functional Debugging

Functional debugging involves the verification of input/output parameters. It is a static process where inputs are supplied, the system is run, and the outputs are compared against the expected results. We will present seven methods of functional debugging.

1. Single Stepping or Trace. Many debuggers allow you to set the program counter to a specific address then execute one instruction at a time. **StepOver** will execute one instruction, unless that instruction is a subroutine call, in which case the simulator will execute the entire subroutine and stop at the instruction following the subroutine call. **StepOut** assumes the execution has already entered a function and will finish execution of the function and stop at the instruction following the function call.

2. Breakpoints without filtering. The first step of debugging is to **stabilize** the system with the bug. In the debugging context, we stabilize the problem by creating a test routine that fixes (or stabilizes) all the inputs. In this way, we can reproduce the exact inputs over and over again. Once stabilized, if we modify the program, we are sure that the change in our outputs is a function of the modification we made in our software and not due to a change in the input parameters. A **breakpoint** is a mechanism to tag places in our software, which when executed will cause the software to stop.

3. Conditional breakpoints. One of the problems with breakpoints is that sometimes we have to observe many breakpoints before the error occurs. One way to deal with this problem is the conditional breakpoint. Add a global variable called `count` and initialize it to zero in the ritual. Add the following conditional breakpoint to the appropriate location, and run the system again (you can change the 32 to match the situation that causes the error).

```
if(++count==32){
   breakpoint();     // <= place breakpoint here
}
```

Notice that the breakpoint occurs only on the 32nd time the break is encountered. Any appropriate condition can be substituted.

4. Instrumentation: print statements. The use of print statements is a popular and effective means for functional debugging. The difficulty with print statements in embedded systems is that a standard "printer" may not be available. Another problem with printing is that most embedded systems involve time-dependent interactions with its external environment. The print statement itself may so slow that the debugging instrument itself causes the system to fail. Therefore, the print statement is usually intrusive. One exception to this rule is if the printing channel occurs in the background using interrupts, and the time between print statements (t_2) is large compared to the time to execution one print (t_1), then the print statements will be minimally intrusive. Nevertheless, this book will focus on debugging methods that do not rely on the availability of a printer.

5. Instrumentation: dump into array without filtering. One of the difficulties with print statements is that they can significantly slow down the execution speed in real-time systems. Many times the bandwidth of the print functions cannot keep pace with data being generated by the debugging process. For example, our system may wish to call a function 1000 times a second (or every 1 ms). If we add print statements to it that require 50 ms to perform, the presence of the print statements will significantly affect the system operation. In this situation, the print statements would be considered extremely intrusive. Another problem with print statements occurs when the system is using the same output hardware for its normal operation, as is required to perform the print function. In this situation, debugger output and normal system output are intertwined.

To solve both these situations, we can add a debugger instrument that dumps strategic information into arrays at run time. We can then observe the contents of the array at a later time. One of the advantages of dumping is that the JTAG debugging allows you to visualize memory even when the program is running.

Assume **Happy** and **Sad** are strategic 32-bit variables. The first step when instrumenting a dump is to define a buffer in RAM to save the debugging measurements. The **Debug_Cnt** will be used to index into the buffers. **Debug_Cnt** must be initialized to zero, before the debugging begins. The debugging instrument, shown in Program 3.13, saves the strategic variables into the buffer.

```
#define SIZE 100
uint32_t Debug_Buffer[SIZE][2];
unsigned int Debug_Cnt=0;
void Debug_Dump(void){ // dump Happy and Sad
  if(Debug_Cnt < SIZE){
    Debug_Buffer[Debug_Cnt][0] = Happy;
    Debug_Buffer[Debug_Cnt][1] = Sad;
    Debug_Cnt++;
  }
}
```

Program 3.12. Instrumentation dump without filtering.

Next, you add **Debug_Dump();** statements at strategic places within the system. You can either use the debugger to display the results or add software that prints the results after the program has run and stopped. In this way, you can collect information in the exact same manner you would if you were using print statements.

6. Instrumentation: dump into array with filtering. One problem with dumps is that they can generate a tremendous amount of information. If you suspect a certain situation is causing the error, you can add a filter to the instrument. A filter is a software/hardware condition that must be true in order to place data into the array. In this situation, if we suspect the error occurs when the pointer nears the end of the buffer, we could add a filter that saves in the array only when data matches a certain condition. In the example shown in Program 3.13, the instrument saves the strategic variables into the buffer only when **Sad** is greater than 100.

```
#define SIZE 100
uint32_t Debug_Buffer[SIZE][2];
unsigned int Debug_Cnt=0;
void Debug_FilteredDump(void){ // dump Happy and Sad
  if((Sad > 100)&&(Debug_Cnt < SIZE)){
    Debug_Buffer[Debug_Cnt][0] = Happy;
    Debug_Buffer[Debug_Cnt][1] = Sad;
    Debug_Cnt ++;
  }
}
```

Program 3.13. Instrumentation dump with filter.

7. Monitor using the LED heartbeat. Another tool that works well for real-time applications is the monitor. A **monitor** is an independent output process, somewhat similar to the print statement, but one that executes much faster and thus is much less intrusive. The OLED or LCD can be an effective monitor for small amounts of information if the time between outputs is much larger than the time to output. Another popular monitor is the LED. You can place one or more LEDs on individual otherwise unused output bits. Software toggles these LEDs to let you know what parts of the program are running. An LED is an example of a Boolean monitor or **heartbeat**. Assume an LED is attached to Port D bit 1. Program 3.14 will toggle the LED.

```
#define PD1 (*((volatile uint32_t *)0x40007008))
#define Debug_HeartBeat() (PD1 ^= 0x02)
```

Program 3.14. An LED monitor.

Next, you add **Debug_HeartBeat();** statements at strategic places within the system. Port D must be initialized so that bit 1 is an output before the debugging begins. You can either observe the LED directly or look at the LED control signals with a high-speed oscilloscope or logic analyzer. When using LED monitors it is better to modify just the one bit, leaving the other 7 as is. In this way, you can have multiple monitors on one port.

Checkpoint 3.18: Write a debugging instrument that toggles Port A bit 3.

3.9.4. Performance Debugging

Performance debugging involves the verification of timing behavior of our system. It is a dynamic process where the system is run, and the dynamic behavior of the system is compared against the expected results. We will present three methods of performance debugging, then apply the techniques to measure execution speed.

1. Counting bus cycles. For simple programs with little and no branching and for simple microcontrollers, we can estimate the execution speed by looking at the assembly code and adding up the time to execute each instruction.

2. Instrumentation measuring with an independent counter. SysTick is a 24-bit counter decremented every bus clock. It automatically rolls over when it gets to 0. If we are sure the execution speed of our function is less than 2^{24} bus cycles, we can use this timer to collect timing information with only a minimal amount of intrusiveness.

3. Instrumentation Output Port. Another method to measure real-time execution involves an output port and an oscilloscope. Connect a microcontroller output bit to your scope. Add debugging instruments that set/clear these output bits at strategic places. Remember to set the port's direction register to 1. Assume an oscilloscope is attached to Port D bit 1. Program 3.15 can be used to set and clear the bit.

```
#define PD1 (*((volatile uint32_t *)0x40007008))
#define Debug_Set()   (PD1 = 0x02)
#define Debug_Clear() (PD1 = 0x00)
```

Program 3.15. Instrumentation output port.

Next, you add **Debug_Set();** and **Debug_Clear();** statements before and after the code you wish to measure. Port D must be initialized so that bit 1 is an output before the debugging begins. You can observe the signal with a high-speed oscilloscope or logic analyzer.

```
Debug_Set();
Stuff();   // User code to be measured
Debug_Clear();
```

To illustrate these three methods, we will consider measuring the execution time of an integer square root function as presented Program 3.16.

The first method is to count bus cycles using the assembly listing. This approach is only appropriate for very short programs, and becomes difficult for long programs with many conditional branch instructions. The time to execute each assembly instruction can be found in the Cortex™-M Technical Reference Manual. Because of the complexity of the ARM® Cortex™-M, this method is only approximate. For example the time to execute a divide depends on the data, and the time to execute a branch depends on the alignment of the instruction pipeline. A portion of the assembly output generated by the ARM Keil™ uVision® compiler is presented on the left side of Program 3.16. Notice that the total cycle count for could range from 155 to 353 cycles. At 16 MHz the execution time could range from 9.69 to 22.1 µs. For most programs it is actually very difficult to get an accurate time measurement using this technique.

```
sqrt MOV    r1,r0             [1]        // Newton's method
     MOVS   r3,#0x01          [1]        // s is an integer
     ADD    r0,r3,r1,LSR #4   [1]        // sqrt(s) is an integer
     MOVS   r2,#0x10          [1]        uint32_t sqrt(uint32_t s){
     B      chck              [2-4]      uint32_t t;   // t*t becomes s
loop MLA    r3,r0,r0,r1       [2]*16     int n;        // loop counter
     UDIV   r3,r3,r0          [2-12]*16    t = s/10+1; // initial guess
     LSRS   r0,r3,#1          [1]*16       for(n = 16; n; --n){  // will finish
     SUBS   r2,r2,#1          [1]*16        t = ((t*t+s)/t)/2;
chck CMP    r2,#0x00          [1]*17     }
     BNE    loop              [2-4]*17   return t;
     BX     lr                [2-4]      }
```

Program 3.16. Assembly listing and C code for a sqrt function.

The second method uses an internal timer called SysTick. The ARM® Cortex™-M microcontrollers provide the 24-bit SysTick register (**NVIC_ST_CURRENT_R**) that is automatically decremented at the bus frequency. When the counter hits zero, it is reloaded to 0xFFFFFF and continues to count down. If we are sure the function will complete in a time less than 2^{24} bus cycles, then the internal timer can be used to measure execution speed empirically. The code in Program 3.17 first reads the SysTick counter, executes the function, and then reads the SysTick counter again. The elapsed time is the difference in the counter before and after. Since the execution speed may be dependent on the input data, it is often wise to measure the execution speed for a wide range of input parameters. There is a slight overhead in the measurement process itself. To be accurate, you could measure this overhead and subtract it off your measurements. In this case, a constant 4 is subtracted so that if the call to the function were completely removed the elapsed time would return 0. Notice that in this example, the total time including parameter passing is measured. Experimental results show this function executes in 204 bus cycles. At 16 MHz, this corresponds to 12.75 µs.

```
uint32_t Before, Elapsed;
void main(void){ volatile uint32_t Out;
   SysTick_Init();              // Program 2.10
   Before = NVIC_ST_CURRENT_R;
   Out = sqrt(230400);
   Elapsed = (Before - NVIC_ST_CURRENT_R - 4)&0x00FFFFFF;
}
```
Program 3.17: Empirical measurement of dynamic efficiency.

The third technique can be used in situations where a timer is unavailable or where the execution time might be larger than 2^{24} counts. In this empirical technique we attach an unused output pin to an oscilloscope or to a logic analyzer. We will set the pin high before the call to the function and set the pin low after the function call. In this way a pulse is created on the digital output with a duration equal to the execution time of the function. We assume Port D is available, and bit 1 is connected to the scope. By placing the function call in a loop, the scope can be triggered. With a storage scope or logic analyzer, the function need be called only once. Together with an oscilloscope or logic analyzer, Program 3.18 measures the execution time of the function **sqrt** (Figure 3.18). We stabilize the system by calling it over and over. Using the scope, we can measure the width of the pulse on PD1, which will be execution time of the function **sqrt**. Running at 16 MHz, the results in Figure 3.18 show it takes 13 μs to execute **sqrt(230400)**, which is 208 bus cycles.

```
int main(void){ uint32_t Out;
   PortD_Init();          // Program 2.6
   while(1){
      Debug_Set();        // Program 3.15
      Out = sqrt(230400);
      Debug_Clear();      // Program 3.15
   }
}
```
Program 3.18. Another empirical measurement of dynamic efficiency.

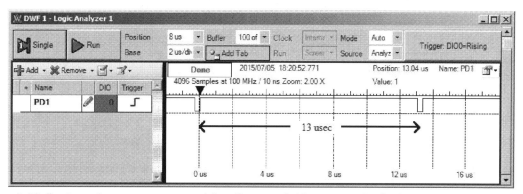

Figure 3.18. Logic analyzer output measured from Program 3.18 using an Analog Discovery.

Checkpoint 3.19: If you were to remove the **Out=sqrt(230400);** line in Program 3.18, what would you expect the pulse width on PD1 to be? Why does Program 3.17 yield a result smaller than Program 3.18?

3.9.5. Profiling

Profiling is a type of performance debugging that collects the time history of program execution. Profiling measures where and when our software executes. It could also include what data is being processed. For example if we could collect the time-dependent behavior of the program counter, then we could see the execution patterns of our software. We can profile the execution of a multiple thread software system to detect reentrant activity.

Profiling using a software dump to study execution pattern. In this section, we will discuss software instruments that study the execution pattern of our software. In order to collect information concerning execution we will add debugging instruments that save the time and location in arrays (Program 3.19). By observing these data, we can determine both a time profile (when) and an execution profile (where) of the software execution. Running this profile revealed the sequence of places as 0, 1, 2, 2, 2, 2, 2, 2, 2, 2, 2, 2, 2, 2, 2, 2, 2, 2, and 3. Each call to **Debug_Profile** requires 32 cycles to execute. Therefore, this instrument is a lot less intrusive than a print statement.

```
uint32_t Debug_time[20];
uint8_t Debug_place[20];
uint32_t n;
void Debug_Profile(uint8_t p){
  if(n < 20){
    Debug_time[n] = NVIC_ST_CURRENT_R; // record current time
    Debug_place[n] = p;
    n++;
  }
}
uint32_t sqrt(uint32_t s){
uint32_t t;   // t*t becomes s
int n;        // loop counter
  Debug_Profile(0);
  t = s/10+1;      // initial guess
  Debug_Profile(1);
  for(n = 16; n; --n){  // will finish
    Debug_Profile(2);
    t = ((t*t+s)/t)/2;
  }
  Debug_Profile(3);
  return t;
}
```
Program 3.19: A time/position profile dumping into a data array.

Profiling using an Output Port. In this section, we will discuss a hardware/software combination to visualize program activity. Our debugging instrument will set output port bits D3–D0 (Program 3.20). We will place these instruments at strategic places in the software. In particular, we will output 1, 2, 4, or 8 to Port D, where each bit uniquely specifies where in the program we are executing (Figure 3.19). We connect the four output pins to a logic analyzer and observe the program activity. Each debugging instrument requires only 4 cycles to execute. So the profile in Program 3.20 is less intrusive than the one in Program 3.19. In particular, notice the execution speed of the sqrt function only increases from 13 to 16 μs. This 3-μs execution penalty is a measure of the intrusiveness of the debugging activity.

```
#define PROFILE (*((volatile uint32_t *)0x4000703C))
uint32_t sqrt(uint32_t s){
uint32_t t;   // t*t becomes s
int n;        // loop counter
  PROFILE = 1;
  t = s/10+1;      // initial guess
  PROFILE = 2;
  for(n = 16; n; --n){  // will finish
    PROFILE = 4;
    t = ((t*t+s)/t)/2;
    PROFILE = 8;
  }
  PROFILE = 0;
  return t;
}
```
Program 3.20: A time/position profile using four output bits.

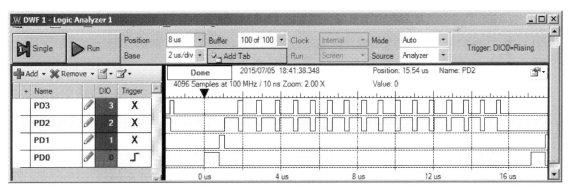

Figure 3.19. Logic analyzer output measured from Program 3.20 using an Analog Discovery.

Thread Profile. When more than one program (multiple threads) is running, you could use the technique in Program 3.20 to visualize the thread that is currently active (the one running). For each thread, we assign one output pin. The debugging instrument would set the corresponding bit high when the thread starts and clear the bit when the thread stops. We would then connect the output pins to a logic analyzer to visualize in real time the thread that is currently running.

3.10. Exercises

3.1 List 3 factors that we can use to evaluate the "quality" of a program.

3.2 In 32 words or less, describe the meaning of each of the following terms.
  a) Dynamic efficiency
  b) Static efficiency
  c) Scope
  d) Cohesion

3.3 Consider the reasons why one chooses which technique to create a variable.
  a) List three reasons why one would implement a variable using a register.
  b) List three reasons why one would implement a variable on the stack and access it using SP indexed mode addressing.
  c) List three reasons why one would implement a variable in RAM and access it using extended mode addressing.

3.4 In 32 words or less, give an example of each of the following terms.
  a) Invocation coupling
  b) Bandwidth coupling
  c) Control coupling

3.5 For each term specify if it is a constant or a variable. For the variables specify if it is permanent or temporary and if it is public or private.
  a) `FIFO_SIZE`
  b) `Fifo_Size`
  c) `FifoSize`
  d) `fifoSize`

3.6 In 32 words or less, give an example of each of the following terms.
  a) Logical cohesion
  b) Temporal cohesion
  c) Procedural cohesion
  d) Communicational cohesion
  e) Sequential cohesion
  f) Functional cohesion

3.7 Describe how you could create an array that is temporary in allocation but public in scope.

3.8 In 32 words or less, explain the differences between a Mealy and Moore FSM. For which types of problems should you implement with Mealy? For which types of problems should you implement with Moore?

3.9 In 32 words or less, describe the meaning of each of the following terms.
  a) Module
  b) BSP
  c) HAL
  d) Device driver

3.10 Give a quantitative measure of modularity. E.g., system A is more modular than system B if…

3.11 In 32 words or less, describe the meaning of each of the following debugging terms.
  a) Profile
  b) Intrusive
  c) Stabilize
  d) Heartbeat
  e) Monitor
  f) Dump
  g) Logic analyzer
  h) Filter

D3.12 Write software to implement the Moore FSM shown in Figure 3.20. Include the FSM state machine, port initialization, timer initialization and the FSM controller. The command sequence will be output, wait the specified time in ms, input, then branch to next state. The 2-bit input is on Port B (PB1 and PB0) and the 3-bit output is also on Port B (PB7, PB6, PB5).

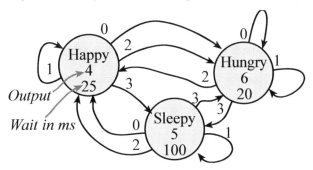

Figure 3.20. FSM for Exercise 3.12.

D3.13 Write software to implement the Mealy FSM shown in Figure 3.21. Include the FSM state machine, port initialization, timer initialization and the FSM controller. The command sequence will be input, output, wait 10 ms, input, and then branch to next state. The 1-bit input is on Port D (PD0) and the 3-bit output is on Port F (PF3, PF2, PF1).

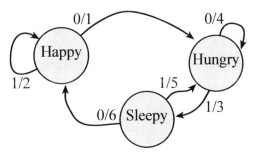

Figure 3.21. FSM for Exercise 3.13.

D3.14 Rewrite the `SysTick_Wait` function in Program 2.10, so that it continuously checks an alarm input on PA7. As long as PA7 is low (normal), it will wait the prescribed time. But, if PA7 goes high (alarm), the wait function returns.

3.11. Lab Assignments

Lab 3.1 The overall objective is to create a **4-key digital lock**. The system has four digital inputs and one digital output. The LED will be initially on, signifying the door is locked. Define two separate key codes, one to lock and one to unlock the door. For example, if the keys are numbers 1, 2, 3 and 4, one possible key code is 23. This means if you push both the 2 and 3 keys (not pushing the 1, 4 keys) the door will unlock. Implement the design such that the unlock function occurs in the software of the microcontroller.

Lab 3.2 The overall objective is to create a **line tracking robot**. The system has two digital inputs and two digital outputs. You can simulate the system with two switches and two LEDs, or build a robot with two DC motors and two optical reflectance sensors. Both sensor inputs will be on if the machine is completely on the line. One sensor input will be on and the other off if the machine is just going off the track. If the machine is totally off the line, then both sensor inputs will be off. Implement the controller using a finite state machine. Choose a Moore or Mealy format as appropriate.

Lab 3.3 The overall objective is to create an **enhanced traffic light controller**. The system has three digital inputs and seven digital outputs. You can simulate the system with three switches and seven LEDs. The inputs are North, East, and Walk. The outputs are six for the traffic light and one for a walk signal. Implement the controller using a finite state machine. Choose a Moore or Mealy format as appropriate.

Lab 3.4 The overall objective is to create an **8-key digital lock**. The system has eight digital inputs and one digital output. The LED will be initially off, signifying the door is locked. Define a key sequence to unlock the door. For example, if the keys are numbers 1, 2, ... and 8, one possible key code is 556. This means if you push the 5, release the 5, push the 5, release the 5 and push the 6, then the door will unlock. The unlock operation will be a two second pulse on the LED.

Lab 3.5 The overall objective is to design a **vending machine controller**. The system has five digital inputs and three digital outputs. You can simulate the system with five switches and three LEDs. The inputs are `quarter`, `dime`, `nickel`, `soda`, and `diet`. The `quarter` input will go high, then go low when a 25¢ coin is added to the machine. The `dime` and `nickel` inputs work in a similar manner for the 10¢ and 5¢ coins. The sodas cost 35¢ each. The user presses the `soda` button to select a regular soda and the `diet` button to select a diet soda. The `GiveSoda` output will release a regular soda if pulsed high, then low. Similarly, the `GiveDiet` output will release a diet soda if pulsed high, then low. The `Change` output will release a 5¢ coin if pulsed high, then low. Implement the controller using a finite state machine. Choose a Moore or Mealy format as appropriate. Since there are so many inputs and at most one is active at a time, you may wish to implement a FSM with a different format than the examples in the book.

4. Hardware-Software Synchronization

Chapter 4 objectives are to:

- Introduce basic performance measures for I/O interfacing
- Formalize timing using equations and graphical diagrams
- Present Petri Nets as a way to describe synchronization
- Introduce Kahn Process Networks (KPN) to describe distributed systems
- Compare and contrast possible approaches to synchronization
- Present the basic hardware/software for a parallel port LCD interface
- Interface a stepper motor using blind-cycle synchronization
- Discuss the basic concepts of busy-wait synchronization
- Introduce the general concept of a handshake interface
- Implement a busy-wait serial port device driver
- Interface a keypad using busy-wait synchronization

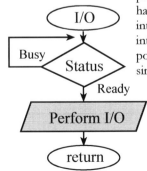

One of the factors that make embedded systems different from regular computers is the specialized input/output devices we attach to our embedded systems. While the entire book addresses the design and analysis of embedded systems, this chapter serves as an introduction to the critical task of I/O interfacing. Interfacing includes both the physical connections of the hardware devices and the software routines that affect information exchange. Key to this task is the need to synchronize the software and hardware components. The chapter begins with performance measures to evaluate the effectiveness of our system (latency, bandwidth, priority). As engineers we are not simply asked to design and build devices, but we are also required to evaluate our products. Latency and bandwidth are two quantitative performance parameters we can measure on our real-time embedded system. A number of formal tools will be presented including timing equations, timing diagrams, Petri Nets, and Kahn Process Networks (KPN). The edge-triggered I/O ports on the TM4C microcontrollers will be presented. Next, five basic approaches to hardware/software synchronization are presented, which include blind cycle, busy wait, interrupts, periodic polling, and direct memory access. A more thorough treatment of interrupts will be presented in Chapter 5. However, the discussions in this chapter will point to situations that require interrupt synchronization. The rest of chapter presents simple examples to illustrate the "blind-cycle" and "busy-wait" approaches to interfacing.

4.1. Introduction

4.1.1. Performance Measures

Latency is the time between when the I/O device indicated service is required and when service is initiated. Latency includes hardware delays in the digital gates plus computer hardware delays. Latency also includes software delays. For an input device, software latency (or software response time) is the time between new input data ready and the software reading the data. For an output device, latency is the delay from output device idle and the software giving the device new data to output. In this book, we will also have periodic events. For example, in our data acquisition systems, we wish to invoke the ADC at a fixed time interval. In this way we can collect a sequence of digital values that approximate the continuous analog signal. Software latency in this case is the time between when the ADC is supposed to be started and when it is actually started. The microcontroller-based control system also employs periodic software processing. Similar to the data acquisition system, the latency in a control system is the time between when the control software is supposed to be run and when it is actually run. A **real-time** system is one that can guarantee a worst-case latency. In other words, there is an upper bound on the software response time. **Throughput** or **bandwidth** is the maximum data flow (bytes per second) that can be processed by the system. Sometimes the bandwidth is limited by the I/O device, while other times it is limited by computer software. Bandwidth can be reported as an overall average or a short-term maximum. **Priority** determines the order of service when two or more requests are made simultaneously. Priority also determines if a high-priority request should be allowed to suspend a low-priority request that is currently being processed. We may also wish to implement equal priority so that no one device can monopolize the computer. The tolerance of a real-time system towards failure to meet the timing requirements determines whether we classify it as **hard real time**, **firm real time**, or **soft real time**. If missing a timing constraint is unacceptable, we call it a hard real-time system. In a firm real-time system, the value of an operation completed past its timing constraint is considered zero but not harmful. In a soft real-time system, the value of an operation diminishes the further it completes after the timing constraint..

4.1.2. Synchronizing the software with the state of the I/O

One can think of the hardware as being in one of three states. The **idle** state occurs when the device is disabled or inactive. No I/O occurs in the idle state. When active (not idle), the hardware toggles between the **busy** and **done** states, as illustrated in Figure 4.1. For an input device, a status flag is set when new input data are available. The busy-to-done state transition will cause a **busy-wait** loop (**gadfly** loop) to complete. This busy-to-done transition could also trigger an interrupt. Once the software recognizes that the input device has new data, it will read the data and ask the input device to create more data. It is the busy-to-done state transition that signals to the computer that service is required. When the hardware is in the done state, the I/O transaction is complete. Often the simple process of reading the data will clear the flag and request another input.

For an output device, a status flag is set when the output is idle and ready to accept more data. The "busy to done" state transition causes a busy-wait loop to complete. Once the software recognizes the output is idle, it gives the output device another piece of data to output. It will be important to make sure the software clears the flag each time new output is started.

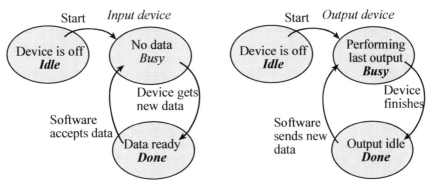

Figure 4.1. Initially the device is idle. While active, it may be in the busy or done state. The input device sets a flag when it has new data. The output device sets a flag when it has finished outputting the last data.

The problem with I/O devices is that they are usually much slower than software execution. Therefore, we need synchronization, which is the process of the hardware and software waiting for each other in a manner such that data is properly transmitted. A way to visualize this synchronization is to draw a state versus time plot of the activities of the hardware and software. For an input device, the software begins by waiting for new input (Figure 4.2). When the input device is busy, it is in the process of creating new input. When the input device is done, new data are available. When the input device makes the transition from busy to done, it releases the software to go forward. In a similar way, when the software accepts the input, it can release the input device hardware. The arrows from one graph to the other represent the synchronizing events. In this example, the time for the software to read and process the data is less than the time for the input device to create new input. This situation is called **I/O bound**.

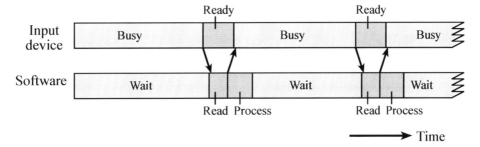

Figure 4.2. The software must wait for the input device to be ready.

If the input device were faster than the software, a situation called **CPU bound**, then the software waiting time would be zero. In general, the bandwidth depends on both the hardware and the software. I/O bound means the overall bandwidth is limited by the speed of the I/O device. Conversely, CPU bound means the overall bandwidth is limited by the execution speed of the software. In a given system, if the producer and consumer rates vary, the system may oscillate between I/O bound and CPU bound.

This configuration is also labeled as unbuffered because the hardware and software must wait for each other during the transmission of each piece of data. A buffered system allows the input device to run continuously, filling a FIFO as fast as it can. In the same way, the software can empty the FIFO whenever it is ready and whenever there is data in the buffer. We will implement buffered interfaces in Chapter 5 using interrupts.

Figure 4.3 contains a state versus time plot of the activities of the output device hardware and software. For an output device, the software begins by generating data then sending it to the output device. When the output device is busy, it is processing the data. Normally when the software writes data to an output port, that only starts the output process. The time it takes an output device to process data is usually longer than the software execution time. When the output device is done, it is ready for new data. When the output device makes the transition from busy to done, it releases the software to go forward. In a similar way, when the software writes data to the output, it releases the output device hardware. The output interface illustrated in Figure 4.3 is also I/O bound because the time for the output device to process data is longer than the time for the software to generate and write it.

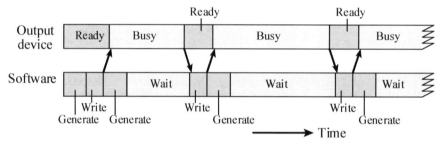

Figure 4.3. The software must wait for the output device to finish the previous operation.

This output interface is also unbuffered, because when the hardware is done, it will wait for the software, and after the software generates data, it waits for the hardware. A buffered system would allow the software to run continuously, filling a FIFO as fast as it wishes. In the same way, the hardware can empty the FIFO whenever it is ready and whenever there is data in the buffer. We will implement buffered interfaces in Chapter 5 using interrupts.

The purpose of our interface is to allow the microprocessor to interact with its external I/O device. There are five mechanisms to synchronize the microprocessor with the I/O device. Each mechanism synchronizes the I/O data transfer to the busy to done transition. The five methods are discussed in the following paragraphs.

Blind cycle is a method where the software simply waits a fixed amount of time and assumes the I/O will complete after that fixed delay. For an input device, the software triggers (starts) the external input hardware, wait a specified time, then reads data from the device (left side of Figure 4.4.) For an output device, the software writes data to the output device, triggers (starts) the device, then waits a specified time (left side of Figure 4.5.) We call this method *blind*, because there is no status information about the I/O device reported to the software. This method is appropriate for situations where the I/O speed is short and predictable.

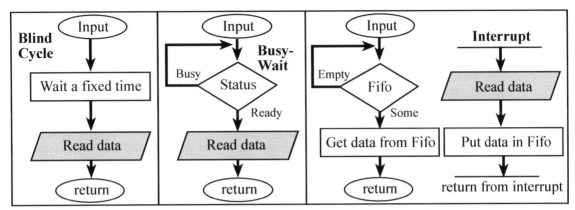

Figure 4.4. Flowcharts showing an input interface using blind-cycle, busy wait and interrupts.

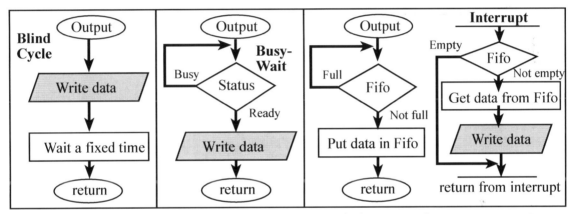

Figure 4.5. Flowcharts showing an output interface using blind-cycle, busy wait and interrupts.

Busy wait or **gadfly** is a software loop that checks the I/O status waiting for the done state. For an input device, the software waits until the input device has new data, and then reads it from the input device (middle of Figure 4.4.) For an output device, the software writes data, triggers the output device then waits until the device is finished. Another approach to output device interfacing is for the software to wait until the output device has finished the previous output, write data, and then trigger the device (middle of Figure 4.5.) We will discuss these two approaches to output device interfacing later in the chapter. Busy-wait synchronization will be used in situations where the software system is relatively simple and real-time response is not important.

An **interrupt** uses hardware to cause special software execution. With an input device, the hardware will request an interrupt when input device has new data. The software interrupt service will read from the input device and save in a global structure (right side of Figure 4.4). With an output device, the hardware will request an interrupt when the output device is idle. The software interrupt service will get data from a global structure, and then write to the device (right side of Figure 4.5).

Sometimes we configure the hardware timer to request interrupts on a periodic basis. The software interrupt service will perform a special function. A data acquisition system needs to read the ADC at a regular rate. Details of data acquisition systems can be found in Chapter 10. Some computers can be configured to request an interrupt on an access to an illegal address or a divide by zero. Interrupt synchronization will be used in situations where the software system is fairly complex or when real-time response is important.

Periodic polling uses a clock interrupt to periodically check the I/O status. With an input device, a ready flag is set when the input device has new data. At the next periodic interrupt, the software will read the data and save them in a global structure. With an output device, a ready flag is set when the output device is idle. At the next periodic interrupt, the software will get data from a global structure and write it. Periodic polling will be used in situations that require interrupts, but the I/O device does not support interrupt requests.

DMA, or direct memory access is an interfacing approach that transfers data directly to/from memory. With an input device, the hardware will request a DMA transfer when the input device has new data. Without the software's knowledge or permission the DMA controller will read from the input device and save in memory. With an output device, the hardware will request a DMA transfer when the output device is idle. The DMA controller will get data from memory, and then write to the device. Sometimes we configure the hardware timer to request DMA transfers on a periodic basis. DMA can be used to implement a high-speed data acquisition system. DMA synchronization will be used when bandwidth and latency are important.

4.1.3. Variety of Available I/O Ports

Microcontrollers perform digital I/O using their ports. In this chapter we will focus on the input and output of digital signals. Microcontrollers have a wide variety of configurations, only few of which are illustrated in Table 4.1. Each microcontroller manufacturer has multiple families consisting of a wide range of parts with varying numbers and types of I/O devices. For example, the TM4C123GH6ZRB has 120 GPIO pins. It is typical for port pins to be programmed via software for alternative functions other than parallel I/O. Each microcontroller family comes with a mechanism to download code and debug. The 9S12 family includes a background debug module. The Texas Instruments MSP430, Stellaris® and Tiva® families include either a JTag or Spy-Bi-Wire debugging interface. The Microchip PIC family debugs in a variety of ways including In-Circuit Serial Programming, In-Circuit Emulator, PICSTART and PROMATE. The Atmel family can be developed with a JTag debugger. Basically, we first choose the processor type (e.g., PIC, MSP430, 9S12, LM3S, or TM4C) depending on our software processing needs. Next, we choose the family depending on our I/O requirements. Lastly, we choose the particular part depending on our memory requirements.

	Port pins	Alternative functions.
PIC12F629	6	Very low cost, ADC, timer
MSP430	Up to 87	Very low power, ADC, SCI, SPI, I2C, USB, timer, and wireless
MC9S12C32	60	Serial, timer, ADC, SPI, CAN
AT91RM	Up to 122	ARM Thumb, ADC, serial, DMA, USB, Ethernet, Smart card
TM4C	Up to 120	Cortex-M, ADC, serial, DMA, USB, Ethernet, LCD, CAN, QEI

Table 4.1. The number of I/O ports and alternative function.

The current trend in the computer industry is customer specific integrated circuits (CSIC). A similar term for this development process is application specific integrated circuits (ASIC). With these approaches, the design engineers (customer) first evaluate the needs of their project. The design engineers working closely with the computer manufacturer make a list of features the microcontroller requires. For example

CPU type	CISC, RISC, multiple buses, multiple cores
Coprocessors	Floating point, DMA, graphics
Memory	RAM, EEPROM, Flash ROM, OTP ROM, ROM
Power	PLL, sleep states, variable supply, reduced output drive
Analog	8 to 16-bit ADC, 8 to 12-bit DAC, analog comparators
Timer	Pulse width modulation, Input capture, Output compare
Parallel Ports	Edge trigger interrupts, pull-up, pull-down, open collector
Serial	Asynchronous (UART), synchronous (SPI), peripheral (I²C)
Sound	Integrated Interchip Sound (I²S)
Motor	Quadrature Encoder Interface (QEI)
Networks	USB, CAN, Ethernet, wireless (wifi, ZigBee, or Bluetooth)

The design engineers either choose a microcontroller from existing products that meets their needs, or the engineers contract with the manufacturer to produce a microcontroller with the exact specifications for that project. Many manufacturers distribute starter code, reference designs, or white papers showing complete implementations using that particular microcontroller to solve actual problems. The availability of such solutions will be extremely helpful, even if the applications are just remotely similar to your problem.

4.2. Timing

4.2.1. Timing Equations

When interfacing devices, it is important to manage when events occur. Typical events include the rise or fall of control signals, when data pins need to be correct, and when data pins actually contain the proper values. In this book, we will use two mechanisms to describe the timing of events. In this section, we present a formal syntax called **timing equations**, which are algebraic mechanisms to describe time. In the next section, we will present graphical mechanisms called timing diagrams.

When using a timing equation, we need to define a **zero-time reference**. For synchronous systems, which are systems based on a global clock, we can define one edge of the clock as time=0. Timing equations can contain number constants typically given in ns, variables, and edges. For example, ↓**A** means the time when signal **A** falls, and ↑**A** means the time when it rises. To specify an interval of time, we give its start and stop times between parentheses separated by a comma. For example, (400, 520) means the time interval begins at 400 ns and ends at 520 ns. These two numbers are relative to the zero-time reference.

We can use algebraic variables, edges, and expressions to describe complex behaviors. Some timing intervals are not dependent on the zero-time reference. For example, (↑**A**-10, ↑**A**+**t**) means the time interval begins 10 ns before the rising edge of signal **A** and ends at time **t** after that same rising edge. Some timing variables we see frequently in data sheets include

t_{pd} propagation delay from a change in input to a change in output
t_{pHL} propagation delay from input to output, as the output goes from high to low
t_{pLH} propagation delay from input to output, as the output goes from low to high
t_{pZL} propagation delay from control to output, as the output goes from floating to low
t_{pZH} propagation delay from control to output, as the output goes from floating to high
t_{pLZ} propagation delay from control to output, as the output goes from low to floating
t_{pHZ} propagation delay from control to output, as the output goes from high to floating
t_{en} propagation delay from floating to driven either high or low, same as t_{pZL} and t_{pZH}
t_{dis} propagation delay from driven high/low to floating, same as t_{pLZ} and t_{pHZ}
t_{su} setup time, the time before a clock input data must be valid
t_h hold time, the time after a clock input data must continue to be valid

Sometimes we are not quite sure exactly when an event starts or stops, but we can give upper and lower bounds. We will use brackets to specify this timing uncertainty. For example, assume we know the interval starts somewhere between 400 and 430 ns, and stops somewhere between 520 and 530 ns, we would then write ([400, 430], [520, 530]).

As examples, we will consider the timing of a not gate, a tristate driver, and an octal D flip-flop, as shown in Figure 4.6. If the input to the 74HC04 is low, its output will be high. Conversely, if the input to the 74HC04 is high, its output will be low. There are eight data inputs to the 74HC244, labeled as **A**. Its eight data outputs are labeled **Y**. The 74HC244 tristate driver has two modes. When the output enable, **OE***, is low, the output **Y** equals the input **A**. When **OE*** is high, the output **Y** floats, meaning it is not driven high or low. The slash with an 8 over top means there are eight signals that all operate in a similar or combined fashion. The 74HC374 octal D flip-flop has eight data inputs (**D**) and eight data outputs (**Q**). A D flip-flip will store or latch its **D** inputs on the rising edge of its **Clk**. The **OE*** signal on the 74HC374 works in a manner similar to the 74HC244. When **OE*** is low, the stored values in the flip-flop are available at its **Q** outputs. When **OE*** is high, the **Q** outputs float. The making **OE*** go high or low does not change the internal stored values. **OE*** only affects whether or not the stored values are driven on the **Q** outputs.

Positive logic means the true or asserted state is a higher voltage than the false or not asserted state. **Negative logic** means the true or asserted state is a lower voltage than the false or not asserted state. The * in the name **OE*** means negative logic. Other syntax styles that mean negative logic include a slash before the symbol (e.g., **\OE**), the letter n in the name (**OEn**), or a line over the top (e.g., $\overline{OE}$).

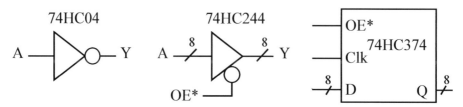

Figure 4.6. A NOT gate, a tristate driver, and an octal D flip-flop.

We will begin with the timing of the 74HC04 not gate. The typical propagation delay time (t_{pd}) for this not gate is 8 ns. Considering just the typical delay, we specify time when **Y** rises in terms of the time when **A** falls. That is

$$\uparrow Y = \downarrow A + t_{pd} = \downarrow A + 8$$

From the 74HC04 data sheet, we see the maximum propagation delay is 15 ns, and no minimum is given. Since the delay cannot be negative, we set the minimum to zero and write

$$\uparrow Y = [\downarrow A, \downarrow A + 15] = \downarrow A + [0, 15]$$

We specify the time interval when **Y** is high as

$$(\uparrow Y, \downarrow Y) = ([\downarrow A, \downarrow A+15], [\uparrow A, \uparrow A+15]) = (\downarrow A+[0,15], \uparrow A+[0,15])$$

When data is transferred from one location (the source) and stored into another (the destination), there are two time intervals that will determine if the transfer will be successful. The **data available interval** specifies when the data driven by the source is valid. The **data required interval** specifies when the data to be stored into the destination must be valid. For a successful transfer the data available interval must overlap (start before and end after) the data required interval. Let **a, b, c, d** be times relative to the same zero-time reference, let the data available interval be (**a, d**), and let the data required interval be (**b, c**), as shown in Figure 4.7. The data will be successfully transferred if

$$a \leq b \quad \text{and} \quad c \leq d$$

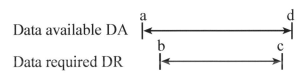

Figure 4.7. The data available interval should overlap the data required interval.

The example shown in Figure 4.8 illustrates the fundamental concept of timing for a digital interface. The objective is to transfer the data from the input, **In**, to the output, **Out**. First, we assume the signal at the **In** input of the 74HC244 is always valid. When the tristate control, **G***, is low then the **In** is copied to the **Bus**. On the rising edge of **C**, the 74HC374 D flip-flop will copy this data to the output **Out**.

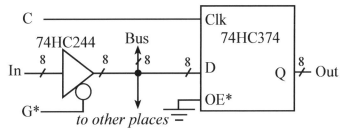

Figure 4.8. Simple circuit to illustrate that the data available interval should overlap the data required interval.

The data available interval defines when the signal **Bus** contains valid data and is determined by the timing of the 74HC244. From its data sheet, the output of the 74HC244 is valid between 0 and 38 ns after the fall of **G***. It will remain valid until 0 to 38 ns after the rise of **G***. The data available interval is

$$DA = (\downarrow G^* + t_{en}, \uparrow G^* + t_{dis}) = (\downarrow G^* + [0, 38], \uparrow G^* + [0, 38])$$

The data required interval is determined by the timing of the 74HC374. The 74HC374 input, **Bus**, must be valid from 25 ns before the rise of **C** and remain valid until 5 ns after that same rise of **C**. The time before the clock the data must be valid is called the **setup time**. The setup time for the 74HC374 is 25 ns. The time after the clock the data must continue to be valid is called the **hold time**. The hold time for the 74HC374 is 5 ns. The data required interval is

$$DR = (\uparrow C - t_{su}, \uparrow C + t_h) = (\uparrow C - 25, \uparrow C + 5)$$

Since the objective is to make the data available interval overlap the data required window, the worst case situation will be the shortest data available and the longest data required intervals. Without loss of information, we can write the shortest data available interval as

$$DA = (\downarrow G^* + 38, \uparrow G^*)$$

Thus the data will be properly transferred if the following are true:

$$\downarrow G^* + 38 \leq \uparrow C - 25 \quad \text{and} \quad \uparrow C + 5 \leq \uparrow G^*$$

Notice in Figure 4.8, the signal between the 74HC244 and 74HC374 is labeled **Bus**. A **bus** is a collection of signals that facilitate the transfer of information from one part of the circuit to another. Consider a system with multiple 74HC244's and multiple 74HC374's. The **Y** outputs of all the 74HC244's and the **D** inputs of all the 74HC374's are connected to this bus. If the system wished to transfer from input 6 to output 5, it would clear **G6*** low, make **C5** rise, and then set **G6*** high. At some point **C5** must fall, but the exact time is not critical. One of the problems with shared bus will be bus arbitration, which is a mechanism to handle simultaneous requests.

4.2.2. Timing Diagrams

An alternative mechanism for describing when events occur uses voltage versus time graphs, called **timing diagrams**. It is very intuitive to describe timing events using graphs because it is easy to visually sort events into their proper time sequence. Figure 4.9 defines the symbols we will use to draw timing diagrams in this book. Arrows will be added to describe the causal relations in our interface. Numbers or variables can be included that define how far apart events will be or should be. It is important to have it clear in our minds whether we are drawing an input or an output signal, because what a symbol means depends on whether we are drawing the timing of an input or an output signal. Many datasheets use the tristate symbol when drawing an input signal to mean "don't care".

Symbol	Input	Output
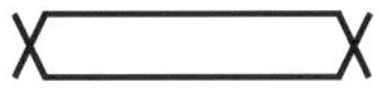 ⟨▬▬⟩	The input must be valid	The output will be valid
 ⎺⎻⎽	If the input were to fall	Then the output will fall
 ⎽⎻⎺	If the input were to rise	Then the output were to rise
 XXXXXXXX	Don't care, will work regardless	Don't know, output value is indeterminate
 ⟩▬▬	Not defined	High impedance, tristate, HiZ, Not driven, floating

Figure 4.9. Nomenclature for drawing timing diagrams.

To illustrate the graphical relationship of dynamic digital signals, we will draw timing diagrams for the three devices presented in the last section, see Figure 4.10. The arrows in the 74HC04 timing diagram describe the causal behavior. If the input were to rise, then the output will fall t_{pHL} time later. The subscript HL refers to the output changing from high to low. Similarly, if the input were to fall, then the output will rise t_{pLH} time later.

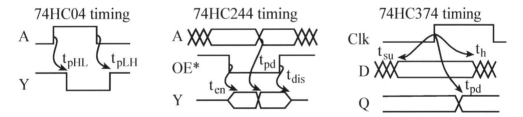

Figure 4.10. Timing diagrams for the circuits in Figure 4.6.

The arrows in the 74HC244 timing diagram also describe the causal behavior. If the input A is valid and if the OE* were to fall, then the output will go from floating to properly driven t_{en} time later. If the OE* is low and if the input A were to change, then the output will change t_{pd} time later. If the OE* were to rise, then the output will go from driven to floating t_{dis} time later.

The parallel lines on the D timing of the 74HC374 mean the input must be valid. "Must be valid" means the D input could be high or low, but it must be correct and not changing. In general, arrows represent causal relationships (i.e., "this" causes "that"). Hence, arrows should be drawn pointing to the right, towards increasing time. The setup time arrow is an exception to the "arrows point to the right" rule. The setup arrow (labeled with t_{su}) defines how long before an edge must the input be stable. The hold arrow (labeled with t_h) defines how long after that same edge the input must continue to be stable.

The timing of the 74HC244 mimics the behavior of devices on the computer bus during a read cycle, and the timing of the 74HC374 clock mimics the behavior of devices during a write cycle. Figure 4.11 shows the timing diagram for the interface problem presented in Figure 4.8. Again we assume the input **In** is valid at all times. The data available (DA) and data required (DR) intervals refer to data on the **Bus**. In this timing diagram, we see graphically the same design constraint developed with timing equations. ↓**G***+38 must be less than or equal to ↑**C**-25 and ↑**C**+5 must be less than or equal to ↑**G***. One of the confusing parts about a timing diagram is that it contains more information than actually matters. For example, notice that the fall of **C** is drawn before the rise of **G***. In this interface, the relative timing of ↑**G*** and ↓**C** does not matter. However, we draw ↓**C** so that we can specify the width of the **C** pulse must be at least 20 ns.

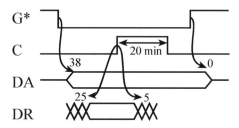

Figure 4.11. Timing diagram of the interface shown in Figure 4.8.

4.3. Petri Nets

In the last chapter, we presented finite state machines as a formal mechanism to describe systems with inputs and outputs. In this chapter, we present two methods to describe synchronization in complex systems: Petri Nets and Kahn Process Networks. Petri Nets can be used to study the dynamic concurrent behavior of network-based systems where there is discrete flow, such as packets of data. A Petri Net is comprised of Places, Transition, and Arcs. **Places**, drawn as circles in Figure 4.12, can contain zero, one, or more tokens. Consider places as variables (or buffers) and **tokens** as discrete packets of data. Tokens are drawn in the net as dots with each dot representing one token. Formally, the tokens need not comprise data and could simply represent the existence of an event. **Transitions**, drawn as vertical bars, represent synchronizing actions. Consider transitions as software that performs work for the system. From a formal perspective, a Petri Net does not model time delay. But, from a practical viewpoint we know executing software must consume time. The **arcs**, drawn as arrows, connect places to transitions. An arc from a place to a transition is an input to the transition, and an arc from a transition to a place is an output of the transition.

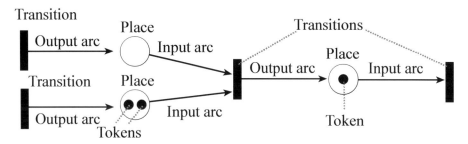

Figure 4.12. Petri Nets are built with places, transitions and arcs. Places can hold tokens.

For example, an input switch could be modeled as a device that inserts tokens into a place. The number of tokens would then represent the number of times the switch has been pressed. An alphanumeric keyboard could also be modeled as an input device that inserts tokens into a place. However, we might wish to assign an ASCII string to the token generated by a keyboard device. An output device in a Petri Net could be modeled as a transition with only input arcs but no output arcs. An output device consumes tokens (data) from the net.

Arcs are never drawn from place to place, nor from transition to transition. Transition node is ready to **fire** if and only if there is at least one token at each of its input places. Conversely, a transition will not fire if one or more input places is empty. Firing a transition produces software action (a task is performed). Formally, firing a transition will consume one token from each of its input places and generate one token for each of its output places. Figure 4.13 illustrates an example firing. In this case, the transition will wait for there to be at least one token in both its input places. When it fires it will consume two tokens and **merge** them into one token added to its output place. In general, once a transition is ready to fire, there is no guarantee when it will fire. One useful extension of the Petri Net assigns a minimum and maximum time delay from input to output for each transition that is ready to fire.

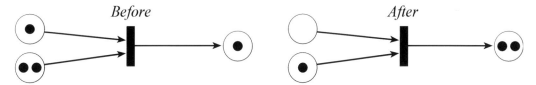

Figure 4.13. Firing a transition consumes one token at each input and produces one token at each output.

Figure 4.14 illustrates a sequential operation. The three transitions will fire in a strictly ordered sequence: first t_1, next t_2, and then t_3.

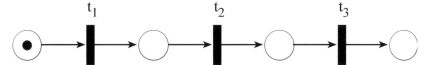

Figure 4.14. A Petri Net used to describe a sequential operation.

Figure 4.15 illustrates concurrent operation. Once transition t_1 fires, transitions t_2 and t_3 are running at the same time. On a distributed system t_2 and t_3 may be running in parallel on separate computers. On a system with one processor, two operations are said to be running concurrently if they are both ready to run. Because there is a single processor, the tasks must run one at a time.

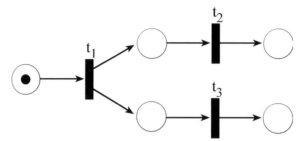

Figure 4.15. A Petri Net used to describe concurrent operations.

Figure 4.16 demonstrates a conflict or **race condition**. Both t_1 and t_2 are ready to fire, but the firing of one leads to the disabling of the other. It would be a mistake to fire them both. A good solution would be to take turns in some fair manner (flip a coin or alternate). A **deterministic model** will always produce the same output from a given starting condition or initial state. Because of the uncertainty when or if a transition will fire, a system described with a Petri Net is not deterministic.

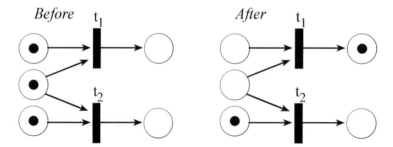

Figure 4.16. If t_1 were to fire, it would disable t_2. If t_2 were to fire, it would disable t_1.

Figure 4.17 describes an assembly line on a manufacturing plant. There are two robots. The first robot picks up one Part1 and one Part2, placing the parts together. After the robot places the two parts, it drills a hole through the combination, and then places the partial assembly into the Partial bin. The second robot first combines Part3 with the partial assembly and screws them together. The finished product is placed into the Done bin. The tokens represent the state of the system, and transitions are actions that cause the state to change.

The three supply transitions are input machines that place parts into their respective parts bins. The tokens in places **Part1**, **Part2**, **Part3**, **Partial**, and **Done** represent the number of components in their respective bins. The first robot performs two operations but can only perform one at a time. The **Rdy-to-P&P** place has a token if the first robot is idle and ready to pick and place. The **Rdy-to-drill** place has a token if the first robot is holding two parts and is ready to drill. The **Pick&Place** transition is the action caused by the first robot as it picks up two parts placing them together. The **Drill** transition is the action caused by the first robot as it drills a hole and places the partial assembly into the Partial bin.

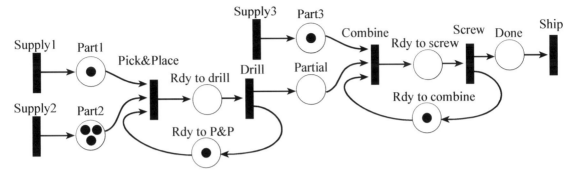

Figure 4.17. A Petri Net used to describe an assembly line.

The first robot performs two operations but can only perform one at a time. The **Rdy-to-P&P** place has a token if the first robot is idle and ready to pick and place. The **Rdy-to-drill** place has a token if the first robot is holding two parts and is ready to drill. The **Pick&Place** transition is the action caused by the first robot as it picks up two parts placing them together. The **Drill** transition is the action caused by the first robot as it drills a hole and places the partial assembly into the Partial bin.

The second robot performs two operations. The **Rdy-to-combine** place has a token if the second robot is idle and ready to combine. The **Rdy-to-screw** place has a token if the second robot is holding two parts and is ready to screw. The **Combine** transition is the action caused by the second robot as it picks up a Part3 and a Partial combining them together. The **Screw** transition is the action caused by the second robot as it screws it together and places the completed assembly into the Done bin. The **Ship** transition is an output machine that sends completed assemblies to their proper destination.

Checkpoint 4.1: Assuming no additional input machines are fired, run the Petri Net shown in Figure 4.17 until it stalls. How many competed assemblies are shipped?

4.4. Kahn Process Networks

Gilles Kahn first introduced the **Kahn Process Network** (KPN). We use KPNs to model distributed systems as well as signal processing systems. Each node represents a computation block communicating with other nodes through unbounded FIFO channels. The circles in Figure 4.18 are computational blocks and the arrows are FIFO queues. The resulting process network exhibits deterministic behavior that does not depend on the various computation or communication delays. As such, KPNs have found many applications in modeling embedded systems, high-performance computing systems, and computational tasks.

▪ 4. Hardware-Software Synchronization

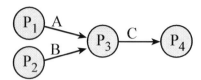

Figure 4.18. A Kahn Process Network consists of process nodes linked by unbounded FIFO queues.

For each FIFO, only one process puts, and only one process gets. Figure 4.18 shows a KPN with four processes and three edges (communication channels). Processes P_1 and P_2 are producers, generating data into channels A and B respectively. Process P_3 consumes one token from channel A and another from channel B (in either order) and then produces one token into channel C. Process P_4 is a consumer because it consumes tokens.

We can use a KPN to describe signal processing systems where infinite streams of data are transformed by processes executing in sequence or parallel. Streaming data means we input/analyze/output one data packet at a time without the desire to see the entire collection of data all at once. Despite parallel processes, multitasking or parallelism are not required for executing this model. In a KPN, processes communicate via unbounded FIFO channels. Processes read and write atomic data elements, or alternatively called **tokens**, from and to channels. The read token is equivalent to a FIFO get and the write token is a FIFO put. In a KPN, writing to a channel is **non-blocking**. This means we expect the put FIFO command to always succeed. In other words, the FIFO never becomes full. From a practical perspective, we can use KPN modeling for situations where the FIFOs never actually do become full. Furthermore, the approximate behavior of a system can be still be deemed for systems where FIFO full errors are infrequent. For these approximations we could discard data with the FIFO becomes full on a put instead of waiting for there to be free space in the FIFO.

On the other hand reading from a channel requires blocking. A process that reads from an empty channel will stall and can only continue when the channel contains sufficient data items (tokens). Processes are not allowed to test an input channel for existence of tokens without consuming them. Given a specific input (token) history for a process, the process must be deterministic so that it always produces the same outputs (tokens). Timing or execution order of processes must not affect the result and therefore testing input channels for tokens is forbidden.

In order to optimize execution some KPNs do allow testing input channels for emptiness as long as it does not affect outputs. It can be beneficial and/or possible to do something in advance rather than wait for a channel. In the example shown in Figure 4.18, process P3 must get from both channel A and channel B. The left side of Program 4.1 shows the process stalls if the AFifo is empty (even if there is data in the BFifo). If the first FIFO is empty, it might be efficient to see if there is data in the other FIFO to save time (right side of Program 4.1).

Processes of a KPN are **deterministic**. For the same input history they must always produce exactly the same output. Processes can be modeled as sequential programs that do reads and writes to ports in any order or quantity as long as the determinism property is preserved.

KPN processes are **monotonic**, which means that they only need partial information of the input stream in order to produce partial information of the output stream. Monotonicity allows parallelism. In a KPN there is a total order of events inside a signal. However, there is no order relation between events in different signals. Thus, KPNs are only partially ordered, which classifies them as an untimed model.

```
void Process3(void){
int32_t inA, inB, out;
  while(1){
    while(AFifo_Get(&inA)){};
    while(BFifo_Get(&inB)){};
    out = compute(inA,inB);
    CFifo_Put(out);
  }
}
```

```
void Process3(void){
int32_t inA, inB, out;
  while(1){
    if(AFifo_Size()==0){
      while(BFifo_Get(&inB)){};
      while(AFifo_Get(&inA)){};
    } else{
      while(AFifo_Get(&inA)){};
      while(BFifo_Get(&inB)){};
    }
    out = compute(inA,inB);
    CFifo_Put(out);
  }
}
```

Program 4.1. Two C implementations of a process on a KPN. The one on the right is optimized.

4.5. Edge-triggered Interfacing

Synchronizing software to hardware events requires the software to recognize when the hardware changes states from busy to done. Many times the busy to done state transition is signified by a rising (or falling) edge on a status signal in the hardware. For these situations, we connect this status signal to an input of the microcontroller, and we use edge-triggered interfacing to configure the interface to set a flag on the rising (or falling) edge of the input. Using edge-triggered interfacing allows the software to respond quickly to changes in the external world. If we are using busy-wait synchronization, the software waits for the flag. If we are using interrupt synchronization, we configure the flag to request an interrupt when set. Each of the digital I/O pins on the TM4C family can be configured for edge triggering. Table 2.14 listed some of the I/O registers associated with the TM4C123. Table 4.2 expands this list to include all the registers available for Port A. The differences between members of the TM4C family include the number of ports (e.g., the MSP432E401Y has ports A – Q, while the TM4C123 has ports A – F) and the number of pins in each port (e.g., the MSP432E401Y has pins 5 – 0 in Port B, while the TM4C123 has pins 7 – 0 in Port B). For more details, refer to the datasheet for your specific microcontroller. Any or all of digital I/O pins can be configured as an edge-triggered input. When writing C code using these registers, include the header file for your particular microcontroller (e.g., **msp432e401y.h**).

To use any of the features for a digital I/O port, we first enable its clock in the **SYSCTL_RCGCGPIO_R**. For each bit we wish to use we must set the corresponding **DEN** (Digital Enable) bit. To use a pin as regular digital input or output, we clear its **AFSEL** (Alternate Function Select) bit. Setting the **AFSEL** will activate the pin's special function (e.g., UART, I^{2}C, CAN etc.) For regular digital input/output, we clear **DIR** (Direction) bits to make them input, and we set **DIR** bits to make them output.

There are four additional registers for the TM4C. We clear bits in the **AMSEL** register to use the port for digital I/O. AMSEL bits exist for those pins which have analog functionality. Which pins have which functionality was shown in Tables 2.4 and 2.5. We set the alternative function using both **AFSEL** and **PCTL** registers. We need to unlock PD7 and PF0 if we wish to use them. Because PC3-0 implements the JTAG debugger, we will never unlock these pins. Pins PC3-0, PD7 and PF0 are the only ones that implement the **CR** bits in their commit registers, where 0 means the pin is locked and 1 means the pin is unlocked. To unlock a pin, we first write 0x4C4F434B to the **LOCK** register, and then we write zeros to the **CR** register.

Address	7	6	5	4	3	2	1	0	Name
$4000.43FC	DATA	DATA	DATA	DATA	DATA	DATA	DATA	DATA	GPIO_PORTA_DATA_R
$4000.4400	DIR	DIR	DIR	DIR	DIR	DIR	DIR	DIR	GPIO_PORTA_DIR_R
$4000.4404	IS	IS	IS	IS	IS	IS	IS	IS	GPIO_PORTA_IS_R
$4000.4408	IBE	IBE	IBE	IBE	IBE	IBE	IBE	IBE	GPIO_PORTA_IBE_R
$4000.440C	IEV	IEV	IEV	IEV	IEV	IEV	IEV	IEV	GPIO_PORTA_IEV_R
$4000.4410	IME	IME	IME	IME	IME	IME	IME	IME	GPIO_PORTA_IM_R
$4000.4414	RIS	RIS	RIS	RIS	RIS	RIS	RIS	RIS	GPIO_PORTA_RIS_R
$4000.4418	MIS	MIS	MIS	MIS	MIS	MIS	MIS	MIS	GPIO_PORTA_MIS_R
$4000.441C	ICR	ICR	ICR	ICR	ICR	ICR	ICR	ICR	GPIO_PORTA_ICR_R
$4000.4420	SEL	SEL	SEL	SEL	SEL	SEL	SEL	SEL	GPIO_PORTA_AFSEL_R
$4000.4500	DRV2	DRV2	DRV2	DRV2	DRV2	DRV2	DRV2	DRV2	GPIO_PORTA_DR2R_R
$4000.4504	DRV4	DRV4	DRV4	DRV4	DRV4	DRV4	DRV4	DRV4	GPIO_PORTA_DR4R_R
$4000.4508	DRV8	DRV8	DRV8	DRV8	DRV8	DRV8	DRV8	DRV8	GPIO_PORTA_DR8R_R
$4000.450C	ODE	ODE	ODE	ODE	ODE	ODE	ODE	ODE	GPIO_PORTA_ODR_R
$4000.4510	PUE	PUE	PUE	PUE	PUE	PUE	PUE	PUE	GPIO_PORTA_PUR_R
$4000.4514	PDE	PDE	PDE	PDE	PDE	PDE	PDE	PDE	GPIO_PORTA_PDR_R
$4000.4518	SLR	SLR	SLR	SLR	SLR	SLR	SLR	SLR	GPIO_PORTA_SLR_R
$4000.451C	DEN	DEN	DEN	DEN	DEN	DEN	DEN	DEN	GPIO_PORTA_DEN_R
$4000.4524	CR	CR	CR	CR	CR	CR	CR	CR	GPIO_PORTA_CR_R
$4000.4528	AMSEL	AMSEL	AMSEL	AMSEL	AMSEL	AMSEL	AMSEL	AMSEL	GPIO_PORTA_AMSEL_R
	31-28	27-24	23-20	19-16	15-12	11-8	7-4	3-0	
$4000.452C	PMC7	PMC6	PMC5	PMC4	PMC3	PMC2	PMC1	PMC0	GPIO_PORTA_PCTL_R
$4000.4520	LOCK (write 0x4C4F434B to unlock, other locks) (reads 1 if locked, 0 if unlocked)								GPIO_PORTA_LOCK_R

Table 4.2. Port A registers for the TM4C.

To configure an edge-triggered pin, we first enable the clock on the port and configure the pin as a regular digital input. Clearing the **IS** (Interrupt Sense) bit configures the bit for edge triggering. If the **IS** bit were to be set, the trigger occurs on the level of the pin. Since most busy to done conditions are signified by edges, we typically trigger on edges rather than levels. Next we write to the **IBE** (Interrupt Both Edges) and **IEV** (Interrupt Event) bits to define the active edge. We can trigger on the rising, falling, or both edges, as listed in Table 4.3. We clear the **IME** (Interrupt Mask Enable) bits if we are using busy-wait synchronization, and we set the **IME** bits to use interrupt synchronization.

DIR	AFSEL	IS	IBE	IEV	IME	Port mode
0	0	0	0	0	0	Input, falling edge trigger, busy wait
0	0	0	0	1	0	Input, rising edge trigger, busy wait
0	0	0	1	-	0	Input, both edges trigger, busy wait
0	0	0	0	0	1	Input, falling edge trigger, interrupt
0	0	0	0	1	1	Input, rising edge trigger, interrupt
0	0	0	1	-	1	Input, both edges trigger, interrupt

Table 4.3. Edge-triggered modes.

The hardware sets an **RIS** (Raw Interrupt Status) bit (called the trigger) and the software clears it (called the acknowledgement). The triggering event listed in Table 4.3 will set the corresponding **RIS** bit in the `GPIO_PORTA_RIS_R` register regardless of whether or not that bit is allowed to request a controller interrupt. In other words, clearing an **IME** bit disables the corresponding pin's interrupt, but it will still set the corresponding **RIS** bit when the interrupt would have occurred. The software can acknowledge the event by writing ones to the corresponding **IC** (Interrupt Clear) bit in the `GPIO_PORTA_IC_R` register. The **RIS** bits are read only, meaning if the software were to write to this registers, it would have no effect. For example, to clear bits 2, 1, and 0 in the `GPIO_PORTA_RIS_R` register, we write a 0x07 to the `GPIO_PORTA_IC_R` register. Writing zeros into **IC** bits will not affect the **RIS** bits.

For input signals we have the option of adding either a pull-up resistor or a pull-down resistor. If we set the corresponding **PUE** (Pull-Up Enable) bit on an input pin, the equivalent of a 50 to 110 kΩ resistor to +3.3 V power is internally connected to the pin. Similarly, if we set the corresponding **PDE** (Pull-Down Enable) bit on an input pin, the equivalent of a 55 to 180 kΩ resistor to ground is internally connected to the pin. We cannot have both pull-up and a pull-down resistor, so setting a bit in one register automatically clears the corresponding bit in the other register.

A typical application of pull-up and pull-down mode is the interface of simple switches. Using these modes eliminates the need for an external resistor when interfacing a switch. Compare the interfaces on Port A to the interfaces on Port B illustrated in Figure 4.19. The Port A interfaces employ software-configured internal resistors, while the Port B interfaces require actual resistors. The PA2 and PB2 interfaces in Figure 4.19a) implement negative logic switch inputs, and the PA3 and PB3 interfaces in Figure 4.19b) implement positive logic switch inputs.

Checkpoint 4.2: What do negative logic and positive logic mean in this context?

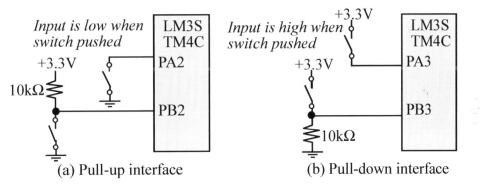

Figure 4.19. Edge-triggered interfaces can generate interrupts on a switch touch.

Checkpoint 4.3: What values to you write into DIR, AFSEL, PUE, and PDE to configure the switch interfaces of PA2 and PA3 in Figure 4.19?

Using edge triggering to synchronize software to hardware centers around the operation of the trigger flags, **RIS**. A busy-wait interface will read the appropriate **RIS** bit over and over, until it is set. When the **RIS** bit is set, the software will clear the **RIS** bit (by writing a one to the corresponding **IC** bit) and perform the desired function. With interrupt synchronization, the initialization phase will arm the trigger flag by setting the corresponding **IME** bit. In this way, the active edge of the pin will set the **RIS** and request an interrupt. The interrupt will suspend the main program and run a special interrupt service routine (ISR). This ISR will clear the **RIS** bit and perform the desired function. At the end of the ISR it will return, causing the main program to resume. In particular, five conditions must be simultaneously true for an edge-triggered interrupt to be requested:

- The trigger flag bit is set (RIS)
- The arm bit is set (IME)
- The level of the edge-triggered interrupt must be less than BASEPRI
- The edge-triggered interrupt must be enabled in the NVIC_EN0_R
- The edge-triggered interrupt must be disabled in the NVIC_DIS0_R
- Bit 0 of the special register PRIMASK is 0

In this chapter we will develop blind-cycle and busy-wait solutions, and then in the next chapter we will redesign the systems using interrupt synchronization.

4.6. Configuring Digital Output Pins

To use a digital port, we must first enable its clock in the **SYSCTL_RCGCGPIO_R** register. Similar to using an input pin, we must set the **DEN** bits and clear the **AFSEL** bits. To make a pin an output, we set the corresponding **DIR** bit. There are a number of choices when configuring digital output pins. The registers for Port A are listed in Table 4.2. The available I_{OH} and I_{OL} for a digital output can be specified with its **DRV2**, **DRV4**, or **DRV8** bits. For each bit, exactly one of the **DRV2**, **DRV4**, or **DRV8** bits is set, specifying the output drive current to be 2 mA, 4 mA, or 8 mA respectively. You can save power by choosing the smallest current required to drive the interface. In particular, determine the I_{IH} and I_{IL} of the device connected to the output pin and make $|I_{OH}| > |I_{IH}|$ and $|I_{OL}| > |I_{IL}|$. The MSP432E401Y microcontroller adds a 12 mA output mode. To activate 12 mA output on PA2 on the MSP432E401Y, we execute

```
GPIO_PORTA_PC_R = (GPIO_PORTA_PC_R&0xFFFFFFCF)+0x0030;
GPIO_PORTA_DR4R_R  |= 0x02;   // 2mA
GPIO_PORTA_DR8R_R  |= 0x02;   // +4mA more
GPIO_PORTA_DR12R_R |= 0x02;   // +4mA more
```

An output pin can be configured as **open drain**, which is similar to open collector, by setting the **ODE** (Open Drain Enable) bit for the pin. In open drain mode, the output states are zero and off. In particular, if we output a 0, the pin will go low (V_{OL}, I_{OL}). If we output a 1, the pin will float, which is neither high nor low. In the floating state, the output will not source or sink any current. We can use open drain mode for interfacing negative logic LEDs and to create a multi-drop network.

An additional configuration bit for output pins is the slew rate control. The **slew rate** of an output signal is defined as the maximum dV/dt occurring when the output is switching. For most digital interfaces, we want the fastest possible slew rate so the digital output spends as little time in the transition range as possible. However, for some network interfaces we will want to limit the slew rate to prevent noise coupling across the wires of a cable. When the output pin is configured at 8 mA drive, we can set **SRL** (Slew Rate Limit) bits in the **GPIO_PORTA_SRL_R** register to limit the slew rate on the pin. When the **SRL** bit is 0, an 8-mA output will rise from 20% to 80% of its voltage in [6, 9 ns] and will fall from 80% to 20% in [6, 10 ns]. When the **SRL** bit is 1, the rise and fall times increase to [10, 12 ns] and [11, 13 ns] respectively.

4.7. Blind-cycle Interfacing

The basic approach for blind-cycle synchronization is to issue an I/O command then wait a fixed time delay for the operation to complete. Its advantage is simplicity. It is appropriate for I/O devices that are fast and predicable. It is robust because it cannot crash (i.e., never returning). Unfortunately, there are several disadvantages of blind-cycle synchronization. If the output rate is variable (like a "carriage return", "tab", "graphics", or "form feed") then this technique is awkward. If the input rate is unknown (like a keyboard) this technique is inappropriate. The delay represents wasted time. For a simple system, this waste usually doesn't matter. But for a more complex system, this time could be used by the software to perform other operations.

4.7.1. HD44780-controlled LCD

Because there is no status feedback from the device, if the device is missing or broken the software will not know. Nevertheless, blind cycle counting can be appropriate for simple high-speed interfaces. In this section, we will use the blind cycle method to interface an LCD because most operations will complete in less than 40 µs. Liquid crystal displays operate at low power and have flexible graphics. Many 5V-powered LCDs use an industry standard HD44780S controller. The HD44780U and ST7066U LCD drivers shown in Figure 4.20 work in a similar way but operate at a supply voltage of 3.3 V. The low-level software initializes and outputs to the LCD controller. The microcontroller simply writes ASCII characters to the LCD controller. Each ASCII character is mapped into a 5 by 8 bit pixel image, called a **font**. A 1 by 8 LCD is 40 pixels wide by 8 pixels tall, and the LCD controller is responsible for refreshing the pixels in a raster-scanned manner similar to maintaining an image on a TV screen.

There are four types of access cycles to the LCD controller depending on RS and R/W as shown in Table 4.4. Normally, you write ASCII characters into the data buffer (called DDRAM in the data sheets) to have them displayed on the screen. However, you can create up to 8 new characters by writing to the character generator RAM (CGRAM). These new characters exist as ASCII data 0 to 7.

RS	R/W	Operation
0	0	Write a command to the LCD instruction register
0	1	Read Busy Flag (bit 7)
1	0	Write data to the LCD data buffer
1	1	Read data from the LCD to the microcontroller

Table 4.4. Two control signals specify the type of access to the LCD.

We could use either blind-cycle or busy-wait synchronization. Most operations require 37 µs to complete while some require 1.52 ms. Programs 4.2 and 4.3 use the SysTick timer to create the blind-cycle wait. Busy-wait synchronization would have provided feedback to detect a faulty interface, but busy wait has the problem of creating a software crash if the LCD never finishes. A better interface would have utilized both busy wait and blind-cycle, so that the software can return with an error code if a display operation does not finish on time (due to a broken display). With blind-cycle synchronization we only use the write command and write data operations. With busy-wait, we would need to connect the **R/W** line to a GPIO output.

216 ■ 4. Hardware-Software Synchronization

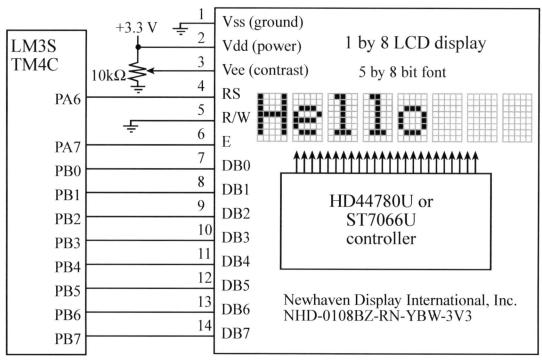

Figure 4.20. Interface of an LCD.

First, we present a low-level private helper function, see Program 4.2. This function would not have a prototype in the **LCD.h** file. The define macros specify the mapping from the logic name to the physical output pin. As shown in Table 4.4, an output command is created with R/W=0 and RS=0. To output a command to the LCD, we first write the 8-bit command to the data lines, then we pulse the **E** pin (**E** = 1, **E**=0.)

```
#define E  0x80 // on PA7
#define RS 0x40 // on PA6
#define LCDDATA (*((volatile uint32_t *)0x400053FC)) // PORTB
#define LCDCMD  (*((volatile uint32_t *)0x40004300))  // PA7-PA6
#define BusFreq 80              // assuming a 80 MHz bus clock
#define T500ns BusFreq/2        // 500ns
#define T40us 40*BusFreq        // 40us
#define T160us 160*BusFreq      // 160us
#define T1600us 1600*BusFreq    // 1.60ms
#define T5ms 5000*BusFreq       // 5ms
#define T15ms 15000*BusFreq     // 15ms
// send a command byte to the LCD
void static OutCmd(uint8_t command){
  LCDDATA = command;
  LCDCMD = 0;           // E=0, R/W=0, RS=0
  SysTick_Wait(T500ns); // wait 500ns
```

```
    LCDCMD = E;              // E=1, R/W=0, RS=0
    SysTick_Wait(T500ns);    // wait 500ns
    LCDCMD = 0;              // E=0, R/W=0, RS=0
    SysTick_Wait(T40us);     // wait 40us
}
```
Program 4.2. Private functions for an LCD. (LCD_xxx). Change 500ns to 6us for a +5V LCD.

Figure 4.21 shows a rough sketch of the E, RS, R/W and data signals as the **OutCmd** function is executed. As time advances, the program executes from top to bottom, and the output signals in the timing diagram progress left to right. The 500-ns waits are to satisfy the setup and hold times for the LCD, and the 40-μs wait is the blind-cycle delay to allow the command to complete. The clear and home commands will require a 1.6-ms blind wait.

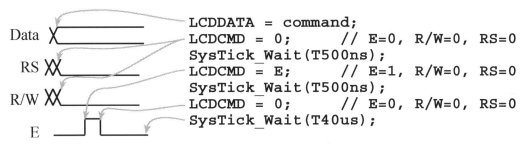

Figure 4.21. Timing diagram of the LCD signals as a command is sent to the LCD.

```
// Initialize LCD
void LCD_Init(void){
  SYSCTL_RCGCGPIO_R |= 0x03;        // 1) activate clock for Ports A and B
  while((SYSCTL_PRGPIO_R&0x03) != 0x03){};// ready?
  GPIO_PORTB_AMSEL_R &= ~0xFF;      // 3) disable analog function on PB7-0
  GPIO_PORTA_AMSEL_R &= ~0xC0;      //    disable analog function on PA7-6
  GPIO_PORTB_PCTL_R = 0x00000000;   // 4) configure PB7-0 as GPIO
  GPIO_PORTA_PCTL_R &= ~0xFF000000; //    configure PA7-6 as GPIO
  GPIO_PORTB_DIR_R = 0xFF;          // 5) set direction register
  GPIO_PORTA_DIR_R |= 0xC0;
  GPIO_PORTB_AFSEL_R = 0x00;        // 6) regular port function
  GPIO_PORTA_AFSEL_R &= ~0xC0;
  GPIO_PORTB_DEN_R = 0xFF;          // 7) enable digital port
  GPIO_PORTA_DEN_R |= 0xC0;
  GPIO_PORTB_DR8R_R = 0xFF;         // enable 8 mA drive
  GPIO_PORTA_DR8R_R |= 0xC0;
  SysTick_Init();          // Program 2.10
  LCDCMD = 0;              // E=0, R/W=0, RS=0
  SysTick_Wait(T15ms);     // see datasheet for specific wait time
  OutCmd(0x30);            // command 0x30 = Wake up
  SysTick_Wait(T5ms);      // must wait 5ms, busy flag not available
  OutCmd(0x30);            // command 0x30 = Wake up #2
  SysTick_Wait(T160us);    // must wait 160us, busy flag not available
  OutCmd(0x30);            // command 0x30 = Wake up #3
```

```
  SysTick_Wait(T160us); // must wait 160us, busy flag not available
  OutCmd(0x38);         // Function set: 8-bit/2-line
  OutCmd(0x10);         // Set cursor
  OutCmd(0x0C);         // Display ON; Cursor ON
  OutCmd(0x06);}        // Entry mode set
// Inputs: letter is ASCII character, 0 to 0x7F    Outputs: none
void LCD_OutChar(char letter){
  LCDDATA = letter;
  LCDCMD = RS;          // E=0, R/W=0, RS=1
  SysTick_Wait(T500ns); // wait 500ns
  LCDCMD = E+RS;        // E=1, R/W=0, RS=1
  SysTick_Wait(T500ns); // wait 500ns
  LCDCMD = RS;          // E=0, R/W=0, RS=1
  SysTick_Wait(T40us);  // wait 40us
}
// Inputs: none                Outputs: none
void LCD_Clear(void){
  OutCmd(0x01);          // Clear Display
  SysTick_Wait(T1600us); // wait 1.6ms
  OutCmd(0x02);          // Cursor to home
  SysTick_Wait(T1600us);} // wait 1.6ms
```

Program 4.3. Public functions for the LCD (LCD_xxx). Change 500ns to 6us for a +5V LCD.

The high-level public functions are shown in Program 4.3. These functions would have prototypes in the **LCD.h** file. The initialization sequence is copied from the data sheet of the LCD controller. Figure 4.22 shows a rough sketch of the E, RS, R/W and data signals as the **LCD_OutChar** function is executed. This interface can operate with an LCD powered with +5 V. For some LCDs, the V_{OH} of the TM4C is not quite high enough for the V_{IH} of the LCD when the LCD is powered with +5 V. In these cases, the **T500ns** delays need to be increased by a factor of 10.

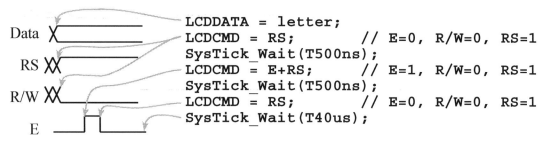

Figure 4.22. Timing diagram of the LCD signals as data is sent to the LCD.

4.7.2. Stepper Motor Interface

Stepper motors are very popular for microcontroller-controlled machines because of their inherent digital interface. It is easy for a microcontroller to control both the position and velocity of a stepper motor in an open-loop fashion. Although the cost of a stepper motor is typically higher than an equivalent DC permanent magnetic field motor, the overall system cost is reduced because stepper motors may not require feedback sensors. For these reasons, they are used in many computer peripherals such as hard disk drives, scanners, and printers. Figure 4.23 shows a stepper motor is made with EM coils and permanent magnets (teeth).

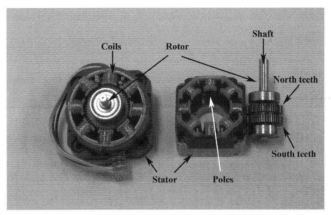

Figure 4.23. Stepper motors have permanent magnets on the rotor and electromagnetics around the stator.

Figure 4.24 shows a simplified stepper motor. The permanent magnet stepper has a rotor and a stator. The **rotor** is manufactured from a gear-shaped permanent magnet. This simple rotor has one North tooth and one South tooth. North and South teeth are equally spaced and offset from each other by half the tooth pitch as illustrated by the rotor with 12 teeth. The **stator** consists of multiple iron-core electromagnets whose poles are also equally spaced. The stator of this simple stepper motor has four electromagnets and four poles. The stepper motors in Figure 4.23 have 50 North teeth and 50 South teeth, resulting in motors with 200 steps per revolution. The stator of a stepper motor with 200 steps per revolution has eight electromagnets each with five poles, making a total of 40 poles.

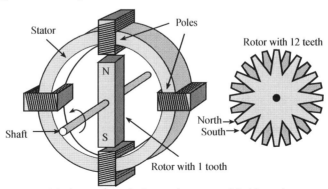

Figure 4.24. Simple stepper motor with 4 steps/revolution and a rotor with 12 teeth.

The operation of this simple stepper motor is illustrated in Figure 4.25. In general, if there are n North teeth and n South teeth, the shaft will rotate $360°/(4 \cdot n)$ per step. For this simple motor in Figure 4.25, each step causes $90°$ rotation.

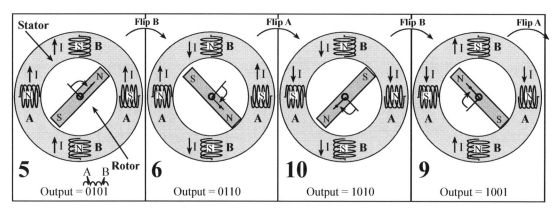

Figure 4.25. The full-step sequence to rotate a stepper motor.

Output=0101. Assume the initial position is the one on the left with the output equal to 0101. There are strong attractive forces between North and South magnets. This is a stable state because the North tooth is equally positioned between the two South electromagnets, and the South tooth is equally positioned between the two North electromagnets. There is no net torque on the shaft, so the motor will stay fixed at this angle. In fact, if there is an attempt to rotate the shaft with an external torque, the stepper motor will oppose that rotation and try to maintain the shaft angle fixed at this position. In fact, stepper motors are rated according to their holding torque. Typical holding torques range from 10 to 300 oz·in.

Output=0110. When the software changes the output to 0110, the polarity of Phase B is reversed. The rotor is in an unstable state, because the North tooth is near the North electromagnet on the top, and the South tooth is near the South electromagnet on the bottom. The rotor will move because there are strong repulsive forces from the top and bottom poles. By observing the left and right poles, the closest stable state occurs if the rotor rotates clockwise, resulting in the stable state illustrated as the picture in Figure 4.25 labeled "Output=0110". The "Output=0110" state is exactly 90° clockwise from the "Output=0101" state. Now, once again the North tooth is near the South poles and the South tooth is near the North poles. This new position has strong attractive forces from all four poles, holding the rotor at this new position.

Output=1010. Next, the software outputs 1010, causing the polarity of Phase A to be reversed. This time, the rotor is in an unstable state because there are strong repulsive forces on the left and right poles. The closest stable state occurs if the rotor rotates clockwise, resulting in the stable state illustrated in the picture Figure 4.25 labeled as "Output=1010". This new state is exactly 90° clockwise from the last state, moving to position the North tooth near the South poles and the South tooth near the North poles.

Output=1001. When the software outputs 1001, the polarity of Phase B is reversed. This causes a repulsive force on the top and bottom poles and the rotor rotates clockwise again by 90°, resulting in the stable state shown as the picture labeled "Output=1001". After each change in software, there are two poles that repel and two poles that attract, causing the shaft to rotate. The rotor moves until it reaches a new stable state with the North tooth close to South poles and the South tooth close to North poles. When the software outputs a 0101, it will rotate 90° resulting in a position similar to the original "Output=0101" state. If the software outputs a new value from the 5,6,10,9 sequence every 250 ms, the motor will spin clockwise at 1 rps. The rotor will spin in a counterclockwise direction if the sequence is reversed.

There is an eight-number sequence called **half-stepping**. In full stepping, the direction of current in one of the coils is reversed in each step. In half-stepping, the coil goes through a no-current state between reversals. The half-stepping sequence is 0101, 0100, 0110, 0010 1010, 1000, 1001, and 0001. If a coil is driven with the 00 command, it is not energized, and the electromagnet applies no force to the rotor. A motor that requires 200 full steps to rotate once will require 400 half-steps to rotate once. In other words, the half-step angle is ½ of a full-step angle.

In a four-wire (or **bipolar**) stepper motor, the electromagnets are wired together, creating two phases. The five- and six-wire (or **unipolar**) stepper motors also have two phases, but each is center-tapped to simplify the drive circuitry. In a bipolar stepper all copper in the windings carries current at all times; whereas in a unipolar stepper, only half the copper in the windings is used at any one time.

To spin the stepper motor at a constant speed the software outputs the 5–6–10–9 sequence separated by a fixed time between outputs. This is an example of blind-cycle interfacing, because there is no feedback from the motor to the software giving the actual speed and/or position of the motor. If Δt is the time between outputs in seconds, and the motor has n steps per revolution, the motor speed will be $60/(n*\Delta t)$ in RPM.

The time between states determines the **rotational speed** of the motor. Let Δt be the time between steps, and let θ be the step angle then the rotational velocity, v, is $\theta/\Delta t$. As long as the load on the shaft is below the holding torque of the motor, the position and speed can be reliably maintained with an open loop software control algorithm. In order to prevent slips (digital commands that produce no rotor motion) it is important to limit the change in acceleration or **jerk**. Let $\Delta t(n-2)$, $\Delta t(n-1)$, $\Delta t(n)$ be the discrete sequence of times between steps. The instantaneous rotational **velocity** is given by

$v(n) = \theta/\Delta t$

The **acceleration** is given by

$a(n) = (v(n) - v(n-1))/\Delta t(n) = (\theta/\Delta t(n) - \theta/\Delta t(n-1))/\Delta t(n) = \theta/\Delta t(n)^2 - \theta/(\Delta t(n-1)\Delta t(n))$

The change in acceleration, or jerk, is given by

$b(n) = (a(n) - a(n-1))/\Delta t(n)$

For example if the time between steps is to be increased from 1000 to 2000 µs, an ineffective approach as shown in Table 4.5 would be simply to go directly from 1000 to 2000. This produces a very large jerk that may cause the motor to slip.

n	Δt (μs)	$v(n)$ (°/sec)	$a(n)$ (°/sec^2)	$b(n)$ (°/sec^3)
1	1000	1800		
2	1000	1800	0.00E+00	
3	2000	900	-2.50E+05	0.00E+00
4	2000	900	0.00E+00	-2.25E+08
5	2000	900	0.00E+00	2.25E+08
6	2000	900	0.00E+00	0.00E+00

Table 4.5. An ineffective approach to changing motor speed.

Table 4.6 shows that a more gradual change from 1000 to 2000 produces a 10 times smaller jerk, reducing the possibility of slips. The optimal solution (the one with the smallest jerk) occurs when $v(t)$ has a quadratic shape. This will make $a(t)$ linear, and $b(t)$ a constant. Limiting the jerk is particularly important when starting to move a stopped motor.

n	Δt (μs)	$v(n)$ (°/sec)	$a(n)$ (°/sec^2)	$b(n)$ (°/sec^3)
1	1000	1800		
2	1000	1800	0.00E+00	
3	1000	1800	0.00E+00	0.00E+00
4	1008	1786	-1.39E+04	-1.38E+07
5	1032	1744	-4.11E+04	-2.64E+07
6	1077	1671	-6.77E+04	-2.47E+07
7	1152	1563	-9.37E+04	-2.26E+07
8	1275	1411	-1.19E+05	-1.96E+07
9	1500	1200	-1.41E+05	-1.48E+07
10	1725	1044	-9.06E+04	2.91E+07
11	1848	974	-3.78E+04	2.86E+07
12	1923	936	-1.96E+04	9.44E+06
13	1968	915	-1.09E+04	4.43E+06
14	1992	904	-5.65E+03	2.63E+06
15	2000	900	-1.77E+03	1.94E+06
16	2000	900	0.00E+00	8.85E+05
17	2000	900	0.00E+00	0.00E+00

Table 4.6. An effective approach to changing motor speed.

The bipolar stepper motor can be controlled with two H-bridge drivers. We could design two H-bridge drivers using individual transistors, but using an integrated driver usually provides better performance at lower cost. In this section, we will present stepper motor interfaces that employ integrated circuits to control the stepper motor. In addition to the devices presented in this section, Allegro, Texas Instruments, and ST Microelectronics offer a variety of stepper motor controllers, e.g., A3972, A3980, UCN5804B, and L6203. The L293 is a simple IC for interfacing stepper motors. It uses Darlington transistors in a double H-bridge configuration (as shown in Figure 4.26), which can handle up to 1 A per channel and voltages from 4 to 36 V. A similar H-bridge interface chip that can drive 5 A is the L6203 from STMicroelectronics.

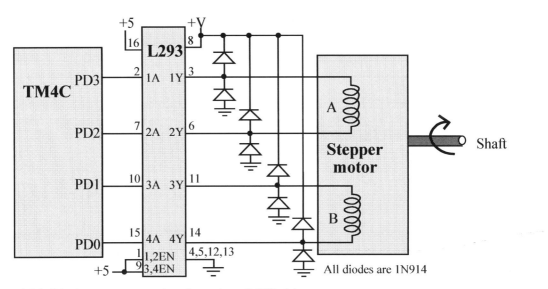

Figure 4.26. Bipolar stepper motor interface using a L293 driver.

The 1N914 snubber diodes protect the electronics from the back EMF generated when currents are switched on and off. The L293D has internal snubber diodes, but can handle only 600 mA. Figure 4.26 shows four digital outputs from the microcontroller connected to the **1A**, **2A**, **3A**, **4A** inputs. The software rotates the stepper motor using either the standard full-step (5–6–10–9...) or half-step (5–4–6–2–10–8–9–1...) sequence.

The unipolar stepper architecture provides for bi-directional currents by using a center tap on each phase, as shown in Figure 4.27. The center tap is connected to the +V power source and the four ends of the phases are controlled with drivers in the L293. Only half of the electromagnets are energized at one time. The L293 provides up to 1A current.

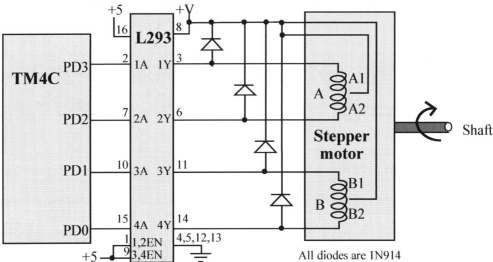

Figure 4.27. Unipolar stepper motor interface.

Checkpoint 4.4: What changes could you make to a stepper motor system to increase torque, increasing the probability that a step command actually rotates the shaft?

Checkpoint 4.5: Do you need a sensor feedback to measure the shaft position when using a stepper motor?

Example 4.1: Interface a 12-V, 200-mA unipolar stepper motor. The motor has 200 steps per revolution. Write software that can set the motor angle.

Solution: The computer can make the motor spin by outputting the sequence ...,10,9,5,6,10,9,5,6,... over and over. For a motor with 200 steps/revolution each new output will cause the motor to rotate 1.8°. If the time in between outputs is fixed at Δt seconds, then the shaft rotation speed will be $0.005/\Delta t$ in rps. In each system, we will connect the stepper motor to the least significant bits of output Port D using Figure 4.27, with the motor voltage $+V$ set to 12 V. A bipolar stepper uses Figure 4.26. The 1N914 diodes will protect the L293 from the back EMF that will develop across the coil when the current is shut off. We define the active state of the coil when current is flowing. The basic operation is summarized in Table 4.7.

Port D output	A1	A2	B1	B2
10	Activate	Deactivate	Activate	Deactivate
9	Activate	Deactivate	Deactivate	Activate
5	Deactivate	Activate	Deactivate	Activate
6	Deactivate	Activate	Activate	Deactivate

Table 4.7. Stepper motor sequence.

We will implement a linked-list data structure to hold the output patterns (Figure 4.28 and Program 4.4). This approach yields a solution that is easy to understand and change. If the computer outputs the sequence backwards then the motor will spin in the other direction. In ensure proper operation, this ...,10,9,5,6,10,9,5,6,... sequence must be followed. For example, assume the computer outputs ..., 9, 5, 6, 10, and 9. Now it wishes to reverse direction, since the output is already at 9, then it should begin at 10, and continue with 6, 5, 9, ... In other words if the current output is "9" then the only two valid next outputs would be "5" if it wanted to spin clockwise or "10" if it wanted to spin counterclockwise. Maintaining this proper sequence will be simplified by implementing a double circular linked-list. For each node in the linked-list there are two valid next states depending upon whether the computer wishes to spin clockwise or counterclockwise.

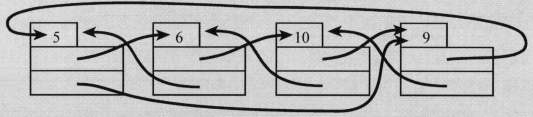

Figure 4.28. A double circular linked-list used to control the stepper motor.

A **slip** is when the computer issues a sequence change, but the motor does not move. A slip can occur if the load on the shaft exceeds the available torque of the motor. A slip can also occur if the computer tries to change the outputs too fast. If the system knows the initial shaft angle, and the motor never slips, then the computer can control both the shaft speed and angle without a position sensor. The routines **CW** and **CCW** will step the motor once in the clockwise and counterclockwise directions respectively. If every time the computer calls **CW** or **CCW** it were to wait for 5 ms, then the motor would spin at 1 rps. The linked data structure will be stored in flash ROM. The variables are allocated into RAM and initialized at run time in the ritual.

```
struct State{
  uint8_t Out;                    // Output
  const struct State *Next[2];    // CW/CCW
};
typedef const struct State StateType;
typedef StateType *StatePtr;
#define clockwise 0              // Next index
#define counterclockwise 1       // Next index
StateType fsm[4]={
  {10,{&fsm[1],&fsm[3]}},
  { 9,{&fsm[2],&fsm[0]}},
  { 5,{&fsm[3],&fsm[1]}},
  { 6,{&fsm[0],&fsm[2]}}
};
uint8_t Pos;         // between 0 and 199
const struct State *Pt; // Current State
```

Program 4.4. A double circular linked list used to control the stepper motor.

The programs that step the motor also maintain the position in the global, **Pos**. If the motor slips, then the software variable will be in error. Also it is assumed the motor is initially in position 0 at the time of the initialization (Program 4.5).

```
#define STEPPER  (*((volatile uint32_t *)0x4000703C))
void Stepper_CW(uint32_t delay){ // Move 1.8 degrees clockwise
  Pt = Pt->Next[clockwise];      // circular
  STEPPER = Pt->Out; // step motor
  if(Pos==199){      // shaft angle
    Pos = 0;         // reset
  }
  else{
    Pos++; // CW
  }
  SysTick_Wait(delay);
}
void Stepper_CCW(uint32_t delay){  // Move 1.8 deg counterclockwise
  Pt = Pt->Next[counterclockwise]; // circular
  STEPPER = Pt->Out; // step motor
  if(Pos==0){        // shaft angle
    Pos = 199;       // reset
  }
  else{
    Pos--; // CCW
  }
  SysTick_Wait(delay); // blind-cycle wait
}
```

```c
void Stepper_Init(void){ // Initialize Stepper interface
  SYSCTL_RCGCGPIO_R |= 0x08;   // activate port D
  SysTick_Init();              // program 2.10
  Pos = 0; Pt = &fsm[0];
  GPIO_PORTD_AFSEL_R &= ~0x0F;     // GPIO function on PD3-0
  GPIO_PORTD_AMSEL_R &= ~0x0F;     // disable analog function on PD3-0
  GPIO_PORTD_PCTL_R &= ~0x0000FFFF;   // configure PD3-0 as GPIO
  GPIO_PORTD_DIR_R |= 0x0F;    // make PD3-0 out
  GPIO_PORTD_DEN_R |= 0x0F;    // enable digital I/O on PD3-0
  GPIO_PORTD_DR8R_R |= 0x0F;} // enable 8 mA drive
```

Program 4.5. Helper and initialization functions used to control the stepper motor.

In Program 4.6, the software will step the motor to the **desired** position. We can use the current position, **pos**, to determine if it would be faster to go CW or CCW. **CWsteps** is calculated as the number of steps from **pos** to **desired** if the motor where to spin clockwise. If it is greater than 100, then it would be faster to get there going counterclockwise.

```c
// Turn stepper motor to desired position (0 <= desired <= 199)
// time is the number of bus cycles to wait after each step
void Stepper_Seek(uint8_t desired, uint32_t time){
int16_t CWsteps;
  if((CWsteps = (desired-Pos))<0){
    CWsteps += 200;
  } // CW steps is 0 to 199
  if(CWsteps > 100){
    while(desired != Pos){
      Stepper_CCW(time);
    }
  }
  else{
    while(desired != Pos){
      Stepper_CW(time);
    }
  }
}
```

Program 4.6. High-level function to control the stepper motor (Stepper_xxx).

There are nine considerations when **selecting a stepper motor**: speed, torque, holding torque, bipolar/unipolar, voltage, current, steps/rotation, size, and weight. The first two parameters are speed and torque. Speed is the rate in rotations per minute (RPM) that the motor will spin, and **torque** is the available force times distance the stepper motor can provide at that speed. Stepper motors also have a **holding torque**, which is the force times distance that the motor will remain stopped when the input pattern is constant. We select the motor voltage to match the available power supply. Unlike LEDs, we MUST NOT use a resistor in series with a motor to reduce the voltage. This is because the impedance ($Z=V/I$) across a motor coil is not constant. In general, we choose the power supply voltage to match the needed motor voltage. We also need to know maximum current. We choose a bipolar stepper for situations where speed, torque, and efficiency are important. Unipolar steppers are appropriate for low-cost systems. If we are trying to control shaft angle it will be better to have a motor with more steps per rotation.

4.8. Busy-Wait Synchronization

To synchronize the software with the I/O device, the microcontroller must be able to recognize the **busy to done** transition. With **busy-wait** synchronization, the software checks a status bit in the I/O device and loops back until the device is ready. Another name for busy-wait is gadfly. The busy-wait loop must precede the data transfer for an input device, but for an output device the busy-wait loop can be either before or after the data transfer (Figure 4.29).

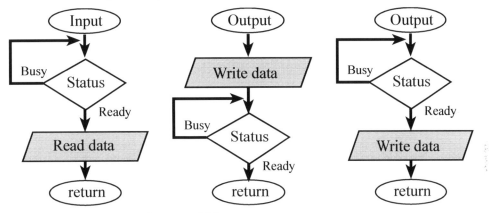

Figure 4.29. Software flowcharts for busy-wait I/O.

The term **gadfly** was chosen as an alternate to busy wait, because it has a negative context outside the computer field. Plato in his book <u>Apology</u> describes Socrates as a gadfly, meaning he is a constant and annoying pest to the Athenian political scene. In biology, a gadfly is a large bug in the horse-fly family. A social gadfly is an irritating person who constantly upsets the status quo. As we will learn in this book, interrupt synchronization will provide more elegant and effective solutions over gadfly synchronization for complex systems when considering issues such as latency, bandwidth, priority, and low-power.

Observation: Busy-wait is appropriate for simple systems that have only a few tasks.

To perform an output the software must transfer the data and wait for completion. The transfer usually executes in a short amount of time because it involves just a few instructions with no backward jumps. On the other hand, the time waiting until the output device is ready is usually long compared to the data transfer. The interface is **I/O bound** when the time for the hardware to perform the output is long compared to the time for the software to produce the data. These two steps can be performed in either order, as long as that order is consistently maintained, and we assume the device is initially ready. Polling before the output, allows the computer to perform additional tasks while the output is occurring. Therefore, polling before the output will have a higher bandwidth than polling after the output. On the other hand, polling after the output, allows the computer to know exactly when the output has been completed.

To illustrate the differences between polling before and after the "write data" operation, consider a system with 3 printers (Figure 4.30). Each printer can print a character in 1ms. In other words, a printer will be ready 1 ms after the "write data" operation. We will also assume all three printers are initially ready. Since the execution speed of the microcontroller is fast compared to the 1 ms it takes to print a character, we will neglect the software execution time

(I/O bound). In the "busy wait before output" system, all three outputs are started together and will operate concurrently. In the "busy wait after output" system, the software waits for the output on printer 1 to finish before starting the output on printer 2. In this system, the three outputs are performed sequentially, that is about three times slower than the first case.

Time(ms)	Busy wait before output	Busy wait after output
0	Start 1,2,3	Start 1
from 0 to 1	Wait for 1	Wait for 1
1	Start 1,2,3	Start 2
from 1 to 2	Wait for 1	Wait for 2
2	Start 1,2,3	Start 3
from 2 to 3	Wait for 1	Wait for 3
3	Start 1,2,3	Start 1
from 3 to 4	Wait for 1	Wait for 1
4	Start 1,2,3	Start 2
from 4 to 5	Wait for 1	Wait for 2

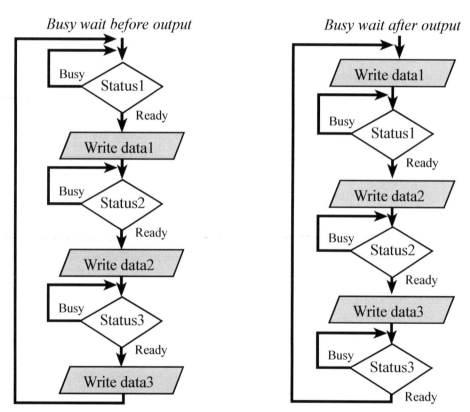

Figure 4.30. Software flowcharts showing busy-wait before output is faster than busy-wait after synchronization.

Performance Tip: Whenever we can establish concurrent I/O operations, we can expect an improvement in the overall system bandwidth.

To implement busy-wait synchronization with multiple I/O devices, simply poll them in sequence and perform service as required. The example in Figure 4.31 implements a fixed polling order, and does not allow high priority devices to suspend the service of lower priority devices. Consider the **interface latency** for device 1, which is the time between when **Status1** become ready until the time the software performs the input/output service on **data1**. The worst case would be for the **Status1** signal to become ready right after the software polled it. For this worst case scenario, the software might have to service device 2, service device 3, and execute other functions before polling device 1 again. When there are multiple I/O devices, the interface latency with busy-wait synchronization will be poor.

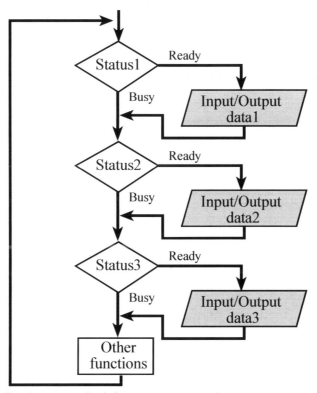

Figure 4.31. A software flowchart for multiple busy-wait inputs and outputs.

A parallel interface encodes the information as separate binary bits on individual port pins, and the information is available simultaneously. For example if we are transmitting ASCII characters, then the ASCII code is represented by digital voltages on 7 (or 8) digital signals. When interfacing a device to a microcontroller, a handshaked interface provides a mechanism for the device to wait for the microcontroller and for the microcontroller to wait for the device.

The device driver for the Kentec EB-LM4F120-L35 graphics module uses a parallel interface, see the example **Kentec_4C123** on the book web site.

Example 4.2: Design a one-directional communication channel between two microcontrollers using a handshaked protocol.

Solution: Handshaking allows each device to wait for the other. This is particularly important when communicating between two computers. The right side of Figure 4.32 shows the timing as one byte is transferred from transmitter to receiver. The key to designing a handshaked protocol is to require each side to wait until the other side is ready. The transmitter will begin communication by putting new data on the **Data** lines and issuing a rising edge on **Ready**. Furthermore, notice the data is properly driven whenever **Ready** is high. Next, the receiver signals it is about to read the data by making its acknowledge, called **Ack**, low. The receiver then reads the result while **Ack** is low. Finally, the receiver acknowledges acceptance of the data by outputting a rising edge on **Ack**, signaling it is ready to accept another. The handshake ends with a falling edge on **Ready**.

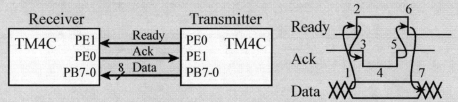

Figure 4.32. Handshaked interface between two microcontrollers.

The initialization for the transmitter is given in Program 4.7. **PE1** is configured as an input. **PE0** and Port B are digital outputs. The initialization for the receiver is given in Program 4.8. **PE1** and Port B are configured as inputs. **PE0** is a simple digital output.

```
#define Ready (*((volatile uint32_t *)0x40024004))  // PE0
#define Ack   (*((volatile uint32_t *)0x40024008))  // PE1
#define Data  (*((volatile uint32_t *)0x400053FC))  // Port B
void Xmt_Init(void){ volatile uint32_t delay;
  SYSCTL_RCGCGPIO_R |= 0x42;  // activate clock for Ports B,E
  while((SYSCTL_PRGPIO_R&0x42) != 0x42){};// ready?
  Ready = 0;
  GPIO_PORTE_DIR_R |= 0x01;  // PE0 is Ready out
  GPIO_PORTE_DIR_R &= ~0x02; // PE1 is Ack in
  GPIO_PORTE_DEN_R |= 0x03;  // enable digital I/O on PE1-0
  GPIO_PORTB_DIR_R = 0xFF;   // make PB7-0 out data
  GPIO_PORTB_DEN_R = 0xFF;   // enable digital I/O on PB7-0
}
```
Program 4.7. Initialization of the transmitter.

```
#define Ack   (*((volatile uint32_t *)0x40024004))  // PE0
#define Ready (*((volatile uint32_t *)0x40024008))  // PE1
#define Data  (*((volatile uint32_t *)0x400053FC))  // Port B
void Rcv_Init(void){
  SYSCTL_RCGCGPIO_R |= 0x42;  // activate clock for Ports B,E
  while((SYSCTL_PRGPIO_R&0x42) != 0x42){};// ready?
  Ack = 1;                    // allow time to finish activating
```

```
  GPIO_PORTE_DIR_R |= 0x01;    // PE0 is Ack out
  GPIO_PORTE_DIR_R &= ~0x02;   // PE1 is Ready in
  GPIO_PORTE_DEN_R |= 0x03;    // enable digital I/O on PE1-0
  GPIO_PORTB_DIR_R  = 0x00;    // make PB7-0 input data
  GPIO_PORTB_DEN_R  = 0xFF;    // enable digital I/O on PB7-0
}
```
Program 4.8. Initialization of the receiver.

This interface is called handshaked or interlocked because each event (1 to 7) follows in sequence one event after the other (Program 4.9). The arrows in Figure 4.32 represent causal events. In a handshaked interface, one event causes the next to occur, and the arrows create a "head to tail" sequence. There is no specific minimum or maximum time delay for these causal events, except they must occur in sequence.

1 Transmitter outputs new **Data**
2 Transmitter makes a rising edge of **Ready** signifying new **Data** available
3 Receiver makes a falling edge of **Ack** signifying it is starting to process data
4 The receiver reads the **Data**
5 Receiver makes a rising edge on **Ack** signifying it has captured the **Data**
6 Transmitter makes a falling edge on **Ready** meaning **Data** is not valid
7 Transmitter no longer needs to maintain **Data** on its outputs

```
// Transmitter                      // Receiver
void Out(uint8_t data){             uint8_t In(void){
  Data = data;      // 1)             uint8_t data;
  Ready = 1;        // 2)             while(Ready == 0){};  // 2)
  while(Ack){};     // 3)             Ack = 0;              // 3)
  while(Ack == 0){};// 5)             data = Data;          // 4)
  Ready = 0;        // 6)             Ack = 1;              // 5)
}                                     while(Ready){};       // 6)
                                      return(data);}
```
Program 4.9. Handshaking routines to initialize and transfer data.

One of the issues involved in handshaked interfaces is whether the receiver should wait for both steps 2 and 6 or just step 2. Similarly, should the transmitter wait for both steps 3 and 5? It is a more robust design to affect all four of these waits. If the receiver did not wait for step 6 (**Ready**=0), then a subsequent call to **In** may find **Ready** still high from the last call, and return the same **Data** a second time. We could have employed edge triggering, but since the interface is interlocked, there is no need to capture the edges. Edge-triggered mode on the PE1 would be required if interrupt synchronization were desired. Program 4.9 shows the solution using busy-wait synchronization.

Checkpoint 4.6: Assume both computers in Program 4.9 have executed the initialization. Assume Out is called before In. List the time-sequenced execution in both computers. Repeat for the case that In is called before Out.

Observation: Programs written for embedded computers are tightly coupled (depend highly) on the hardware, therefore it is good programming practice to document the hardware configuration in the software.

Performance tip: When initializing output pins, it is better to first write the desired initial output value, and then set the direction register to output. This way the pin goes from input to output of the correct value, rather than from input to output of the wrong value, and then output with the correct value.

Handshaking is a very reliable synchronization method when connecting devices from different manufacturers and different speeds. It also allows you to upgrade one device (e.g., get a newer and faster sensor) without redesigning both sides of the interface. Handshaking is used for the Small Computer Systems Interface (SCSI) and the IEEE488 instrumentation bus.

4.9. UART Interface

In this section we will develop a simple device driver using the Universal Asynchronous Receiver/Transmitter (UART). This serial port allows the microcontroller to communicate with devices such as other computers, printers, input sensors, and LCDs. Serial transmission involves sending one bit a time, such that the data is spread out over time. The total number of bits transmitted per second is called the **baud rate**. The reciprocal of the baud rate is the **bit time**, which is the time to send one bit. Most microcontrollers have at least one UART. Before discussing the detailed operation on the TM4C, we will begin with general features common to all devices. Each UART will have a baud rate control register, which we use to select the transmission rate. Each device is capable of creating its own serial clock with a transmission frequency approximately equal to the serial clock in the computer with which it is communicating. A **frame** is the smallest complete unit of serial transmission. Figure 4.33 plots the signal versus time on a serial port, showing a single frame, which includes a **start bit** (which is 0), 8 bits of data (least significant bit first), and a **stop bit** (which is 1). There is always only one start bit, but the Tiva UARTs allow us to select the 5 to 8 data bits and 1 or 2 stop bits. The UART can add even, odd, or no parity bit. However, we will employ the typical protocol of 1 start bit, 8 data bits, no parity, and 1 stop bit. This protocol is used for both transmitting and receiving. The information rate, or **bandwidth**, is defined as the amount of data or useful information transmitted per second. From Figure 4.33, we see that 10 bits are sent for every byte of usual data. Therefore, the bandwidth of the serial channel (in bytes/second) is the baud rate (in bits/sec) divided by 10.

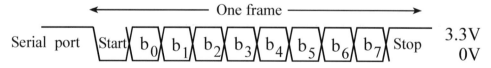

Figure 4.33. *A serial data frame with 8-bit data, 1 start bit, 1 stop bit, and no parity bit.*

Common Error: If you change the bus clock frequency without changing the baud rate register, the UART will operate at an incorrect baud rate.

Checkpoint 4.7: Assuming the protocol drawn in Figure 4.33 and a baud rate of 9600 bits/sec, what is the bandwidth in bytes/sec?

Table 4.8 shows the three most commonly used RS232 signals. The RS232 standard uses a DB25 connector that has 25 pins. The EIA-574 standard uses RS232 voltage levels and a DB9 connector that has only 9 pins. The most commonly used signals of the full RS232 standard are available with the EIA-574 protocols. Only **TxD**, **RxD**, and **SG** are required to implement a simple bidirectional serial channel, thus the other signals are not shown (Figure 4.34). We define the **data terminal equipment** (DTE) as the computer or a terminal and the **data communication equipment** (DCE) as the modem or printer.

DB25 Pin	RS232 Name	DB9 Pin	EIA-574 Name	Signal	Description	True	DTE	DCE
2	BA	3	103	TxD	Transmit Data	-12V	out	in
3	BB	2	104	RxD	Receive Data	-12V	in	out
7	AB	5	102	SG	Signal Ground			

Table 4.8. The commonly-used signals on the RS232 and EIA-574 protocols.

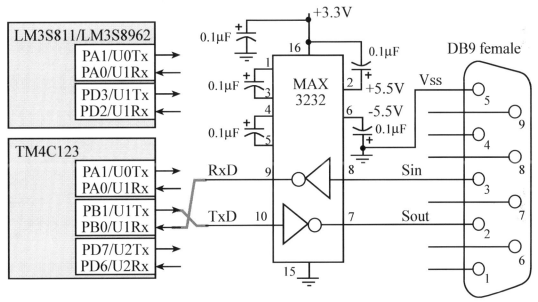

Figure 4.34. Hardware interface implementing an asynchronous RS232 channel. The LM3S1968 has three UART ports. The TM4C123 and MSP432E401Y have eight UART ports.

Observation: Most LM3S/TM4C development kits send one UART channel through the USB cable, so the circuit shown in Figure 4.34 will not be needed. On the PC side of the cable, the serial channel becomes a virtual COM port. After you have installed the drivers for your development kit, look in the Windows Device Manager to see which COM port it is.

RS232 is a non-return-to-zero (NRZ) protocol with true signified as a voltage between -5 and -15 V. False is signified by a voltage between +5 and +15 V. A MAX3232 converter chip is used to translate between the +5.5/-5.5 V RS232 levels and the 0/+3.3 V digital levels, as shown in Figure 4.34. The capacitors in this circuit are important, because they form a charge pump used to create the ±5.5 voltages from the +3.3 V supply. The RS232 timing is generated automatically by the UART. During transmission, the Maxim chip translates a digital high on microcontroller side to -5.5V on the RS232/EIA-574 cable, and a digital low is translated to +5.5V. During receiving, the Maxim chip translates negative voltages on RS232/EIA-574 cable to a digital high on the microcontroller side, and a positive voltage is translated to a digital low. The computer is classified as DTE, so its serial output is pin 3 in the EIA-574 cable, and its serial input is pin 2 in the EIA-574 cable. When connecting a DTE to another DTE, we use a cable with pins 2 and 3 crossed. I.e., pin 2 on one DTE is connected to pin 3 on the other DTE and pin 3 on one DTE is connected to pin 2 on the other DTE. When connecting a DTE to a DCE, then the cable passes the signals straight across. In all situations, the grounds are connected together using the SG wire in the cable. This channel is classified as **full duplex**, because transmission can occur in both directions simultaneously.

4.9.1. Transmitting in asynchronous mode

We will begin with transmission, because it is simple. The transmitter portion of the UART includes a data output pin, with digital logic levels as drawn in Figure 4.33. The transmitter has a 16-element FIFO and a 10-bit shift register, which cannot be directly accessed by the programmer (Figure 4.35). The FIFO and shift register in the transmitter are separate from the FIFO and shift register associated with the receiver. To output data using the UART, the software will first check to make sure the transmit FIFO is not full (it will wait if **TXFF** is 1) and then write to the transmit data register (e.g., **UART0_DR_R**). The bits are shifted out in this order: start, b_0, b_1, b_2, b_3, b_4, b_5, b_6, b_7, and then stop, where b_0 is the LSB and b_7 is the MSB. The transmit data register is write only, which means the software can write to it (to start a new transmission) but cannot read from it. Even though the transmit data register is at the same address as the receive data register, the transmit and receive data registers are two separate registers.

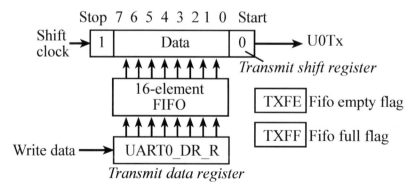

Figure 4.35. Data and shift registers implement the serial transmission.

When a new byte is written to **UART0_DR_R**, it is put into the transmit FIFO. Byte by byte, the UART gets data from the FIFO and loads them into the 10-bit transmit shift register. The 10-bit shift register includes a start bit, 8 data bits, and 1 stop bit. Then, the frame is shifted out one bit at a time at a rate specified by the baud rate register. If there are already data in the FIFO or in the shift register when the **UART0_DR_R** is written, the new frame will wait until the previous frames have been transmitted, before it too is transmitted. The FIFO guarantees the data are transmitted in the order they were written. The serial port hardware is actually controlled by a clock that is 16 times faster than the baud rate, referred to in the datasheet as **Baud16**. When the data are being shifted out, the digital hardware in the UART counts 16 times in between changes to the **U0Tx** output line.

The software can actually write 16 bytes to the **UART0_DR_R**, and the hardware will send them all one at a time in the proper order. This FIFO reduces the software response time requirements of the operating system to service the serial port hardware. Unfortunately, it does complicate the hardware/software timing. At 9600 bits/sec, it takes 1.04 ms to send a frame. Therefore, there will be a delay ranging from 1.04 and 16.7 ms between writing to the data register and the completion of the data transmission. This delay depends on how much data are already in the FIFO at the time the software writes to **UART0_DR_R**.

4.9.2. Receiving in asynchronous mode

Receiving data frames is a little trickier than transmission because we have to synchronize the receive shift register with the incoming data. The receiver portion of the UART includes an **U0Rx** data input pin with digital logic levels. At the input of the microcontroller, true is 3.3V and false is 0V. There is also a 16-element FIFO and a 10-bit shift register, which cannot be directly accessed by the programmer (Figure 4.36). Again the receive shift register and receive FIFO are separate from those in the transmitter. The receive data register, **UART0_DR_R**, is read only, which means write operations to this address have no effect on this register (recall write operations activate the transmitter). The receiver obviously cannot start a transmission, but it recognizes a new frame by its start bit. The bits are shifted in using the same order as the transmitter shifted them out: start (0), b_0, b_1, b_2, b_3, b_4, b_5, b_6, b_7, and then stop (1).

There are six status bits generated by receiver activity. The Receive FIFO empty flag, **RXFE**, is clear when new input data are in the receive FIFO. When the software reads from **UART0_DR_R**, data are removed from the FIFO. When the FIFO becomes empty, the **RXFE** flag will be set, meaning there are no more input data. There are other flags associated with the receiver. There is a Receive FIFO full flag **RXFF**, which is set when the FIFO is full. There are four status bits associated with each byte of data. For this reason, the receive FIFO is 12 bits wide. The overrun error, **OE**, is set when input data are lost because the FIFO is full and more input frames are arriving at the receiver. An overrun error is caused when the receiver interface latency is too large. The break error, **BE**, is set when the input is held low for more than a frame. The **PE** bit is set on a parity error. Because the error rate is so low, most systems do not implement parity. The framing error, **FE**, is set when the stop bit is incorrect. Framing errors are probably caused by a mismatch in baud rate.

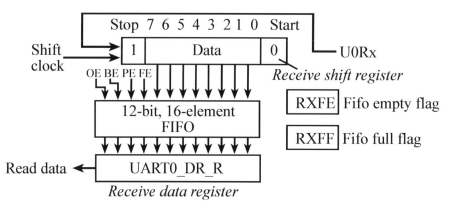

Figure 4.36. Data register shift registers implement the receive serial interface.

The receiver waits for the 1 to 0 edge signifying a start bit, then shifts in 10 bits of data one at a time from the **U0Rx** line. The internal **Baud16** clock is 16 times faster than the baud rate. After the 1 to 0 edge, the receiver waits 8 **Baud16** clocks and samples the start bit. 16 **Baud16** clocks later it samples b_0. Every 16 **Baud16** clocks it samples another bit until it reaches the stop bit. The UART needs an internal clock faster than the baud rate so it can wait the half a bit time between the 1 to 0 edge beginning the start bit and the middle of the bit window needed for sampling. The start and stop bits are removed (checked for framing errors), the 8 bits of data and 4 bits of status are put into the receive FIFO. The FIFO implements hardware buffering so data can be safely stored if the software is performing other tasks.

> **Observation:** If the receiving UART device has a baud rate mismatch of more than 5%, then a framing error can occur when the stop bit is incorrectly captured.

An overrun occurs when there are 16 elements in the receive FIFO, and a 17th frame comes into the receiver. In order to avoid overrun, we can design a real-time system, i.e., one with a maximum latency. The latency of a UART receiver is the delay between the time when new data arrives in the receiver (RXFE=0) and the time the software reads the data register. If the latency is always less than 160 bit times, then overrun will never occur.

> **Observation:** With a serial port that has a shift register and one data register (no FIFO buffering), the latency requirement of the input interface is the time it takes to transmit one data frame.

In the example illustrated in Figure 4.37, assume the UART receive shift register and receive FIFO are initially empty (**RXFE**=1). 17 incoming serial frames occur one right after another (letters A – Q), but the software does not respond. At the end of the first frame, the 0x41 goes into the receive FIFO, and the **RXFE** flag is cleared. Normally, the **UART_InChar** function would respond to **RXFE** being clear and read the data from the UART. In this scenario however, the software is busy doing other things and does not respond to the presence of data in the receive FIFO. Next, 15 more frames are shifted in and entered into the receive FIFO. At the end of the 16th frame, the FIFO is full (**RXFF**=1). If the software were to respond at this point, then all 16 characters would be properly received. If the 17th frame occurs before the first is read by the software, then an overrun error occurs, and a frame is lost. We can see from this worst case scenario that the software must read the data from the UART within 160 bit times of the clearing of **RXFE**.

Figure 4.37. Seventeen receive data frames result in an overrun (OE) error.

4.9.3. TM4C UART Details

Next we will overview the specific UART functions on the TM4C microcontrollers. This section is intended to supplement rather than replace the Texas Instruments manuals. When designing systems with any I/O module, you must also refer to the reference manual of your specific microcontroller. It is also good design practice to review the errata for your microcontroller to see if any quirks (mistakes) exist in your microcontroller that might apply to the system you are designing.

The TM4C microcontrollers have one to eight UARTs. The specific port pins used to implement the UARTs vary from one chip to the next. To find which pins your microcontroller uses, you will need to consult its datasheet. Table 4.9 shows some of the registers for the UART0. If the microcontroller has multiple UARTs, the register names will replace the 0 with a 1 – 7. For the exact register addresses, you should include the appropriate header file (e.g., **msp432e401y.h**). To activate a UART you will need to turn on the UART clock in the **SYSCTL_RCGCUART_R** register. You should also turn on the clock for the digital port in the **SYSCTL_RCGCGPIO_R** register. You need to enable the transmit and receive pins as digital signals. The alternative function for these pins must also be selected.

The OE, BE, PE, and FE are error flags associated with the receiver. You can see these flags in two places: associated with each data byte in **UART0_DR_R** or as a separate error register in **UART0_RSR_R**. The overrun error (**OE**) is set if data has been lost because the input driver latency is too long. **BE** is a break error, meaning the other device has sent a break. **PE** is a parity error (however, we will not be using parity). The framing error (**FE**) will get set if the baud rates do not match. The software can clear these four error flags by writing any value to **UART0_RSR_R**.

The status of the two FIFOs can be seen in the **UART0_FR_R** register. The **BUSY** flag is set while the transmitter still has unsent bits. It will become zero when the transmit FIFO is empty and the last stop bit has been sent. If you implement busy-wait output by first outputting then waiting for **BUSY** to become 0 (middle flowchart of Figure 4.29), then the routine will write new data and return after that particular data has been completely transmitted.

The **UART0_CTL_R** control register contains the bits that turn on the UART. **TXE** is the Transmitter Enable bit, and **RXE** is the Receiver Enable bit. We set **TXE**, **RXE**, and **UARTEN** equal to 1 in order to activate the UART device. However, we should clear **UARTEN** during the initialization sequence.

Address	31–12	11	10	9	8	7–0	Name
$4000.C000		OE	BE	PE	FE	DATA	UART0_DR_R

Address	31–3	3	2	1	0	Name
$4000.C004		OE	BE	PE	FE	UART0_RSR_R

Address	31–8	7	6	5	4	3	2–0	Name
$4000.C018		TXFE	RXFF	TXFF	RXFE	BUSY		UART0_FR_R

Address	31–16	15–0	Name
$4000.C024		DIVINT	UART0_IBRD_R

Address	31–6	5–0	Name
$4000.C028		DIVFRAC	UART0_FBRD_R

Address	31–8	7	6–5	4	3	2	1	0	Name
$4000.C02C		SPS	WLEN	FEN	STP2	EPS	PEN	BRK	UART0_LCRH_R

Address	31–10	9	8	7	6–3	2	1	0	Name
$4000.C030		RXE	TXE	LBE		SIRLP	SIREN	UARTEN	UART0_CTL_R

Address	31–6	5–3	2–0	Name
$4000.C034		RXIFLSEL	TXIFLSEL	UART0_IFLS_R

Address	31–11	10	9	8	7	6	5	4	Name
$4000.C038		OEIM	BEIM	PEIM	FEIM	RTIM	TXIM	RXIM	UART0_IM_R
$4000.C03C		OERIS	BERIS	PERIS	FERIS	RTRIS	TXRIS	RXRIS	UART0_RIS_R
$4000.C040		OEMIS	BEMIS	PEMIS	FEMIS	RTMIS	TXMIS	RXMIS	UART0_MIS_R
$4000.C044		OEIC	BEIC	PEIC	FEIC	RTIC	TXIC	RXIC	UART0_ICR_R

Table 4.9. Some UART registers. Each register is 32 bits wide. Shaded bits are zero.

The **UART0_IBRD_R** and **UART0_FBRD_R** registers specify the baud rate. The baud rate **divider** is a 22-bit binary fixed-point value with a resolution of 2^{-6}. The **Baud16** clock is created from the system bus clock, with a frequency of (Bus clock frequency)/**divider**. The baud rate is 16 times slower than **Baud16**

Baud rate = **Baud16**/16 = (Bus clock frequency)/(16***divider**)

For example, if the bus clock is 8 MHz and the desired baud rate is 19200 bits/sec, then the **divider** should be 8,000,000/16/19200 or 26.04167. As a binary fixed-point number, this number is about 11010.000011. We can establish this baud rate by putting the 11010 into **UART0_IBRD_R** and the 000011 into **UART0_FBRD_R**. In reality, 11010.000011 is equal to 1667/64 or 26.046875. The baud rates in the transmitter and receiver must match within 5% for the channel to operate properly. The error for this example is 0.02%.

The three registers **UART0_LCRH_R**, **UART0_IBRD_R**, and **UART0_FBRD_R** form an internal 30-bit register. This internal register is only updated when a write operation to **UART0_LCRH_R** is performed, so any changes to the baud-rate divisor must be followed by a write to the **UART0_LCRH_R** register for the changes to take effect. Out of reset, both FIFOs are disabled and act as 1-byte-deep holding registers. The FIFOs are enabled by setting the **FEN** bit in **UART0_LCRH_R**.

Checkpoint 4.8: Assume the bus clock is 6 MHz. What is the baud rate if UART0_IBRD_R equals 10 and UART0_FBRD_R equals 20?

Checkpoint 4.9: Assume the bus clock is 50 MHz. What values should you put in UART0_IBRD_R and UART0_FBRD_R to make a baud rate of 38400 bits/sec?

To use interrupts we will enable the FIFOs by setting the **FEN** bit in the **UART0_LCRH_R** register. **RXIFLSEL** specifies the receive FIFO level that causes an interrupt. **TXIFLSEL** specifies the transmit FIFO level that causes an interrupt.

RXIFLSEL	RX FIFO	Set RXMIS interrupt trigger when
0x0	$\geq$ ⅛ full	Receive FIFO goes from 1 to 2 characters
0x1	$\geq$ ¼ full	Receive FIFO goes from 3 to 4 characters
0x2	$\geq$ ½ full	Receive FIFO goes from 7 to 8 characters
0x3	$\geq$ ¾ full	Receive FIFO goes from 11 to 12 characters
0x4	$\geq$ ⅞ full	Receive FIFO goes from 13 to 14 characters

TXIFLSEL	TX FIFO	Set TXMIS interrupt trigger when
0x0	$\leq$ ⅞ empty	Transmit FIFO goes from 15 to 14 characters
0x1	$\leq$ ¾ empty	Transmit FIFO goes from 13 to 12 characters
0x2	$\leq$ ½ empty	Transmit FIFO goes from 9 to 8 characters
0x3	$\leq$ ¼ empty	Transmit FIFO goes from 5 to 4 characters
0x4	$\leq$ ⅛ empty	Transmit FIFO goes from 3 to 2 characters

There are seven possible interrupt trigger flags that are in the **UART0_RIS_R** register. The setting of the **TXRIS** and **RXRIS** flags is defined above. The **OERIS** flag is set on an overrun, new incoming frame received but the receive FIFO is full. The **BERIS** flag is set on a break error. The **PERIS** flag is set on a parity error. The **FERIS** flag is set on a framing error (stop bit is not high). The **RTRIS** is set on a receiver timeout, which is when the receiver FIFO is not empty and no incoming frames have occurred in a 32-bit time period. Each of the seven trigger flags has a corresponding arm bit in the **UART0_IM_R** register. A bit in the **UART0_MIS_R** register set if the trigger flag is both set and armed. To acknowledge an interrupt (make the trigger flag become zero), software writes a 1 to the corresponding bit in **UART0_IC_R**.

4.9.4. UART device driver

Software that sends and receives data must implement a mechanism to synchronize the software with the hardware. In particular, the software should read data from the input device only when data is indeed ready. Similarly, software should write data to an output device only when the device is ready to accept new data. With busy-wait synchronization, the software continuously checks the hardware status waiting for it to be ready. In this section, we will use busy-wait synchronization to write I/O programs that send and receive data using the UART. After a frame is received, the receive FIFO will be not empty (**RXFE** becomes 0) and the 8-bit data is available to be read. To get new data from the serial port, the software first waits for **RXFE** to be zero, then reads the result from **UART0_DR_R**. Recall that when the software reads **UART0_DR_R** it gets data from the receive FIFO. This operation is illustrated in Figure 4.38 and shown in Program 4.10. In a similar fashion, when the software wishes to output via the serial port, it first waits for **TXFF** to be clear, then performs the output. When the software writes **UART0_DR_R** it puts data into the transmit FIFO. An interrupt synchronization method will be presented in Chapter 5.

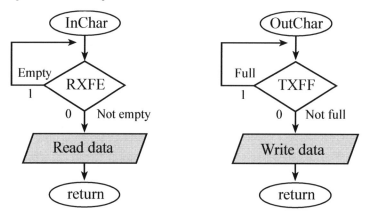

Figure 4.38. Flowcharts of InChar and OutChar using busy-wait synchronization.

The initialization program, **UART_Init**, enables the UART device and selects the baud rate. The input routine waits in a loop until **RXFE** is 0 (FIFO not empty), then reads the data register. The output routine first waits in a loop until **TXFF** is 0 (FIFO not full), then writes data to the data register. Polling before writing data is an efficient way to perform output.

```
// Assumes an 80 MHz bus clock, creates 115200 baud rate
void UART_Init(void){         // should be called only once
  SYSCTL_RCGCUART_R |= 0x0001; // activate UART0
  SYSCTL_RCGCGPIO_R |= 0x0001; // activate port A
  UART0_CTL_R &= ~0x0001;      // disable UART
  UART0_IBRD_R = 43; // IBRD=int(80000000/(16*115,200)) = int(43.40278)
  UART0_FBRD_R = 26; // FBRD = round(0.40278 * 64) = 26
  UART0_LCRH_R = 0x0070;       // 8-bit word length, enable FIFO
  UART0_CTL_R = 0x0301;        // enable RXE, TXE and UART
  GPIO_PORTA_PCTL_R = (GPIO_PORTA_PCTL_R&0xFFFFFF00)+0x00000011; // UART
  GPIO_PORTA_AMSEL_R &= ~0x03; // TM4C, disable analog on PA1-0
  GPIO_PORTA_AFSEL_R |= 0x03;  // enable alt funct on PA1-0
  GPIO_PORTA_DEN_R |= 0x03;    // enable digital I/O on PA1-0
}
// Wait for new input, then return ASCII code
char UART_InChar(void){
  while((UART0_FR_R&0x0010) != 0); // wait until RXFE is 0
  return((char)(UART0_DR_R&0xFF));
}
// Wait for buffer to be not full, then output
void UART_OutChar(char data){
  while((UART0_FR_R&0x0020) != 0);  // wait until TXFF is 0
  UART0_DR_R = data;
}
```

Program 4.10. Device driver functions that implement serial I/O (UART_xxx).

Checkpoint 4.10: How does the software clear RXFE?

Checkpoint 4.11: How does the software clear TXFF?

Checkpoint 4.12: Describe what happens if the receiving computer is operating on a baud rate that is twice as fast as the transmitting computer?

Checkpoint 4.13: Describe what happens if the transmitting computer is operating on a baud rate that is twice as fast as the receiving computer?

4.10. Keyboard Interface

In this section we attempt to interface switches to digital I/O pins and will consider three interfacing schemes, as shown in Figure 4.39. In a **direct interface** we connect each switch up to a separate microcontroller input pin. For example using just one 8-bit parallel port, we can connect 8 switches using the direct scheme. An advantage of this interfacing approach is that the software can recognize all 256 (2^8) possible switch patterns. If the switches were remote from the microcontroller, we would need a 9-wire cable to connect it to the microcontroller. In general, if there are n switches, we would need $n/8$ parallel ports and $n+1$ wires in the cable. This method will be used when there are a small number of switches, or when we must recognize multiple simultaneous key presses. We will use the direct approach for music keyboards and for modifier keys such as shift, control, and alt.

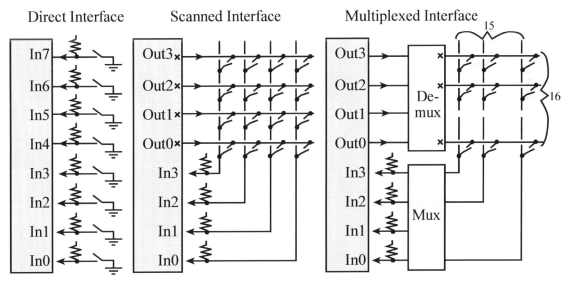

Figure 4.39. Three approaches to interfacing multiple keys.

In a **scanned interface** the switches are placed in a row/column matrix. The × at the four outputs signifies open drain (an output with two states: HiZ and low.) The software drives one row at a time to zero, while leaving the other rows at HiZ. By reading the column, the software can detect if a key is pressed in that row. The software "scans" the device by checking all rows one by one. The Table 4.10 illustrates the sequence to scan the 4 rows.

Row	Out3	Out2	Out1	Out0
3	0	HiZ	HiZ	HiZ
2	HiZ	0	HiZ	HiZ
1	HiZ	HiZ	0	HiZ
0	HiZ	HiZ	HiZ	0

Table 4.10. Scanning patterns for a 4 by 4 matrix keyboard.

For computers without an open drain output mode, the direction register can be toggled to simulate the two output states, HiZ/0, or open drain logic. This method can interface many switches with a small number of parallel I/O pins. In our example situation, the single 8-bit I/O port can handle 16 switches with only an 8-wire cable. The disadvantage of the scanned approach over the direct approach is that it can only handle situations where 0, 1 or 2 switches are simultaneously pressed. This method is used for most of the switches in our standard computer keyboard. The shift, alt, and control keys are interfaced with the direct method. We can "arm" this interface for interrupts by driving all the rows to zero. The edge-triggered input can be used to generate interrupts on touch and release. Because of the switch bounce, an edge-triggered interrupt will occur when any of the keys change. In this section we will interface the keypad using busy-wait synchronization.

With a scanned approach, we give up the ability to detect three or more keys pressed simultaneously. If three keys are pressed in an "L" shape, then the fourth key that completes the rectangle will appear to be pressed. Therefore, special keys like the shift, control, option, and alt are not placed in the scanned matrix, but rather are interfaced directly, each to a separate input port. In general, an n by m matrix keypad has $n*m$ keys, but requires only $n+m$ I/O pins. You can detect any 0, 1, or 2 key combinations, but it has trouble when 3 or more are pressed. The scanned keyboard operates properly if

1. No key is pressed
2. Exactly one key is pressed
3. Exactly two keys are pressed.

In a **multiplexed interface**, the computer outputs the binary value defining the row number, and a hardware decoder (or demultiplexer) will output the zero on the selected row and HiZ's on the other rows. The decoder must have open collector outputs (illustrated again by the × in the above circuit.) The computer simply outputs the sequence 0x00,0x10,0x20,0x30,...,0xF0 to scan the 16 rows, as shown in the Table 4.11.

	Computer output				Decoder			
Row	Out3	Out2	Out1	Out0	15	14	...	0
15	1	1	1	1	0	HiZ		HiZ
14	1	1	1	0	HiZ	0		HiZ
...								
1	0	0	0	1	HiZ	HiZ		HiZ
0	0	0	0	0	HiZ	HiZ		0

Table 4.11. Scanning patterns for a multiplexed 16 by 16 matrix keyboard.

In a similar way, the column information is passed to a hardware encoder that calculates the column position of any zero found in the selected row. One additional signal is necessary to signify the condition that no keys are pressed in that row. Since this interface has 16 rows and 16 columns, we can interface up to 256 keys! We could sacrifice one of the columns to detect the no key pressed in this row situation. In this way, we can interface 240 (15•16) keys on the single 8-bit parallel port. If more than one key is pressed in the same row, this method will only detect one of them. Therefore we classify this scheme as only being able to handle zero or one key pressed.

Applications that can utilize this approach include touch screens and touch pads because they have a lot of switches but are only interested in the 0 or 1 touch situation. Implementing an interrupt driven interface would require too much additional hardware. In this case, periodic polling interrupt synchronization would be appropriate. In general, an *n* by *m* matrix keypad has *n*m* keys, but requires only $x+y+1$ I/O pins, where $2^x = n$ and $2^y = m$. The extra input is used to detect the condition when no key is pressed in that

Example 4.3: Interface a 16-key matrix keyboard. There will be either one key touched or no keys touched.

Solution: This matrix keyboard divides the sixteen keys into four rows and four columns, as shown in Figure 4.40. Each key exists at a unique row/column location. It will take eight I/O pins to interface the rows and columns. Any output port on the TM4C could have been used to interface the rows. To scan the matrix, the software will drive the rows one at a time with open collector logic then read the columns. The open collector logic, with outputs HiZ and 0, will be created by toggling the direction register on the four rows. Actual 10 kΩ pull-up resistors will be placed on the column inputs (PA5-PA2) rather than configured internally, because the internal pull-ups are not fast enough to handle the scanning procedure.

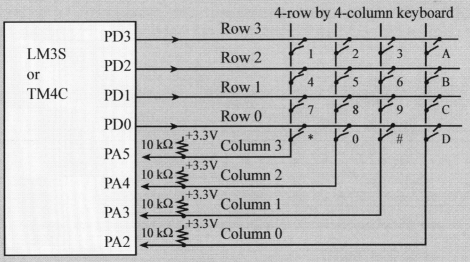

Figure 4.40. A matrix keyboard interfaced to the microcontroller.

Program 4.11 shows the initialization software. The data structure will assist in the scanning algorithm, and it provides a visual mapping from the physical layout of the keys to the ASCII code produced when touching that key. The structure also makes it easy to adapt this solution to other keyboard interfaces. A periodic interrupt can be used to debounce the switches. The key to debouncing is to not observe the switches more frequently than once every 10 ms.

```
void MatrixKeypad_Init(void){
  SYSCTL_RCGCGPIO_R |= 0x09;        // 1) activate clock for Ports A and D
  while((SYSCTL_PRGPIO_R&0x09) != 0x09){};// ready?
  GPIO_PORTA_AFSEL_R &= ~0x3C;      // GPIO function on PA5-2
  GPIO_PORTA_AMSEL_R &= ~0x3C;      // disable analog function on PA5-2
  GPIO_PORTA_PCTL_R &= ~0x00FFFF00; // configure PA5-2 as GPIO
  GPIO_PORTA_DEN_R |= 0x3C;         // enable digital I/O on PA5-2
  GPIO_PORTA_DIR_R &= ~0x3C;        // make PA5-2 in (PA5-2 columns)
  GPIO_PORTD_AFSEL_R &= ~0x0F;      // GPIO function on PD3-0
  GPIO_PORTD_AMSEL_R &= ~0x0F;      // disable analog function on PD3-0
  GPIO_PORTD_PCTL_R &= ~0x0000FFFF; // configure PD3-0 as GPIO
  GPIO_PORTD_DATA_R &= ~0x0F;       // DIRn=0, OUTn=HiZ; DIRn=1, OUTn=0
  GPIO_PORTD_DEN_R |= 0x0F;         // enable digital I/O on PD3-0
  GPIO_PORTD_DIR_R &= ~0x0F;        // make PD3-0 in (PD3-0 rows)
  GPIO_PORTD_DR8R_R |= 0x0F;}       // enable 8 mA drive
```

Program 4.11. Initialization software for a matrix keyboard.

Program 4.12 shows the scanning software. The scanning sequence is listed in Table 4.12. There are two steps to scan a particular row:

1. Select that row by driving it low, while the other rows are HiZ,
2. Read the columns to see if any keys are pressed in that row,
   0 means the key is pressed
   1 means the key is not pressed

It is important to observe column and row signals on a dual trace oscilloscope while running the software at full speed, because it takes time for correct signal to appear on the column after the row is changed. In some cases, a software delay should be inserted between setting the row and reading the column. The length of the delay you will need depends on the size of the pull-up resistor and any stray capacitance that may exist in your circuit.

direction	PD0	PD1	PD2	PD3	PA2	PA3	PA4	PA5
0x01	0	HiZ	HiZ	HiZ	1	2	3	A
0x02	HiZ	0	HiZ	HiZ	4	5	6	B
0x04	HiZ	HiZ	0	HiZ	7	8	9	C
0x08	HiZ	HiZ	HiZ	0	*	0	#	D

Table 4.12. Patterns for a 4 by 4 matrix keyboard.

```
struct Row{
  uint8_t direction;
  char keycode[4];};
typedef const struct Row RowType;
RowType ScanTab[5]={
{   0x01, "123A" }, // row 0
```

```c
{   0x02, "456B" }, // row 1
{   0x04, "789C" }, // row 2
{   0x08, "*0#D" }, // row 3
{   0x00, "    " }};

/* Returns ASCII code for key pressed,
   Num is the number of keys pressed
   both equal zero if no key pressed */
char MatrixKeypad_Scan(int32_t *Num){
  RowType *pt;
  uint32_t column; char key;
  uint32_t j;
  (*Num) = 0;
  key = 0;    // default values
  pt = &ScanTab[0];
  while(pt->direction){
    GPIO_PORTD_DIR_R = pt->direction; // one output
    GPIO_PORTD_DATA_R &= ~0x0F;   // DIRn=0, OUTn=HiZ; DIRn=1, OUTn=0
    for(j=1; j<=10; j++);         // very short delay
    column = ((GPIO_PORTA_DATA_R&0x3C)>>2);// read columns
    for(j=0; j<=3; j++){
      if((column&0x01)==0){
        key = pt->keycode[j];
        (*Num)++;
      }
      column>>=1;  // shift into position
    }
    pt++;
  }
  return key;
}
/* Waits for a key be pressed, then released
   Returns ASCII code for key pressed,
   Num is the number of keys pressed
   both equal zero if no key pressed */
char MatrixKeypad_In(void){ int32_t n;
char letter;
  do{
    letter = MatrixKeypad_Scan(&n);
  } while (n != 1); // repeat until exactly one
  do{
    letter = MatrixKeypad_Scan(&n);
  } while (n != 0); // repeat until release
  return letter;
}
```
Program 4.12. Scanning software for a matrix keyboard (MatrixKeypad_xxx).

Two-key **rollover** occurs when the operator is typing quickly. For example, if the operator is typing the A, B, then C, he/she might type A, AB, B, BC, C, and then release. With rollover, the keyboard does not go through a no-key state in between typing. The hardware interface in Figure 4.40 could handle two-key rollover, but the software solution in Program 4.12 does not. An interrupt version of this interface will be presented in Chapter 5.

4.11. Exercises

4.1 In 16 words or less, describe the meaning of each of the following terms.
 a) Latency
 b) Real-time
 c) Bandwidth
 d) I/O bound
 e) CPU bound
 f) Blind-cycle synchronization
 g) Busy-wait synchronization
 h) Interrupt synchronization
 i) Monotonic
 j) Non-blocking
 k) Deterministic
 l) Slew rate
 m) Open drain
 n) Handshake

4.2 In 16 words or less, describe the meaning of each of the following terms.
 a) Frame
 b) Baud rate
 c) Bandwidth
 d) Full duplex
 e) DTE
 f) DCE
 g) Framing error
 h) Overrun error
 i) Device driver
 j) NRZ
 k) Bit time
 l) Start and stop bits

4.3 List five different methods for I/O synchronization. In 16 words or less, describe each method.

4.4 List one advantage of blind-cycle synchronization. List one disadvantage.

4.5 What is difference between busy-wait and gadfly synchronization?

4.6 In 16 words or less, define setup time. In 16 words or less, define hold time.

4.7 Simplify $\downarrow A = 5 + [10, 20] - [5, 15]$

4.8 Consider a circuit like Figure 4.8, except using 74LS244 and 74LS374. Rework the timing equations to find the appropriate timing relationship between $\downarrow G^*$, $\uparrow C$, $\uparrow C$ and $\uparrow G^*$ so the interface still works. Look up the data sheets on **www.ti.com**.

4.9 Consider a circuit in Figure 4.8 running at voltage supply V_{CC} of 2 V at 25C. Rework the timing equations to find the appropriate timing relationship between $\downarrow G^*$, $\uparrow C$, $\uparrow C$ and $\uparrow G^*$ so the interface still works. Look up the data sheets on **www.ti.com**.

4.10 High speed CMOS logic will run at a wide variety of supply voltages. Look in the data sheets for the 74HC04 74HC244 and 74HC374. Make a general observation about the relationship between how fast the chip operates and the power supply voltage for HC logic.

4.11 What is the effect of capacitance added to a digital line?

4.12 Write code to configure Port D bit 5 as a rising edge-triggered input. Write code that waits for a rising edge on PD5.

4.13 Write code to configure Port A bit 2 as a falling edge edge-triggered input. Write code that waits for a falling edge on PA2.

4.14 Write code to initialize all of Port D as outputs, in power saving mode.

4.15 Interface a positive logic switch to PA5 without an external resistor. Write code to configure Port A bit 5 as needed. Write code that waits for the switch to be touched, then released. You may assume the switch does not bounce.

4.16 Interface a negative logic switch to PB6 without an external resistor. Write code to configure Port B bit 6 as needed. Write code that waits for the switch to be touched, then released. You may assume the switch does not bounce.

4.17 Assume you are using the FIFO implementations in Program 3.7. What happens in the KPN of Program 4.1 if a FIFO becomes empty? What happens in Program 4.1 if a FIFO becomes full?

4.18 Assuming the bus clock to be 8 MHz, write code to make the baud rate 19200 bits/sec.

4.19 Assuming the bus clock to be 24 MHz, what value goes into **UART0_IBRD_R** and **UART0_FBRD_R** to make the baud rate 9600 bits/sec?

4.20 Look up the UART module in the data sheet and describe the similarities and differences between BUSY and TXFF.

4.21 Look up the UART module in the data sheet and describe the two bits of WLEN.

4.22 Look up the UART module in the data sheet.
    a) Redraw Figure 4.33 with 1 start, 6 data, no parity and 1 stop bit.
    b) Redraw Figure 4.33 with 1 start, 8 data, no parity and 2 stop bits.
    c) Redraw Figure 4.33 with 1 start, 8 data, even parity and 1 stop bit.

4.23 Assume you are given a working system with a UART connected to an external device. To save power, you decide to slow down the bus clock by a factor of 4. Briefly explain the changes required for the UART software driver.

4.24 Look up in the data sheet how the UART operated when the FIFOs are disabled. Would the solution in Program 4.10 still work if the FIFO enable bit (**UART0_LCRH_R** bit 4) were cleared? If it does work, in what way is it less efficient?

4.25 Is the write access to Port A in Program 4.2 friendly? What would you change if you wished to switch the two pins on Port A to Port E?

D4.26 Draw the equivalent of the left side of the KPN in Program 4.1 using a Petri Net.

D4.27 Look at the data sheet for the Newhaven Display NHD-0108BZ-RN-YBW-3V3. Write a software function using the LCD hardware in Section 4.7.1 that reads the LCD status bit. Use this function to convert **LCD_OutChar** from blind cycle to busy wait.

D4.28 Write the **Stepper_Seek** function in Program 4.6 to minimize jerk. Slowly ramp up the acceleration as the motor goes from stopped to full speed (**time** parameter). Similarly, slowly ramp down the acceleration as the motor goes from full speed to stopped.

D4.29 Rewrite the programs in Example 4.2 to use UART0 instead of the parallel ports. Assume the bus clock is 50 MHz. Do it is such a way that the function prototypes are identical. Look up the maximum allowable baud rate for the UART. Create a project for Example 4.2 and count the number of assembly instructions in both **Out** and **In**. Add these two counts together (because they wait for each other). Multiply the total number of instructions by 2 bus periods (40ns). Use this number to estimate how long it takes to send one byte. Compare the bandwidths of the serial and parallel approaches. Most high speed interfaces use a serial protocol (e.g., SATA, Ethernet.) Most processors communicate with memory using a parallel protocol. List three reasons other than bandwidth for using a serial link over a parallel port when communicating between two computers.

D4.30 Rewrite **UART_OutChar** in Program 4.10 to use busy-wait synchronization on the **BUSY** bit. Use the middle flowchart in Figure 4.29 so the routine returns after the data has been transmitted.

D4.31 Rewrite the matrix keypad driver in Programs 4.12 and 4.13 to use open drain mode on pins PD3 – PD0. In other words, the direction register is set once during initialization, and the data register is modified during scanning.

4.12. Lab Assignments

Lab 4.1 The overall objective is to create a serial port device driver that supports fixed-point input/output. The format will be 32-bit unsigned decimal fixed-point with a resolution of 0.001. You will be able to find on the implementations of a serial port driver that supports character, string, and integer I/O, **UART_xxx**. In particular, you will design, implement, and test two routines called **UART_FixIn** and **UART_FixOut**. **UART_FixIn** will accept input from the UART similar to the function **UART_InUDec**. **UART_FixOut** will transmit output to the UART similar to the function **UART_OutUDec**. During the design phase of this lab, you should define the range of features available. You should design, implement, and test a main program that illustrates the range of capabilities.

Lab 4.2 The same as Lab 4.1 except the format will be 32-bit signed binary fixed-point with a resolution of 2^{-8}.

Lab 4.3 Design a four function (add, subtract, multiply, divide) calculator using fixed-point math and the UART for input/output. The format should be 32-bit signed decimal fixed-point with a resolution of 0.01. You are free to design the syntax of the calculator however you wish.

Lab 4.4 Design a four function (add, subtract, multiply, divide) calculator using fixed-point math. Take input from a matrix keypad and put output on an LCD. The format should be 32-bit signed decimal fixed-point with a resolution of 0.001. You are free to design the syntax of the calculator however you wish.

Lab 4.5 The overall objective of this lab is to design, implement and test a parallel port expander. Using less than 16 I/O pins of your microcontroller, you will design hardware and software that supports two 8-bit latched input ports and two strobed output ports. Each input port has 8 data lines and one **latch** signal. On the rising each of the **latch**, your system should capture (latch), the data lines. Each output port has 8 data lines and one **strobe** signal. The hardware/software system should generate a pulse out on **strobe** whenever new output is sent. The output ports do not need to be readable.

Lab 4.6 The overall objective of this lab is to design, implement and test a parallel output port expander. Using just three I/O pins of your microcontroller, you will design hardware and software that supports four 8-bit output ports. The output ports do not need to be readable.

Lab 4.7 The overall objective of this lab is to design, implement and test a parallel input port expander. Using just three I/O pins of your microcontroller, you will design hardware and software that supports four 8-bit input ports. The input ports do not need to be latched.

Lab 4.8 The overall objective is to create a HD44780-controlled LCD device driver that supports fixed-point output. The format will be 32-bit unsigned decimal fixed-point with resolutions of 0.1, 0.01 and 0.001. You will be able to find on the book web site implementations of a HD44780-controlled LCD device driver that supports character, string, and integer I/O. In particular, you will design, implement, and test three routines called **LCD_FixOut1 LCD_FixOut2** and **LCD_FixOut3**, implementing fixed-point resolutions 0.1, 0.01 and 0.001 respectively. During the design phase of this lab, you should define the range of features available. You should design, implement, and test a main program that illustrates the range of capabilities.

Lab 4.9 The same as Lab 4.8 except the format will be 32-bit signed decimal fixed-point with resolutions of 0.1, 0.01 and 0.001. You will design, implement, and test three routines.

Lab 4.10 The same as Lab 4.8 except the format will be 32-bit signed binary fixed-point with resolutions of 2^{-4}, 2^{-8} and 2^{-12}. You will design, implement, and test three routines.

Lab 4.11 The same as Lab 4.8 except the formats will be 12, 24 and 32-bit unsigned integers. You will design, implement, and test three routines.

Lab 4.12 The overall objective is to create a HD44780-controlled LCD device driver that supports voltage versus time graphical output. The display will be 8 pixels high by 40 pixels wide, using 8 character positions on the LCD. You will need to continuously write to the CGRAM creating new fonts, then output the new font as data to create the images. The initialization routine will clear the graph and set the minimum and maximum range scale on the voltage axis (y-axis). The plot routine takes a voltage data point between minimum and maximum and draws one pixel on the 8 by 40 display. It takes 40 calls to plot to complete one image on the display as illustrated in Figure 4.41.

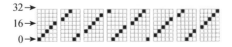

Figure 4.41. Image after 40 calls to plot with the LCD is initialized to Max=32, Min=0. The software outputs this repeating pattern 0,4,8,12,16,20,24,28,0,4,8,12,16...

Subsequent calls to plot should remove the point from 40 calls ago and draw the new point. If the plot is called at a fixed period, the display should show a continuous sweep of the voltage versus time data, like an untriggered oscilloscope. During the design phase of this lab, you should define the range of features available. You should design, implement, and test a main program that illustrates the range of capabilities.

Lab 4.13 You will be able to find on the implementations of a LCD driver that supports the printf I/O, **ST7735_4C123**. You will design, implement, and test three routines called **ST7735_FixOut1**, **ST7735_FixOut2**, and **ST7735_FixOut3**. These three functions should implement 16-bit unsigned fixed-point resolutions 0.1, 0.01 and 0.001 respectively. Create functions like **Fixed_uDecOut2s** that convert the integer portion of the fixed-point number with Δ=0.01 to an ASCII string. You should be able to control the position and color of these outputs in a convenient manner. During the design phase of this lab, you should define the range of features available. You should design, implement, and test a main program that illustrates the range of capabilities.

Lab 4.14 The same as Lab 4.13 except the format will be 16-bit signed decimal fixed-point with resolutions of 0.1, 0.01 and 0.001. Create functions like **Fixed_sDecOut2s** that convert the integer portion of the fixed-point number with Δ=0.01 to an ASCII string. You will design, implement, and test three routines.

Lab 4.15 The same as Lab 4.13 except the format will be 16-bit unsigned binary fixed-point with resolutions of 2^{-4}, 2^{-8} and 2^{-12}. You will design, implement, and test three routines. For example the function , **Fixed_uBinOut8s** converts the integer portion of the unsigned 16-bit fixed-point number with Δ=2^{-8} to an ASCII string.

Lab 4.16 The same as Lab 4.13 except the output will go through the UART to a COM port on the PC. Run a terminal program like PuTTY on the PC.

Lab 4.17 The same as Lab 4.14 except the output will go through the UART to a COM port on the PC. Run a terminal program like PuTTY on the PC.

Lab 4.18 The same as Lab 4.15 except the output will go through the UART to a COM port on the PC. Run a terminal program like PuTTY on the PC.

5. Interrupt Synchronization

Chapter 5 objectives are to:

- Introduce the concept of interrupt synchronization
- Discuss the issues involved in reentrant programming
- Use the first in first out circular queue for buffered I/O
- Discuss the specific details of using interrupts on Cortex-M
- Interface devices using edge-triggered interrupts
- Create periodic interrupts using SysTick
- Implement background I/O for simple devices using periodic polling

There are many reasons to consider interrupt synchronization. The first consideration is that the software in a real-time system must respond to hardware events within a prescribed time. Given a change in input, it is not only necessary to get the correct response, but it will be necessary to get the correct response at the correct time. To illustrate the need for interrupts, consider a keyboard interface where the time between new keyboard inputs might be as small as 10ms. In this situation, the software latency is the time from when the new keyboard input is ready until the time the software reads the new data. In order to prevent loss of data in this case, the software latency must be less than 10ms. We can implement real-time software using busy-wait polling only when the size and complexity of the system is very small. Interrupts are important for real-time systems because they provide a mechanism to guarantee an upper bound on the software response time. Interrupts also give us a way to respond to infrequent but important events. Alarm conditions like low battery power and error conditions can be handled with interrupts. Periodic interrupts, generated by the timer at a regular rate, will be necessary to implement data acquisition and control systems. In the unbuffered interfaces of the previous chapter, the hardware and software took turns waiting for each other. Interrupts provide a way to buffer the data, so that the hardware and software spend less time waiting. In particular, the buffer we will use is a first in first out queue placed between the interrupt routine and the main program to increase the overall bandwidth. We will begin our discussion with general issues, then present the specific details about the TM4C microcontrollers. After that, a number of simple interrupt examples will be presented, and at the end of the chapter we will discuss some advanced concepts like priority and periodic polling.

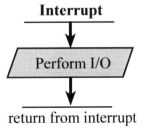

5.1. Multithreading

An **interrupt** is the automatic transfer of software execution in response to a hardware event that is asynchronous with the current software execution. This hardware event is called a **trigger**. The hardware event can either be a busy to ready transition in an external I/O device (like the UART input/output) or an internal event (like bus fault, memory fault, or a periodic timer.) When the hardware needs service, signified by a busy to ready state transition, it will request an interrupt by setting its trigger flag. A **thread** is defined as the path of action of software as it executes. The execution of the interrupt service routine is called a background thread. This thread is created by the hardware interrupt request and is killed when the interrupt service routine returns from interrupt (e.g., executing a `BX LR`). A new thread is created for each interrupt request. It is important to consider each individual request as a separate thread because local variables and registers used in the interrupt service routine are unique and separate from one interrupt event to the next interrupt. In a **multithreaded** system, we consider the threads as cooperating to perform an overall task. Consequently we will develop ways for the threads to communicate (e.g., FIFO) and synchronize with each other. Most embedded systems have a single common overall goal. On the other hand, general-purpose computers can have multiple unrelated functions to perform. A **process** is also defined as the action of software as it executes. Processes do not necessarily cooperate towards a common shared goal. Threads share access to I/O devices, system resources, and global variables, while processes have separate global variables and system resources. Processes do not share I/O devices.

There are no standard definitions for the terms mask, enable and arm in the professional, Computer Science, or Computer Engineering communities. Nevertheless, in this book we will adhere to the following specific meanings. To **arm (disarm)** a device means to enable (shut off) the source of interrupts. Each potential interrupting device has a separate arm bit. One arms (disarms) a device if one is (is not) interested in interrupts from this source. To **enable (disable)** means to allow interrupts at this time (postponing interrupts until a later time). On the ARM® Cortex™-M processor there is one interrupt enable bit for the entire interrupt system. We disable interrupts if it is currently not convenient to accept interrupts. In particular, to disable interrupts we set the I bit in **PRIMASK**.

The software has dynamic control over some aspects of the interrupt request sequence. First, each potential interrupt trigger has a separate **arm** bit that the software can activate or deactivate. The software will set the arm bits for those devices it wishes to accept interrupts from, and will deactivate the arm bits within those devices from which interrupts are not to be allowed. In other words it uses the arm bits to individually select which devices will and which devices will not request interrupts. The second aspect that the software controls is the interrupt enable bit. Specifically, bit 0 of the special register **PRIMASK** is the interrupt mask bit, **I**. If this bit is 1 most interrupts and exceptions are not allowed, we will define as **disabled**. If the bit is 0, then interrupts are allowed, we will define as **enabled**. The BASEPRI register prevents interrupts with lower priority interrupts, but allows higher priority interrupts. For example if **BASEPRI** equals 3, then requests with level 0, 1, and 2 can interrupt, while requests at levels 3 and higher will be postponed. The software can specify the priority level of each interrupt request. If **BASEPRI** is zero, then the priority feature is disabled and all interrupts are allowed. Four conditions must be true for an interrupt to be generated: arm, enable, level, and trigger. A device must be armed; interrupts must be enabled (I=0); the level of the requested interrupt must be less than **BASEPRI**; and an external event must occur setting a trigger flag (e.g., new UART input is ready). An interrupt causes the following sequence of events. First, the current instruction is finished. Second, the execution of the currently running program is suspended, pushing eight registers on the stack (**R0**, **R1**, **R2**, **R3**, **R12**, **LR**, **PC**, and **PSR** with the **R0** on top). Third, the **LR** is set to a specific value signifying an interrupt service routine (ISR) is being run (bits [31:8] to 0xFFFFFF, bits [7:1] specify the type of interrupt return to perform, bit 0 will always be 1 on the Cortex-M meaning Thumb mode). Fourth, the **IPSR** is set to the interrupt number being processed. Lastly, the **PC** is loaded with the address of the ISR (vector). These five steps, called a **context switch**, occur automatically in hardware as the context is switched from foreground to background. Next, the software executes the ISR.

If a trigger flag is set, but the processor is disabled, the interrupt level is not high enough, or the flag is disarmed, the request is not dismissed. Rather the request is held **pending**, postponed until a later time, when the system deems it convenient to handle the requests. We will pay special attention to these enable/disable software actions. In particular we will need to disable interrupts when executing nonreentrant code but disabling interrupts will have the effect of increasing the response time of software to external events. Clearing a trigger flag is called **acknowledgement**, which occurs only by specific software action. Each trigger flag has a specific action software must perform to clear that flag. The SysTick periodic interrupt will be the only example of an automatic acknowledgement. For SysTick, the periodic timer requests an interrupt, but the trigger flag will be automatically cleared. For all the other trigger flags, the ISR must explicitly execute code that clears the flag.

The **interrupt service routine** (ISR) is the software module that is executed when the hardware requests an interrupt. There may be one large ISR that handles all requests (polled interrupts), or many small ISRs specific for each potential source of interrupt (vectored interrupts). The Cortex-M has both polled and vectored interrupts. The design of the interrupt service routine requires careful consideration of many factors. The ISR must acknowledge the trigger flag that caused the interrupt. After the ISR provides the necessary service, it will execute **BX LR**. Because LR contains a special value, this instruction pulls the 8 registers from the stack, which returns control to the main program. There are two stack pointers: PSP and MSP. The software in this book will exclusively use the MSP. It is imperative that the ISR software balance the stack before exiting. Execution of the main program will then continue with the exact stack and register values that existed before the interrupt. Although interrupt handlers can create and use local variables, parameter passing between threads must be implemented using shared global memory variables. A private global variable can be used if an interrupt thread wishes to pass information to itself, e.g., from one interrupt instance to another. The execution of the main program is called the foreground thread, and the executions of the various interrupt service routines are called background threads.

An axiom with interrupt synchronization is that the ISR should execute as fast as possible. The interrupt should occur when it is time to perform a needed function, and the interrupt service routine should perform that function, and return right away. Placing backward branches (busy-wait loops, iterations) in the interrupt software should be avoided if possible. The percentage of time spent executing any one ISR should be minimized. For an input device, the **interface latency** is the time between when new input is available, and the time when the software reads the input data. We can also define **device latency** as the response time of the external I/O device. For example, if we request that a certain sector be read from a disk, then the device latency is the time it take to find the correct track and spin the disk (seek) so the proper sector is positioned under the read head. For an output device, the interface latency is the time between when the output device is idle, and the time when the software writes new data. A **real-time** system is one that can guarantee a worst case interface latency.

Many factors should be considered when deciding the most appropriate mechanism to synchronize hardware and software. One should not always use busy wait because one is too lazy to implement the complexities of interrupts. On the other hand, one should not always use interrupts because they are fun and exciting. Busy-wait synchronization is appropriate when the I/O timing is predicable, and when the I/O structure is simple and fixed. Busy wait should be used for dedicated single thread systems where there is nothing else to do while the I/O is busy. Interrupt synchronization is appropriate when the I/O timing is variable, and when the I/O structure is complex. In particular, interrupts are efficient when there are I/O devices with different speeds. Interrupts allow for quick response times to important events. In particular, using interrupts is one mechanism to design real-time systems, where the interface latency must be short and bounded.

Interrupts can also be used for infrequent but critical events like power failure, memory faults, and machine errors. Periodic interrupts will be useful for real-time clocks, data acquisition systems, and control systems. For extremely high bandwidth and low latency interfaces, DMA should be used.

An **atomic** operation is a sequence that once started will always finish, and cannot be interrupted. All instructions on the ARM® Cortex™-M processor are atomic except store and load multiple, **STM LDM**. If we wish to make a section of code atomic, we can run that code with I=1. In this way, interrupts will not be able to break apart the sequence. Again, requested that are triggered while I=1 are not dismissed, but simply postponed until I=0. In particular, to implement an atomic operation we will 1) save the current value of the **PRIMASK**, 2) disable interrupts, 3) execute the operation, and 4) restore the **PRIMASK** back to its previous value.

Checkpoint 5.1: What four conditions must be true for an interrupt to occur?

Checkpoint 5.2: How do you enable interrupts?

Checkpoint 5.3: What are the steps that occur when an interrupt is processed?

As you develop experience using interrupts, you will come to notice a few common aspects that most computers share. The following paragraphs outline three essential mechanisms that are needed to utilize interrupts. Although every computer that uses interrupts includes all three mechanisms there are a wide spectrum of implementation methods.

All interrupting systems must have the **ability for the hardware to request action from computer**. The interrupt requests can be generated using a separate connection to processor for each device, or using a shared negative logic wire-or requests using open collector logic. The TM4C microcontrollers use separate connections to request interrupts.

All interrupting systems must have the **ability for the computer to determine the source**. A vectored interrupt system employs separate connections for each device so that the computer can give automatic resolution. You can recognize a vectored system because each device has a separate interrupt vector address. With a polled interrupt system, the interrupt software must poll each device, looking for the device that requested the interrupt. Most interrupts on the TM4C microcontrollers are vectored, but there are some triggers, like edge-triggered interrupts on the GPIO pins, that share the same vector. For these interrupts the ISR must poll to see which trigger caused the interrupt.

The third necessary component of the interface is the **ability for the computer to acknowledge the interrupt.** Normally there is a trigger flag in the interface that is set on the busy to ready state transition, i.e., when the device needs service. In essence this trigger flag is the cause of the interrupt. Acknowledging the interrupt involves clearing this flag. It is important to shut off the request, so that the computer will not mistakenly request a second (and inappropriate) interrupt service for the same condition. The first Intel x86 processors used a hardware acknowledgment that automatically clears the request. Except for periodic SysTick, TM4C microcontrollers use software acknowledge. So when designing an interrupting interface, it will be important to know exactly what hardware conditions will set the trigger flag (and request an interrupt) and how the software will clear it (acknowledge) in the ISR.

Common Error: The system will crash if the interrupt service routine doesn't either acknowledge or disarm the device requesting the interrupt.

Common Error: The ISR software should not disable interrupts at the beginning nor should it reenable interrupts at the end. Which interrupts are allowed to run is automatically controlled by the NVIC.

5.2. Interthread Communication and Synchronization

For regular function calls we use the registers and stack to pass parameters, but interrupt threads have logically separate resisters and stack. In particular, registers are automatically saved by the processor as it switches from main program (foreground thread) to interrupt service routine (background thread). Exiting an ISR will restore the registers back to their previous values. Thus, all parameter passing must occur through global memory. One cannot pass data from the main program to the interrupt service routine using registers or the stack.

In this chapter, multithreading means one main program (foreground thread) and multiple ISRs (background threads). An operating system allows multiple foreground threads. Synchronizing threads is a critical task affecting efficiency and effectiveness of systems using interrupts. In this section, we will present in general form three constructs to synchronize threads: binary semaphore, mailbox, and FIFO queue.

A **binary semaphore** is simply a shared flag, as described in Figure 5.1. There are two operations one can perform on a semaphore. **Signal** is the action that sets the flag. **Wait** is the action that checks the flag, and if the flag is set, the flag is cleared and important stuff is performed. This flag must exist as a private global variable with restricted access to only the **Wait** and **Signal** functions. In C, we add the qualifier `static` to a global variable to restrict access to software within the same file. In order to reduce complexity of the system, it will be important to limit the access to this flag to as few modules as possible.

The history of semaphores dates back to the invention of fire. Man has used optical telegraphs such as fire, smoke, and flags to communicate over short distances. The transmitter encodes the information as recognizable images. The observer at a distant location sees (receives) the signals produced by the transmitter. Semaphores were used in the early railroads to allow multiple trains to share access to a common track. The position of the semaphore flag described the status of the upcoming track. The basic idea of these physical semaphores is used in I/O interfacing. A shared flag describes the status, and threads can observe or change the flag.

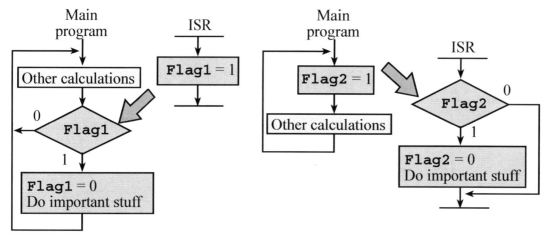

Figure 5.1. A semaphore can be used to synchronize threads.

With binary semaphores, the flag has two states: 0 and 1. It is good design to assign a meaning to this flag. For example, 0 might mean the switch has not been pressed, and 1 might mean the switch has been pressed. Figure 5.1 shows two examples of the binary semaphore. The big arrows in this figure signify the synchronization link between the threads. In the example on the left, the ISR signals the semaphore and the main program waits on the semaphore. Notice the "important stuff" is run in the foreground once per execution of the ISR. In the example on the right, the main program signals the semaphore and the ISR waits. It is good design to have NO backwards jumps in an ISR. In this particular application, if the ISR is running and the semaphore is 0, the action is just skipped and the computer returns from the interrupt.

A **counting semaphore** is a shared counter, representing the number of objects. When a new object is created, the counter is incremented, and when an object is destroyed the counter is decremented. A counter describes the status, and threads can observe, increment or decrement.

The second interthread synchronization scheme is the **mailbox**. The mailbox is a binary semaphore with associated data variable. Figure 5.2 illustrates an input device interfaced using interrupt synchronization.

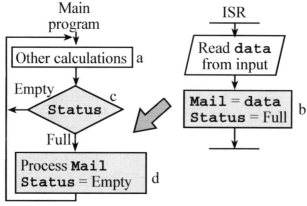

Figure 5.2. A mailbox can be used to pass data between threads.

The big arrow in Figure 5.2 signifies the communication and synchronization link between the background and foreground. The mailbox structure is implemented with two shared global variables. **Mail** contains data, and **Status** is a binary semaphore specifying whether the mailbox is full or empty. The interrupt is requested on its hardware trigger, signifying new data are ready from the input device. The ISR will read the data from the input device and store it in the shared global variable **Mail**, then update its status to full. The main program will perform other calculations, while occasionally checking the status of the mailbox. When the mailbox has data, the main program will process it. This approach is adequate for situations where the input bandwidth is slow compared to the software processing speed.

One way to visualize the interrupt synchronization is to draw a state versus time plot of the activities of the hardware, the mailbox, and the two software threads. Figure 5.3 shows that at time (a) the mailbox is empty, the input device is idle and the main program is performing other tasks, because mailbox is empty. When new input data are ready, the trigger flag will be set, and an interrupt will be requested. At time (b) the ISR reads data from input device and saves it in **Mail**, and then it sets **Status** to full. At time (c) the main program recognizes **Status** is full. At time (d) the main program processes data from **Mail**, sets **Status** to empty. Notice that even though there are two threads, only one is active at a time. The interrupt hardware switches the processor from the main program to the ISR, and the return from interrupt switches the processor back.

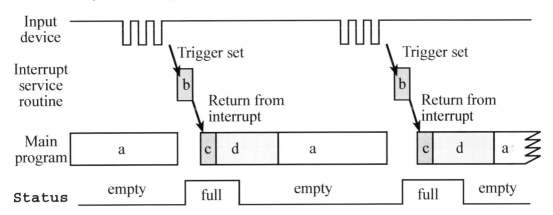

Figure 5.3. Hardware/software timing of an input interface using a mailbox.

The third synchronization technique is the FIFO queue, shown in Figure 5.4. With mailbox synchronization, the threads execute in lock-step: one, the other, one, the other... However, with the FIFO queue execution of the threads is more loosely coupled. The classic producer/consumer problem has two threads. One thread produces data and the other consumes data. For an input device, the background thread is the producer because it generates new data, and the foreground thread is the consumer because it uses up the data. For an output device, the data flows in the other direction so the producer/consumer roles are reversed. It is appropriate to pass data from the producer thread to the consumer thread using a FIFO queue. Figure 5.4 shows one producer linked to one consumer. However, it is possible to have multiple producers connected to multiple consumers. An example of multiple producers is a USB hub or Ethernet router, where packets can arrive from multiple essentially equivalent ports. Another name for this is **client-server**, where the server produces data and the client consumes data.

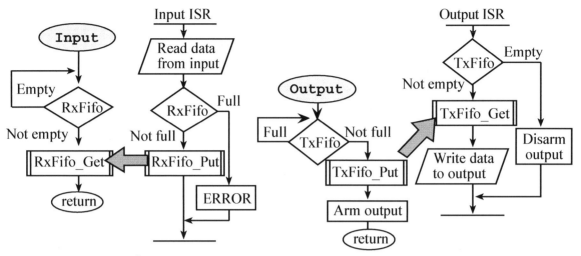

Figure 5.4. In a producer/consumer system, FIFO queues can be used to pass data between threads.

Observation: For systems with interrupt-driven I/O on multiple devices, there will be a separate FIFO for each device.

We could process the data within the ISR itself, and just report the results of the processing to the main program using the mailbox. Processing data in the ISR is usually poor design because we try to minimize the time running in the ISR, in order to minimize latency of other interrupts.

An input device needs service (busy to done state transition) when new data are available, see Figures 5.4 and 5.5. The interrupt service routine (background) will accept the data and put it into a FIFO. Typically, the ISR will restart the input hardware, causing a done to busy transition.

An output device needs service (busy to done state transition) when the device is idle, ready to output more data. The interrupt service routine (background) will get more data from the FIFO and output it. The output function will restart the hardware causing a done to busy transition. Two particular problems with output device interrupts are

1. How does one generate the first interrupt?
    In other words, how does one start the output thread? and
2. What does one do if an output interrupt occurs (device is idle)
    but there is no more data currently available (e.g., FIFO is empty)?

The foreground thread (main program) executes a loop, and accesses the appropriate FIFO when it needs to input or output data. The background threads (interrupts) are executed when the hardware needs service. Use Little's Theorem to estimate the size of the fifo needed.

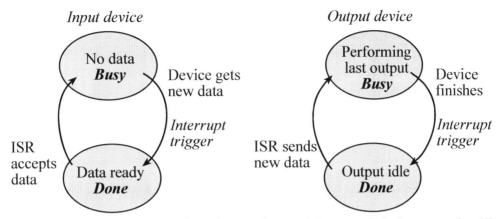

Figure 5.5. *The input device interrupts when it has new data, and the output device interrupts when idle.*

One way to visualize the interrupt synchronization is to draw a state versus time plot of the activities of the hardware and the two software modules. Figure 5.6 is drawn to describe a situation where the time between inputs is about twice as long as it takes the software to process the data. For this example, the main thread begins by waiting because the FIFO is empty (a). When the input device is busy it is in the process of creating new input. When the input device is done, new data are available and an interrupt is requested. The interrupt service routine will read the data and put it into the FIFO (b). Once data are in the FIFO, the main program is released to go on because the get function will return with data (c). The main program processes the data (d) and then waits for more input (a). The arrows from one graph to the other represent the synchronizing events. Because the time for the software to read and process the data is less than the time for the input device to create new input, this situation is called **I/O bound**. A system is I/O bound if the overall bandwidth is limited by the speed of the I/O device. In this situation, the FIFO has either 0 or 1 entry, and the use of interrupts does not enhance the bandwidth over the busy-wait implementations presented in the previous chapter. Even with an I/O bound device it may be more efficient to utilize interrupts because it provides a straight-forward approach to servicing multiple devices.

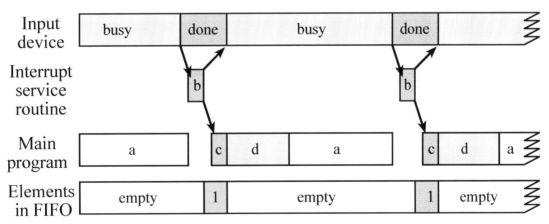

Figure 5.6. *Hardware/software timing of an I/O bound input interface.*

In this second example, the input device starts with a burst of high bandwidth activity. Figure 5.7 is drawn to describe a situation where the input rate is temporarily two to three times faster than the software can handle. As long as the interrupt service routine is fast enough to keep up with the input device, and as long as the FIFO does not become full during the burst, no data are lost. The software waits for the first data (a), but then does not have to wait until the burst is over. In this situation, the overall bandwidth is higher than it would be with a busy-wait implementation, because the input device does not have to wait for each data byte to be processed (b). This is the classic example of a "buffered" input, because the ISR puts data the FIFO. The main program gets data from the FIFO (c), and then processes it (d). When the bandwidth is limited by software execution speed, the system is called **CPU bound**. As we will see later, this system will work only if the producer rate temporarily exceeds the consumer rate (a short burst of high bandwidth input). If the external device sustained the high bandwidth input rate, then the FIFO would become full and data would be lost.

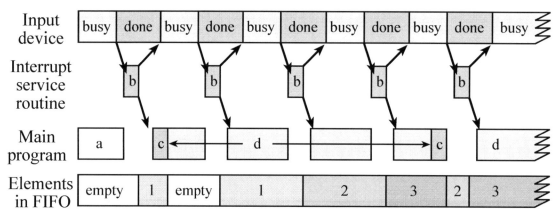

Figure 5.7. Hardware/software timing of an input interface during a high bandwidth burst.

For an input device, if the FIFO is usually empty, the interface is I/O bound. During times when there are many elements, the interface is CPU bound.

For an output device, the interrupt is requested when the output is idle and ready to accept more data. The "busy to done" state transition causes an interrupt. The interrupt service routine gives the output device another piece of data to output. Again, we can visualize the interrupt synchronization by drawing a state versus time plot of the activities of the hardware and the two software modules. Figure 5.8 is drawn to describe a situation where the time between outputs is about half as long as it takes the software to generate new data. For an output device interface, the output device is initially disarmed and the FIFO is empty. The main thread begins by generating new data (a). After the main program puts the data into the FIFO it arms the output interrupts (b). This first interrupt occurs immediately and the ISR gets some data from the FIFO and outputs it to the external device (c). The output device becomes busy because it is in the process of outputting data. It is important to realize that it only takes the software on the order of 1 μsec to write data to one of its output ports, but usually it takes the output device much longer to fully process the data. When the output device is done, it is ready to accept more data and an interrupt is requested. If the FIFO is empty at this point, the ISR will disarm the output device (d). If the FIFO is not empty, the interrupt service routine will get from the FIFO, and write it out to the output port. Once data are written to the output port, the output device is released to go on. In this first example, the time for the software to generate data is larger than the time for the external device to output it. This is an example of a **CPU bound** system. In this situation, the FIFO has either 0 or 1 entry, and the use of interrupts does not enhance the bandwidth over the busy-wait implementations presented in the previous chapter. Nevertheless interrupts provide a well-defined mechanism for dealing with complex systems.

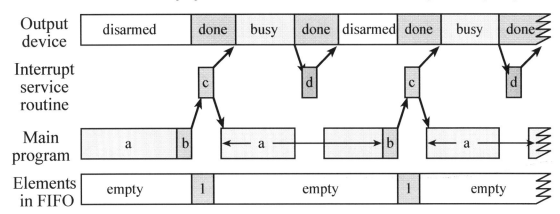

Figure 5.8. Hardware/software timing of a CPU bound output interface.

In this second output example, the software starts with a burst of high bandwidth activity. Figure 5.9 is drawn to describe a situation where the software produces data at a rate that is temporarily much faster than the hardware can handle. As long as the FIFO does not become full, no data are lost. In this situation, the overall bandwidth is higher than it would be with a busy-wait implementation, because the software does not have to wait for each data byte to be processed by the hardware. The software generates data (a) and puts it into the FIFO (b). When the output is idle, it generates an interrupt. The ISR gets data and restarts the output device (c).

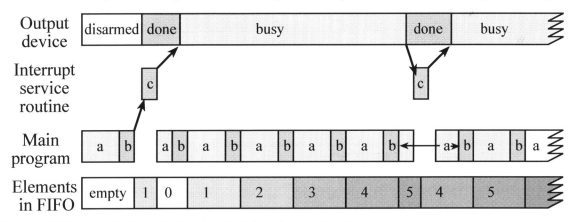

Figure 5.9. Hardware/software timing of an I/O bound output interface.

This is the classic example of a "buffered" output, because data enters the system (via the main program) is temporarily stored in a buffer (put into the FIFO) and the data are processed later (by the ISR, get from the FIFO, write to external device.) When the I/O device is slower than the software, the system is called I/O bound. Just like the input scenario, the FIFO might become full if the producer rate is too high for too long.

There are other types of interrupt that are not an input or output. For example, we will configure the computer to request an interrupt on a periodic basis. This means an interrupt handler will be executed at fixed time intervals. This periodic interrupt will be essential for the implementation of real-time data acquisition systems and real-time control systems. For example if we are implementing a digital controller that executes a control algorithm 100 times a second, then we will set up the internal timer hardware to request an interrupt every 10 ms. The interrupt service routine will execute the digital control algorithm and return to the main thread. We will use periodic interrupts to output to a DAC and input from an ADC.

> **Performance Tip:** It is poor design to employ backward jumps in an ISR, because they may affect the latency of other interrupt requests. Whenever you are thinking about using a backward jump, consider redesigning the system with more or different triggers to reduce the number of backward jumps.

As you recall, the FIFO passes the data from the producer to the consumer. In general, the rates at which data are produced and consumed can vary dynamically. Humans do not enter data into a keyboard at a constant rate. Even printers require more time to print color graphics versus black and white text. Let t_p be the time (in sec) between calls to **Fifo_Put**, and r_p be the arrival rate (producer rate in bytes/sec) into the system, so $r_p = 1/t_p$. Similarly, let t_g be the time (in sec) between calls to **Fifo_Get**, and r_g be the service rate (consumer rate in bytes/sec) out of the system, so $r_g = 1/t_g$.

If the minimum time between calls to **Fifo_Put** is greater than the maximum time between calls to **Fifo_Get**, then a FIFO is not necessary and the data flow could be solved with a mailbox. I.e., no FIFO is needed if $\min(t_p) \geq \max(t_g)$. On the other hand, if the time between calls to **Fifo_Put** becomes less than the time between calls to **Fifo_Get** because either

- The arrival rate temporarily increases
- The service rate temporarily decreases

then information will be collected in the FIFO. For example, a person might type very fast for a while, followed by long pause. The FIFO could be used to capture without loss all the data as it comes in very fast. Clearly on average the system must be able to process the data (the consumer thread) at least as fast as the average rate at which the data arrives (producer thread). If the average producer rate is larger than the average consumer rate

$$\mathrm{Ave}(r_p) > \mathrm{Ave}(r_g)$$

then the FIFO will eventually overflow no matter how large the FIFO. If the producer rate is temporarily high, and that causes the FIFO to become full, then this problem can be solved by increasing the FIFO size.

There is fundamental difference between an empty error and a full error. Consider the application of using a FIFO between your computer and its printer. This is a good idea because the computer can temporarily generate data to be printed at a very high rate followed by long pauses. The printer is like a turtle. It can print at a slow but steady rate. The computer will put a byte into the FIFO that it wants printed. The printer will get a byte out of the FIFO when it is ready to print another character. A full error occurs when the computer calls **Fifo_Put** at too fast a rate. A full error is serious, because if ignored data will be lost. Recall that one of the definitions of a Kahn Process Network is that the FIFOs are never full. So, implementing the data flow in such a way that the FIFOs never become full allows us to model the system as a

KPN. On the other hand, an empty error occurs when the printer is ready to print but the computer has nothing in mind. An empty error is not serious, because in this case the printer just sits there doing nothing.

> **Checkpoint 5.4:** If the FIFO becomes full, can the situation always be solved by increasing the size?

Consider a FIFO that has a feature where we can determine the number of elements by calling **Fifo_Size**. If we place this debugging instrument inside the producer, we can measure a histogram of FIFO sizes telling us 1) if the FIFO ever became full; 2) if the interface is CPU bound; or 3) if the interface is I/O bound.

```
uint32_t Histogram[FIFOSIZE];
#define Collect() (Histogram[Fifo_Size()]++;)
```

5.3. Critical Sections

In general, if two threads access the same global memory and one of the accesses is a write, then there is a **causal dependency** between the threads. This means, the execution order may affect the outcome. Shared global variables are very important in multithreaded systems because they are required to pass data between threads, but they exhibit complex behavior, and it is hard to find bugs that result with their use.

A program segment is **reentrant** if it can be concurrently executed by two (or more) threads. To implement reentrant software, we place variables in registers or on the stack, and avoid storing into global memory variables. When writing in assembly, we use registers, or the stack for parameter passing to create reentrant subroutines. Typically each thread will have its own set of registers and stack. A nonreentrant subroutine will have a section of code called a **vulnerable window** or **critical section**. An error occurs if

> 1) One thread calls the function in question
> 2) It is executing in the critical section when interrupted by a second thread
> 3) The second thread calls the same function.

There are a number of scenarios that can happen next. In the most common scenario, the second thread is allowed to complete the execution of the function, control is then returned to the first thread, and the first thread finishes the function. This first scenario is the usual case with interrupt programming. In the second scenario, the second thread executes part of the critical section, is interrupted and then re-entered by a third thread, the third thread finishes, the control is returned to the second thread and it finishes, lastly the control is returned to the first thread and it finishes. This second scenario can happen in interrupt programming if the second interrupt has higher priority than the first. A critical section may exist when two different functions that access and modify the same memory-resident data structure.

Program 5.1 shows a C function and the assembly code generated by the ARM Keil™ uVision® compiler. The function is nonreentrant because of the read-modify-write nonatomic access to the global variable, **num**.

5. Interrupt Synchronization

```
num     SPACE   4
Count   LDR     r0,[pc,#116]   ; R0= &num
;*******start of critical section***
        LDR     r0,[r0,#0x00]  ; R0=num
;could be bad if interrupt occurs here
        ADDS    r0,r0,#1
;could be bad if interrupt occurs here
        LDR     r1,[pc,#108]   ; R1=&num
;could be bad if interrupt occurs here
        STR     r0,[r1,#0x00]  ; update num
;*******end of critical section***
        BX      lr
ptr     DCD     num
```

```c
uint32_t volatile num;
void Count(void){
   num = num + 1;
}
```

Program 5.1. This function is nonreentrant because of the read-modify-write access to a global.

Assume there are two concurrent threads (the main program and a background ISR) that both call this function. Concurrent means that both threads are ready to run. Because there is only one computer, exactly one thread will be running at a time. Typically, the operating system switches execution control back and forth using interrupts. There are three places in the assembly code at which if an interrupt were to occur and the ISR called the same function, the end result would be **num** would be incremented only once, even though the function was called twice. Assume for this example **num** is initially 100. An error occurs if:

1. The main program calls **Count**
2. The main executes **LDR r0,[r0,#0x00]** making R0 = 100
3. The OS halts the main (using an interrupt) and starts the ISR
4. the ISR calls **Count**
   Executes **num=num+1;** making equal to 101
5. The OS returns control back to the main program
   R0 is back to its original value of 100
6. The main program finished the function (adding 1 to R0)
   Making **num** equal to 101

An **atomic operation** is one that once started is guaranteed to finish. In most computers, once an instruction has begun, the instruction must be finished before the computer can process an interrupt. In general, nonreentrant code can be grouped into three categories all involving 1) nonatomic sequences, 2) writes and 3) global variables. We will classify I/O ports as global variables for the consideration of critical sections. We will group registers into the same category as local variables because each thread will have its own registers and stack.

The first group is the **read-modify-write** sequence:

1. The software reads the global variable producing a copy of the data
2. The software modifies the copy (original variable is still unmodified)
3. The software writes the modification back into the global variable.

In the second group, we have a **write followed by read**, where the global variable is used for temporary storage:

1. The software writes to the global variable (only copy of the information)
2. The software reads from the global variable expecting the original data to be there.

In the third group, we have a **non-atomic multi-step write** to a global variable:

1. The software writes part of the new value to a global variable
2. The software writes the rest of the new value to a global variable.

Observation: When considering reentrant software and vulnerable windows we classify accesses to I/O ports the same as accesses to global variables.

Observation: Sometimes we store temporary information in global variables out of laziness. This practice is to be discouraged because it wastes memory and may cause the module to not be reentrant.

Sometime we can have a critical section between two different software functions (one function called by one thread, and another function called by a different thread). In addition to above three cases, a **non-atomic multi-step read** will be critical when paired with a **multi-step write**. For example, assume a data structure has multiple components (e.g., hours, minutes, and seconds). In this case, the write to the data structure will be atomic because it occurs in a high priority ISR. The critical section exists in the foreground between steps 1 and 3. In this case, a critical section exists even though no software has actually been reentered.

Foreground thread	Background thread
1. The main reads some of the data	
	2. ISR writes to the data structure
3. The main reads the rest of the data	

In a similar case, a **non-atomic multi-step write** will be critical when paired with a **multi-step read**. Again, assume a data structure has multiple components. In this case, the read from the data structure will be atomic because it occurs in a high priority ISR. The critical section exists in the foreground between steps 1 and 3.

Foreground thread	Background thread
1. The main writes some of the data	
	2. ISR reads from the data structure
3. The main writes the rest of the data	

When multiple threads are active, it is possible for two threads to be executing the same program. For example, the system may be running in the foreground and calls `Func`. Part way through execution the `Func`, an interrupt occurs. If the ISR also calls `Func`, two threads are simultaneously executing the function. To experimentally determine if a function has been reentered, we could use two flags or two output pins. Set one of them (**PD1**, `Entered`) at the start and clear it at the end. The thread has been re-entered if this flag or pin is set at the start of the function, as shown in Program 3.7. In this example, Port D bits 1,0 are not part of the original code, but rather used just for the purpose of debugging. **PD1** is 1 when one thread starts executing the function. However, if **PD0** becomes 1, then the function has been reentered.

```
#define PD0     (*((volatile uint32_t *)0x40007004))
#define PD1     (*((volatile uint32_t *)0x40007008))
```

`// function to be tested`
`volatile int Entered=0,Flag=0;`
`void Func(void){`
`  if(Entered) Flag = 1;`
`  Entered = 1;`
`// the regular function`
`  Entered = 0;}`
	`// function to be tested`
	`void Func(void){`
	`  if(PD1) PD0 = 1;`
	`  PD1 = 2;`
	`// the regular function`
	`  PD1 = 0;`
	`}`

Program 5.2. Detection of re-entrant behavior using two flags or two output pins.

If critical sections do exist, we can either eliminate it by removing the access to the global variable or implement **mutual exclusion**, which simply means only one thread at a time is allowed to execute in the critical section. In general, if we can eliminate the global variables, then the subroutine becomes reentrant. Without global variables there are no "vulnerable" windows because each thread has its own registers and stack. Sometimes one must access global memory to implement the desired function. Remember that all I/O ports are considered global. Furthermore, global variables are necessary to pass data between threads.

```
;********** DisableInterrupts **************
; disable interrupts
; inputs:  none
; outputs: none
DisableInterrupts
        CPSID   I
        BX      LR
;********** EnableInterrupts **************
; disable interrupts
; inputs:  none
; outputs: none
EnableInterrupts
        CPSIE   I
        BX      LR
;********** StartCritical **********************
; make a copy of previous I bit, disable interrupts
; inputs:  none
; outputs: previous I bit
StartCritical
        MRS     R0, PRIMASK   ; save old status
        CPSID   I             ; mask all (except faults)
        BX      LR
;********** EndCritical ***********************
; using the copy of previous I bit, restore I bit to previous value
; inputs:  previous I bit
; outputs: none
EndCritical
        MSR     PRIMASK, R0
        BX      LR
```

Program 5.3. Assembly functions needed for interrupt enabling and disabling.

A simple way to implement mutual exclusion is to disable interrupts while executing the critical section. It is important to disable interrupts for as short a time as possible, so as to minimize the effect on the dynamic performance of the other threads. While we are running with interrupts disabled, time-critical events like power failure and danger warnings cannot be processed. Notice also that the interrupts are not simply disabled then enabled. Before the critical section, the interrupt status is saved, and the interrupts disabled. After the critical section, the interrupt status is restored. You cannot save the interrupt status in a global variable, rather you should save it either on the stack or in a register. We will add the assembly code of Program 5.3 to the **Startup.s** file in our projects that use interrupts. Program 5.4 illustrates how to implement mutual exclusion and eliminate the critical section.

```
uint32_t volatile num;
void Count(void){ int32_t sr;
  sr = StartCritical();
  num = num + 1;
  EndCritical(sr);
}
```

Program 5.4. This function is reentrant because of the read-modify-write access to the global is atomic.

Checkpoint 5.5: Consider the situation of nested critical sections. For example, a function with a critical section calls another function that also has a critical section. What would happen if you simply added disable interrupt at the beginning and a reenable interrupt at the end of each critical section?

Another category of timing-dependent bugs, similar to critical sections, is called a **race condition**. A race condition occurs in a multi-threaded environment when there is a causal dependency between two or more threads. In other words, different behavior occurs depending on the order of execution of two threads. In this example of a race condition, Thread-1 initializes Port B bits 3 – 0 to be output using **GPIO_PORTB_DIR_R = 0x0F;** Thread-2 initializes Port B bits 6 – 4 to be output using **GPIO_PORTB_DIR_R = 0x70;** In particular, if Thread-1 runs first and Thread-2 runs second, then Port B bits 3 – 0 will be set to inputs. Conversely, if Thread-2 runs first and Thread-1 runs second, then Port B bits 6 – 4 will be set to inputs. This is a race condition caused by unfriendly code. The solution to this problem is to write the two initializations in a friendly manner, and make both initializations atomic.

In a second example, assume two threads are trying to get data from the same input device. Both call the function **UART_InChar** given in Program 4.10. When data arrives at the input, the thread that executes first will capture the data. This example is equivalent to the Petri Net conflict drawn in Figure 4.16.

5.4. NVIC on the ARM® Cortex-M Processor

On the ARM® Cortex™-M processor, exceptions include resets, software interrupts and hardware interrupts. Each exception has an associated 32-bit vector that points to the memory location where the ISR that handles the exception is located. Vectors are stored in ROM at the beginning of memory. Program 5.5 shows the first few vectors as defined in the **Startup.s** file.

DCD is an assembler pseudo-op that defines a 32-bit constant. ROM location 0x0000.0000 has the initial stack pointer, and location 0x0000.0004 contains the initial program counter, which is called the reset vector. It points to a function called the reset handler, which is the first thing executed following reset. There are up to 240 (77 on the TM4C123 microcontroller) possible interrupt sources and their 32-bit vectors are listed in order starting with location 0x0000.0008. From a programming perspective, we can attach ISRs to interrupts by writing the ISRs as regular C functions with no input or output parameters and editing the **Startup.s** file to specify those functions for the appropriate interrupt. For example, if we wrote a Port C interrupt service routine named **PortCISR**, then we would replace **GPIOPortC_Handler** with **PortCISR**. In this book, we will write our ISRs using standard function names so that the **Startup.s** file need not be edited. I.e., we will simply call the ISR for edge-triggered interrupts on Port C as **GPIOPortC_Handler**. For more details see the **Startup.s** files within the interrupt examples posted on the book web site.

```
        EXPORT    __Vectors
__Vectors                                ; address        ISR
        DCD    StackMem + Stack          ; 0x00000000 Top of Stack
        DCD    Reset_Handler             ; 0x00000004 Reset Handler
        DCD    NMI_Handler               ; 0x00000008 NMI Handler
        DCD    HardFault_Handler         ; 0x0000000C Hard Fault Handler
        DCD    MemManage_Handler         ; 0x00000010 MPU Fault Handler
        DCD    BusFault_Handler          ; 0x00000014 Bus Fault Handler
        DCD    UsageFault_Handler        ; 0x00000018 Usage Fault Handler
        DCD    0                         ; 0x0000001C Reserved
        DCD    0                         ; 0x00000020 Reserved
        DCD    0                         ; 0x00000024 Reserved
        DCD    0                         ; 0x00000028 Reserved
        DCD    SVC_Handler               ; 0x0000002C SVCall Handler
        DCD    DebugMon_Handler          ; 0x00000030 Debug Monitor Handler
        DCD    0                         ; 0x00000034 Reserved
        DCD    PendSV_Handler            ; 0x00000038 PendSV Handler
        DCD    SysTick_Handler           ; 0x0000003C SysTick Handler
        DCD    GPIOPortA_Handler         ; 0x00000040 GPIO Port A
        DCD    GPIOPortB_Handler         ; 0x00000044 GPIO Port B
        DCD    GPIOPortC_Handler         ; 0x00000048 GPIO Port C
        DCD    GPIOPortD_Handler         ; 0x0000004C GPIO Port D
        DCD    GPIOPortE_Handler         ; 0x00000050 GPIO Port E
        DCD    UART0_Handler             ; 0x00000054 UART0
        DCD    UART1_Handler             ; 0x00000058 UART1
        DCD    SSI0_Handler              ; 0x0000005C SSI
        DCD    I2C0_Handler              ; 0x00000060 I2C
        DCD    PWMFault_Handler          ; 0x00000064 PWM Fault
        DCD    PWM0_Handler              ; 0x00000068 PWM Generator 0
```
Program 5.5. Software syntax to set the interrupt vectors for the TM4C.

Table 5.1 explains where to find the priority bits for some of the interrupts on the TM4C. In particular, this table shows the vector address, interrupt number, IRQ number, ISR name as defined in the file **Startup.s**, which register contains priority bits and which bits to modify when configuring priority. Each processor is a little different so check the data sheet.

Vector address	Number	IRQ	ISR name in **Startup.s**	NVIC	Priority bits
0x00000038	14	-2	`PendSV_Handler`	`NVIC_SYS_PRI3_R`	23 – 21
0x0000003C	15	-1	`SysTick_Handler`	`NVIC_SYS_PRI3_R`	31 – 29
0x00000040	16	0	`GPIOPortA_Handler`	`NVIC_PRI0_R`	7 – 5
0x00000044	17	1	`GPIOPortB_Handler`	`NVIC_PRI0_R`	15 – 13
0x00000048	18	2	`GPIOPortC_Handler`	`NVIC_PRI0_R`	23 – 21
0x0000004C	19	3	`GPIOPortD_Handler`	`NVIC_PRI0_R`	31 – 29
0x00000050	20	4	`GPIOPortE_Handler`	`NVIC_PRI1_R`	7 – 5
0x00000054	21	5	`UART0_Handler`	`NVIC_PRI1_R`	15 – 13
0x00000058	22	6	`UART1_Handler`	`NVIC_PRI1_R`	23 – 21
0x0000005C	23	7	`SSI0_Handler`	`NVIC_PRI1_R`	31 – 29
0x00000060	24	8	`I2C0_Handler`	`NVIC_PRI2_R`	7 – 5
0x00000064	25	9	`PWMFault_Handler`	`NVIC_PRI2_R`	15 – 13
0x00000068	26	10	`PWM0_Handler`	`NVIC_PRI2_R`	23 – 21
0x0000006C	27	11	`PWM1_Handler`	`NVIC_PRI2_R`	31 – 29
0x00000070	28	12	`PWM2_Handler`	`NVIC_PRI3_R`	7 – 5
0x00000074	29	13	`Quadrature0_Handler`	`NVIC_PRI3_R`	15 – 13
0x00000078	30	14	`ADC0_Handler`	`NVIC_PRI3_R`	23 – 21
0x0000007C	31	15	`ADC1_Handler`	`NVIC_PRI3_R`	31 – 29
0x00000080	32	16	`ADC2_Handler`	`NVIC_PRI4_R`	7 – 5
0x00000084	33	17	`ADC3_Handler`	`NVIC_PRI4_R`	15 – 13
0x00000088	34	18	`WDT_Handler`	`NVIC_PRI4_R`	23 – 21
0x0000008C	35	19	`Timer0A_Handler`	`NVIC_PRI4_R`	31 – 29
0x00000090	36	20	`Timer0B_Handler`	`NVIC_PRI5_R`	7 – 5
0x00000094	37	21	`Timer1A_Handler`	`NVIC_PRI5_R`	15 – 13
0x00000098	38	22	`Timer1B_Handler`	`NVIC_PRI5_R`	23 – 21
0x0000009C	39	23	`Timer2A_Handler`	`NVIC_PRI5_R`	31 – 29
0x000000A0	40	24	`Timer2B_Handler`	`NVIC_PRI6_R`	7 – 5
0x000000A4	41	25	`Comp0_Handler`	`NVIC_PRI6_R`	15 – 13
0x000000A8	42	26	`Comp1_Handler`	`NVIC_PRI6_R`	23 – 21
0x000000AC	43	27	`Comp2_Handler`	`NVIC_PRI6_R`	31 – 29
0x000000B0	44	28	`SysCtl_Handler`	`NVIC_PRI7_R`	7 – 5
0x000000B4	45	29	`FlashCtl_Handler`	`NVIC_PRI7_R`	15 – 13
0x000000B8	46	30	`GPIOPortF_Handler`	`NVIC_PRI7_R`	23 – 21
0x000000BC	47	31	`GPIOPortG_Handler`	`NVIC_PRI7_R`	31 – 29
0x000000C0	48	32	`GPIOPortH_Handler`	`NVIC_PRI8_R`	7 – 5
0x000000C4	49	33	`UART2_Handler`	`NVIC_PRI8_R`	15 – 13
0x000000C8	50	34	`SSI1_Handler`	`NVIC_PRI8_R`	23 – 21
0x000000CC	51	35	`Timer3A_Handler`	`NVIC_PRI8_R`	31 – 29
0x000000D0	52	36	`Timer3B_Handler`	`NVIC_PRI9_R`	7 – 5
0x000000D4	53	37	`I2C1_Handler`	`NVIC_PRI9_R`	15 – 13
0x000000D8	54	38	`Quadrature1_Handler`	`NVIC_PRI9_R`	23 – 21
0x000000DC	55	39	`CAN0_Handler`	`NVIC_PRI9_R`	31 – 29
0x000000E0	56	40	`CAN1_Handler`	`NVIC_PRI10_R`	7 – 5
0x000000E4	57	41	`CAN2_Handler`	`NVIC_PRI10_R`	15 – 13
0x000000E8	58	42	`Ethernet_Handler`	`NVIC_PRI10_R`	23 – 21
0x000000EC	59	43	`Hibernate_Handler`	`NVIC_PRI10_R`	31 – 29
0x000000F0	60	44	`USB0_Handler`	`NVIC_PRI11_R`	7 – 5
0x000000F4	61	45	`PWM3_Handler`	`NVIC_PRI11_R`	15 – 13
0x000000F8	62	46	`uDMA_Handler`	`NVIC_PRI11_R`	23 – 21
0x000000FC	63	47	`uDMA_Error`	`NVIC_PRI11_R`	31 – 29

Table 5.1. Some of the interrupt vectors for the TM4C (goes to number 154 on the M4).

Interrupts on the Cortex™-M processor are controlled by the Nested Vectored Interrupt Controller (NVIC). To activate an interrupt source we need to set its priority and enable that source in the NVIC. This activation is in addition to the arm and enable steps. Table 5.1 lists the interrupt sources available on the LM3S/LM4F/TM4C family of microcontrollers. Interrupt numbers 0 to 15 contain the faults, software interrupt and SysTick; these interrupts will be handled differently from interrupts 16 to 63.

Table 5.2 shows the twelve priority registers on the NVIC. There are 35 such registers on the TM4C. Each register contains an 8-bit priority field for four devices. On the TM4C microcontrollers, only the top three bits of the 8-bit field are used. This allows us to specify the interrupt priority level for each device from 0 to 7, with 0 being the highest priority. The interrupt number (number column in Table 5.1) is loaded into the **IPSR** register. The servicing of interrupts does not set the I bit in the **PRIMASK**, so a higher priority interrupt can suspend the execution of a lower priority ISR. If a request of equal or lower priority is generated while an ISR is being executed, that request is postponed until the ISR is completed. In particular, those devices that need prompt service should be given high priority.

Address	31 – 29	23 – 21	15 – 13	7 – 5	Name
0xE000E400	GPIO Port D	GPIO Port C	GPIO Port B	GPIO Port A	NVIC_PRI0_R
0xE000E404	SSI0, Rx Tx	UART1, Rx Tx	UART0, Rx Tx	GPIO Port E	NVIC_PRI1_R
0xE000E408	PWM Gen 1	PWM Gen 0	PWM Fault	I2C0	NVIC_PRI2_R
0xE000E40C	ADC Seq 1	ADC Seq 0	Quad Encoder	PWM Gen 2	NVIC_PRI3_R
0xE000E410	Timer 0A	Watchdog	ADC Seq 3	ADC Seq 2	NVIC_PRI4_R
0xE000E414	Timer 2A	Timer 1B	Timer 1A	Timer 0B	NVIC_PRI5_R
0xE000E418	Comp 2	Comp 1	Comp 0	Timer 2B	NVIC_PRI6_R
0xE000E41C	GPIO Port G	GPIO Port F	Flash Control	System Control	NVIC_PRI7_R
0xE000E420	Timer 3A	SSI1, Rx Tx	UART2, Rx Tx	GPIO Port H	NVIC_PRI8_R
0xE000E424	CAN0	Quad Encoder 1	I2C1	Timer 3B	NVIC_PRI9_R
0xE000E428	Hibernate	Ethernet	CAN2	CAN1	NVIC_PRI10_R
0xE000E42C	uDMA Error	uDMA Soft Tfr	PWM Gen 3	USB0	NVIC_PRI11_R
0xE000ED20	SysTick	PendSV	--	Debug	NVIC_SYS_PRI3_R

Table 5.2. Some of the TM4C NVIC registers. Bits not shown are zero.

There are five enable registers **NVIC_EN0_R** to **NVIC_EN4_R** on the TM4C. The 32 bits in **NVIC_EN0_R** control the IRQ numbers 0 to 31 (interrupt numbers 16 – 47). In Table 5.1 we see UART0 is IRQ=5. To enable UART0 interrupts we set bit 5 in **NVIC_EN0_R**. The bottom 16 bits in **NVIC_EN1_R** control the IRQ numbers 32 to 47 (interrupt numbers 48 – 63). In Table 5.1 we see Timer 3A is IRQ=35. To enable Timer 3A interrupts we set bit 3 (35-32=3) in **NVIC_EN1_R**. Not every interrupt source is available on every TM4C microcontroller, so you will need to refer to the data sheet for your microcontroller when designing I/O interfaces. Writing zeros to the **NVIC_EN0_R** to **NVIC_EN4_R** registers has no effect. To disable interrupts we write ones to the corresponding bit in the **NVIC_DIS0_R** to **NVIC_DIS4_R** register.

Figure 5.10 shows the context switch from executing in the foreground to running an edge-triggered ISR from Port C. Assume Port C interrupts are configured for a priority level of 5. The I bit in the PRIMASK is 0 signifying interrupts are enabled. The interrupt number (ISRNUM) in the **IPSR** register is 0, meaning we are running in **Thread mode** (i.e., the main program, and not an ISR). **Handler mode** is signified by a nonzero value in **IPSR**. When **BASEPRI** register is zero, all interrupts are allowed and the **BASEPRI** register is not active.

When a Port C interrupt is triggered, the current instruction is finished. (a) Eight registers are pushed on the stack with **R0** on top. These registers are pushed onto the stack using whichever stack pointer is active: either the **MSP** or **PSP**. (b) The vector address is loaded into the **PC** ("Vector address" column in Table 5.2). (c) The **IPSR** register is set to 18 ("Number" column in Table 5.2) (d) The top 24 bits of **LR** are set to 0xFFFFFF, signifying the processor is executing an ISR. Bits [7:1] specify how to return from interrupt. Bit 0 means Thumb mode.

0xE1 Return to Handler mode MSP (using floating point state on TM4C)
0xE9 Return to Thread mode MSP (using floating point state on TM4C)
0xED Return to Thread mode PSP (using floating point state on TM4C)
0xF1 Return to Handler mode MSP
0xF9 Return to Thread mode MSP
0xFD Return to Thread mode PSP

After pushing the registers, the processor always uses the main stack pointer (**MSP**) during the execution of the ISR. Events b, c, and d can occur simultaneously

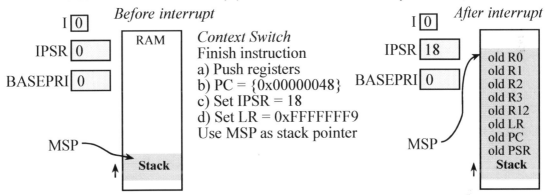

Figure 5.10. Stack before and after an interrupt, in this case a Port C edge-triggered interrupt.

To **return from an interrupt**, the ISR executes the typical function return **BX LR**. However, since the top 24 bits of **LR** are 0xFFFFFF, it knows to return from interrupt by popping the eight registers off the stack. Since the bottom eight bits of **LR** in this case are 0xF9, it returns to thread mode using the **MSP** as its stack pointer, in Thumb mode. Since the **IPSR** is part of the **PSR** that is pulled, so it is automatically reset its previous state.

A **nested interrupt** occurs when a higher priority interrupt suspends an ISR. The lower priority interrupt will finish after the higher priority ISR completes. When one interrupt preempts another, the **LR** is set to 0xFFFFFFF1, so it knows to return to handler mode. **Tail chaining** occurs when one ISR executes immediately after another. Optimization occurs because the eight registers need not be popped only to be pushed once again. If an interrupt is triggered and is in the process of stacking registers when a higher priority interrupt is requested, this **late arrival interrupt** will be executed first.

On the Cortex™-M4, if an interrupt occurs while in floating point state, an additional 18 words are pushed on the stack. These 18 words will save the state of the floating point processor. Bits 7-4 of the LR will be 0b1110 (0xE), signifying it was interrupted during a floating point state. When the ISR returns, it knows to pull these 18 words off the stack and restore the state of the floating point processor.

Priority determines the order of service when two or more requests are made simultaneously. Priority also allows a higher priority request to suspend a lower priority request currently being processed. Usually, if two requests have the same priority, we do not allow them to interrupt each other. NVIC assigns a priority level to each interrupt trigger. This mechanism allows a higher priority trigger to interrupt the ISR of a lower priority request. Conversely, if a lower priority request occurs while running an ISR of a higher priority trigger, it will be postponed until the higher priority service is complete.

5.5. Edge-triggered Interrupts

Table 4.2 listed the registers for Port A. The other ports have similar registers. We will begin with a simple example that counts the number of rising edges on Port C bit 4 (Program 5.6). The initialization requires many steps. (a) The clock for the port must be enabled. (b) The global variables should be initialized. (c) The appropriate pins must be enabled as inputs. (d) We must specify whether to trigger on the rise, the fall, or both edges. In this case we will trigger on the rise of PC4. (e) It is good design to clear the trigger flag during initialization so that the first interrupt occurs due to the first rising edge after the initialization has been run. We do not wish to trigger on a rising edge that might have occurred during the power up phase of the system. (f) We arm the edge-trigger by setting the corresponding bits in the **IM** register. (g) We establish the priority of Port C by setting bits 23 – 21 in the **NVIC_PRI0_R** register as listed in Table 5.2. We activate Port C interrupts in the NVIC by writing a one to bit 2 in the **NVIC_EN0_R** register ("IRQ number" in Table 5.1). This initialization is shown to enable interrupts in step (i). However, in most systems we would not enable interrupts in the device initialization. Rather, it is good design to initialize all devices in the system, then enable interrupts. On the TM4C we also clear the corresponding bits in **AMSEL** and **PCTL**.

```
volatile uint32_t FallingEdges = 0;
void EdgeCounter_Init(void){
  SYSCTL_RCGCGPIO_R |= 0x04;      // (a) activate clock for Port C
  FallingEdges = 0;               // (b) initialize counter
  GPIO_PORTC_DIR_R &= ~0x10;      // (c) make PC4 in
  GPIO_PORTC_DEN_R |= 0x10;       //     enable digital I/O on PC4
  GPIO_PORTC_IS_R &= ~0x10;       // (d) PC4 is edge-sensitive
  GPIO_PORTC_IBE_R &= ~0x10;      //     PC4 is not both edges
  GPIO_PORTC_IEV_R &= ~0x10;      //     PC4 falling edge event
  GPIO_PORTC_ICR_R = 0x10;        // (e) clear flag4
  GPIO_PORTC_IM_R |= 0x10;        // (f) arm interrupt on PC4
  NVIC_PRI0_R = (NVIC_PRI0_R&0xFF00FFFF)|0x00A00000; // (g) priority 5
  NVIC_EN0_R = 4;                 // (h) enable interrupt 2 in NVIC
  EnableInterrupts();             // (i) Program 5.3
}
void GPIOPortC_Handler(void){
  GPIO_PORTC_ICR_R = 0x10;        // acknowledge flag4
  FallingEdges = FallingEdges + 1;
}
```
Program 5.6. Interrupt-driven edge-triggered input that counts rising edges of PC4 (EdgeInterrupt_xxx).

All ISRs must acknowledge the interrupt by clearing the trigger flag that requested the interrupt. For edge-triggered PC4, the trigger flag is bit 4 of the `GPIO_PORTC_RIS_R` register. This flag can be cleared by writing a 0x10 to `GPIO_PORTC_ICR_R`.

If two or more triggers share the same vector, these requests are called **polled interrupts**, and the ISR must determine which trigger generated the interrupt. If the requests have separate vectors, then these requests are called **vectored interrupts** and the ISR knows which trigger caused the interrupt. Example 5.1 illustrates these differences.

Example 5.1. Interface two switches and signal associated semaphores when each switch is pressed.

Solution: We will assume the switches do not bounce (interfacing switches that bounce will be covered later in the chapter). The semaphore SW1 will be signaled when switch SW1 is pressed, and similarly, semaphore SW2 will be signed when switch SW2 is pressed. In the first solution, we will use vectored interrupts by connecting one switch to Port C and the other switch to Port E. Since the two sources have separate vectors, the switch on Port C will automatically activate `GPIOPortC_Handler` and switch on Port E will automatically activate `GPIOPortE_Handler`. The left side of Figures 5.11 and 5.12 show the solution with vectored interrupts.

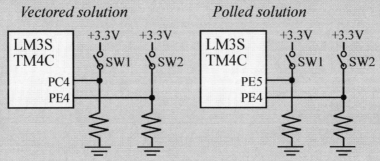

Figure 5.11. Two solutions of switch-triggered interrupts.

The software solution using vectored interrupts is in Program 5.7. We initialize two I/O pins as inputs with rising edge interrupt triggers. In this way, we get an interrupt request when the switch is touched. I.e., an interrupt occurs on the 0 to 1 rising edge either of PC4 or PE4. To acknowledge an interrupt we clear the trigger flag. Writing a 0x10 to the flag register, `GPIO_PORTn_ICR_R`, will clear bit 4 without affecting the other bits in the register. Notice that the acknowledgement uses an "=" instead of an "|=".

```
volatile uint8_t SW1, SW2; // semaphores
void VectorButtons_Init(void){
  SYSCTL_RCGCGPIO_R |= 0x14;   // activate clock for Ports C and E
  SW1 = 0;                     // clear semaphores
  SW2 = 0;
  GPIO_PORTC_DIR_R &= ~0x10;   // make PC4 in (PC4 built-in button)
  GPIO_PORTC_DEN_R |= 0x10;    // enable digital I/O on PC4
  GPIO_PORTC_IS_R &= ~0x10;    // PC4 is edge-sensitive (default setting)
```

```
  GPIO_PORTC_IBE_R &= ~0x10;    // PC4 is not both edges (default setting)
  GPIO_PORTC_IEV_R |= 0x10;     // PC4 rising edge event
  GPIO_PORTC_ICR_R = 0x10;      // clear flag4
  GPIO_PORTC_IM_R  |= 0x10;     // arm interrupt on PC4
  NVIC_PRI0_R = (NVIC_PRI0_R&0xFF00FFFF)|0x00400000; // PortC=priority 2
  GPIO_PORTE_DIR_R &= ~0x10;    // make PE4 in (PE4 button)
  GPIO_PORTE_DEN_R |= 0x10;     // enable digital I/O on PE4
  GPIO_PORTE_IS_R  &= ~0x10;    // PE4 is edge-sensitive (default setting)
  GPIO_PORTE_IBE_R &= ~0x10;    // PE4 is not both edges (default setting)
  GPIO_PORTE_IEV_R |= 0x10;     // PE4 rising edge event
  GPIO_PORTE_ICR_R = 0x10;      // clear flag4
  GPIO_PORTE_IM_R  |= 0x10;     // arm interrupt on PE4
  NVIC_PRI1_R = (NVIC_PRI1_R&0xFFFFFF00)|0x00000040; // PortE=priority 2
  NVIC_EN0_R = (NVIC_EN0_INT2+NVIC_EN0_INT4); // enable interrupts 2,4
  EnableInterrupts();
}
void GPIOPortC_Handler(void){
  GPIO_PORTC_ICR_R = 0x10;      // acknowledge flag4
  SW1 = 1;                      // signal SW1 occurred
}
void GPIOPortE_Handler(void){
  GPIO_PORTE_ICR_R = 0x10;      // acknowledge flag4
  SW2 = 1;                      // signal SW2 occurred
}
```

Program 5.7. Example of a vectored interrupt (TwoButtonVector_xxx).

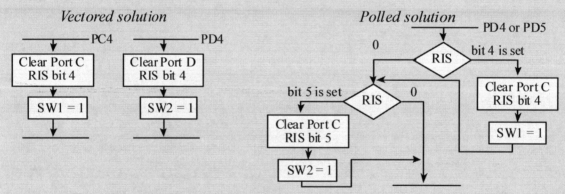

Figure 5.12. Flowcharts for a vectored and polled interrupt.

The right sides of Figures 5.11 and 5.12 show the solution with polled interrupts. Touching either switch will cause a Port E interrupt. The ISR must poll to see which one or possibly both caused the interrupt. Fortunately, even though they share a vector, the acknowledgements are separate. The code `GPIO_PORTE_ICR_R=0x10;` will clear bit 4 in the status register without affecting bit 5, and the code `GPIO_PORTE_ICR_R=0x20;` will clear bit 5 in the status register without affecting bit 4. This means the timing of one switch does not affect whether or not pushing the other switch will signal its semaphore. On the other hand, whether we are using polled or vectored interrupt, because there is only one processor, the timing of one interrupt may delay the servicing of another interrupt.

The polled solution is Program 5.8. It takes three conditions to cause an interrupt. 1) The PE4 and PE5 are armed in the initialization; 2) The LM3S/LM4F/TM4C is enabled for interrupts with the **EnableInterrupts()** function; 3) The trigger **GPIO_PORTE_RIS_R** is set on the rising edge of PE4 or the trigger **GPIO_PORTE_RIS_R** is set on the rising edge of PE5. Because the two triggers have separate acknowledgments, if both triggers are set, both will get serviced. Furthermore, the polling sequence does not matter.

```
volatile uint8_t SW1, SW2;
void PolledButtons_Init(void){
  SYSCTL_RCGCGPIO_R |= 0x10;     // activate clock for Port E
  SW1 = 0; SW2 = 0;              // clear semaphores
  GPIO_PORTE_DIR_R &= ~0x30;     // make PE5-4 in (PE5-4 buttons)
  GPIO_PORTE_DEN_R |= 0x30;      // enable digital I/O on PE5-4
  GPIO_PORTE_IS_R &= ~0x30;      // PE5-4 is edge-sensitive
  GPIO_PORTE_IBE_R &= ~0x30;     // PE5-4 is not both edges
  GPIO_PORTE_IEV_R |= 0x30;      // PE5-4 rising edge event
  GPIO_PORTE_ICR_R = 0x30;       // clear flag5-4
  GPIO_PORTE_IM_R |= 0x30;       // arm interrupts on PE5-4
  NVIC_PRI1_R = (NVIC_PRI1_R&0xFFFFFF00)|0x00000040; // PortE=priority 2
  NVIC_EN0_R = NVIC_EN0_INT4;    // enable interrupt 4 in NVIC
  EnableInterrupts();
}
void GPIOPortE_Handler(void){
  if(GPIO_PORTE_RIS_R&0x10){    // poll PE4
    GPIO_PORTE_ICR_R = 0x10;    // acknowledge flag4
    SW1 = 1;                     // signal SW1 occurred
  }
  if(GPIO_PORTE_RIS_R&0x20){    // poll PE5
    GPIO_PORTE_ICR_R = 0x20;    // acknowledge flag5
    SW2 = 1;                     // signal SW2 occurred
  }
}
```
Program 5.8. Example of a polled interrupt (TwoButtonPoll_xxx).

5.6. Interrupt-Driven UART

Figure 5.13 shows a data flow graph with buffered input and buffered output. FIFOs implemented in this section are statically allocated global structures. Because they are global variables, it means they will exist permanently and can be carefully shared by the foreground and background threads. The advantage of using a FIFO structure for a data flow problem is that we can decouple the producer and consumer threads. Without the FIFO we would have to produce one piece of data, then process it, produce another piece of data, then process it. With the FIFO, the producer thread can continue to produce data without having to wait for the consumer to finish processing the previous data. This decoupling can significantly improve system performance.

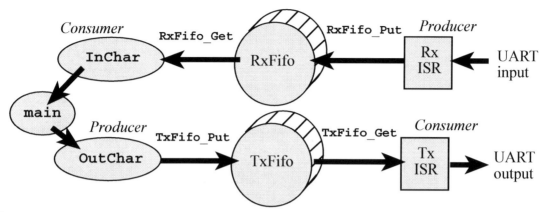

Figure 5.13. A data flow graph showing two FIFOs that buffer data between producers and consumers.

Checkpoint 5.6: What does it mean if the RxFifo in Figure 5.13 is empty?

Checkpoint 5.7: What does it mean if the TxFifo in Figure 5.13 is empty?

The system shown in Figure 5.13 has two channels, one for input and one for output, and each channel employs a separate FIFO queue. Program 5.9 shows the interrupt-driven UART device driver. The flowchart for this interface was shown previously as Figure 5.4. During initialization, Port A pins 0 and 1 are enabled as alternate function digital signals. The two software FIFOs of Program 3.9 are initialized. The baud rate is set at 115200 bits/sec, and the hardware FIFOs are enabled. A transmit interrupt will occur as the transmit FIFO goes from 2 elements down to 1 element. Not waiting until the hardware FIFO is completely empty allows the software to refill the hardware FIFO and maintain a continuous output stream, achieving maximum bandwidth. There are two conditions that will request a receive interrupt. First, if the receive FIFO goes from 2 to 3 elements a receive interrupt will be requested. At this time there is still 13 free spaces in the receive FIFO so the latency requirement for this real-time input will be 130 bit times (about 1 ms). The other potential source of receiver interrupts is the receiver time out. This trigger will occur if the receiver becomes idle and there are data in the receiver FIFO. This trigger will allow the interface to receive input data when it comes just one or two frames at a time. In the NVIC, the priority is set at 2 and UART0 (IRQ=5) is activated. Normally, one does not enable interrupts in the individual initialization functions. Rather, interrupts should be enabled in the main program, after all initialization functions have completed.

When the main thread wishes to output it calls **UART_OutChar**, which will put the data into the software FIFO. Next, it copies as much data from the software FIFO into the hardware FIFO and arms the transmitter. The transmitter interrupt service will also get as much data from the software FIFO and put it into the hardware FIFO. The **copySoftwareToHardware** function has a critical section and is called by both **UART_OutChar** and the ISR. To remove the critical section the transmitter is temporarily disarmed in the **UART_OutChar** function when **copySoftwareToHardware** is called. This helper function guarantees data is transmitted in the same order it was produced.

When input frames are received they are placed into the receive hardware FIFO. If this FIFO goes from 2 to 3 elements, or if the receiver becomes idle with data in the FIFO, a receive interrupt occurs. The helper function **copyHardwareToSoftware** will get from the receive hardware FIFO and put into the receive software FIFO. When the main thread wished to input data it calls **UART_InChar**. This function simply gets from the software FIFO. If the receive software FIFO is empty, it will spin.

```
#define FIFOSIZE   16        // size of the FIFOs (must be power of 2)
#define FIFOSUCCESS 1        // return value on success
#define FIFOFAIL    0        // return value on failure
AddIndexFifo(Rx, FIFOSIZE, char, FIFOSUCCESS, FIFOFAIL)
AddIndexFifo(Tx, FIFOSIZE, char, FIFOSUCCESS, FIFOFAIL)
// Assumes an 80 MHz bus clock, creates 115200 baud rate
void UART_Init(void){              // should be called only once
  SYSCTL_RCGCUART_R |= 0x0001;     // activate UART0
  SYSCTL_RCGCGPIO_R |= 0x0001;     // activate port A
  RxFifo_Init();                   // initialize empty FIFOs
  TxFifo_Init();
  UART0_CTL_R &= ~UART_CTL_UARTEN;       // disable UART
  UART0_IBRD_R = 43; // IBRD=int(80000000/(16*115,200)) = int(43.40278)
  UART0_FBRD_R = 26; // FBRD = round(0.40278 * 64) = 26
  UART0_LCRH_R = 0x0070; // 8-bit word length, enable FIFO
  UART0_IFLS_R &= ~0x3F; // clear TX and RX interrupt FIFO level fields
                         // configure interrupt for TX FIFO <= 1/8 full
                         // configure interrupt for RX FIFO >= 1/8 full
  UART0_IFLS_R += (UART_IFLS_TX1_8|UART_IFLS_RX1_8);
          // enable TX and RX FIFO interrupts and RX time-out interrupt
  UART0_IM_R |= (UART_IM_RXIM|UART_IM_TXIM|UART_IM_RTIM);
  UART0_CTL_R |= 0x0301;                 // enable RXE TXE UARTEN
  GPIO_PORTA_PCTL_R = (GPIO_PORTA_PCTL_R&0xFFFFFF00)+0x00000011; // UART
  GPIO_PORTA_AMSEL_R &= ~0x03; // disable analog on PA1-0
  GPIO_PORTA_AFSEL_R |= 0x03;  // enable alt funct on PA1-0
  GPIO_PORTA_DEN_R |= 0x03;             // enable digital I/O on PA1-0
  NVIC_PRI1_R = (NVIC_PRI1_R&0xFFFF00FF)|0x00004000; // UART0=priority 2
  NVIC_EN0_R = NVIC_EN0_INT5;           // enable interrupt 5 in NVIC
  EnableInterrupts();
}
// copy from hardware RX FIFO to software RX FIFO
// stop when hardware RX FIFO is empty or software RX FIFO is full
void static copyHardwareToSoftware(void){ char letter;
  while(((UART0_FR_R&UART_FR_RXFE)==0)&&(RxFifo_Size() < (FIFOSIZE-1))){
    letter = UART0_DR_R;
    RxFifo_Put(letter);
  }
}
// copy from software TX FIFO to hardware TX FIFO
// stop when software TX FIFO is empty or hardware TX FIFO is full
void static copySoftwareToHardware(void){ char letter;
  while(((UART0_FR_R&UART_FR_TXFF) == 0) && (TxFifo_Size() > 0)){
    TxFifo_Get(&letter);
    UART0_DR_R = letter;
  }
```

```c
}
// input ASCII character from UART
// spin if RxFifo is empty
char UART_InChar(void){
  char letter;
  while(RxFifo_Get(&letter) == FIFOFAIL){}; // spin if no data
  return(letter);
}
// output ASCII character to SCI
// spin if TxFifo is full
void UART_OutChar(char data){
  while(TxFifo_Put(data) == FIFOFAIL){}; // spin if TxFIFO has no room
  UART0_IM_R &= ~UART_IM_TXIM;           // disable TX FIFO interrupt
  copySoftwareToHardware();
  UART0_IM_R |= UART_IM_TXIM;            // enable TX FIFO interrupt
}
// at least one of three things has happened:
// hardware TX FIFO goes from 3 to 2 or less items
// hardware RX FIFO goes from 1 to 2 or more items
// UART receiver has timed out with one or more items available
void UART0_Handler(void){
  if(UART0_RIS_R&UART_RIS_TXRIS){        // hardware TX FIFO <= 2 items
    UART0_ICR_R = UART_ICR_TXIC;         // acknowledge TX FIFO
    // copy from software TX FIFO to hardware TX FIFO
    copySoftwareToHardware();
    if(TxFifo_Size() == 0){              // software TX FIFO is empty
      UART0_IM_R &= ~UART_IM_TXIM;       // disable TX FIFO interrupt
    }
  }
  if(UART0_RIS_R&UART_RIS_RXRIS){        // hardware RX FIFO >= 2 items
    UART0_ICR_R = UART_ICR_RXIC;         // acknowledge RX FIFO
    // copy from hardware RX FIFO to software RX FIFO
    copyHardwareToSoftware();
  }
  if(UART0_RIS_R&UART_RIS_RTRIS){        // receiver timed out
    UART0_ICR_R = UART_ICR_RTIC;         // acknowledge receiver time out
    // copy from hardware RX FIFO to software RX FIFO
    copyHardwareToSoftware();
  }
}
```

Program 5.9. Interrupt-driven device driver for the UART uses two hardware FIFOs and two software FIFOs to buffer data (UARTint_xxx).

5.7. Periodic Interrupts using SysTick

The SysTick timer is a simple way to create periodic interrupts. A periodic interrupt is one that is requested on a fixed time basis. This interfacing technique is required for data acquisition and control systems, because software servicing must be performed at accurate time intervals. For a data acquisition system, it is important to establish an accurate sampling rate. The time in between ADC samples must be equal (and known) in order for the digital signal processing to function properly. Similarly for microcontroller-based control systems, it is important to maintain both the ADC and DAC timing. Later in Chapter 6, we will see the general purpose timers can also create periodic interrupts.

Another application of periodic interrupts is called "intermittent polling" or "periodic polling". Figure 5.14 shows busy wait side by side with periodic polling. In busy-wait synchronization, the main program polls the I/O devices continuously. With periodic polling, the I/O devices are polled on a regular basis (established by the periodic interrupt.) If no device needs service, then the interrupt simply returns. If the polling period is Δt, then on average the interface latency will be $\frac{1}{2}\Delta t$, and the worst case latency will be Δt. Periodic polling is appropriate for low bandwidth devices where real-time response is not necessary. This method frees the main program from the I/O tasks.

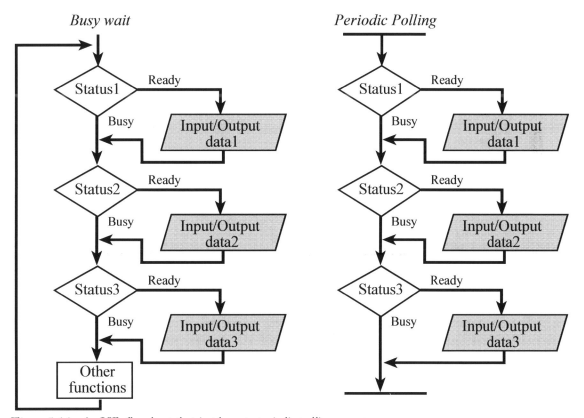

Figure 5.14. An ISR flowchart that implements periodic polling.

We use periodic polling if the following two conditions apply:

1. The I/O hardware cannot generate interrupts directly
2. We wish to perform the I/O functions in the background

Table 5.3 shows the SysTick registers used to create a periodic interrupt. SysTick has a 24-bit counter that decrements at the bus clock frequency. Let f_{BUS} be the frequency of the bus clock, and let n be the value of the **RELOAD** register. The frequency of the periodic interrupt will be $f_{BUS}/(n+1)$. First, we clear the **ENABLE** bit to turn off SysTick during initialization. Second, we set the **RELOAD** register. Third, we write to the **NVIC_ST_CURRENT_R** value to clear the counter. Lastly, we write the desired mode to the control register, **NVIC_ST_CTRL_R**. We must set **CLK_SRC**=1, because **CLK_SRC**=0 external clock mode is not implemented.

Address	31-24	23-17	16	15-3	2	1	0	Name
$E000E010	0	0	COUNT	0	CLK_SRC	INTEN	ENABLE	NVIC_ST_CTRL_R
$E000E014	0			24-bit RELOAD value				NVIC_ST_RELOAD_R
$E000E018	0			24-bit CURRENT value of SysTick counter				NVIC_ST_CURRENT_R

Address	31-29	28-24	23-21	20-8	7-5	4-0	Name
$E000ED20	TICK	0	PENDSV	0	DEBUG	0	NVIC_SYS_PRI3_R

Table 5.3. SysTick registers.

We set **INTEN** to enable interrupts. We establish the priority of the SysTick interrupts using the TICK field in the **NVIC_SYS_PRI3_R** register. We need to set the **ENABLE** bit so the counter will run. When the **CURRENT** value counts down from 1 to 0, the **COUNT** flag is set. On the next clock, the **CURRENT** is loaded with the **RELOAD** value. In this way, the SysTick counter (**CURRENT**) is continuously decrementing. If the **RELOAD** value is n, then the SysTick counter operates at modulo $n+1$ (…n, n-1, n-2 … 1, 0, n, n-1, …). In other words, it rolls over every $n+1$ counts. Thus, the **COUNT** flag will be configured to trigger an interrupt every $n+1$ counts.

```
volatile uint32_t Counts;
#define PD0     (*((volatile uint32_t *)0x40007004))
// period has units of the bus clock (e.g., 12.5ns or 20ns)
void SysTick_Init(uint32_t period){
  SYSCTL_RCGCGPIO_R |= 0x00000008; // activate port D
  Counts = 0;
  GPIO_PORTD_DIR_R |= 0x01;    // make PD0 out
  GPIO_PORTD_PCTL_R &= ~0x0000000F; // GPIO
  GPIO_PORTD_AMSEL_R &= ~0x01; // disable analog on PD0
  GPIO_PORTD_AFSEL_R &= ~0x01; // disable alt funct on PD0
  GPIO_PORTD_DEN_R |= 0x01;    // enable digital I/O on PD0
  NVIC_ST_CTRL_R = 0;          // disable SysTick during setup
  NVIC_ST_RELOAD_R = period-1; // reload value
  NVIC_ST_CURRENT_R = 0;       // any write to current clears it
  NVIC_SYS_PRI3_R = (NVIC_SYS_PRI3_R&0x00FFFFFF)|0x40000000; //priority 2
  NVIC_ST_CTRL_R = 0x00000007;// enable with core clock and interrupts
  EnableInterrupts();
}
void SysTick_Handler(void){
```

```
    PD0 ^= 0x01;           // toggle PD0
    Counts = Counts + 1;
}
```
Program 5.10. Implementation of a periodic interrupt using SysTick (PeriodicSysTickInts_xxx).

One of the problems with switches is called **switch bounce**. Many inexpensive switches will mechanically oscillate for up to a few milliseconds when touched or released. It behaves like an underdamped oscillator. These mechanical oscillations cause electrical oscillations such that a port pin will oscillate high/low during the bounce. In some cases this bounce should be removed.

Example 5.2. Redesign Example 4.3 of a matrix keyboard interface using interrupt synchronization. The ISR will put ASCII data into the FIFO. The main will get data from the FIFO.

Solution: Figure 4.40 is redrawn here as Figure 5.15. There are two good solutions to using interrupt synchronization for the keyboard. The approach implemented here uses periodic polling, because it affords a simple solution to both bouncing and two-key rollover. The time between interrupts is selected to be longer than the maximum bounce time, but shorter than the minimum time between key strikes. If you type ten characters per second, the minimum time between rising and falling edges is about 50 ms. Since switch bounce times are less than 10 ms, we will poll the keyboard every 25 ms. This means the average latency will be 12.5 ms, and the maximum latency will be 25 ms.

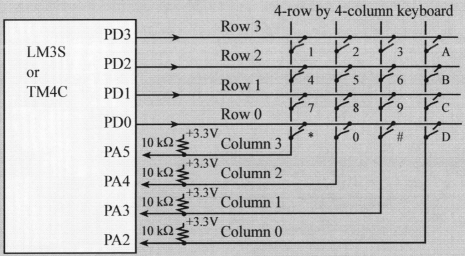

Figure 5.15. A matrix keyboard interfaced to the microcontroller (same as Figure 4.40).

The initialization combines Program 4.11 with Program 5.10 with the additional initialization of the FIFO. The PD0 debugging will have to be removed from SysTick because PD0 is used to interrupt the keyboard. A key is recognized if the scanning returns one key found, and this key is different from what it scanned 25 ms ago.

```
AddIndexFifo(Matrix, 16, char, 1, 0) // create a FIFO
char static LastKey;
```

5. Interrupt Synchronization

```
void Matrix_Init(void){
  LastKey = 0;              // no key typed
  MatrixFifo_Init();
  MatrixKeypad_Init();      // Program 4.11
  SysTick_Init(1250000);    // Program 5.10, 25 ms polling
}
void SysTick_Handler(void){ char thisKey; int32_t n;
  thisKey = MatrixKeypad_Scan(&n); // scan
  if((thisKey != LastKey) && (n == 1)){
    MatrixFifo_Put(thisKey);
    LastKey = thisKey;
  } else{
    LastKey = 0; // invalid
  }
}
char Matrix_InChar(void){  char letter;
  while(MatrixFifo_Get(&letter) == FIFOFAIL){};
  return(letter);
}
```
Program 5.11. Periodic polling interface of a scanned keyboard (MatrixKeypadPeriodic_xxx).

One of the advantages of Program 5.11 is **two-key rollover**. When people type very fast, they sometimes type the next key before the release the first key. For example, when the operator types the letters "BCD" slowly with one finger, the keyboard status goes in this sequence

<none>, <B>, <none>, <C>, <none>, <D>, <none>

Conversely, if the operator types quickly, there can be two-key rollover, which creates this sequence

<none>, <B>, <BC>, <C>, <CD>, <D>, <none>

where <BC> means both keys 'B' and 'C' are touched. Two-key rollover means the keyboard does not go through a state where no keys are touched between typing the 'B' and the 'C'. Since each of the keys goes through a state where exactly one key is pressed and is different than it was 25 ms ago, Program 5.11 will handle two-key rollover.

A second approach is to arm the device for interrupts by driving all rows to zero. In this manner, we will receive a falling edge on one of the Port A inputs when any key is touched. During the ISR we could scan the keyboard and put the key into the FIFO. To solve the bounce problem this solution implements a time delay from key touch to when the software scans for keys. This approach uses a combination of edge-triggered inputs and output compare interrupts to perform input in the background. When arming for interrupts, we set all four rows to output zero. In this way, a falling edge interrupt will occur on any key touched. When an edge-triggered interrupt occurs we will disarm this input and arm an output compare to trigger in 10 ms. It is during the output compare ISR we scan the matrix. If there is exactly one key, we enter it into the FIFO. An interrupt may occur on release due to bounce. However, 10 ms after the release, when we scan during the Timer0A_Handler the **MatrixKeypad_Scan** function will return a **Num** of zero, and we will ignore it. This solution solves switch bounce, but not two-key rollover. You can find this solution on the web site as **MatrixKeypadInt_xxx**.

There are three solutions to debounce an individual switch posted on the example page
1) **Blind** synchronization: read switch, and then wait 10 ms
2) **Periodic polling**: use a periodic interrupt at 10 ms
3) **Interrupt**: edge-triggered interrupt, then time delay interrupt

5.8. Low-Power Design

Reducing the amount of power used by an embedded system will save money and extend battery life. In CMOS digital logic, power is required to make signals rise or fall. Most microcontrollers allow you to adjust the frequency of the bus clock. Most microcontrollers allow you to change the bus clock using a PLL. Selecting the bus clock frequency is a tradeoff between power and performance. To optimize for power, we choose the slowest bus clock frequency that satisfies the minimum requirements of the system. When we implement the software system with interrupts, it allows us to focus the processor on executing tasks, when they need to run. Because there are fewer backward jumps wasting time, interrupt-driven systems typically will be able to perform the same functions at a slower bus clock.

Capacitance plays a major role in high speed digital systems. The current/voltage relationship across a capacitor is $I = C\, dV/dt$. For a given digital circuit the capacitances are approximately fixed. As we increase the bus frequency using the PLL, the slew rates in the signals must increase (larger dV/dt), requiring more current to operate. Furthermore, CMOS logic requires charge to create a transition, so as the frequency increases the number of transitions per second increases, requiring more current (charge per second is current).

A second factor in low-power design follows the axiom "turn the light off when you leave the room." Basically, we turn off devices when they are not being used. Most I/O devices on the microcontroller are initialized as off, so we have to turn them on to use them. However, rather than turning it on once in the initialization and leaving it on continuously, we could dynamically turn it on when needed then turn it off when done. In Chapter 9, we will learn ways to turn off external analog circuits when they are not needed.

In general, as we reduce the voltage, the power is reduced. Some microcontroller families have versions that will run at different voltages. For example the MC9S12C32 can operate at either 3.3 or 5 V. If a device has a fixed resistance from supply voltage to ground, reducing from 5 to 3.3 V will drop the power by a factor of $(5^2-3.3^2)/5^2 = 56\%$. The MSP430F2012 can operate on a supply voltage ranging from 1.8 to 3.6 V. Reducing the voltage on the MSP430 will decrease the maximum speed at which we can run the bus clock.

Many microcontrollers allow us to control the current output on the output pins. Reducing the current on an output pin will make it operate slower but draw less current. Any unused input pins should have internal or external pull-up (or pull-down). An input pin not connected to anything may oscillate at a frequency of an external field, wasting power unnecessarily. Unused pins can also be made outputs. On TM4C microcontrollers we can turn off individual pins that we are not using by clearing bits in the DEN register (e.g., `GPIO_PORTD_DEN_R`). If we have an entire port not being used, we keep it completely off.

One way to save power is to perform all operations as background tasks and put the processor to sleep when in the foreground. The `WFI` instruction stops the bus clock, and the processor stops executing instructions. One way to use it is to make a function that C programs can call.

```
WaitForInterrupt
        WFI
        BX      LR
```

An interrupt will wake up the sleeping processor. A TM4C123 consumes about 25 mA while running at 80 MHz. However, in sleep mode, the supply current drops to about 12 mA, depending on what else is active. The TM4C family also has a deep sleep mode, where it consumes about 2.5 mA. To illustrate this approach, consider the two systems in Program 5.12. Both systems execute **Stuff** at a fixed rate, but the one with the wait for interrupt requires less power.

```
int main(void){                         int main(void){
  SysTick_Init(50000);                    SysTick_Init(50000);
  while(1){                               while(1){
    WaitForInterrupt();                     // runs at full power
  }                                       }
}                                       }
void SysTick_Handler(void){             void SysTick_Handler(void){
  Stuff();                                Stuff();
}                                       }
```

Program 5.12. Example showing how to save power by putting the processor to sleep.

Maintenance Tip: Whenever your software performs a backward jump (e.g., waiting for an event), it may be possible to put the processor to sleep, thus saving power.

5.9. Debugging Profile

One way to see both how long it takes to execute an ISR and how often it executes is to toggle an output pin three times (triple toggle technique), as shown in Program 5.13. Figure 5.16 shows the time between interrupts is 1ms, and the time to execute one ISR is 650ns.

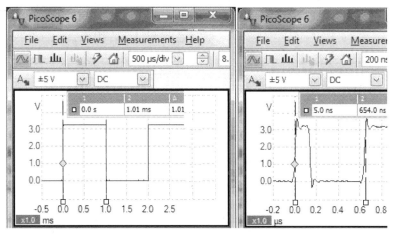

Figure 5.16. Profile showing both time between interrupts and time within the interrupt (TM4C123, 50 MHz).

```
#define PF2    (*((volatile uint32_t *)0x40025010))
void SysTick_Handler(void){
  PF2 ^= 0x04;        // toggle PF2
  PF2 ^= 0x04;        // toggle PF2
  Stuff();
  PF2 ^= 0x04;        // toggle PF2
}
```
Program 5.13. The triple toggle technique shows both the execution time for Stuff() and the time between ISRs.

5.10. Exercises

5.1 Syntactically, I/O ports are public globals. In order to separate mechanisms from policies (i.e., improve the quality of the software system), how should I/O be actually used?
    a) Local in allocation   **b)** Private in scope
    c) Global in allocation   **d)** Like volatile memory
    e) Public in scope   **f)** Like nonvolatile memory

5.2 Why do we add the **volatile** qualifier in all I/O port definitions?

5.3 What happens if an interrupt service routine does not acknowledge or disarm?
    a) Software crashes because no more interrupts will be requested
    b) The next interrupt is lost
    c) This interrupt is lost
    d) Software crashes because interrupts are requested over and over.

5.4 The main program synthesizes data and a periodic interrupt will output the data separated by a fixed time. A FIFO queue is used to buffer data between a main program (e.g., main program calls **Fifo_Put**). The interrupt service routine calls **Fifo_Get** and actually outputs. Experimental observations show this FIFO is usually empty, and has at most 3 elements. What does it mean? Choose a-f.
    a) The system is CPU bound   **b)** Bandwidth could be increased by increasing FIFO size
    c) The system is I/O bound   **d)** The FIFO could be replaced by a global variable
    e) The latency is small and bounded   **f)** Interrupts are not needed in this system

5.5 Answer question 5.4, under the condition that the FIFO often becomes full.

5.6 An edge-triggered input is armed so that interrupts occur when new data arrives into the microcontroller. Consider the situation in which a FIFO queue is used to buffer data between the ISR and the main program. The ISR inputs the data and saves the data by calling **Fifo_Put**. When the main program wants input it calls **Fifo_Get**. Experimental observations show this FIFO is usually empty, having at most 3 elements. What does it mean? Choose a-f.
    a) The system is CPU bound
    b) Bandwidth could be increased by increasing FIFO size
    c) The system is I/O bound
    d) The FIFO could be replaced by a global variable

e) The latency is small and bounded
f) Interrupts are not needed in this system

5.7 Answer question 5.6, under the condition that the FIFO often becomes full.

5.8 Consider the following interrupt service routine. The goal is to measure the elapsed time from one interrupt call to the other. What qualifier do you place in the **XX** position to make this measurement operational? Choose from **volatile static float const** or **public**.

```
uint32_t Elapsed;   // time between interrupt
void handler(void){
XX uint32_t last=0;
  Elapsed = (NVIC_ST_CURRENT_R - last)&0x00FFFFFF;
  last = NVIC_ST_CURRENT_R; }
```

5.9 What purpose might there be to use the PLL and slow down the microcontroller?
  a) The system is CPU bound
  b) To make the batteries last longer on a battery-powered system
  c) In order to adjust the baud rate to a convenient value
  d) In order to balance the load between foreground and background threads
  e) To reduce latency
  f) None of the above, because there is never a reason to run slower

5.10 This is a *functional debugging* question. However, the debugging instrument still needs to be *minimally intrusive*. Assume **y=Function(x)** is a function with 16-bit input **x** and 16-bit output **y** and is called from an ISR as part of a real-time system. The UART, Port F and Port G are unused by the system, and Port F and Port G are digital outputs. The debugging code will be placed at the end just before the return, unless otherwise stated. **UART_OutSDec** outputs a 16-bit signed integer. **BufX** and **BufY** are 16-bit signed global buffers of length 100, **n** is a global variable initialized to 0. Which debugging code would you add to verify the correctness of this function?

  a)   `GPIO_PORTF_DATA_R =x; GPIO_PORTG_DATA_R =y;`
  b)   `UART_OutSDec(x); UART_OutSDec(y);   // busy wait`
  c)   `if(n<100){BufX[n]=x; BufY[n]=y; n++;}`
  d)   `GPIO_PORTF_DATA_R |= 0x01;    // at beginning`
          `GPIO_PORTF_DATA_R &= ~0x01;   // at end`
  e)   `if(n<100){BufX[n]=x; BufY[n]=NVIC_CURRENT_R; n++;}`

5.11 Four events must occur for an edge-triggered interrupt on Port E bit 1 to be generated:
1) Software sets the arm bit (IM bit 1)
2) Software enables interrupts (I=0)
3) Hardware sets the flag bit (RIS bit 1)
4) Software configures the NVIC to allow Port E input interrupts
Which time sequence of these four events cause the interrupt to be generated?
  a) Only the order 1,2,3,4
  b) Only the order 1,2,4,3
  c) Only the order 4,1,2,3
  d) Any order will generate an interrupt

5.12 The following multithreaded system has a critical section. Modify these programs to remove the error. You may assume the interrupts are enabled and are running.

```
uint32_t Sec,Min;                        void InterruptHandler(void){
void main(void){                           if(Sec == 59){
  while(1){                                  Sec = 0; Min++;
    if((Min==10)&&(Sec==0)){                } else{
      UART_OutString("done");                 Sec++;
    }                                       }
  }                                        }
}
```

5.13 The following multithreaded system uses Port F as a debugging profile. One interrupt has higher priority than the other. **Stuff1** and **Stuff2** are unrelated. Is there a critical section? If yes, edit the code to remove the critical section. If no, justify your answer in 16 words or less.

```
void InterruptHandler1(void){            void InterruptHandler2(void){
  GPIO_PORTF_DATA_R ^= 0x02;               GPIO_PORTF_DATA_R ^= 0x04;
  Stuff1();                                Stuff2();
}                                        }
```

The assembly code for the **InterruptHandler1** is
```
PUSH  {lr}
LDR   r0,[pc,#180]     ; R0 points to Port F
LDR   r0,[r0,#0x00]    ; R0 is the value from Port F
EOR   r0,r0,#0x02      ; toggle bit 1
LDR   r1,[pc,#152]     ; R1 points to Port F
STR   r0,[r1,#0x3FC]   ; write new value to Port F
BL.W  Stuff0           ; call function
POP   {pc}             ; return from interrupt
```

5.14 What happens if two interrupt requests are made during the same instruction? Is one lost? If both are serviced, which one goes first?

5.15 Specify whether each statement is TRUE or FALSE.
a) A **signed char** or **int8_t** variable can store values from –128 to +128.
b) Consider an **input** device. The **interface latency** is the time from when the software asks for new data until the time new data are ready.
c) Consider an **output** device. The **interface latency** is the time from when the software sends new data to the output device until the time the output operation is complete.
d) The **static** qualifier is used with functions to specify the function is permanent, created at compile time and is never destroyed. E.g.,
  static int16_t AddTwo(int16_t in){ return in+2;}
e) The **static** qualifier is used with a variable defined inside a function to specify the variable is permanent, created at compile time, initialized to 0, and is never destroyed. E.g.,
  void function(int16_t in){ static int16_t myData;
f) The **const** qualifier is used with a global variable to specify the variable should be allocated in ROM. E.g., **const int16_t myData=5;**
g) Code that is **friendly** means it can be executed by more than one thread without causing a crash or loss of data.

h) The `volatile` qualifier is used with variables to tell the compiler that code that accesses this variable should be optimized as much as possible.

i) A read-modify-write access to a shared global variable always creates a **critical** section.

j) The compiler automatically sets I=1 at the beginning of the **interrupt service routine** and clears I=0 at the end so that the computer runs with interrupts disabled while servicing the interrupt.

5.16 Modify Program 5.9 assuming the bus clock is 8 MHz and want to a baud rate of 9600 bits/sec.

5.17 Modify Program 5.9 assuming the bus clock is 6 MHz and want to a baud rate of 38400 bits/sec.

5.18 Modify Program 5.9 assuming the bus clock is 25 MHz and want to a baud rate of 19200 bits/sec.

5.19 Modify Program 5.9 to use UART1, assuming the bus clock is 50 MHz and want to a baud rate of 115200 bits/sec.

5.20 Which of the following debugging instruments is faster?
```
        GPIO_PORTD0 ^= 0x01;
```
or
```
        GPIO_PORTD0 = 0x01; GPIO_PORTD0 = 0x00;
```

5.21 Modify Program 5.10 assuming the bus clock is 8 MHz and you want to interrupt every 1 ms.

5.22 Modify Program 5.10 assuming the bus clock is 50 MHz and you want to interrupt every 100 μs.

5.23 Modify Program 5.10 assuming the bus clock is 6 MHz and you want to interrupt every 1 sec.

5.24 These seven events all occur during each SysTick interrupt. Order these events into a proper time sequence. More than one answer may be correct.
1) The **CURRENT** equals **0** and the hardware sets the SysTick trigger flag
2) The SysTick vector address is loaded into the PC
3) The SysTick trigger flag is automatically cleared by hardware
4) The software executes the SysTick ISR
5) The R0, R1, R2, R3, R12, LR, PC and PSR are pushed on the stack
6) The R0, R1, R2, R3, R12, LR, PC and PSR are pulled from the stack
7) The software executes `bx lr`

D5.25 Design and implement a FIFO that can hold up to 4 elements. Each element is 10 bytes. There will be three functions: `Init`, `Put` one element into FIFO and `Get` one element from the FIFO. Show the private RAM-based variables. Write a function that initializes the FIFO. Write a function that puts one 10-byte element into the FIFO. Write a function that gets one 10-byte element from the FIFO. Document how parameters are passed in each function.

D5.26 Design and implement a FIFO that uses the memory manager to allocate space. Each element is 32-bits. Call malloc and free on blocks of 1024 bytes (256 words). If the FIFO needs more space allocate it by calling malloc(1024). After all the data from one block is returned via Get, then deallocate that block by calling Free. You are allowed to permanently define pointers and counters, but no permanent buffers are allowed. There will be three functions: **Fifo_Init**, **Fifo_Put** one element into FIFO and **Fifo_Get** one element from the FIFO, using the same prototypes as Program 3.7.

D5.27 Solve the FSM described in Exercise D3.12 using SysTick interrupts. Include three components: the FSM structure, an initialization function, and the SysTick ISR. There should be no backwards jumps.

D5.28 Solve the FSM described in Exercise D3.13 using SysTick interrupts. Run one pass through the FSM every 10 ms. Include three components: the FSM structure, an initialization function, and the SysTick ISR. There should be no backwards jumps.

D5.29 Solve the FSM described in Example 3.1 using SysTick interrupts. Run the FSM in the ISR. Include three components: the FSM structure, an initialization function, and the SysTick ISR. There should be no backwards jumps.

D5.30 You will design and implement the Example 3.2 FSM using the *edge-triggered interrupts*. One of the limitations of the FSM implementations in Chapter 3 is that they require 100% of the processor time and run in the foreground. In this system, there are two inputs and two outputs. We will implement a FSM where state transitions only occur on the edges of one of the two inputs. These edges should cause an interrupt, and the FSM controller will be run in the interrupt handler.

5.11. Lab Assignments

Lab 5.1 The overall objective is to create an alarm clock. A periodic interrupt establishes the time of day. Input/output of the system uses an interrupting UART port. Input is used to set the time, and set the alarm. Design a command interpreter that performs the necessary operations. Output is used to display the current time. A flashing LED or sound buzzer can be used to signify the alarm.

Lab 5.2 The overall objective is to create a software-driven variable-frequency digital square wave output. A periodic interrupt will set an output pin high/low. Input/output of the system uses an interrupting UART port. Input is used to set the frequency of the wave. Design a command interpreter that performs the necessary operations. Connect the output to an oscilloscope.

Lab 5.3 The overall objective is to create an interrupt-driven LED light display. Interface 8 to 16 colored LEDs to individual output pins. A periodic interrupt will change the LED pattern. Connect one or two switches to the system, and use them to control which LED light pattern is being displayed. Design a linked data structure that contains the light patterns. Create device drivers for the LED outputs, switch inputs, and periodic interrupt. The main program will initialize the LED outputs, switch inputs, and periodic interrupt. All of the input/output will be performed in the ISR. You should create a general purpose timer system that accepts a function to execute and a period. E.g.,

```
void main(void){
  LED_Init();      // Initialize LED output system
  Switch_Init();   // Initialize switch input system
  SysTick_Init(&FSMcontroller,250);   // Run FSMcontroller() every 250ms
  while(1){}
}
```

Lab 5.4 The overall objective is to create an **interrupt-driven traffic light controller**. The system has three digital inputs and seven digital outputs. You can simulate the system with three switches and seven LEDs. The inputs are North, East, and Walk. The outputs are six for the traffic light and one for a walk signal. Implement the controller using a finite state machine. Choose a Moore or Mealy data structure as appropriate. A periodic interrupt will run the FSM. The main program will initialize the LED outputs, switch inputs, and periodic interrupt. All of the input/output will be performed in the ISR. You should create a general purpose timer system that accepts a function to execute and a period. E.g.,

```
void main(void){
  Traffic_Init();    // Initialize switches and LED
  Timer_Init(&Traffic_Controller,100);   // Run FSM every 100ms
  while(1){}
}
```

6. Time Interfacing

Chapter 6 objectives are to:

- Use input capture to generate interrupts and measure period or pulse width
- Use output compare to create periodic interrupts and generate square waves
- Use both input capture and output compare to measure frequency
- Interface coil-activated devices like a DC motor, solenoid, and relay
- Interface an optical tachometer
- Generate waveforms using the pulse-width modulator

The timer systems on state-of-the-art microcontrollers are very versatile. Over the last 30 years, the evolution of these timer functions has paralleled the growth of new applications possible with these embedded computers. In other words, inexpensive yet powerful embedded systems have been made possible by the capabilities of the timer system. In this chapter we will introduce these functions, then use them throughout the remainder of the book. To adjust power to a DC motor we will use a time-based method called pulse-width modulation. To measure motor speed we will use a tachometer an employ the timer to measure period. We will introduce the real-time clock (RTC) in Chapter 9 as part of low power systems design. This chapter introduces fundamental principles of time used as an input and as an output, and presents a few examples. So it is important when designing to use this book in conjunction with the datasheet for your specific microcontroller.

6.1. Input Capture or Input Edge Time Mode

6.1.1. Basic Principles

The Texas Instruments microcontrollers have timers that are separate and distinct from SysTick, see Figure 6.1. Input edge time mode (or input capture mode) is used to make time measurements on input signals. We can use input capture to measure the period or pulse width of digital-level signals. The input capture system can also be used to trigger interrupts on rising or falling transitions of external signals. A General Purpose Timer Module (GPTM) has two 16-bit timers, which can be extended to 24 bits. Each GPTM input capture module has

    An external input pin, e.g., **CCP0**
    A trigger flag bit, called Raw Interrupt Status, e.g., **CAERIS**
    Two edge control bits, Event bits, e.g., **TAEVENT**
    An arm bit, called interrupt mask, e.g., **CAEIM**
    A 16-bit or 24-bit input capture register, e.g., **TAR**

The various members of the TM4C family have from zero to twelve input capture modules. Figure 6.1 shows the port pins used for input capture vary from microcontroller to microcontroller. The input capture and output compare pins are labeled **CCP0**, **CCP1**, ... A timer module has associated I/O pins. The even CCP pin is attached to Submodule A and the odd pin to Submodule B. Some timer modules are not attached to any I/O pins For example, the MSP432E401Y has eight timers, but Timer 6 and Timer 7 do not have I/O pins. Timers without pins can be used to generate periodic interrupts, but not for input capture. Tables 2.4 and 2.5 describe how to attach I/O pins to the timer modules on the TM4C123 and MSP432E401Y.

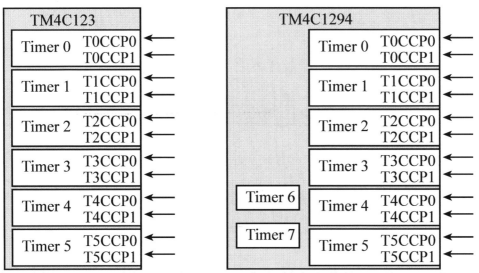

Figure 6.1. Input capture pins on the TM4C123/MSP432E401Y are called TnCCP0 and TnCCP1.

In this book we use the term **arm** to describe the bit that allows/denies a specific flag from requesting an interrupt. The Texas Instruments manuals refer to this bit as a **mask**. I.e., the device is armed when the mask bit is 1. Typically, there is a separate arm bit for every flag that can request an interrupt. An external digital signal is connected to the input capture pin. During initialization we specify whether the rising or falling edge of the external signal will trigger an input capture event. The 16-bit counter decrements at the rate of the bus clock, when it hits 0, it automatically rolls over to 0xFFFF and continues to count down (Figure 6.2). Two or three actions result from an input capture event: 1) the current timer value is copied into the input capture register (TAR or TBR), 2) the input capture flag is set (RIS) and 3) an interrupt is requested if armed (IM). This means an interrupt can be requested on a capture event. When using the prescaler on the TM4C, the counter can be extended to 24 bits (not on LM3S parts).

The input capture mechanism has many uses. For input capture, an external digital signal is connected to an input (TnCCP0 or TnCCP1) of the microcontroller. Three of common applications are:

1. An ISR is executed on the active edge of the external signal
2. Perform two rising edge input captures and subtract the two to get period
3. Perform a rising edge and then a falling edge capture and
       subtract the two measurements to get pulse width

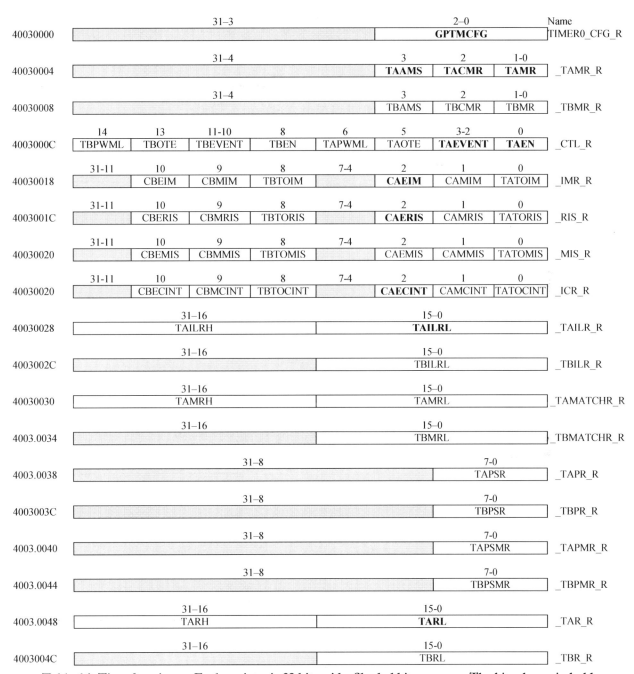

Table 6.1. Timer0 registers. Each register is 32 bits wide. Shaded bits are zero. The bits shown in bold will be used in this section. Timers 1, 2, ... have the same formats.

There are also 32-bit timers, called wide timers, which can be extended to 48 bits. However, in this book, we will only present the 32-bit timers.

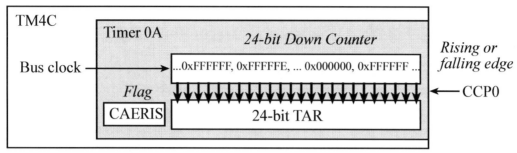

Figure 6.2. Rising or falling edge of CCP0 causes the counter to be latched into TAR, setting CAERIS.

Checkpoint 6.1: When does an input capture event occur?

Checkpoint 6.2: What happens during an input capture event?

Observation: The timer is very accurate because of the stability of the crystal clock.

Observation: When measuring period or pulse-width, the measurement resolution will equal the bus clock period.

6.1.2. Input Capture Details on the TM4C

Next we will overview the specific input capture functions on the TM4C family. This section is intended to supplement rather than replace the data sheets. When designing systems with input capture, please refer to the reference manual of your specific microcontroller. Table 6.1 shows some of the registers for Timer 0. We begin initialization by enabling the clock for the timer and for the digital port we will be using. We enable the digital pin and select its alternative function. We will disable the timer during initialization by clearing the **TAEN** (or **TBEN**) bit in the `TIMER0_CTL_R` register. To use 16-bit mode, we set **GPTMCFG** field to 4. We clear the **TAAMS** (or **TBAMS**) bit for capture mode. We set the **TACMR** (or **TBCMR**) bit for input edge time mode. The **TAMR** (or **TBMR**) field is set to 3 for capture mode. In summary, we write a 0x0007 to the `TIMER0_TAMR_R` register to select input capture mode. Table 6.2 lists the edge capture modes for **TAEVENT** (or **TBEVENT**.)

TAEVENT	Active edge
00	Capture on rising
01	Capture on falling
10	Reserved
11	Capture on both rising and falling

Table 6.2. Two control bits define the active edge used for input capture (TBEVENT is the same).

When we are measuring time with prescaler, such as period measurement and pulse width measurement, we set the 24-bit reload value to 0xFFFFFF. In this way, the 24-bit subtraction of two capture events yields the time difference between events. In particular, we will initialize `TIMER0_TAILR_R` to 0xFFFF and `TIMER0_TAPR_R` to 0xFF. We arm the input capture by setting the **CAEIM** (or **CBEIM**) bit in the `TIMER0_IMR_R` register. It is good practice to clear the trigger flag in the initialization so that the first interrupt occurs do to actions occurring after the initialization, and not due to edges that might have occurred during power up. The trigger flags are in the **TIMER0_RIS_R** register. These flags are cleared by writing 1's into corresponding bits in the **TIMER0_ICR_R** register. After all configuration bits are set, the Timer can be enabled by setting the **TAEN** (or **TBEN**) bit in the **TIMER0_CTL_R** register. If interrupts are required then the NVIC must be configured by setting the priority and enabling the appropriate interrupt number.

There is an 8-bit prescaler defined for each submodules A and B: **TIMER0_TAPMR_R** and **TIMER0_TBPMR_R**. These prescalers are not active during input capture mode on LM3S, but the prescalers on the TM4C are used to extend the 16-bit timer to 24 bits.

The **TAEVENT** bits of **TIMER0_CTL_R** register specify whether the rising or falling edge of **CCP0** will trigger an input capture event on Timer 0A. Two or three actions result from an input capture event: 1) the current timer value is copied into the input capture register, **TIMER0_TAR_R**, 2) the input capture flag (**CAERIS**) is set, and 3) an interrupt is requested if the mask bit (**CAEIM**) is 1. The **CAERIS** and **CBERIS** flag bits in the **TIMER0_RIS_R** register do not behave like a regular memory location. In particular, the flag cannot be set by software. Rather, an input capture or output compare hardware event will set the flag. The other peculiar behavior of the flag is that the software must write a one to the **TIMER0_ICR_R** register in order to clear the flag. If the software writes a zero to the **TIMER0_ICR_R** register, no change will occur. From Table 6.1, we see the **CAERIS** trigger flag is in bit 2 of the **TIMER0_RIS_R** register. The proper way to clear this trigger flag is

```
TIMER0_ICR_R = 0x0004;
```

Writes the **TIMER0_RIS_R** register have no effect. No effect occurs in the bits to which we write a zero in the **TIMER0_ICR_R** register.

Observation: The phase-lock-loop (PLL) on the ARM will affect the timer period.

Example 6.1. Design a measurement system for the robot that counts the number of times a wheel turns. This count will be a measure of the total distance travelled. The desired resolution is 1/32 of a turn

Solution: Whenever you measure something, it is important to consider the resolution and range. The basic idea is to use an optical sensor (QRB1134) to visualize the position of the wheel. A black/white striped pattern is attached to the wheel, and an optical reflective sensor is placed near the stripes. The sensor has an infrared LED output and a light sensitive transistor. The resolution will be determined by the number of stripes and the ability of the sensor to distinguish one stripe from another. The range will be determined by the precision of the software global variable used to count edges.

The basic idea is to trigger an interrupt on each stripe and then count the stripes in the interrupt service routine. If there are 32 stripes on the pattern then the number of times the wheel has turned will be

Count/32

The anode and cathode leads of the sensor control the amount of emitted IR light. The operating point for the LED in the QRB1134 is 15 mA at 1.8 V. The current to the LED is controlled by the R_1 resistor. In this circuit, the LED current will be $(3.3-1.8V)/R_1$, which we set to 15 mA by making R_1 equal to 100Ω. The R_2 pull-up resistor on the transistor creates an output swing at V_1 depending on whether the sensor sees a black stripe or white stripe. Unfortunately, the signal V_1 is not digital. The rail-to-rail op amp, in open loop mode, creates a clean digital signal at V_2, which has the same frequency as V_1. The negative terminal is set to a voltage approximately in the center of V_1, shown as +2V in Figure 6.3. Slew rate is defined as dV/dt. An uncertainty in voltage δV will translate to an uncertainty in time, $\delta t = \delta V/(dV/dt)$. Thus to minimize time error, we choose a place with maximum slew rate. In other words, we should select the threshold at the place in the wave where the slope is at maximum.

We then interface V_2 to an input capture pin, and configure the system to trigger an interrupt on each rising edge. Several of the GPIO pins could have been used, but we selected PB6 (T0CCP0) as shown in Figure 6.3. This means that Timer0A will be used.

Because there are 32 stripes on the wheel, there will be 32 interrupts each time the wheel rotates once. A 32-bit global variable will be used to count the number of rising edges. This count is a binary fixed-point number with a resolution of 2^{-5} revolutions. E.g., if the count is 100, this means 100/32 or 3.125 revolutions. If the circumference of the wheel is fixed, and if the wheel does not slip, then this count is also a measure of distance traveled. We solved a similar problem in Program 5.6 using edge-triggered inputs. However in this solution, we will use input capture.

The initialization sets the direction register bit 6 to 0, so PB6 is an input. Bit 6 in **GPIO_PORTB_AFSEL_R** is set, making timer channel 4 an input capture. Bits 2 and 3 (**TAEVENT**) in **TIMER0_CTL_R** specify we want Timer0A to capture on the rising edge of PB6. We arm the input capture channel by setting bit 2 (CAEIM) in **TIMER0_IMR_R**. It is good design practice to clear trigger flags in the initialization, so the first interrupt is due to a rising edge on the input occurring after the initialization and not due to events occurring during power up or before initialization.

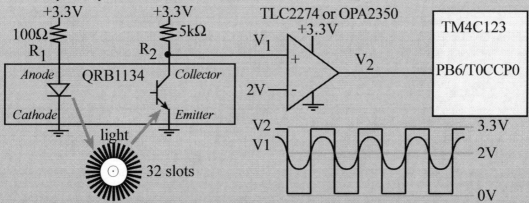

Figure 6.3. An external signal is connected to the input capture.

```
volatile uint32_t Count;        // incremented on interrupt
void TimerCapture_Init(void){
  SYSCTL_RCGCTIMER_R |= 0x01;   // activate timer0
```

```
  SYSCTL_RCGCGPIO_R |= 0x00000002;  // activate port B
  Count = 0;                        // allow time to finish activating
  GPIO_PORTB_DEN_R |= 0x40;         // enable digital I/O on PB6
  GPIO_PORTB_AFSEL_R |= 0x40;       // enable alt funct on PB6
  GPIO_PORTB_PCTL_R = (GPIO_PORTB_PCTL_R&0xF0FFFFFF)+0x07000000;
  TIMER0_CTL_R &= ~0x00000001;      // disable timer0A during setup
  TIMER0_CFG_R = 0x00000004;        // configure for 16-bit timer mode
  TIMER0_TAMR_R = 0x00000007;       // configure for input capture mode
  TIMER0_CTL_R &= ~(0x000C);        // TAEVENT is rising edge
  TIMER0_TAILR_R = 0x0000FFFF;      // start value
  TIMER0_IMR_R |= 0x00000004;       // enable capture match interrupt
  TIMER0_ICR_R = 0x00000004;        // clear timer0A capture flag
  TIMER0_CTL_R |= 0x00000001;       // enable timer0A
  NVIC_PRI4_R =(NVIC_PRI4_R&0x00FFFFFF)|0x40000000; // Timer0A=priority 2
  NVIC_EN0_R = 0x00080000;          // enable interrupt 19 in NVIC
  EnableInterrupts();
}
void Timer0A_Handler(void){
  TIMER0_ICR_R = 0x00000004;        // acknowledge timer0A capture match
  Count = Count + 1;
}
```

Program 6.1. Counting interrupt using input capture (InputCapture_xxx).

An input capture interrupt occurs on each rise of **T0CCP0**. The latency of the system is defined as the time delay between the rise of the input capture signal to the increment of **Count**. Assuming there are no other interrupts, and assuming the main program does not disable interrupts, the delay will be on the order of 1µs. The latency may be larger if there are other sections of code that execute with the interrupts disabled, or if there are higher priority interrupts active. The ritual, **TimerCapture_Init()**, sets input capture to interrupt on the rise, and initializes the global, **Count**. The interrupt software performs a poll, acknowledges the interrupt and increments the global variable. Actual measurements collected with this interface are shown in Figure 6.4.

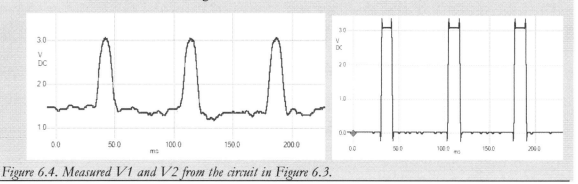

Figure 6.4. Measured V1 and V2 from the circuit in Figure 6.3.

On the TM4C123, the T0CCP0 input could be connected to either PB6 or to PF0 as described in Table 2.4. We used PB6 so we set **PCTL** bits for PB6 to 7. To use PF0, we set **PCTL** bits for PF0 to 7. E.g.,

```
SYSCTL_RCGCGPIO_R |= 0x00000020; // activate port F
Count = 0;
GPIO_PORTF_DEN_R |= 0x01;        // enable digital I/O on PF0
GPIO_PORTF_AFSEL_R |= 0x01;
GPIO_PORTF_PCTL_R = (GPIO_PORTF_PCTL_R&0xFFFFFFF0)+0x00000007;
```

Checkpoint 6.3: Explain how to change Program 6.1 to still run on a TM4C123 with the input connected to T3CCP0/PB2.

Checkpoint 6.4: Explain how to change Program 6.1 to run on a TM4C1294 / MSP432E401Y with the input connected to T0CCP0.

Checkpoint 6.5: Write code to clear CBMRIS in Timer 1.

6.1.3. Period Measurement

The basic idea of period measurement is to generate two input captures on the same edge (both rise or both fall), record the times of each edge, and calculate period as the difference between those two times. Before one implements a system that measures period, it is appropriate to consider the issues of resolution, precision and range. The **resolution** of a period measurement is defined as the smallest change in period that can reliably be detected. In Example 6.2, the bus clock is 80 MHz. This means, if the period increases by 12.5 ns, then there will be one more Timer clock between the first rising edge and the second rising edge. In this situation, the 24-bit subtraction will increase by 1, therefore the period measurement resolution is 12.5 ns. The resolution is the smallest measurable change. Resolution defines the units of the measurement. In this first example, if the calculation of **Period** results in 1000, then it represents a period of 1000•12.5ns or 12.5µs. The **precision** of the period measurement is defined as the number of separate and distinguishable measurements. If the 24-bit counter is used, there are about 16 million different periods that can be measured. We can specify the precision in alternatives, e.g., 2^{24}, or in bits, e.g., 24 bits. The last issue to consider is the **range** of the period measurement, which is defined as the minimum and maximum values that can reliably be measured. We are concerned what happens if the period is too small or too large. A good measurement system should be able to detect overflows and underflows. In addition, we would not like the system to crash, or hang-up if the input period is out of range. Similarly, it is desirable if the system can detect when there is no period. For edge detection, the input must be high for at least two system clock periods and low for at least two system clock periods.

Example 6.2. Design a system that measures the rotational speed of a motor shaft using period measurement with a precision of 24 bits and a resolution of 12.5 ns.

Solution: For details of the sensor refer back to Figure 6.3. In this example, the digital input signal is connected to an input capture pin, CCP0. The diodes, 47k, and 220nF create a 0 to 3.3V signal on V_1. The 10k-4.7k create a reference voltage V_t, and the 10k positive feedback resistor removes glitches. V_2 is a squarewave at the same frequency as the input. Let N be the number of rising edges as the shaft rotates once. We will make the bus clock equal 80 MHz. Each rising edge will cause Timer0A to generate an input capture interrupt (Figure 6.5).

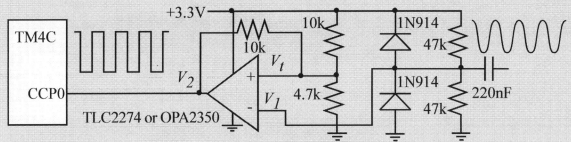

Figure 6.5. To measure period we connect the external signal an input capture, PB6 on the TM4C123.

The period is calculated as the difference in **TIMER0_TAR_R** latch values from one rising edge to the other. If N=100, and the motor is spinning at 300 RPM, then the period will be [(60000ms/min)/(300RPM)/100edges/rotation)], which will be 2.00 ms/edge, see Figure 6.6.

For example, if the period is 2000 µs, the Timer0A interrupts will be requested every 160,000 cycles, and the 24-bit difference between **TIMER0_TAR_R** latch values will be 160,000. This subtraction remains valid even if the timer reaches zero and wraps around in between Timer0A interrupts. On the other hand, this method will not operate properly if the period is larger than 2^{24} cycles, or about 209 ms.

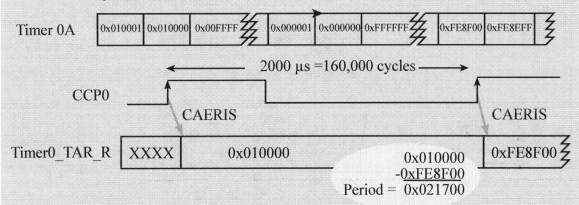

Figure 6.6. Timing example showing counter rollover during 24-bit period measurement.

The resolution is 12.5 ns because the period must increase by at least this amount before the difference between Timer0A measurements will reliably change. Even though a 24-bit counter is used, the precision is a little less than 24 bits, because the shortest period that can be handled with this interrupt-driven approach is about 1 µs. It takes about 1 µs to complete the context switch, execute the ISR software, and return from interrupt. This factor is determined by experimental measurement. In other words, as the period approaches 1 µs, a higher and higher percentage of the computer execution is utilized just in the handler itself.

Because the input capture interrupt has a separate vector the software does not poll. An interrupt is requested on each rising edge of the input signal. In this situation we count all the cycles required to process the interrupt. The period measurement system written for the TM4C123 is presented in Program 6.2. The 24-bit subtraction is produced by anding the difference with 0x0FFFFFF, calculating the number of bus clocks between rising edges. The first period measurement will be incorrect and should be neglected.

```c
uint32_t Period;              // 24-bit, 12.5 ns units
uint32_t static First;        // Timer0A first edge, 12.5 ns units
int32_t Done;                 // mailbox status set each rising
void PeriodMeasure_Init(void){
  SYSCTL_RCGCTIMER_R |= 0x01;        // activate timer0
  SYSCTL_RCGCGPIO_R |= 0x02;         // activate port B
  First = 0;                         // first will be wrong
  Done = 0;                          // set on subsequent
  GPIO_PORTB_DIR_R &= ~0x40;         // make PB6 input
  GPIO_PORTB_AFSEL_R |= 0x40;        // enable alt funct on PB6
  GPIO_PORTB_DEN_R |= 0x40;          // configure PB6 as T0CCP0
  GPIO_PORTB_PCTL_R = (GPIO_PORTB_PCTL_R&0xF0FFFFFF)+0x07000000;
  TIMER0_CTL_R &= ~0x00000001;       // disable timer0A during setup
  TIMER0_CFG_R = 0x00000004;         // configure for 16-bit capture mode
  TIMER0_TAMR_R = 0x00000007;        // configure for rising edge event
  TIMER0_CTL_R &= ~0x0000000C;       // rising edge
  TIMER0_TAILR_R = 0x0000FFFF;       // start value
  TIMER0_TAPR_R = 0xFF;              // activate prescale, creating 24-bit
  TIMER0_IMR_R |= 0x00000004;        // enable capture match interrupt
  TIMER0_ICR_R = 0x00000004;         // clear timer0A capture match flag
  TIMER0_CTL_R |= 0x00000001;        // timer0A 24-b, +edge, interrupts
  NVIC_PRI4_R = (NVIC_PRI4_R&0x00FFFFFF)|0x40000000; //Timer0A=priority 2
  NVIC_EN0_R = 1<<19;                // enable interrupt 19 in NVIC
  EnableInterrupts();
}
void Timer0A_Handler(void){
  TIMER0_ICR_R = 0x00000004;         // acknowledge timer0A capture
  Period = (First - TIMER0_TAR_R)&0x00FFFFFF; // 12.5ns resolution
  First = TIMER0_TAR_R;              // setup for next
  Done = 1;                          // set semaphore
}
```

Program 6.2. 24-bit period measurement (PeriodMeasure_xxx).

Example 6.3. Design an ohmmeter with a range of 0 to 250 kΩ and a resolution of 10Ω.

Solution: One way to measure resistance or capacitance is convert the electrical parameter to time. In particular, we will use a TLC555 to convert the input resistance (R_{in}) to a time period (P). Figure 6.7 shows the hardware circuit. The 555 is called an astable multivibrator, created a signal wave with a period equal to $C_T*(R_A + 2*R_B)*\ln(2)$. C_T and R_A are constants, but R_B is 10kΩ plus the unknown input, R_{in}. We can then use the period measurement software from Program 6.2. The resistance measurement resolution (ΔR) depends on the period measurement resolution (Δt) and the capacitor C_T, $\Delta t = C_T *\Delta R*\ln(2)$. At a bus frequency of 50 MHz, the period measurement resolution is 20 ns. If we desire a 10Ω resistance resolution, the capacitor will be about 3nF. In this circuit, a 3.3nF capacitor was used because it is a standard value. As the input resistance varies from 0 to 250 kΩ, the period (in 20ns units) varies from about 3000 to 60000. A linear function is used to convert the measured period (20 ns units) to resistance (10Ω units). The same initialization of Program 6.2 can be used.

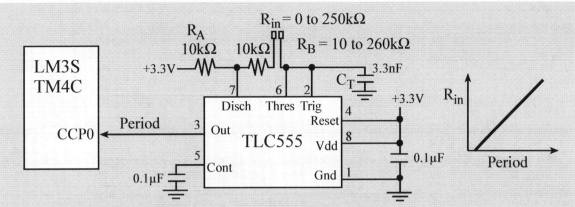

Figure 6.7. The TLC555 outputs a square wave with a period depending on R_A R_B and C_T.

Program 6.3 shows the ISR of Program 6.2 extended to include the period to resistance conversion. The coefficients of this linear fit were determined by empirical calibration R_{in} = 0.4817*Period*-1466.5. This equation is implemented with fixed-point math.

Although mathematically, it seems like the system has a 10 Ω resolution and a precision of 25000 alternatives, the actual resolution will be limited by the stability of the TLC555, the stability of capacitor, and any added noise into the circuit.

```
uint32_t Period;                  // 24-bit, 20 ns units
uint32_t Resistance;              // 10 ohm units, 0 to 25000
uint32_t static First;            // Timer0A first edge
uint32_t Done;                    // mailbox status set each rising
void Timer0A_Handler(void){
  TIMER0_ICR_R = 0x00000004;                       // acknowledge timer0A capture
  Period = (First - TIMER0_TAR_R)&0xFFFF;          // 20 ns resolution
  Resistance = (1973*Period-6006784)>>12;          // conversion
  First = TIMER0_TAR_R;                            // setup for next
  Done = 1;                                        // set mailbox flag
}
```
Program 6.3. 16-bit period measurement used to measure resistance.

6.1.4. Pulse Width Measurement

The basic idea of pulse width measurement is to cause an input capture event on both the rising and falling edges of an input signal. Each edge captures a timer value. The difference between these two captured times will be the pulse width. Just like period measurement, the resolution is determined by the rate at which the timer is decremented. The maximum pulse width is 2^{16} times the resolution, and is limited by the 16-bit timer. When considering measurement resolution it is important to consider voltage noise as well. For example, in Figures 6.3 and 6.5, any voltage noise on the sensor will cause a time-jitter, which is a noise in the time measurement. If the slew rate of the sensor at the threshold voltage is dV/dt, then a voltage noise of δV will cause a time error of δt, according to $\delta t = \delta V/(dV/dt)$.

Example 6.4. Design a system to measure resistance and use it to interface a 10 kΩ joystick.

Solution: Clearly, you could just connect the potentiometer across the power rails and measure the voltage drop using the ADC (which will be described later). However, it is much cheaper and easier to measure time precisely than to measure voltage precisely. This makes sense, considering that an inexpensive clock can run for months before it needs to be reset, but even a high quality voltmeter measures to only a few digits of precision. Therefore, we will convert the resistance to a pulse width using external circuitry and measure the pulse using input capture. The TM4C123 microcontroller runs at 80 MHz, so the pulse width resolution will be 12.5 ns. The range will be 1 µs to 209 ms. With the minimum determined by the time to execute software and the maximum determined by the 24-bit counter (12.5 ns *2^{24}.)

The objective is to use *input capture pulse width measurement* to measure resistance. This basic approach is employed by most joystick interfaces. The resistance measurement range is $0 \leq R \leq 10$ kΩ. The desired resolution is 1 Ω. We will use busy-wait synchronization.

Most joysticks have two variable resistances, but we will show the solution for just one of the potentiometers. The variable resistance R in Figure 6.8 is one channel of the joystick. We use a monostable to convert unknown resistance, R, to time difference t. To perform high quality measurements we will need a high quality capacitor, because of the basic conversion follows $\Delta t = R*C$. PB5 is a digital output and PB6 is an input capture input. A rising edge on PB5 causes a monostable positive logic pulse on the "Q" output of the 74HC123. We choose R_1 and C, so that the resistance resolution maps into a pulse width measurement resolution of 12.5 ns, and the resistance range $0 \leq R \leq 10$ kΩ maps into $125 \leq t \leq 250$ µs. The following equation describes the pulse width generated by the 74HC123 monostable as a function of the resistances and capacitance.

$$t = 0.45 \cdot (R + R_1) \cdot C$$

For a linear system, with x as input and y as output, we can use calculus to relate the measurement resolution of the input and output.

$$\Delta y = \frac{dy}{dx} \Delta x$$

Therefore, the relationship between the pulse width measurement resolution, Δt, and the resulting resistance measurement resolution is determined by the value of the capacitor.

$$\Delta t = 0.45 \cdot \Delta R \cdot C$$

To make a Δt of 12.5 ns correspond to a ΔR of 1 Ω, we choose

$C = \Delta t/(0.45 \Delta R) = 12.5\text{ns}/(0.45 \cdot 1\Omega) = 27.8\text{nF}$

We will use a C0G high-stability 27nF ceramic capacitor. To design for the minimum pulse width, we set R=0, $t = 0.45 \cdot R_1 \cdot C$. We choose R_1 to make the minimum pulse width 125 µs,

$R_1 = t/(0.45 \cdot C) = 125\text{µs}/(0.45 \cdot 27\text{nF}) = 10.288$ kΩ

We will use a 1% metal film 10 kΩ resistor for R_1. To check the minimum and maximum pulse widths we set $R=0$ and $R_1=10$kΩ, and calculate $t = 0.45 \cdot (10\text{k}\Omega) \cdot 27\text{nF} = 121.5$µs (close to 125), and $t = 0.45 \cdot (20\text{k}\Omega) \cdot 27\text{nF} = 243$µs (close to 250). The importance parameters for the 74HC123, R_1 and C are the long term stability. I.e., their performance should be constant over time. Any differences between assumed values and real values for the capacitor and the 0.45 constant can be compensated for with software calibration.

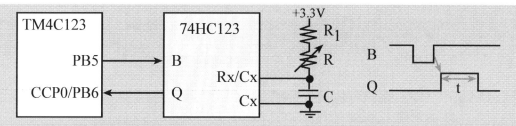

Figure 6.8. To measure resistance using pulse width we connect the external signal an input capture.

The measurement function returns the resistance, R, in Ω. For example, if the resistance, R, is 1234 Ω, then return parameter will be 1234. We will not worry about resistances, R, greater than 55535 Ω or if R is disconnected. The solution is shown as Program 6.4.

```
#define CALIBRATION    0
void ResistanceMeasure_Init(void){
  SYSCTL_RCGCTIMER_R |= 0x01;        // activate timer0
  SYSCTL_RCGCGPIO_R |= 0x02;         // activate port B
  while((SYSCTL_PRGPIO_R&0x02) != 0x02){};
  GPIO_PORTB_DATA_R |= 0x20;         // set PB5 high
  GPIO_PORTB_DIR_R = (GPIO_PORTB_DIR_R& ~0x40)|0x20;  // PB5 out, PB6 in
  GPIO_PORTB_DEN_R |= 0x60;          // enable digital PB5 and PB6
  GPIO_PORTB_AFSEL_R |= 0x40;        // enable alt funct on PB6
  GPIO_PORTB_PCTL_R = (GPIO_PORTB_PCTL_R&0xF0FFFFFF)+0x07000000;
  TIMER0_CTL_R &= ~0x00000001;       // disable timer0A during setup
  TIMER0_CFG_R = 0x00000004;         // configure for 16-bit timer mode
  TIMER0_TAMR_R = 0x00000007;        // configure for input capture mode
  TIMER0_TAILR_R = 0x0000FFFF;       // start value
  TIMER0_IMR_R &= ~0x7;              // disable all interrupts for timer0A
  TIMER0_ICR_R = 0x00000004;         // clear timer0A capture match flag
  TIMER0_CTL_R |= 0x00000001;        // enable timer0A edge, no interrupts
}
// return resistance in ohms, range is 0 to 10000 ohm
uint16_t ResistanceMeasure(void){ uint16_t rising;
  GPIO_PORTB_DATA_R &= ~0x20;        // turn off PB5
  TIMER0_CTL_R &= ~(0x000C);         // rising edge
  TIMER0_ICR_R = 0x00000004;         // clear timer0A capture flag
  GPIO_PORTB_DATA_R |= 0x20;         // turn on PB5, trigger 74HC123
  while((TIMER0_RIS_R&0x00000004)==0){};// wait for rise
  rising = TIMER0_TAR_R;             // timerA0 at rising edge
  TIMER0_ICR_R = 0x00000004;         // clear timer0A flag
  TIMER0_CTL_R &= ~(0x000C);
  TIMER0_CTL_R += 0x00000004;        // falling edge
  while((TIMER0_RIS_R&0x00000004)==0){}; // wait for fall
  TIMER0_ICR_R = 0x00000004;         // clear timer0A flag
  return(rising-TIMER0_TAR_R-CALIBRATION)&0xFFFF;
}
```
Program 6.4. Measuring resistance using pulse-width measurement.

The difficulty with pulse width measurement in the previous example was the need to switch from rising to falling edge during each measurement. It was not a problem with this problem because the smallest pulse width was 125 μs. However, to handle shorter pulses we will need to use two input capture pins. One pin measures the time of the rise and the other pin measures the time of the fall. In order for input capture to operate, the input must be high for at least two bus clocks and low for at least two bus clocks. Otherwise the minimum pulse width does not depend on software execution time or interrupt latency. However the minimum period will depend on software speed.

Example 6.5. Design a system to measure pulse width using interrupts, with a precision of 24 bits and a resolution of 12.5 ns.

Solution: In this example, the digital-level input signal is connected to two input capture pins, CCP0 and CCP1 (Figure 6.9). The bus clock is selected to be 80 MHz so the measurement resolution will be 12.5 ns. The rising edge time will be measured by Timer0B without the need of an interrupt and the falling edge interrupts will be handled by Timer0A. The pulse width is calculated as the difference in **TIMER0_TBR_R-TIMER0_TAR_R** latch values. In this example the Timer0A interrupt handler simply sets the global variable, **PW**, at the time of the falling edge. Because no software is required to process the Timer0B measurement, there is no software limit to the minimum pulse width. There is the hardware limit requiring at least two bus clock periods while high and two bus clock periods while low. On the other hand, software processing is required to handle the Timer0A signal, so there is a minimum period. E.g., there must be more than 2 μs from one falling edge to the next falling edge. This time depends on software execution speed in the ISR, and the context switch. This minimum period will be larger for systems with higher priority interrupts. Again, the first measurement may or may not be accurate.

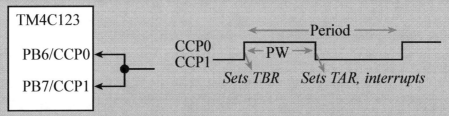

Figure 6.9. The rising edge is measured with Timer0B, and falling edge is measured with Timer0A.

The pulse width measurement is performed from rising edge to falling edge. The resolution is 12.5 ns, determined by the system bus clock. The range is about 25 ns to 209ms with no overflow checking. Timer0A interrupts only occur on the falling edges. The global, **PW**, contains the most recent measurement. **Done** is set at the falling edge by the ISR signifying a new measurement is available. If the first edge after the **PWMeasure2_Init();** is executed is a falling edge, then the first measurement will be incorrect (because **TIMER0_TBR_R** is incorrect). If the first edge after the **PWMeasure2_Init();** is executed is a rising edge, then the first measurement will be correct. Notice how little software overhead is required to perform these measurements (Program 6.5).

```
uint32_t PW;                    // 24 bits, 12.5 ns units
int Done;                       // set each falling
void PWMeasure2_Init(void){     // TM4C123 code
```

```c
  SYSCTL_RCGCTIMER_R |= 0x01;       // activate timer0
  SYSCTL_RCGCGPIO_R |= 0x02;        // activate port B
  Done = 0;                         // allow time to finish activating
  GPIO_PORTB_DIR_R &= ~0xC0;        // make PB6, PB7 inputs
  GPIO_PORTB_DEN_R |= 0xC0;         // enable digital PB6, PB7
  GPIO_PORTB_AFSEL_R |= 0xC0;       // enable alt funct on PB6, PB7
  GPIO_PORTB_PCTL_R = (GPIO_PORTB_PCTL_R&0x00FFFFFF)+0x77000000;
  TIMER0_CTL_R &= ~0x00000003;
  TIMER0_CFG_R = 0x00000004;        // configure for 16-bit timer mode
// **** timer0A initialization ****
  TIMER0_TAMR_R = 0x00000007;
  TIMER0_CTL_R = (TIMER0_CTL_R&(~0x0C))+0x04; // falling edge
  TIMER0_TAILR_R = 0x0000FFFF;      // start value
  TIMER0_TAPR_R = 0xFF;             // activate prescale, creating 24-bit
  TIMER0_IMR_R |= TIMER_IMR_CAEIM;  // enable capture match interrupt
  TIMER0_ICR_R = TIMER_ICR_CAECINT; // clear timer0A capture match flag
// **** timer0B initialization ****
  TIMER0_TBMR_R = 0x00000007;
  TIMER0_CTL_R = (TIMER0_CTL_R&(~0x0C00))+0x00; // rising edge
  TIMER0_TBILR_R = 0x0000FFFF;      // start value
  TIMER0_TBPR_R = 0xFF;             // activate prescale, creating 24-bit
  TIMER0_IMR_R &= ~0x700;           // disable all interrupts for timer0B
  TIMER0_CTL_R |= 0x00000003;       // enable timers
// **** interrupt initialization ****
  NVIC_PRI4_R = (NVIC_PRI4_R&0x00FFFFFF)|0x40000000; // Timer0=priority 2
  NVIC_EN0_R = 1<<19;               // enable interrupt 19 in NVIC
  EnableInterrupts();
}
void Timer0A_Handler(void){
  TIMER0_ICR_R = 0x00000004;// acknowledge timer0A capture flag
  PW = (TIMER0_TBR_R-TIMER0_TAR_R)&0x00FFFFFF;// from rise to fall
  Done = 1;
}
```

Program 6.5. Pulse-width measurement using two input captures.

6.2. Output Compare or Periodic Timer

In output compare (periodic timer) mode the timer is configured as a 16-bit down-counter with an optional 8-bit prescaler that effectively extends the counting range of the timer to 24 bits. When the timer counts from 1 to 0 it sets the trigger flag. On the next count, the timer is reloaded with the value in **TIMER0_TAILR_R** (or **TIMER0_TBILR_R**). We select periodic timer mode by setting the 2-bit TAMR (or TBMR) field of the **TIMER0_TAMR_R** (or **TIMER0_TBMR_R**) to 0x02. If we set this field to 0x01, the timer is in one shot mode. In periodic mode the timer runs continuously, and in one shot mode, it runs once and stops. The periodic mode can also be used to create pulse width modulated outputs.

We will use output compare to create time delays, trigger periodic interrupts, and control ADC sampling. We will also use output compare together with input capture to measure frequency. Output compare and input capture can also be combined to measure period and frequency over a wide range of ranges and resolutions. We may run the output compare modes with or without an external output pin attached. Each periodic timer module has

>An external output pin, e.g., CCP0,
>A flag bit, e.g., TATORIS
>A control bit to connect the output to the ADC as a trigger, e.g., TAOTE,
>An interrupt arm bit, e.g., TATOIM
>A 16-bit reload register, e.g., **TIMER0_TAILR_R**
>A 8-bit prescale register, e.g., **TIMER0_TAPR_R**
>A 8-bit prescale match register, e.g., **TIMER0_TAPMR_R**

The members of the TM4C family have varying number of timers. When designing a system using the timers, you will need to consult the datasheet for your particular microcontroller. In particular, some of the channels do not have an associated output pin. For example, the MSP432E401Y has eight timers, creating up to eight periodic interrupts, but has only six I/O pins, as shown in Figure 6.10.

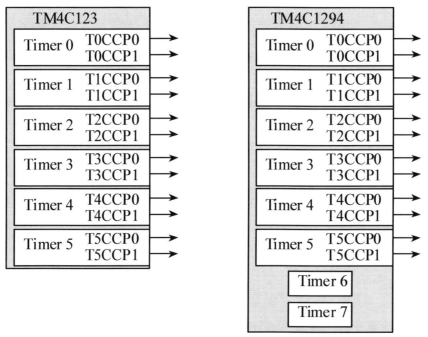

Figure 6.10. Output compare pins on the TM4C123, and the MSP432E401Y are called TnCCP0 and TnCCP1. Notice the MSP432E401Y has two timers not connected to any I/O pins.

To connect the output pin to the timer, we must set the alternative function bit for that pin. Tables 2.4 and 2.5 describe how to attach I/O pins to the timer module on the TM4C123 and MSP432E401Y. The output compare pin is an output of the computer, hence can be used for debugging or to control an external device. An output compare event occurs, changing the state of the output pin, when the 16-bit timer matches the 16-bit **TIMER0_TAILR_R** register. The timer will be used without the output pin if the corresponding alternative function bit is clear.

The output compare event occurs when a timer counts down to zero. The timer mode specifies what effect the output compare event will have on the output pin or the rest of the system. If the timer is in one-shot or periodic timer mode, the **TATORIS** (or **TBTORIS**) bit of the Raw Interrupt Status register (**TIMER0_RIS_R**) is set. If the arm bit **TATOIM** (or **TBTOIM**) in the **TIMER0_IMR_R** register is set, a timer interrupt is requested. The hardware can also trigger an ADC conversion at this time. If the timer is in one-shot mode, it stops counting after the first output compare event. In periodic timer mode, the timer continues counting indefinitely until explicitly disabled by clearing the **TAEN** (or **TBEN**) enable bit in the **TIMER0_CTL_R** register. Just like the input capture, the output compare flag is cleared by writing a 1 to its corresponding bit in the Interrupt Clear Register (**TIMER0_ICR_R**).

One simple application of output compare is to create a fixed time delay. Let `delay` be the number of bus cycles you wish to wait, up to 65,535. The steps to create the delay are:

0) Enable the General-Purpose Timer Module in of **RCGCTIMER**
1) Ensure that the timer is disabled before making any changes (clear **TAEN**)
2) Put the timer module in 16-bit mode by writing 0x4 to **TIMER0_CFG_R**
3) Write 0x1 to **TAMR**
4) Load the desired `delay` into **TIMER0_TAILR_R**
5) Write a 1 to **TATOCINT** of **TIMER0_ICR_R** to clear the time-out flag
6) Set the **TAEN** bit to start the timer and begin counting down from `delay`
7) Poll **TATORIS** of **TIMER0_RIS_R**, wait is over when this bit is set

A second application of output compare is to create a periodic interrupt. Let `prescale` be an 8-bit number loaded into **TIMER0_TAPR_R**. The timer frequency will be bus frequency divided by prescale+1. The default prescale is 0, meaning the timer frequency equals the bus frequency. Let `period` be the 32-bit value loaded into **TIMER0_TAILR_R**. The steps to create the periodic interrupt are:

0) Enable the General-Purpose Timer Module in of **RCGCTIMER**
1) Ensure that the timer is disabled before making any changes (clear **TAEN**)
2) Put the timer module in 32-bit mode by writing 0x00 to **TIMER0_CFG_R**
3) Write 0x2 to **TAMR** to configure for periodic mode
4) Load `period` into **TIMER0_TAILR_R**
5) Load `prescale` into **TIMER0_TAPR_R**
6) Write a 1 to **TATOCINT** of **TIMER0_ICR_R** to clear the time-out flag
7) Set **TATOIEN** of **TIMER0_IMR_R** to arm the time-out interrupt
8) Set the priority in the correct NVIC Priority register
9) Enable the correct interrupt in the correct NVIC Interrupt Enable register
10) Set the **TAEN** bit to start the timer and begin counting down from `period`

If the bus period is Δ*t*, then the timer interrupt period will be

$$\Delta t * (\texttt{prescale}+1) * (\texttt{period}+1).$$

A few cycles of instructions should separate Steps 0 and 1 to ensure that the timer is receiving a clock before the program attempts to use it. Move Step 0 earlier in your program or insert dummy instructions between Steps 0 and 1 if you ever get a Hardware Fault. The maximum **period** can be 24 bits using prescaler or 32 bits using TimerA and TimerB together in 32-bit mode. Resolution is 1 bus cycle.

> **Checkpoint 6.6**: When does an output compare event occur when in one-shot/periodic timer mode?
>
> **Checkpoint 6.7**: What happens during an output compare event in one-shot/periodic timer mode?

Example 6.6. Design a system to execute a user task at a periodic rate with units of 12.5 ns.

Solution: We will generate a periodic interrupt and call the user task from the ISR. Assuming a 80 MHz bus clock, we disable the prescale, meaning the timer counts every 12.5ns. To define the user task we will create a private global variable containing a pointer to the user's function. We will set the variable during initialization and call that function at run time. Another name for a dynamically set function pointer is a **hook**. The maximum possible value for **period** is $12.5\text{ns} * 2^{32}$, which is about 53 seconds.

```
void (*PeriodicTask)(void);   // user function
```

The initialization sequence follows the 1 – 10 outline listed above (Program 6.6).

```c
void Timer2A_Init(void(*task)(void), uint32_t period){
  SYSCTL_RCGCTIMER_R |= 0x04;     // 0) activate timer2
  PeriodicTask = task;             // user function
  TIMER2_CTL_R = 0x00000000;       // 1) disable timer2A during setup
  TIMER2_CFG_R = 0x00000000;       // 2) configure for 32-bit mode
  TIMER2_TAMR_R = 0x00000002;      // 3) configure for periodic mode
  TIMER2_TAILR_R = period-1;       // 4) reload value
  TIMER2_TAPR_R = 0;               // 5) bus clock resolution
  TIMER2_ICR_R = 0x00000001;       // 6) clear timer2A timeout flag
  TIMER2_IMR_R = 0x00000001;       // 7) arm timeout interrupt
  NVIC_PRI5_R = (NVIC_PRI5_R&0x00FFFFFF)|0x80000000; // 8) priority 4
  NVIC_EN0_R = 1<<23;              // 9) enable IRQ 23 in NVIC
  TIMER2_CTL_R = 0x00000001;       // 10) enable timer2A
  EnableInterrupts();
}
void Timer2A_Handler(void){
  TIMER2_ICR_R = 0x00000001;       // acknowledge timer2A timeout
  (*PeriodicTask)();               // execute user task
}
```

Program 6.6. Implementation of a periodic interrupt using Timer2A (PeriodicTimer0AInts_xxx).

Example 6.7. Design a system that generates a 50% duty cycle 100 Hz square wave on PB5.

Solution: This example generates a 50% duty cycle square wave using output compare. The output is high for `Period` cycles then low for `Period` cycles. Program 6.6 will be used to request interrupts at a rate twice as fast as the resulting square wave frequency. One interrupt is required for the rising edge and another for the falling edge. The output compare interrupt handler simply acknowledges the interrupt and toggles the output pin.

```
#define PB5 (*((volatile uint32_t *)0x40005080))
void TogglePB5(void){ PB5 ^= 0x20;}
```

There will be some software jitter due to the latency in processing the interrupt. We add these lines to the initialization

```
SYSCTL_RCGCGPIO_R |= 0x00000002; // activate clock for Port B
while((SYSCTL_PRGPIO_R&0x00000002) == 0){};// ready?
GPIO_PORTB_DEN_R  |= 0x20;       // enable digital I/O on PB5
GPIO_PORTB_DIR_R  |= 0x20;       // PB5 is an output
```

To start the square wave we call initialization in Program 6.6 so interrupts occur 200 times a second (every 5ms), creating the 100 Hz wave.

```
Timer2A_Init(&TogglePB5,400000); //400,000*12.5ns = 5ms
```

6.3. Pulse Width Modulation

The problem with the solution to Example 6.7 is that the minimum period is determined by software execution speed, the time running with interrupts disabled, and the existence of higher priority interrupts. Generating output waves is an essential task for real-time systems, so the microcontrollers have multiple methods to create output waves.

6.3.1. Pulse Width Modulation using the Timer Module

Pulse width modulation is an effective and thus popular mechanism for the embedded microcontrollers to control external devices. The timer can create PWM outputs by setting the **TAAMS** bit and selecting periodic mode in **TAMR** field in the **TIMER0_TAMR_R** register. The output is one for `High` cycles then zero for `Low` cycles. This example generates a variable duty cycle square wave using output compare. Output compare events will again be requested at a rate twice as fast as the resulting square wave frequency. One event is required for the rising edge and another for the falling edge. In the examples below, we make `High` plus `Low` be a constant. By adjusting the ratio of `High` and `Low` the software can control the duty cycle.

$$Dutycycle = \frac{\texttt{High}}{\texttt{High}+\texttt{Low}} = \frac{\texttt{High}}{\texttt{Period}}$$

This implementation occurs in hardware and does not require interrupts. Therefore, it can generate waves close to 0 or 100% duty cycle. If we clear the **TAPWML** bit of the control register we select normal PWM mode. In normal mode, the corresponding output pin is set when the timer is loaded with the value in the **TIMER0_TAILR_R** register. When it reaches the value stored in the **TIMER0_TAMATCHR_R** register, the pin is cleared. The **TAPWML** bit of the control register inverts this behavior. In all modes, the timer is reloaded with the value in the **TIMER0_TAILR_R** register on the cycle after it reaches 0x0000. In PWM output mode, the timer continues counting indefinitely until explicitly disabled by clearing the **TAEN** (or **TBEN**) bit in the **TIMER0_CTL_R** register. Figure 6.11 shows the PWM output can be used to interface a DC motor to the microcontroller.

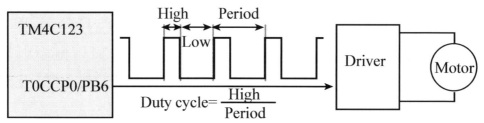

Figure 6.11. The PWM output of Timer 0A can adjust the power to the DC motor.

Program 6.7 configures Timer 0A for PWM output. The user calls **PWM_Init** once to turn it on, and then calls **PWM_Duty** to adjust the duty cycle.

```
// period is number of bus clock cycles in the PWM period
// high is number of bus clock cycles the signal is high
void PWM_Init(uint16_t period, uint16_t high){
  SYSCTL_RCGCTIMER_R |= 0x01;         // activate timer0
  SYSCTL_RCGCGPIO_R  |= 0x02;         // activate port B
  while((SYSCTL_PRGPIO_R&0x02) == 0){};
  GPIO_PORTB_AFSEL_R |= 0x40;         // enable alt funct on PB6
  GPIO_PORTB_DEN_R   |= 0x40;         // enable digital I/O on PB6
  GPIO_PORTB_PCTL_R = (GPIO_PORTB_PCTL_R&0xF0FFFFFF)+0x07000000;
  TIMER0_CTL_R &= ~0x00000001;        // disable timer0A during setup
  TIMER0_CFG_R = 0x00000004;          // configure for 16-bit timer mode
  TIMER0_TAMR_R = 0x0000000A;         // PWM and periodic mode
  TIMER0_TAILR_R = period-1;          // timer start value
  TIMER0_TAMATCHR_R = period-high-1;  // duty cycle = high/period
  TIMER0_CTL_R |= 0x00000001;         // enable timer0A 16-b, PWM
}
void PWM_Duty(uint16_t high){ // duty cycle is high/period
  TIMER0_TAMATCHR_R = TIMER0_TAILR_R-high; // duty cycle = high/period
}
```
Program 6.7. Software to generate a PWM output using Timer 0A (Timer0APWM_xxx).

Checkpoint 6.8: When does an output compare event occur when in PWM mode?

Checkpoint 6.9: What happens during an output compare event in PWM mode?

6.3.2. Pulse Width Modulation using the PWM Module

PWM outputs are so important that many microcontrollers have dedicated PWM modules. The number of PWMs and associated pins vary from one microcontroller to the next, see Figure 6.12. The LM4F120 has none, the TM4C123 has sixteen and the MSP432E401Y has eight. Refer to Tables 2.4 and 2.5.

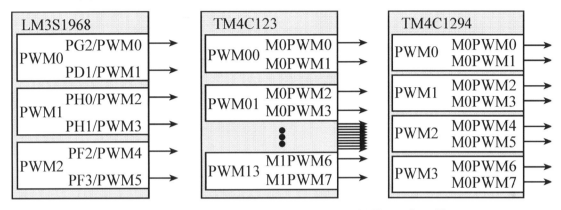

Figure 6.12. PWM pins on the LM3S1968, the TM4C123, and the MSP432E401Y.

The PWM0 block produces the PWM0 and PWM1 outputs, the PWM1 block produces the PWM2 and PWM3 outputs, the PWM2 block produces the PWM4 and PWM5 outputs, and the PWM3 on the TM4C123 block produces the PWM6 and PWM7 outputs. The TM4C123 has a second block providing for an additional eight PWM outputs. The design of a PWM system considers three factors. The first factor is period of the PWM output. Most applications choose a period, initialize the waveform at that period, and adjust the duty cycle dynamically. The second factor is precision, which is the total number of duty cycles that can be created. A 16-bit channel can potentially create up to 65536 different duty cycles. However, since the duty cycle register must be less than or equal to the period register, the precision of the system is determined by the value written to the period register. The last consideration is the number of channels. Different members of the LM3S/LM4F/TM4C family have from zero to sixteen PWM outputs. Refer to the data sheet for your specific microcontroller to determine how many PWM outputs it has and to which ports the PWM is connected.

Program 6.8 shows the initialization on a TM4C123 for generating a PWM on the PB6/PWM0 pin. 1) First, we activate the clock for the PWM module. 2) Second, we activate the output pin as a digital alternate function. 3) Next, we select the clock to be used for the PWM in RCC register. If we do not use the PWM divider, then it is clocked from the bus clock. With the divider we can choose /2, /4, /8, /16, /32, or /64. Assuming the TM4C123 is running at 50 MHz, this program specifies the PWM clock to be 25 MHz. 4) We set the PWM to countdown mode. We specify in the **PWM_0_GENA_R** register that the comparator action is to set to one, and the load action is set to zero. 5) We specify the period in the **PWM_0_LOAD_R** register. 6) We specify the duty cycle in the **PWM_0_CMPA_R** register. 7) Lastly, we start and enable the PWM. We call **PWM0_Init** once to turn it on, and then call **PWM0_Duty** to adjust the duty cycle. Assume the bus clock is 6 MHz, we call **PWM0_Init(30000,15000);** to create a 10 ms period 50 % duty cycle output on PWM0 (PB6).

```c
// period is 16-bit number of PWM clock cycles in one period (3<=period)
// duty is number of PWM clock cycles output is high  (2<=duty<=period-1)
// PWM clock rate = processor clock rate/SYSCTL_RCC_PWMDIV
//                = BusClock/2 (in this example)
void PWM0_Init(uint16_t period, uint16_t duty){
  SYSCTL_RCGCPWM_R  |= 0x01;             // 1) activate PWM0
  SYSCTL_RCGCGPIO_R |= 0x02;             // 2) activate port B
  while((SYSCTL_PRGPIO_R&0x02) == 0){};
  GPIO_PORTB_AFSEL_R |= 0x40;            // enable alt funct on PB6
  GPIO_PORTB_PCTL_R = (GPIO_PORTB_PCTL_R&0xF0FFFFFF)+0x04000000; // PWM0
  GPIO_PORTB_AMSEL_R &= ~0x40;           // disable analog fun on PB6
  GPIO_PORTB_DEN_R |= 0x40;              // enable digital I/O on PB6
  SYSCTL_RCC_R = 0x00100000 |            // 3) use PWM divider
      (SYSCTL_RCC_R & (~0x000E0000));    //    configure for /2 divider
  PWM0_0_CTL_R = 0;                      // 4) re-loading down-counting mode
  PWM0_0_GENA_R = 0xC8;                  // low on LOAD, high on CMPA down
  PWM0_0_LOAD_R = period - 1;            // 5) cycles needed to count down to 0
  PWM0_0_CMPA_R = duty - 1;              // 6) count value when output rises
  PWM0_0_CTL_R |= 0x00000001;            // 7) start PWM0
  PWM0_ENABLE_R |= 0x00000001;           // enable PB6/M0PWM0
}
void PWM0_Duty(uint16_t duty){
  PWM_0_CMPA_R = duty - 1;               // 6) count value when output rises
}
```

Program 6.8. Implementation of a 16-bit PWM output. low on LOAD, high on CMPA (PWM_xxx).

6.4. Frequency Measurement

6.4.1. Frequency Measurement Concepts

The direct measurement of frequency involves counting input pulses for a fixed amount of time. The basic idea is to use input capture to count pulses, and use output compare to create the fixed time interval, Δt. For example, we could initialize input capture to decrement a counter in `TIMER0_TAR_R` on every rising edge of our input signal. At the beginning of our fixed time interval, `TIMER0_TAR_R` is initialized to `TIMER0_TAILR_R`, and at the end of the interval, we can calculate frequency:

f = (`TIMER0_TAILR_R` - `TIMER0_TAR_R`)/Δt

The frequency resolution, Δf, is defined to be the smallest change in frequency that can be reliably measured by the system. In order for the system to detect a change, the frequency must increase (or decrease) enough so that there is one more (or one less) pulse during the fixed time interval. Therefore, the frequency resolution is

$$\Delta f = 1/\Delta t$$

This frequency resolution also specifies the units of the measurement.

Example 6.8. Design a system that measures frequency with a resolution of 100 Hz.

Solution: If we count pulses in a 10ms time interval, then the number of pulses represents the signal frequency with units 1/10 ms or 100 Hz. E.g., if there are 7 pulses during the 10 ms interval then the frequency is 700 Hz. For this system, the measurement resolution is 100 Hz, so the frequency would have to increase to 800 Hz (or decrease to 600 Hz) for the change to be detected (Figure 6.13.)

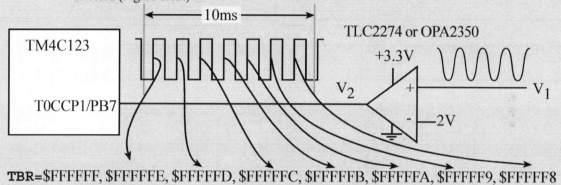

Figure 6.13. Frequency measurement using both input capture and output compare.

The highest frequency that can be measured will be determined by how fast the Input Capture hardware can count pulses. Since there must be two clock periods while it is high and two clock pulses while it is low, the fastest frequency that can be measured is the bus frequency divided by four. The precision of the measurement is determined by the number of bits in the input capture register, which in this case is 16 bits. In this example, the digital logic input signal is connected to PB7 (T0CCP1, attached to Timer0B). The rising edge will decrement the counter in **TIMER0_TBR_R**.

The frequency measurement software is given as Program 6.9. The frequency measurement counts the number of rising edges in a 10 ms interval. The measurement resolution is 100 Hz (determined by the 10 ms interval). A Timer0B input capture event occurs on each rising edge, and a Timer0A periodic interrupt occurs every 10 ms. The foreground/background threads communicate via a mailbox. The background thread will update the data in **Freq** with a new measurement and set the flag **Done**. When **Done** is set, the foreground thread will read the global **Freq** and clear **Done**.

```
uint32_t Freq;    // Frequency with units of 100 Hz
int Done;         // Set each measurement, every 10 ms
```

If the bus clock is 16 MHz, the largest frequency we can measure will be 8 MHz. Therefore, the maximum values of the frequency measurement will be 80000, which will not overflow the 24-bit counter. A frequency of 0 will result in no input capture events and the system will properly report the frequency of 0. The precision of this system is 80001 alternatives or about 16 bits.

```c
void FreqMeasure_Init(void){
  SYSCTL_RCGCTIMER_R |= 0x01;     // activate timer0
  SYSCTL_RCGCGPIO_R |= 0x02;      // activate port B
  Freq = 0;                        // allow time to finish activating
  Done = 0;
  GPIO_PORTB_DIR_R &= ~0x80;       // make PB7 inputs
  GPIO_PORTB_DEN_R |= 0x80;        // enable digital PB7
  GPIO_PORTB_AFSEL_R |= 0x80;      // enable alt funct on PB7
  GPIO_PORTB_PCTL_R = (GPIO_PORTB_PCTL_R&0x0FFFFFFF)+0x70000000;
  TIMER0_CTL_R = 0x00000000;       // disable TIMER0A TIMER0B during setup
  TIMER0_CFG_R = 0x00000004;       // configure for 16-bit mode
  TIMER0_TAMR_R = 0x00000002;      // TIMER0A for periodic mode
  TIMER0_TAILR_R = 9999;           // 4) reload value 100 Hz
  TIMER0_TAPR_R = 15;              // 1us resolution
  TIMER0_TBMR_R = 0x00000003;      // edge count mode rising edge on PB7
  TIMER0_TBILR_R = 0xFFFFFFFF;     // start value
  TIMER0_TBPR_R = 0xFF;            // activate prescale, creating 24-bit
  TIMER0_IMR_R &= ~0x700;          // disable all interrupts for timer0B
  TIMER0_ICR_R = 0x00000001;       // clear TIMER0A timeout flag
  TIMER0_IMR_R = 0x00000001;       // arm timeout interrupt
  NVIC_PRI4_R = (NVIC_PRI4_R&0x00FFFFFF)|0x80000000; // priority 4
  NVIC_EN0_R = 1<<19;              // enable IRQ 19 in NVIC
  TIMER0_CTL_R |= 0x00000101;      // enable TIMER0A TIMER0B
}
// called every 10ms to collect a new measurement
void Timer0A_Handler(void){
  TIMER0_ICR_R = 0x00000001;         // acknowledge timer0A timeout
  Freq = (0xFFFFFF-TIMER0_TBR_R);    // f = (pulses)/(fixed time)
  Done = -1;
  TIMER0_CTL_R &= ~0x00000100;       // disable TIMER0B
  TIMER0_TBILR_R = 0xFFFFFFFF;       // start value
  TIMER0_TBPR_R = 0xFF;              // activate prescale, creating 24-bit
  TIMER0_CTL_R |= 0x00000100;        // enable TIMER0B
}
```

Program 6.9. Software to measure frequency with a resolution of 100 Hz (FreqMeasure_xxx).

6.4.2. Using Period Measurement to Calculate Frequency

Period and frequency are obviously related, so when faced with a problem that requires frequency information we could measure period, and calculate frequency from the period.

If we have a bus clock of 8 MHz, the a period measurement system will have a resolution of 125 ns. Assume p is 16-bit period measurement. With a resolution of 125ns, the period can range from about 40 to 8192 µs. This corresponds to a frequency range of 122 Hz to 25 kHz. We can calculate frequency f from this period measurement, $f = 8000000/p$

It is easy to see how the 40 to 8192 µs period range maps into the 122 Hz to 25 kHz frequency range, but mapping the 125 ns period resolution into an equivalent frequency resolution is a little trickier. If the frequency is f, then the frequency must change to $f+\Delta f$ such that the period changes by at least Δp = 125 ns. $1/f$ is the initial period, and $1/(f+\Delta f)$ is the new period. These two periods differ by 125 ns. In other words,

$$\Delta p = \frac{1}{f} - \frac{1}{f + \Delta f}$$

We can rearrange this equation to relate Δf as a function of Δp and f.

$$\Delta f = \frac{1}{1/f - \Delta p} - f$$

This very nonlinear relationship, shown in Table 6.3, illustrates that although the period resolution is fixed at 125 ns, the equivalent frequency resolution varies from 500 Hz to 0.0005 Hz. If the signal frequency is restricted to the range from 125 to 2828 Hz, then we can say the frequency resolution will be better than 1 Hz.

Frequency (Hz)	Period (µsec)	Δf (Hz)
25000	40	78.370
16000	63	32.064
8000	125	8.008
2000	500	0.500
1000	1000	0.125
500	2000	0.031
250	4000	0.008
125	8000	0.002

Table 6.3. Relationship between frequency resolution and frequency when calculated using period measurement.

6.4.3. Using Frequency Measurement to Calculate Period

Similarly, when faced with a problem that requires a period measurement we could measure frequency, and calculate period from the frequency measurement. A similar nonlinear relationship exists between the frequency resolution and period resolution. In general, the period measurement approach will be faster, but the frequency measurement approach will be more robust in the face of missed edges or extra pulses.

$$\Delta p = \frac{1}{1/p - \Delta f} - p$$

6.5. Binary Actuators

6.5.1. Electrical Interface

Relays, solenoids, and DC motors are grouped together because their electrical interfaces are similar. We can add speakers to this group if the sound is generated with a square wave. In each case, there is a coil, and the computer must drive (or not drive) current through the coil. To interface a coil, we consider **voltage**, **current** and **inductance**. We need a power supply at the desired voltage requirement of the coil. If the only available power supply is larger than the desired coil voltage, we use a voltage regulator (rather than a resistor divider to create the desired voltage.) We connect the power supply to the positive terminal of the coil, shown as **+V** in Figure 6.14. We will use a transistor device to drive the negative side of the coil to ground. The computer can turn the current on and off using this transistor. The second consideration is current. In particular, we must however select the power supply and an interface device that can support the coil current. The 7406 is an open collector driver capable of sinking up to 40 mA. The 2N2222 is a **bipolar junction transistor** (BJT), NPN type, with moderate current gain. The TIP120 is a **Darlington transistor**, also NPN type, which can handle larger currents. The IRF540 is a **MOSFET** transistor that can handle even more current. BJT and Darlington transistors are current-controlled (meaning the output is a function of the input current), while the MOSFET is voltage-controlled (output is a function of input voltage). When interfacing a coil to the microcontroller, we use information like Table 6.4 to select an interface device capable the current necessary to activate the coil. It is a good design practice to select a driver with a maximum current at least twice the required coil current. When the digital **Port** output is high, the interface transistor is active and current flows through the coil. When the digital **Port** output is low, the transistor is not active and no current flows through the coil.

The third consideration is inductance in the coil. The 1N914 diode in Figure 6.14 provides protection from the **back emf** generated when the switch is turned off, and the large dI/dt across the inductor induces a large voltage (on the negative terminal of the coil), according to $V=L \cdot dI/dt$. For example, if you are driving 0.1A through a 0.1 mH coil (**Port** output = 1) using a 2N2222, then disable the driver (**Port** output = 0), the 2N2222 will turn off in about 20ns. This creates a dI/dt of at least $5 \cdot 10^6$ A/s, producing a back emf of 500 V! The 1N914 diode shorts out this voltage, protecting the electronics from potential damage. The 1N914 is called a **snubber diode**.

Device	Type	Maximum current
TM4C123	CMOS	8 mA
MSP432E401Y	CMOS	12 mA
7406	TTL logic	40 mA
PN2222	BJT NPN	150 mA
2N2222	BJT NPN	500 mA
TIP120	Darlington NPN	5 A
IRF540	power MOSFET	28 A

Table 6.4. Possible devices that can be used to interface a coil compared to the microcontroller.

Observation: It is important to realize that many devices cannot be connected directly up to the microcontroller. In the specific case of motors, we need an interface that can handle the voltage and current required by the motor.

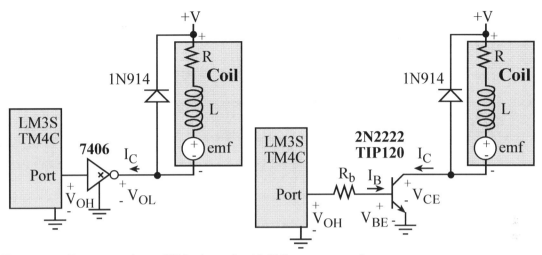

Figure 6.14. Binary interface to EM relay, solenoid, DC motor or speaker.

If you are sinking 16 mA (I_{OL}) with the 7406, the output voltage (V_{OL}) will be 0.4V. However, when the I_{OL} of the 7406 equals 40 mA, its V_{OL} will be 0.7V. 40 mA is not a lot of current when it comes to typical coils. However, the 7406 interface is appropriate to control small relays.

Checkpoint 6.10: A relay is interfaced with the 7406 circuit in Figure 6.14. The positive terminal of the coil is connected to +5V, and the coil requires 40 mA. What will be the voltage across the coil when active?

When designing an interface, we need to know the desired coil voltage (V_{coil}) and coil current (I_{coil}). Let V_{be} be the base-emitter voltage that activates the NPN transistor and let h_{fe} be the current gain. There are three steps when interfacing an N-channel (right side of Figure 6.14.)

1) Choose the interface voltage V equal to V_{coil} (since V_{CE} is close to zero)
2) Calculate the desired base current $I_b = I_{coil}/h_{fe}$ (since I_C equals I_{coil})
3) Calculate the interface resistor $R_b \leq (V_{OH} - V_{be})/I_b$ (choose a resistor 2 to 5 times smaller)

With an N-channel switch, like Figure 6.14, current is turned on and off by connecting/disconnecting one side of the coil to ground, while the other side is fixed at the voltage supply. A second type of binary interface uses P-channel switches to connect/disconnect one side of the coil to the voltage supply, while the other side fixed at ground, as shown in Figure 6.15. In other to activate a PNP transistor (e.g., PN2907 or TIP125), there must be a V_{EB} greater than 0.7 V. In order to deactivate a PNP transistor, the V_{EB} voltage must be 0. Because the transistor is a current amplifier, there must be a resistor into the base in order to limit the base current.

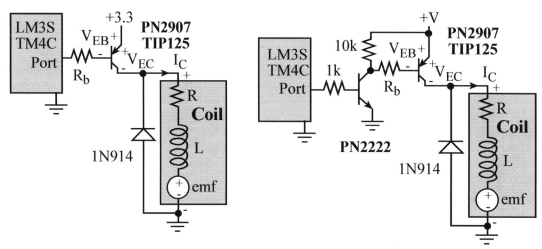

Figure 6.15. PNP interface to EM relay, solenoid, DC motor or speaker.

To understand how the PNP interface on the right of Figure 6.15 operates, consider the behavior for the two cases: the Port output is high and the Port output is low. If the Port output is high, its output voltage will be between 2.4 and 3.3 V. This will cause current to flow into the base of the PN2222, and its V_{be} will saturate to 0.7 V. The base current into the PN2222 could be from (2.4-0.7)/1000 to (3.3-0.7)/1000, or 1.7 to 2.6 mA. The microcontroller will be able to source this current. This will saturate the PN2222 and its V_{CE} will be 0.3 V. This will cause current to flow out of the base of the PN2907, and its V_{EB} will saturate to 0.7 V. If the supply voltage is V, then the PN2907 base current is $(V-0.7-0.3)/R_b$. Since the PNP transistor is on, V_{EC} will be small and current will flow from the supply to the coil. If the port output is low, the voltage output will be between 0 and 0.4V. This not high enough to activate the PN2222, so the NPN transistor will be off. Since there is no I_C current in the PN2222, the 10k and R_b resistors will place +V at the base of the PN2907. Since the V_{EB} of the PN2907 is 0, this transistor will be off, and no current will flow into the coil. For parameter values refer back to Table 1.6.

MOSFETs can handle significantly more current than BJT or Darlington transistors. MOSFETs are voltage controlled switches. The difficulty with interfacing MOSFETs to a microcontroller is the large gate voltage needed to activate it. The left side of Figure 6.16 is an N-channel interface. The IRF540 N-channel MOSFET can sink up to 28A when the gate-source voltage is above 7V. This circuit is negative logic. When the port pin is high, the 2N2222 is active making the MOSFET gate voltage 0.3V (V_{CE} of the PN2222). A V_{GS} of 0.3V turns off the MOSFET. When the port pin is low, the 2N2222 is off making the MOSFET gate voltage +V (pulled up through the 10kΩ resistor). The V_{GS} is +V, which turns the MOSFET on.

The right side of Figure 6.16 shows a P-channel MOSFET interface. The IRF9540 P-channel MOSFET can source up to 20A when the source-gate voltage is above 7V. The FQP27P06 P-channel MOSFET can source up to 27A when the source-gate voltage is above 6V. This circuit is positive logic. When the port pin is high, the 2N2222 is active making the MOSFET gate voltage 0.3V. This makes V_{SG} equal to +V-0.3, which turns on the MOSFET. When the port pin is low, the 2N2222 is off. Since the 2N2222 is off, the 10kΩ pull-up resistor makes the MOSFET gate voltage +V. In this case V_{SG} equals 0, which turns off the MOSFET.

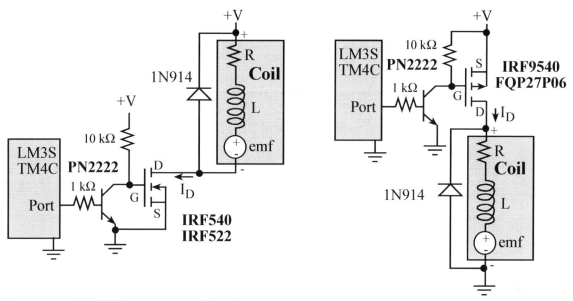

Figure 6.16. MOSFET interfaces to EM relay, solenoid, DC motor or speaker.

An H-bridge uses four transistors, allowing current to flow in either direction. Figures 4.26 and 4.27 show applications of the L293 H-bridge, while Figure 6.17 shows one of the H-bridge circuits internal to the L293. Each output is a totem-pole drive circuit with a Darlington transistor sink and a pseudo-Darlington source. If 1A is high, Q_1 is on and Q_2 is off. If 1A is low, Q_1 is off and Q_2 is on. 2A controls Q_3 and Q_4 in a similar fashion. If 1A is high and 2A is low, then Q_1 Q_4 are on and current flows left to right across coil A. If 1A is low and 2A is high, then Q_2 Q_3 are on and current flows right to left across coil A.

Observation: We used the L293 to interface unipolar and bipolar stepper motors back in Section 4.7.2.

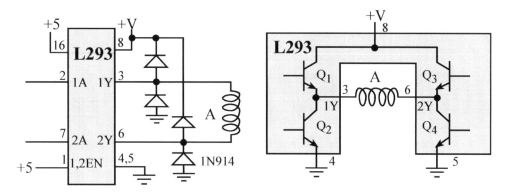

Figure 6.17. An H-bridge can drive current in either direction. Also consider integrated H-bridge drivers like the DRV8838 and DRV8848.

6.5.2. Electromagnetic and Solid State Relays

A relay is a device that responds to a small current or voltage change by activating switches or other devices in an electric circuit. It is used to remotely switch signals or power. The input control is usually electrically isolated from the output switch. The input signal determines whether the output switch is open or closed. Relays are classified into three categories depending upon whether the output switches power (i.e., high currents through the switch) or electronic signals (i.e., low currents through the switch). Another difference is how the relay implements the switch. An electromagnetic (EM) relay uses a coil to apply EM force to a contact switch that physically opens and closes. The solid state relay (SSR) uses transistor switches made from solid state components to electronically allow or prevent current flow across the switch). The three types are

1. The classic general purpose relay has an EM coil and can switch AC power
2. The reed relay has an EM coil and can switch low-level DC electronic signals
3. The solid state relay (SSR) has an input triggered semiconductor power switch

Three solid state relays are shown in Figure 6.18. Interfacing a SSR is similar to interfacing an LED. A SSR interface was developed as Example 2.1 and Figure 2.27. SSRs allow the microcontroller to switch AC loads from 1 to 30A. They are appropriate in situations where the power is turned on and off many times.

The input circuit of an EM relay is a coil with an iron core. The output switch includes two sets of silver or silver-alloy contacts (called **poles**.) One set is fixed to the relay **frame**, and the other set is located at the end of leaf spring poles connected to the **armature**. The contacts are held in the "normally closed" position by the armature return spring. When the input circuit energizes the EM coil, a "pull in" force is applied to the armature and the "normally closed" contacts are released (called **break**) and the "normally open" contacts are connected (called **make**.) The armature pull in can either energize or de-energize the output circuit depending on how it is wired. Relays are mounted in special sockets, or directly soldered onto a PC board.

The number of poles (e.g., single pole, double pole, 3P, 4P etc.) refers to the number of switches that are controlled by the input. The relay shown below is a double pole because it has two switches. **Single-throw** means each switch has two contacts that can be open or closed. **Double-throw** means each switch has three contacts. The common contact will be connected to one of the other two contacts (but not both at the same time.) The parameters of the output switch include maximum AC (or DC) power, maximum current, maximum voltage, on resistance, and off resistance. A DC signal will weld the contacts together at a lower current value than an AC signal, therefore the maximum ratings for DC are considerable smaller than for AC. Other relay parameters include turn on time, turn off time, life expectancy, and input/output isolation. **Life expectancy** is measured in number of operations. Figure 6.19 illustrates the various configurations available. The sequence of operation is described in Table 6.5.

Figure 6.18. Solid state relays can be used to control power to an AC appliance.

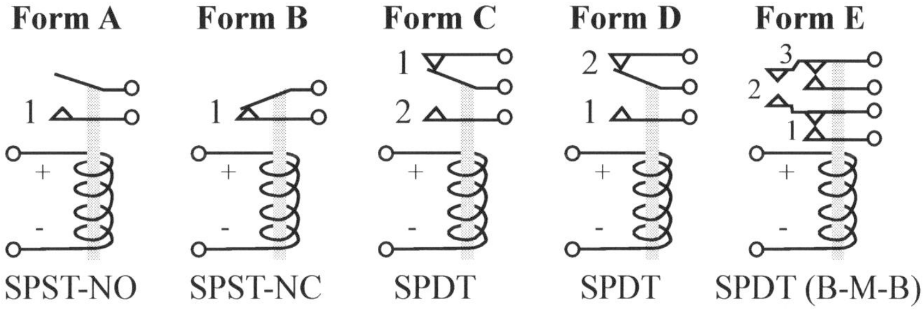

Figure 6.19. Standard relay configurations.

Form	Activation Sequence	Deactivation Sequence
A	Make 1	Break 1
B	Break 1	Make 1
C	Break 1, Make 2	Break 2, Make 1
D	Make 1, Break 2	Make 2, Break 1
E	Break 1, Make 2, Break 3	

Table 6.5. Standard definitions for five relay configurations.

6.5.3. Solenoids

Solenoids are used in discrete mechanical control situations such as door locks, automatic disk/tape ejectors, and liquid/gas flow control valves (on/off type). Much like an EM relay, there is a frame that remains motionless, and an armature that moves in a discrete fashion (on/off). A solenoid has an electro-magnet. When current flows through the coil, a magnetic force is created causing a discrete motion of the armature. Each of the solenoids shown Figure 6.20 has a cylindrically-shaped armature the moves in the horizontal direction relative to the photograph. The solenoid on the top is used in a door lock, and the second from top is used to eject the tape from a video cassette player. When the current is removed, the magnetic force stops, and the armature is free to move. The motion in the opposite direction can be produced by a spring, gravity, or by a second solenoid.

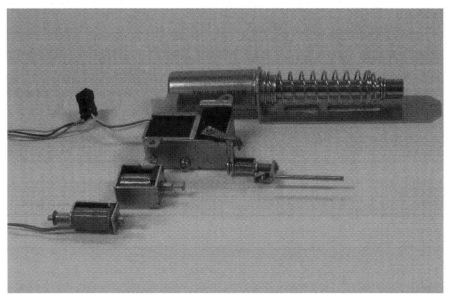

Figure 6.20. Photo of four solenoids.

6.5.4. DC Motor Interface with PWM

Similar to the solenoid and EM relay, the DC motor has a frame that remains motionless, and an armature that moves. In this case, the armature moves in a circular manner (shaft rotation).

In the previous interfaces the microcontroller was able to control electrical power to a device in a binary fashion: either all on or all off. Sometimes it is desirable for the microcontroller to be able to vary the delivered power in a variable manner. One effective way to do this is to use pulse width modulation (PWM). The basic idea of PWM is to create a digital output wave of fixed frequency, but allow the microcontroller to vary its duty cycle. The system is designed in such a way that **High+Low** is constant (meaning the frequency is fixed). The **duty cycle** is defined as the fraction of time the signal is high:

$$duty\ cycle\ =\ \frac{High}{High+Low}$$

Hence, duty cycle varies from 0 to 1. We interface this digital output wave to an external actuator (like a DC motor), such that power is applied to the motor when the signal is high, and no power is applied when the signal is low. We purposely select a frequency high enough so the DC motor does not start/stop with each individual pulse, but rather responds to the overall average value of the wave. The average value of a PWM signal is linearly related to its duty cycle and is independent of its frequency. Let P ($P=V*I$) be the power to the DC motor, shown in Figure 6.21, when the PP0 signal is high. Notice the circuit in Figure 6.21 is one of the examples previously described in Figure 6.14. Under conditions of constant speed and constant load, the delivered power to the motor is linearly related to duty cycle.

$$\text{Delivered Power} = \text{duty cycle} * P = \frac{High}{High + Low} * P$$

Unfortunately, as speed and torque vary, the developed emf will affect delivered power. Nevertheless, PWM is a very effective mechanism, allowing the microcontroller to adjust delivered power.

A DC motor has an electro-magnet as well. When current flows through the coil, a magnetic force is created causing a rotation of the shaft. Brushes positioned between the frame and armature are used to alternate the current direction through the coil, so that a DC current generates a continuous rotation of the shaft. When the current is removed, the magnetic force stops, and the shaft is free to rotate. The resistance in the coil (R) comes from the long wire that goes from the + terminal to the – terminal of the motor. The inductance in the coil (L) arises from the fact that the wire is wound into coils to create the electromagnetics. The coil itself can generate its own voltage (emf) because of the interaction between the electric and magnetic fields. If the coil is a DC motor, then the emf is a function of both the speed of the motor and the developed torque (which in turn is a function of the applied load on the motor.) Because of the internal emf of the coil, the current will depend on the mechanical load. For example, a DC motor running with no load might draw 50 mA, but under load (friction) the current may jump to 500 mA.

There are six considerations when **selecting a DC motor**: speed, torque, voltage, current, size, and weight. Speed is the rate in rotations per minute (RPM) that the motor will spin, and torque is the available force times distance the motor can provide at that speed. We select the motor voltage to match the available power supply. Unlike LEDs, we MUST not use a resistor in series with a motor to reduce the voltage. In general, the motor voltage matches the power supply voltage. When interfacing we will need to know maximum current.

There are lots of motor driver chips, but they are fundamentally similar to the circuits shown in Figure 6.14. For the 2N2222 and TIP120 NPN transistors, if the port output is low, no current can flow into the base, so the transistor is off, and the collector current, I_C, will be zero. If the port output is high, current does flow into the base and V_{BE} goes above V_{BEsat} turning on the transistor. The transistor is in the linear range if $V_{BE} \leq V_{BEsat}$ and $I_c = h_{fe} \cdot I_b$. The transistor is in the saturated mode if $V_{BE} \geq V_{BEsat}$, $V_{CE} = 0.3\text{V}$ and $I_c < h_{fe} \cdot I_b$. We select the resistor for the NPN transistor interfaces to operate right at the transition between linear and saturated mode. We start with the desired coil current, I_{coil} (the voltage across the coil will be $+V-V_{CE}$ which will be about +V-0.3V). Next, we calculate the needed base current (I_b) given the current gain of the NPN

$$I_b = I_{coil} / h_{fe}$$

knowing the current gain of the NPN (h_{fe}), see Table 6.6. Finally, given the output high voltage of the microcontroller (V_{OH} is about 3.3 V) and base-emitter voltage of the NPN (V_{BEsat}) needed to activate the transistor, we can calculate the desired interface resistor.

$$R_b \leq (V_{OH} - V_{BEsat})/ I_b = h_{fe} *(V_{OH} - V_{BEsat})/ I_{coil}$$

The inequality means we can choose a smaller resistor, creating a larger I_b. Because the of the transistors can vary a lot, it is a good design practice to make the R_b resistor about ½ the value shown in the above equation. Since the transistor is saturated, the increased base current produces the same V_{CE} and thus the same coil current.

Parameter	PN2222 (I_C=150mA)	2N2222 (I_C=500mA)	TIP120 (I_C=3A)
h_{fe}	100	40	1000
h_{ie}	60 Ω	250 to 8000 Ω	70 to 7000 Ω
V_{BEsat}	0.6	2	2.5 V
V_{CE} at saturation	0.3	1	2 V

Table 6.6. Design parameters for the 2N2222 and TIP120.

The IRF540 MOSFET is a voltage-controlled device, if the **Port** output is high, the 2N2222 is on, the MOSFET is off, and the coil current will be zero. If the **Port** output is low, the 2N2222 is off, the gate voltage of the MOSFET will be +V, the MOSFET is on, and the V_{DS} will be very close to 0. The IRF540 needs a large gate voltage (> 10V) to fully turn so the drain will be able to sink up to 28 A.

Because of the resistance of the coil, there will not be significant dI/dt when the device is turned on. Consider a DC motor as shown in Figure 6.21 with V= 12V, R = 50 Ω and L = 100 µH. Assume we are using a 2N2222 with a V_{CE} of 1 V at saturation. Initially the motor is off (no current to the motor). At time t=0, the digital port goes from 0 to +3.3 V, and transistor turns on. Assume for this section, the emf is zero (motor has no external torque applied to the shaft) and the transistor turns on instantaneously, we can derive an equation for the motor (I_c) current as a function of time. The voltage across both LC together is 12-V_{CE} = 11 V at time = 0^+. At time = 0^+, the inductor is an open circuit. Conversely, at time = ∞, the inductor is a short circuit. The I_c at time 0^- is 0, and the current will not change instantaneously because of the inductor. Thus, the I_c is 0 at time = 0^+. The I_c is 11V/50Ω= 220mA at time = ∞.

$$11\ V = I_c * R + L * dI_c/dt$$

General solution to this differential equation is

$$I_c = I_0 + I_1 e^{-t/\tau} \quad dI_c/dt = -(I_1/\tau)e^{-t/\tau}$$

We plug the general solution into the differential equation and boundary conditions.

$$11\ V = (I_0 + I_1 e^{-t/\tau})*R - L*(I_1/\tau)e^{-t/\tau}$$

To solve the differential equation, the time constant will be $\tau = L/R$ = 2 µsec. Using initial conditions, we get

$$I_c = 220mA*(1 - e^{-t/2\mu s})$$

Example 6.10. Design an interface for two +12V 1A geared DC motors. These two motors will be used to propel a robot with two independent drive wheels.

Solution: We will use two copies of the TIP120 circuit in Figure 6.21 because the TIP120 can sink at least three times the current needed for this motor. We select a +12V supply and connect it to the +V in the circuit. The needed base current is

$$I_b = I_{coil}/h_{fe} = 1A/1000 = 1mA$$

The desired interface resistor.

$$R_b \leq (V_{OH} - V_{be})/I_b = (3.3-2.5)/1mA = 800\ \Omega$$

To cover the variability in h_{fe}, we will use a 330 Ω resistor instead of the 800 Ω. The actual voltage on the motor when active will be +12-2 = 10V.

The coils and transistors can vary a lot, so it is appropriate to experimentally verify the design by measuring the voltages and currents. Two copies of Program 6.8 are used to control the robot. The period of the PWM output is chosen to be about 10 times shorter than the time constant of the motor. The electronic driver will turn on and off at this rate, but the motor only responds to the average level. The software sets the duty cycle of the PWM to adjust the delivered power. When active, the interface will drive +10 V across the motor. The current will be a function of the friction applied to the shaft.

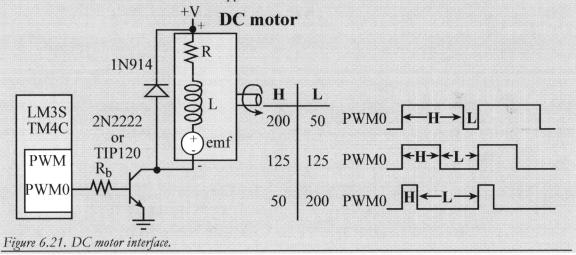

Figure 6.21. DC motor interface.

Similar to the solenoid and EM relay, the DC motor has a frame that remains motionless (called the **stator**), and an armature that moves (called the **rotor**). A **brushed DC motor** has an electromagnetic coil as well, located on the rotor, and the rotor is positioned inside the stator. In Figure 6.22, North and South refer to a permanent magnet, generating a constant B field from left to right. In this case, the rotor moves in a circular manner. When current flows through the coil, a magnetic force is created causing a rotation of the shaft. A brushed DC motor uses commutators to flip the direction of the current in the coil. In this way, the coil on the right always has an up force, and the one on the left always has a down force. Hence, a constant current generates a continuous rotation of the shaft. When the current is removed, the magnetic force stops, and the shaft is free to rotate. In a pulse-width modulated DC motor, the computer activates the coil with a current of fixed magnitude but varies the duty cycle in order to adjust the power delivered to the motor.

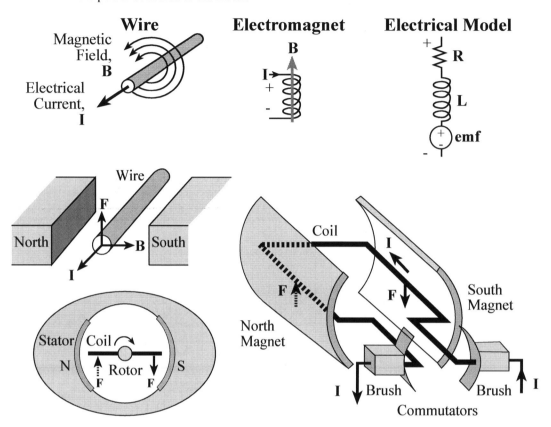

Figure 6.22. A brushed DC motor uses a commutator to flip the coil current.

A **brushless DC motor** (BLDC), as the name implies, does not have mechanical commutators or brushes to flip the currents. It is a synchronous electric motor powered by direct current and has an electronic commutation system, rather than a mechanical commutator and brushes. In BLDC motors, current-to-torque and voltage-to-rpm are linear relationships. The controller uses either the back emf of the motor itself or Hall-effect sensors to know the rotational angle of the shaft. The controller uses this angle to set the direction of the currents in the electromagnets, shown as the six step sequence in Figure 6.23. Other differences from a brushed DC motor are that the BLDC permanent magnets are in the rotor and the electromagnets are in the stator. Typically, there are three electromagnetic coils, labeled Phase A, Phase B, and Phase C, which are arranged in a Wye formation. Each coil can be modeled as a resistance, inductance, and emf, as previously shown in Figure 6.14. The Hall sensor goes through the sequence 001, 000, 100, 110, 111, 011 each time the shaft rotates once. It is a synchronous motor because the controller adjusts the phase current according to the six-step sequence. For example, if the Hall sensor reads 001, then the controller places +V on Phase A and ground on Phase C (step 1). In other words, the phase currents are synchronized to the shaft position. To rotate the motor in the other direction, we reverse the currents in each step. We will see later for stepper motors that the process is reversed. For stepper motors, the controller sets the phase currents and the motor moves to that position. To adjust the power to a BLDC motor, we change the voltage, V, or use PWM on the control signals themselves. The PWM period should be at least 10 times shorter than the time for each of the six steps. In other words, the PWM frequency should be 60 times faster than the shaft rotational frequency.

BLDC motors have many advantages and few disadvantages when compared to brushed DC motors. Because there are no brushes, they require less maintenance and hence have a longer life. Therefore, they are appropriate for applications where servicing is inconvenient or expensive. BLDC motors produce more output torque per weight than brushed DC motors and hence are used for pilotless airplanes and helicopters. Because the rotor is made of permanent magnets, the rotor inertia is less, allowing it to spin faster and to change quicker. In other words, it has faster acceleration and deceleration. Removing the brushes reduces friction, which also contributes to the improved speed and acceleration. It has a linear speed/torque relationship. Because there is no brush contact, BLDC motors operate more quietly and have less **Electromagnetic Interference** (EMI). The only disadvantages are the complex controller and increased cost.

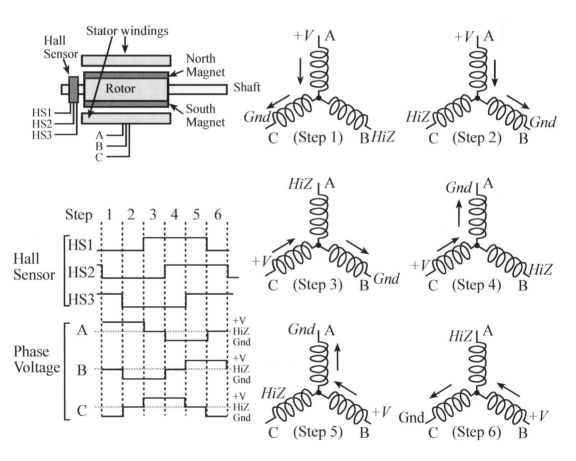

Figure 6.23. A brushless DC motor uses an electronic commutator.

Example 6.11. Interface a 24-V 2-A brushless DC motor.

Solution: A brushless DC motor has three coils connected in a Wye pattern. Each of the phases can be driven into one of three states: 24 V, ground, or floating. We will use MOSFETs to source and sink the current required by the motor (Figure 6.16). Remember, when the motor is under load, the current will increase. The P-channel MOSFET will connect the 24 V to the phase when its gate voltage is below 24 V. The N-channel MOSFET will drive the phase to ground when its gate is above zero. It will be important to prevent turning on both MOSFETs at the same time. For safety reasons, we will use digital logic in the interface so the driver can only be in the three valid states. Table 6.7 shows the design specification for Phase A. When **EnA** is low, both MOSFETs are off and the phase will float (HiZ). When **EnA** is high, the **InA** determines whether the phase is high or low. The six gate voltages are labeled in Figure 6.24 as **G1** to **G6**. These gate voltages are 24 V, produced by the 10 kΩ pull-up, when the corresponding 7406 driver output is floating. Alternatively, these gate voltages are 0.5 V when the 7406 driver output is low. It is good design to use integrated drivers like the ULN2074, L293, TPIC0107, and MC3479 rather than individual transistors. In particular, the entire interface circuit in Figure 6.24 could be replaced with three L6203 full bridge drivers.

EnA	InA	G1	G2	P-chan	N-chan	Phase A
1	1	Low	Low	On	Off	+24 V
1	0	High	High	Off	On	Ground
0	X	High	Low	Off	Off	HiZ

Table 6.7. Control signals for one phase of the brushless DC motor.

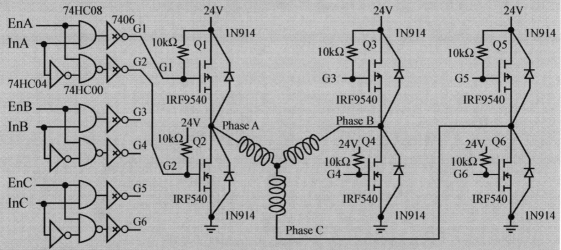

Figure 6.24. Brushless DC motor interface.

The **InA**, **InB**, and **InC** signals in Figure 6.24 are connected to any output ports, whereas the **EnA**, **EnB**, and **EnC** signals will be attached to PWM outputs. The PWM period will be selected 60 times faster than the motor speed in rps. The three Hall-effect sensor signals will be attached to input capture pins. Interrupts will be armed for both the rise and fall of these three sensors. In this way, an ISR will be run at the beginning of each of the six steps. The BLDC motor is a synchronous motor, so the six control signals are a function of the shaft position. In particular, the ISR will look up the Hall sensors and output the pattern, as shown in Table 6.8.

Step	HS1	HS2	HS3	EnA	InA	EnB	InB	EnC	InC	A	B	C
1	0	0	1	PWM	1	0	X	PWM	0	24V	HiZ	0V
2	0	0	0	PWM	1	PWM	0	0	X	24V	0V	HiZ
3	1	0	0	0	X	PWM	0	PWM	1	HiZ	0V	24V
4	1	1	0	PWM	0	0	X	PWM	1	0V	HiZ	24V
5	1	1	1	PWM	0	PWM	1	0	X	0V	24V	HiZ
6	0	1	1	0	X	PWM	1	PWM	0	HiZ	24V	0V

Table 6.8. Input-output relationships for the synchronous controller.

Furthermore, when any of the enable signals are scheduled to be high, they will be pulsed using positive logic PWM. The software can adjust the delivered power to the BLDC motor by setting the duty cycle of the PWM. The software implementation has been left as Lab 6.4.

6.6. Integral Control of a DC Motor

A **control system** is a collection of mechanical and electrical devices connected for the purpose of commanding, directing, or regulating a **physical plant.** In this section the physical plant is a DC motor as described previously in Section 6.5.4. The **real state variables** are the properties of the physical plant that are to be controlled. In this example, we wish to spin the motor at 1500 RPM, or 25 rps. Thus, the state variable in this case will be motor speed. The **sensor** and **state estimator** comprise a data acquisition system. The goal of this data acquisition system is to estimate the state variables. We will attach a tachometer to the motor so the system can measure speed. Figures 6.4 and 6.4 illustrate one type of tachometer and how the signal is interfaced to the microcontroller. The **estimated state variables**, $X'(t)$, in this system will be the measured speed in 0.1 rps. The **actuator** is a transducer that converts the control system commands, $U(t)$, into driving forces, $V(t)$, that are applied the physical plant. In this example we will use the circuit in Figure 6.21, which allows the microcontroller to adjust power to the motor using the functions in Program 6.8. In this example, the bus clock is 80 MHz, so the PWM clock is 40 MHz. We will fix the PWM period 1ms, and initialize the PWM output at 0.1% by calling `PWM0_Init(40000,40)`. We define the actuator command, $U(t)$, as the parameter 40 to 39960 that we pass in when we call `PWM0_Duty`. Therefore we can adjust the power from 0.1% to 99.9%.

In general, the goal of the control system is to drive the real state variables to equal the desired state variables. In actuality though, the controller attempts to drive the estimated state variables to equal the desired state variables. It is important to have an accurate state estimator, because any differences between the estimated state variables and the real state variables will translate directly into controller errors. If we define the error as the difference between the desired and estimated state variables:

$$e(t) = X^*(t) - X'(t)$$

A **closed loop** control system uses the output of the state estimator in a feedback loop to drive the errors to zero. The control system compares $X'(t)$, to the **desired state variables**, $X^*(t)$, in order to decide appropriate action, $U(t)$. See Figure 6.25.

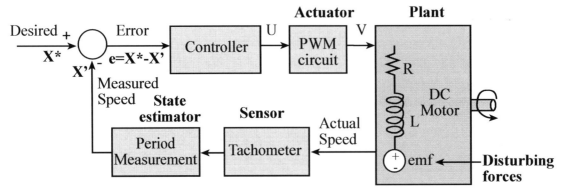

Figure 6.25. Block diagram of a microcomputer-based closed loop control system.

We can combine the period measurement from Section 6.1, the PWM output of Section 6.3, and the DC motor interface of Section 6.5.4 to build a motor controller. One effective yet simple control algorithm is an integral controller. We specify the actuator output as the integral of the accumulated errors.

$$U(t) = \int_0^t K_i E(\tau) d\tau$$

where K_i is a controller constant. For this controller, if the error is zero the actuator command remains constant. If the motor is spinning too slowly, the controller will increase power. If the motor is spinning too quickly, it will decrease power. For an integral controller, the amount of increase or decrease is linearly related to the error. So if the error is large it adds (or subtracts) a lot, and if the error is small it adds (or subtracts) a little. If the time constant of the motor is 100ms, then we will run the controller 10 times faster. The Timer2A ISR in Program 6.11 is set to execute every 10 ms. The software in Program 6.2 will set the global variable **Period** with the measured period in 12.5 ns units.

```
uint32_t Period;              // 24-bit, 12.5 ns units
uint32_t Speed;               // motor speed in 0.1 rps
int32_t E;                    // speed error in 0.1 rps
int32_t U;                    // duty cycle 40 to 39960
void Timer2A_Handler(void){
  TIMER2_ICR_R = 0x01;        // acknowledge timer2A timeout
  Speed = 800000000/Period;   // 0.1 rps
  E = 250-Speed;              // 0.1 rps
  U = U+(3*E)/64;             // discrete integral
  if(U < 40) U=40;            // Constrain output
  if(U>39960) U=39960;        // 40 to 39960
  PWM0_Duty(U);               // output
}
```
Program 6.11. ISR to implement an integral controller

6.7. Exercises

6.1 Show the changes you need to make to Program 6.1 to run on a TM4C123 with the input connected to T1CCP0/PB4.

6.2 Show the changes you need to make to Program 6.1 to run on a TM4C123 with the input connected to T2CCP0/PF4.

6.3 If Timer 0A and Timer 0B are both armed for interrupt, are the two sources polled or vectored? If Timer 0A and Timer 1A are both armed for interrupt, are the two sources polled or vectored?

6.4 How do you change the resolution of the period measurement resolution in Program 6.2?

6.5 If the bus clock were changed to 50 MHz without any changes to Program 6.2, what would the new period measurement range be?

6.6 Show the changes you need to make to Program 6.6 to run Timer 1A.

6.7 Show the changes you need to make to Program 6.6 to run Timer 2B.

6.8 Show the changes you need to make to Program 6.7 to run on a TM4C123 with the output connected to T1CCP0/PB4.

6.9 Show the changes you need to make to Program 6.8 to run on a TM4C123 with the output connected to PWM1/PB8.

D6.10 Create two synchronized 50% duty cycle square waves using PWM channels. Interface the two PWM outputs to a stepper motor using a L293. If you use just 2 PWM outputs you will need to add digital logic inverters. Write software that initialized the PWM given the desired shaft speed in 1 RPM, with a range of speeds from 0 to 100 RPM. Assume there are 200 steps/revolution. The stepper will spin without software overhead until the function is called again. If the desired speed is zero, stop the motor at one of the valid states (5, 6, 10 or 9).

D6.11 The objective of this problem is to measure the frequency of a square wave connected to CCP2. The frequency range is 0 to 2000 Hz and the *resolution is 0.1Hz*. For example, if the frequency is 567.83 Hz, then your software will set the global **Freq** to 5678. Don't worry about frequencies above 2000 Hz.

D6.12 The objective of this problem is to measure body temperature using input capture (period measurement). A shunt resistor is placed in parallel with a thermistor. The thermistor-shunt combination, R, has the following linear relationship for temperatures from 90 to 110 °F.

$R = 100\ k\Omega - (T - 90°F) \cdot 1k\Omega/°F$ where R is the resistance of the thermistor-shunt

In other words, the resistance varies from 100 kΩ to 80 kΩ as the temperature varies from 90 to 110°F. The range of your system is 90 to 110 °F and the resolution should be better than 0.01°F. You will use a TLC555 to convert the resistance to a period. The period of a TLC555 timer is $0.693 \cdot C_T \cdot (R_A + 2R_B)$. Set $R_A = R$, and $R_B = 50k\Omega$.

a) The pulse width measurement resolution will be 125ns because the bus clock is 8 MHz. Choose the capacitor value so that this pulse width measurement resolution matches the desired temperature resolution of 0.01°F.

b) Given this value of C, what is the pulse width at 90 °F? Give the answer both in μs and bus clock cycles.
c) Given this value of C, what is the pulse width at 110 °F? Give the answer both in μs and bus clock cycles.
d) Write the initialization that configures the input capture interrupts.
e) Show the interrupt handler that performs the temperature measurement tasks in the background and sets a global, **Temperature**. **Temperature** will vary from 9000 to 11000 as temperature varies from 90 to 110 °F.

D6.13 Design a wind direction measurement instrument using the input capture technique. Again, you are given a transducer that has a resistance that is linearly related to the wind direction. As the wind direction varies from 0 to 360 degrees, the transducer resistance varies from 0 to 1000 Ω. The frequencies of interest are 0 to 0.5 Hz, and the sampling rate will be 1 Hz. One way to interface the transducer to the computer is to use an astable multivibrator like the TLC555. The period of a TLC555 timer is $0.693 \cdot C_T \cdot (R_A + 2R_B)$.
a) Show the hardware interface
b) Write the initialization and interrupt service routine that measures the wind direction and creates a 16-bit unsigned result with units of degrees. I.e., the value varies from 0 to 359. You do not have to write software that samples at 1 Hz, simply a function that measures wind direction in the background.

D6.14 The objective of this problem is to design an underwater ultrasonic ranging system. The distance to the object, d, can vary from 1 to 100 m. The ultrasonic transducer will send a short 5 μs sound pulse into the water in the direction of interest. The sound wave will travel at 1500m/sec and reflect off the first object it runs into. The reflected wave will also travel at 1500m/sec back to the transducer. The reflected pulse is sensed by the same transducer. Your system will trigger the electronics (give a 5μs digital pulse), measure the time of flight, then calculate the distance to the object. Using periodic interrupts, the software will issue a 5 μs pulse out once a second. Using interrupting input capture, the software will measure the time of flight, Δt. The input capture interrupt handler will calculate distance, d, as a decimal fixed-point value with units of 0.01m, and enter it into a FIFO queue. The main program will call the ritual, then get data out of the FIFO queue. The main program will call **Alarm()** if the distance is less than 15 m. You do not have to give the implementation of **Alarm()**. You may use any of the FIFOs in Chapter 3 without showing its implementation. Assume the bus clock is 8 MHz.

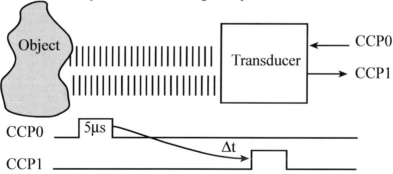

Figure 6.25. Interface for Question D6.14.

a) Derive an equation that relates the distance, d, to the time of flight, Δt.

b) Use this equation to calculate the minimum and maximum possible time of flight, Δt.
c) Give the initialization routine.
d) Give the ISRs that measure distance.

6.8. Lab Assignments

Lab 6.1 The overall objective is to interface a joystick to the microcontroller. The joystick is made with two potentiometers. You can use two astable multivibrators (TLC555) to convert the two resistances into two periods. Use two period measurement channels to estimate the X-Y position of the joystick. Organize the software interface into a device driver, and write a main program to test the interface.

Lab 6.2 The overall objective is to measure temperature. A thermistor is a transducer with a resistance that is a function of its temperature. You can use an astable multivibrator (TLC555) to convert the resistance into a period. Use a period measurement channel to estimate the resistance of the thermistor. Use a table lookup with linear interpolation to convert resistance to temperature. Organize the software interface into a device driver, and write a main program that outputs temperature to the UART channel.

Lab 6.3 The overall objective is to measure linear position. A slide-pot is a transducer with a resistance that is a function of the linear position of the slide. You can use an astable multivibrator (TLC555) to convert the resistance into a period. Use a period measurement channel to estimate the resistance of the potentiometer. Use a table lookup with linear interpolation to convert resistance to position. Organize the software interface into a device driver, and write a main program that outputs position to the UART channel.

Lab 6.4 The objective of this lab is to control a servo motor. Interface a servo motor to the PWM output pin of the microcontroller. The desired angle is input from the UART channel (connected to PC running PuTTY.) Organize the software interface into a device driver, and write a main program that inputs from the UART channel and maintains the PWM output to the servo. Servos are a popular mechanism to implement steering in robotics. Ranging from micro servos with 15oz-in torque to powerful heavy-duty sailboat servos, they all share several common characteristics. A servo is essentially a positionable motor. The servo "knows" two things: where it is (the actual position) and where it wants to be (the desired position). When the servo receives a position, it attempts to move the servo horn to the desired position. The task of the servo, then, is to make the actual position the desired position. The first step to understanding how servos work is to understand how to control them. Power is usually between 4V and 6V and should be separate from system power (as servos are electrically noisy). Even small servos can draw over an amp under heavy load so the power supply should be appropriately rated. Though not recommended, servos may be driven to higher voltages to improve torque and speed characteristics. Servos are commanded through "Pulse Width Modulation," or PWM, signals sent through the command wire. Essentially, the width of a pulse defines the position. For example, sending a 1.5ms pulse to the servo, tells the servo that the desired position is 90 degrees. In order for the servo to hold this position, the command must be sent at about 50 Hz, or every 20 ms. If you were to send a pulse longer than 2.5 ms or shorter than 0.5 ms, the servo would attempt to overdrive (and possibly damage) itself. Once the servo has received the desired position (via the PWM signal) the servo must attempt to match the desired and actual positions. It does this by turning a small, geared motor left or right. If, for example, the desired position is less than the actual position, the servo will turn to the left. On the other hand, if the desired position is greater than the actual position, the servo will turn to the right. In this manner, the servo "zeros-in" on the correct position. Should a load force the servo horn to the right or left, the servo will attempt to compensate. Note that there is no control mechanism for the speed of movement and, for most servos, the speed is specified in degrees/second. For more information refer to the data sheet of your servo.

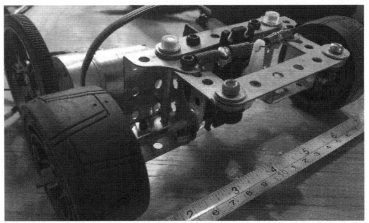

Figure 6.26. Servo motor used to create rack and pinion steering.

Lab 6.5 The overall objective is to measure capacitance. The system will measure from 100 pF to 1uF with a resolution of 100 pF. You can use an astable multivibrator (TLC555) to convert the capacitance into a period. Use a period measurement channel to estimate the capacitance. Use a table lookup with linear interpolation to convert period to capacitance. Organize the software interface into a device driver, and write a main program that outputs position to the UART channel or on a local display.

Lab 6.6. The objective of this lab is to interface a brushless DC motor. Rather than use discrete transistors like Figure 6.21, use an integrated motor interface chip like the L293 or L6103. Using three power resistors in a Wye shape, test the motor drive circuit. Write two software functions: Initialization and PowerSet. The second function can be used to adjust the power to the motor from 0 to 100%. Add period measurements to the input capture ISRs, and use them to measure shaft speed in RPM. Use a logic analyzer to take measurements of shaft speed versus duty cycle. Take transient measurements of speed versus time, as the duty cycle changes from 25 to 75%.

7. Serial Interfacing

Chapter 7 objectives are to:

- Present the physical drivers and protocols for serial interfacing
- Describe the RS232 standard
- Design interfaces that convert between +5 and +3.3 V digital logic
- Present differential line protocols such as RS422 and USB
- Describe and employ the SSI synchronous serial protocol
- Describe and employ the I²C serial bus
- Introduce the USB protocol

In many applications, a single dedicated microcontroller is insufficient to perform all the tasks of the embedded system. One solution would be to use a larger and more powerful microcontroller, and another approach would be to distribute the tasks among multiple microcontrollers. This second approach has the advantages of modularity and expandability. To implement a distributed system, we need a mechanism to send and receive information between the microcontrollers. A second scenario that requires communication is a central general-purpose computer linked to multiple remote embedded systems for the purpose of distributed data collection or distributed control. For situations where the required bandwidth is less than about 100,000 bytes/sec, the built-in serial ports of the microcontroller can be used.

In this chapter we will interface DAC and ADC devices to the synchronous serial port. In Chapters 8 and 10 we continue by showing how to employ the DAC and ADC for signal generation and data acquisition. We can use the I²C bus to interface sensors and actuators. In this chapter we will focus on serial channels that employ a direct physical connection between the microcontrollers, and later in Chapter 11 we expand the communication system to include networks.

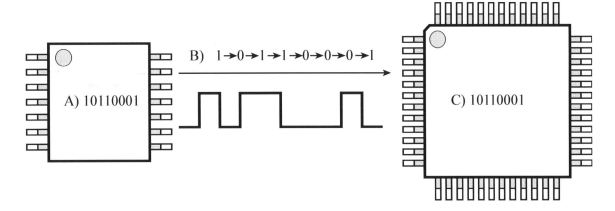

7.1. Introduction to Serial Communication

Serial communication involves the transmission of one bit of information at a time. One bit is sent, a time delay occurs, then the next bit is sent. This section will introduce the use of serial communication as an interfacing technique for various microcontroller peripherals. Since many peripheral devices such as printers, keyboards, scanners, and mice operate have their own computers, the communication problem can be generalized to one of transmitting information between two computers. The **universal asynchronous receiver/transmitter (UART)** is the interface device that implements the serial data transmission. The **serial channel** is the collection of signals (or wires) that implement the communication, see Figure 7.1. To improve bandwidth, remove noise and increase range, we place interface logic between the digital logic UART device and the serial channel. We define the **data terminal equipment (DTE)** as the computer or a terminal and the **data communication equipment (DCE)** as the modem or printer.

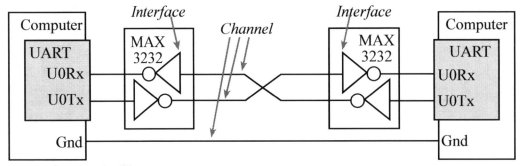

Figure 7.1. A serial channel connects two data terminal equipment (DTE) devices.

When transmitting in a serial fashion, there are many ways to encode binary information on the line, as shown in Figure 7.2. The goal is to maximize bandwidth and minimize errors. **Non-return-to-zero** (NRZ) encoding is a binary code in which the signal is never at zero voltage. There is more energy in the wire compared to typical V_{OL}/V_{OH} digital logic, because there is always a voltage causing current to flow continuously. Energy is the fundamental property needed to communicate information over a distance. On a single wire, voltages are measured relative to ground (e.g., RS232). On a twisted pair, voltages are differential (e.g., RS422, USB). The binary values of 1 and 0 are encoded as positive or negative voltage differences. In positive logic, 1 is a positive voltage difference and 0 is a negative voltage difference. In negative logic, the voltage representing 0 is higher than the voltage representing 1.

Non-return-to-zero-inverted (NRZI) is a method of encoding binary signal as transitions or changes in the signal. Similar to NRZ, the signal in a NRZI protocol is never zero. The binary information is encoded as the presence or absence of a transition at a clock boundary, illustrated by the arrows in Figure 7.2. Both the transmitter and receiver will synchronize their clocks so the receiver knows when to look for the transition. A transition is a change from positive to negative or from negative to positive. For example, we could send a 1 by placing a transition on the signal, or send a 0 by causing no transition. Also, NRZI might take the opposite convention, as in Universal Serial Bus (USB), where a transition means 0 and a steady level means 1.

Manchester encoding code encodes binary bits as either a low-to-high transition, or a high-to-low transition. Manchester encoding is a type of phase encoding, which means the information transmitted as phase shifts. It is used with the IEEE 802.3 Ethernet protocol. There is a fixed time period inside which one bit is transmitted. The original Manchester encoding scheme defined 0 as a low-to-high, and 1 as a high-to-low. However, the IEEE 802.3 convention defines the bits in the opposite manner. The transitions that signify 0 or 1 occur at the midpoint of a period, shown as arrows in Figure 7.2. There may be an additional transitions at the start of a period; these are extra and do not signify data. Since every bit has at least one transition, it is easier for the receiver to align correctly or to synchronize its clock with the transmitter clock. However, the cost of this ease of synchronization is a doubling of the bandwidth requirement of the physical channel as compared to NRZ or NZRI encoding. Consider the case where a system is communicating a long sequence of zeros at 1 Mb/s. Using the NRZI USB encoding, the line will be a square wave at 1 MHz. However, using IEEE 802.3 encoding, the line will be a square wave at 2 MHz.

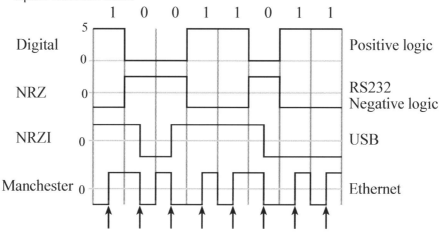

Figure 7.2. Four encodings of the binary 10011011.

Figure 7.3 shows the UART pins available on some LM3S/TM4C microcontrollers. The descriptions and software drivers were presented earlier as Program 4.10 and 5.11. Tables 2.4 and 2.5 describe how to attach I/O pins to the UART modules on the TM4C123/MSP432E4.

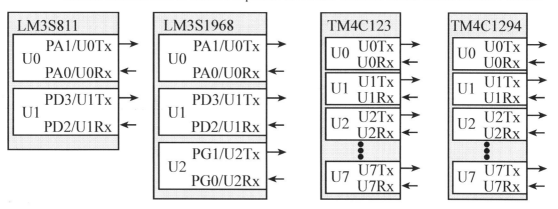

Figure 7.3. Serial port pins available on various LM3S/TM4C/MSP432E microcontrollers.

The interface logic (e.g., MAX3232) converts between digital logic levels and RS232 logic levels. In this protocol, a typical bidirectional channel requires 3 wires (Rx, Tx, Ground.) We use RS422 voltage levels when we want long cable lengths and high bandwidths. The binary signal is encoded on the RS422 line as a voltage difference (e.g., MAX485). In this protocol, a typical bidirectional channel requires 5 wires (RxD$^+$, RxD$^-$, TxD$^+$, TxD$^-$, Ground.) Typical voltage levels are shown in Table 7.1. If the two computers are in the same box, we can implement the serial channel without the interface drivers.

		+3.3V logic	+5V logic	RS232 level	RS422 level
True	Mark	+3V	+5V	TxD = -5.5V	(TxD$^+$ - TxD$^-$) = -3V
False	Space	+0.1V	+0.1V	TxD = +5.5V	(TxD$^+$ - TxD$^-$) = +3V

Table 7.1. Typical voltage levels for the digital logic, RS232 and RS422 protocols.

A **frame** is a complete and nondivisible packet of bits. A frame includes both information (e.g., data, characters) and overhead (start bit, error checking, and stop bits.) A frame is the smallest packet that can be transmitted. The RS232 and RS422 protocols have 1 **start bit**, 7/8 data bits, no/even/odd **parity**, and 1/1.5/2 **stop bits**. The RS232 idle level is true (-5V). The start bit is false (+5V.) A true data bit is -5V, and a false data bit is +5V (Figure 7.4).

Observation: RS232 and RS422 data channels are in negative logic because the true voltage is less than the false voltage.

Observation: The RS232 protocol always has one start bit and at least one stop bit.

Checkpoint 7.1: If the RS232 protocol has eight data bits, no parity, and one stop bit, how many total bits are in a frame?

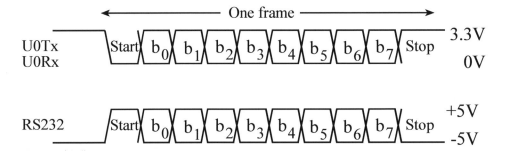

Figure 7.4. A RS232 frame showing 1 start, 8 data, no parity, and 1 stop bit.

Parity can be used to detect errors. Parity is generated by the transmitter and checked by the receiver. For **even parity**, the number of ones in the data plus parity is an even number. For **odd parity**, the number of ones in the data plus parity is an odd number. If errors are unlikely, then operating without parity is faster and simpler.

The **bit time** is the basic unit of time used in serial communication. It is the time between each bit. The transmitter outputs a bit, waits one bit time, and then outputs the next bit. The start bit is used to synchronize the receiver with the transmitter. The receiver waits on the idle line until a start bit is first detected. After the true to false transition, the receiver waits a half of a bit time. The half of a bit time wait places the input sampling time in the middle of each data bit, giving the best tolerance to variations between the transmitter and receiver clock rates. In order to operate properly the data available interval must overlap the data required interval (Section 4.2). Next, the receiver reads one bit every bit time. The **baud rate** is the total number of bits (information, overhead, and idle) per time that are transmitted in the serial communication.

$$\text{baud rate} = 1/(\text{bit time})$$

We will define **information** as the data that the "user" intends to be transmitted by the communication system. Examples of information include
- Characters to be printed on your printer
- A picture file to be transmitted to another computer
- A digitally encoded voice message communicated to your friend
- The object code file to be downloaded from the PC to the microcontroller

We will define **overhead** as signals added to the communication in order to affect reliable transmission. Examples of overhead include
- Start bit(s) start byte(s) or start code(s)
- Stop bit(s) stop byte(s) or stop code(s)
- Error checking bits like parity, CRC (cyclic redundancy check), and checksum
- Synchronization messages like ACK, NAK, XON, XOFF

Bandwidth, latency, and reliability are the fundamental performance measures for a communication system. Although, in a general sense overhead signals contain "information", overhead signals are not included when calculating bandwidth or considering full duplex, half duplex, and simplex. In similar way, if we are sending 2 bits of data, but add 6 bits of zeros to fill the byte field in the frame, we consider that there are 2 bits of information per frame (not 8 bits.) We will use the three terms **bandwidth**, **bit rate** and **throughput** interchangeably to specify the number of information bits per time that are transmitted. These terms apply to all forms of communication:

- Parallel
- Serial
- Mixed parallel/serial

For serial communication systems, we can calculate:

$$\text{Bandwidth} = \frac{\text{number of information bits/frame}}{\text{total number of bits/frame}} \cdot \text{Baud rate}$$

Latency is the time delay between when a message is sent and when it is received. For the simple systems in this chapter, at the physical layer, latency can be calculated as the frame size in bits divided by the baud rate in bits/sec. For example a RS232 protocol with 10-bit frames running at 9600-bps baud rate will take 1.04 ms to go from transmitter to receiver.

Reliability is defined as the probability of corrupted data or the **mean time between failures** (MTBF). One of the confusing aspects of bandwidth is that it could mean two things. The **peak bandwidth** is the maximum achievable data transfer rate over short periods during times when nothing else is competing for resources. When we say the bandwidth of a serial channel with 10-bit frames and a baud rate of 9600 bps is 960 bytes/s, we are defining peak bandwidth. At the component level, it is appropriate to specify peak bandwidth. However, on a complex system, there will be delays caused by the time it takes software to run, and there will be times when the transmission will be stalled due to conditions like full or empty FIFOs. The **sustained bandwidth** is the achievable data transfer rate over long periods of time and under typical usage and conditions. At the system level, it is appropriate to specify sustained bandwidth.

The design parameters that affect bandwidth are resistance, capacitance and power. It takes energy to encode each bit, therefore the bandwidth in bits per second is related to the power, which is energy per second. Capacitance exists because of the physical proximity of the wires in the cable. The time constant τ of a simple RC circuit is $R \cdot C$. An increase in capacitance will decrease the slew rate of the signal (see Figure 1.14), limiting the rate at which signals can change, thereby reducing the bandwidth of the digital transmission. However, we can increase the slew rate by using more power. We can increase the energy over the same time period by increasing voltage, increasing current, or decreasing resistance.

A **full duplex communication system** allows information (data, characters) to transfer simultaneously in both directions. A **full duplex channel** allows bits (information, error checking, synchronization or overhead) to transfer simultaneously in both directions, shown previously in Figure 7.1.

A **half duplex communication** system allows information to transfer in both directions, but in only one direction at a time. Half duplex is a term usually defined for modem communications, but in this book we will expand its meaning to include any serial protocol that allows communication in both directions, but only one direction at a time. A fundamental problem with half duplex is the detection and recovery from a collision. A collision occurs when both computers simultaneously transmit data. Fortunately, every transmission frame is echoed back into its own receiver. The transmitter program can output a frame, wait for the frame to be transmitted (which will be echoed into its own receiver) then check the incoming parity and compare the data in order to detect a collision. If a collision occurs, then it will probably be detected by both computers. After a collision, the transmitter can wait awhile and retransmit the frame. The two computers need to decide which one will transmit first after a collision so that a second collision can be avoided.

Observation: Most people communicate in half duplex.

A common hardware mechanism for half duplex utilizes open drain logic. The microcontroller open drain mode has two output states: zero and HiZ (Figure 7.5). The logic high is created with the passive pull-up. With open drain, the half duplex channel is the logical AND of the two TxD outputs. In this system, the transmitter simply transmits its frame without needing to enable or disable a driver. If both microcontrollers transmit at the same time, the zeros on the channel will dominate the ones, and the data will be corrupted.

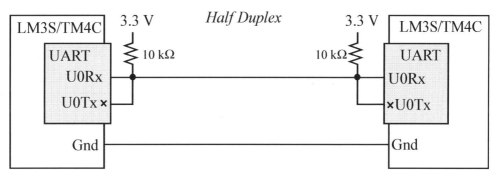

Figure 7.5. A half duplex serial channel can be implemented with open drain logic.

Checkpoint 7.2: What is the difference between full duplex and half duplex?

A **simplex communication** system allows information to transfer only in one direction. The XON/XOFF protocol that we will cover later is an example of a communication system that has a full duplex channel, but implements simplex communication. This is simplex because information is transmitted from the computer to the printer, but only XON/XOFF (error checking flags) are sent from the printer back to computer. In this case, no data is sent from printer to computer.

In order to transfer information correctly, both sides of the channel must operate at the same baud rate. In an **asynchronous** communication system, the two devices have separate and distinct clocks. Because these two clocks are generated separately (one on each side), they will not have exactly the same frequency or be in phase. If the two baud rate clocks have different frequencies, the phase between the clocks will also drift over time. Transmission will occur properly as long as the periods of the two baud rate clocks are close enough. The -5V to +5V edge at the beginning of the start bit is used to synchronize the receiver with the transmitter. If the two baud rate clock periods in a RS232 system differ by less than 5%, then after 10 bits the receiver will be off by less than a half a bit time (and no error will occur.) Any larger difference between the two periods may cause an error.

In a **synchronous** communication system, the two devices share the same clock (Figure 7.6). Typically a separate wire in the serial cable carries the clock. In this way, very high baud rates can be obtained. Another advantage of synchronous communication is that very long frames can be transmitted. Larger frames reduce the operating system overhead for long transmissions because fewer frames need be processed per message. Even though in this chapter, we will design various low bandwidth synchronous systems using the SSI, synchronous communication is best applied to systems that require bandwidths above 1 Mbps. The cost of this increased performance is the additional wire in the cable. The clock must be interfaced with channel drivers (e.g., RS232, RS422, optocouplers) similar to the transmit and receive data signals. If the two computers are in the same box, we can implement communication without the interface drivers.

Checkpoint 7.3: What is the difference between synchronous and asynchronous communication?

Observation: Self-centered people employ simplex communication.

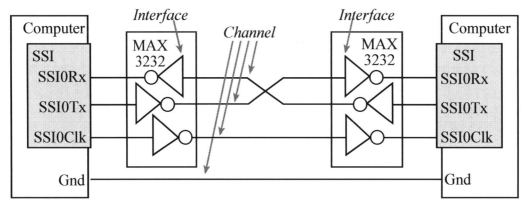

Figure 7.6. Nodes on a synchronous channel operate off a common clock.

Checkpoint 7.4: How does the MAX3232 in Figures 7.1 and 7.6 make it better?

7.2. RS232 Interfacing

The baud rate in a RS232 system can be as high as 115,200 bits/sec. Because a typical cable has 50pF/foot, the maximum distance for RS232 transmission is limited to 50 feet. There are 21 signals defined for full MODEM (MOdulate/DEModulate) communication (Figure 7.7.) The RS232 standard uses a DB25 connector that has 25 pins. The EIA-574 standard uses RS232 voltage levels and a DB9 connector that has only 9 pins. The EIA-561 standard also uses RS232 voltage levels but with a RJ45 connector that has only 8 pins.

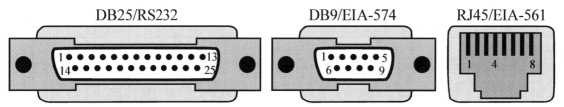

Figure 7.7. RS232, EIA-574 and EIA-561 connectors used in many serial applications.

Table 7.2 shows the entire set of RS232 signals. The most commonly used signals of the full RS232 standard are available with the EIA-561/EIA-574 protocols.

The frame ground is connected on one side to the **ground shield** of the cable. The shield will provide protection from electric field interference. The **twisted cable** has a small area between the wires. The smaller the area, the less the magnetic field pickup. There is one disadvantage to reducing the area between the connectors. The capacitance to ground is inversely related to the separation distance between the wires. Thus as the area decreases, the capacitance will increase. This increased capacitive load will limit both the distance and the baud rate.

DB25 Pin	RS232 Name	DB9 Pin	EIA-574 Name	RJ45 Pin	EIA-561 Name	Signal	Description	True	DTE	DCE
1						FG	Frame Ground/Shield			
2	BA	3	103	6	103	TxD	Transmit Data	-5V	out	in
3	BB	2	104	5	104	RxD	Receive Data	-5V	in	out
4	CA	7	105/133	8	105/133	RTS	Request to Send	+5V	out	in
5	CB	8	106	7	106	CTS	Clear to Send	+5V	in	out
6	CC	6	107			DSR	Data Set Ready	+5V	in	out
7	AB	5	102	4	102	SG	Signal Ground			
8	CF	1	109	2	109	DCD	Data Carrier Detect	+5V	in	out
9							Positive Test Voltage			
10							Negative Test Voltage			
11							Not Assigned			
12						sDCD	secondary DCD	+5V	in	out
13						sCTS	secondary CTS	+5V	in	out
14						sTxD	secondary TxD	-5V	out	in
15	DB					TxC	Transmit Clk (DCE)		in	out
16						sRxD	secondary RxD	-5V	in	out
17	DD					RxC	Receive Clock		in	out
18	LL						Local Loopback			
19						sRTS	secondary RTS	+5V	out	in
20	CD	4	108	3	108	DTR	Data Terminal Rdy	+5V	out	in
21	RL					SQ	Signal Quality	+5V	in	out
22	CE	9	125	1	125	RI	Ring Indicator	+5V	in	out
23						SEL	Speed Selector DTE		in	out
24	DA					TCK	Speed Selector DCE		out	in
25	TM					TM	Test mode	+5V	in	out

Table 7.2. Pin assignments for the RS232 EIA-574 and EIA-561 protocols.

The signal ground is connected on both sides to the supply return. A separate wire should be used for signal ground (do not use the ground shield to connect the two signal grounds.) The noise immunity will be degraded if the ground shield is connected on both sides. There are many available RS232 driver chips (e.g., Maxim MAX3232). These chips employ a charge pump (using the 100 nF capacitors) to create the standard +5 and -5 output voltages with only a +3.3 V supply. The following overviews the RS232 specifications for an output signal

- Must be able to withstand a short to ground or short to another wire
- True voltage is between -5 and -15 V (operating range)
- False voltage is between +5 and +15 V (operating range)
- Short circuit current less than 0.5 A
- Spends less than 4% of the time in the transition range from -3 to +3 V

Similarly, the following overviews the RS232 specifications for an input signal

- The maximum slew rate is 30 V/μs
- True input is -3 to -15 V
- False input is +3 to +15 V
- The transition range is -3 to +3 V
- The input impedance between 3000 and 7000 Ω
- The input capacitance including cable must be less than 2500 pF
- The input open circuit voltage must be less than 2 V

On the data lines, a true or mark voltage is negative. Conversely for the control lines, a true signal has a positive voltage. **Request to Send (RTS)** is a signal from the computer to the MODEM requesting transmission be allowed. **Clear to Send (CTS)** is the acknowledge signal back from the MODEM signifying transmission can proceed. **Data Set Ready (DSR)** is a MODEM signal specifying that input/output can occur. **Data Carrier Detect (DCD)** is a modem signal specifying that the carrier frequencies have been established on its telephone line. **Ring Indicator (RI)** is a modem signal that is true when the phone rings. The **Receive Clock** and **Transmit Clock** are used to establish synchronous serial communication. **Data Terminal Ready (DTR)** is a printer signal specifying the status of the printer. When DTR is +5V, the printer is ready and can accept more characters. Conversely when DTR is -5V, the printer is busy and cannot accept more characters. Most microcontrollers do not support these control lines explicitly. On the other hand, it is straight forward to implement these hardware handshaking signals using simple input/output lines. If we wish to generate interrupts on edges of these lines, then we would use edge-triggered input features.

Example 7.1. Design a simplex printer driver with DTR synchronization.

Solution: One problem with printers is that the printer bandwidth (the actual number of characters per second that can be printed) may be less than the maximum bandwidth supported by the serial channel. There are five conditions that might lead to a situation where the computer outputs serial data to the printer, but the printer isn't ready to accept the data. First, special characters may require more time to print (e.g., carriage return, line feed, tab, formed, and graphics). Second, most printers have internal FIFOs that could get full. If the FIFO is not full, then it can accept data as fast as the channel will allow, but when the FIFO becomes full, the computer should stop sending data. Third, the printer cable may be disconnected. Fourth, the printer may be deselected or attached to different computer. Fifth, the printer power may be off. The output interfaces shown previously provide no feedback from the printer that could be used to detect/correct these five problems. There are two mechanisms, called flow control, to synchronize the computer with a variable rate output device. These two flow control protocols are called DTR and XON/XOFF.

The first method uses a hardware signal, DTR (pin 4 on the DB9 connector), as feedback from the printer to the microcontroller, see Figure 7.8. DTR is –5V if the printer is busy and is not currently able to accept transmission. DTR is +5V if the printer is ready and able to accept transmission. This mechanism can handle all five of the above conditions where the printer is not ready. The computer input mechanism will handle the DTR protocol using additional software checking.

With a standard RS232 interface when DTR is –5V, PC4 will be +3.3V (which will stop the output even if the UART transmit FIFO is not full). Thus, when DTR is –5V, transmission is temporarily suspended. When DTR is +5V, PC4 will be 0V and transmission can proceed normally. In this design, edge-triggered input on PC4 will be used to detect changes in the DTR signal.

The DTR signal from the printer provides feedback information about the printer status. When DTR is -5V (PC4 is high), the printer is not ready to accept more data. In this case, our computer will postpone transmitting more frames. When DTR is +5V (PC4 is low), the printer is ready to accept more data. At this point, the computer will resume transmission. The `Printer_OutChar(data)` in Program 7.1 is called by the main program to print.

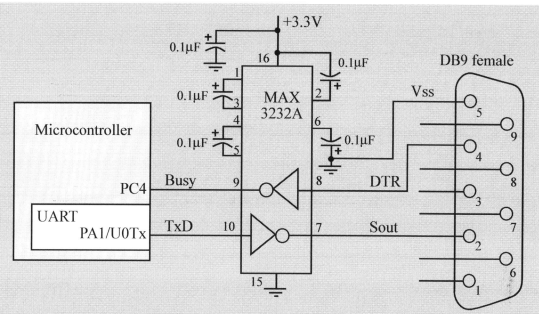

Figure 7.8. Hardware interface implementing a RS232 simplex channel with DTR handshaking.

```
void Printer_Init2(void){
  SYSCTL_RCGCUART_R |= 0x01;          // activate UART0
  SYSCTL_RCGCGPIO_R |= 0x05;          // activate ports A, C
  TxFifo_Init();                      // initialize empty FIFOs
  GPIO_PORTC_DIR_R &= ~0x10;          // make PC4 in
  GPIO_PORTC_DEN_R |= 0x10;           // enable digital I/O on PC4
  GPIO_PORTC_IS_R &= ~0x10;           // PC4 is edge-sensitive
  GPIO_PORTC_IBE_R |= 0x10;           // PC4 is both edges
  GPIO_PORTC_ICR_R = 0x10;            // clear flag4
  GPIO_PORTC_IM_R |= 0x10;            // enable interrupt on PC4
  UART0_CTL_R &= ~UART_CTL_UARTEN;    // disable UART
  UART0_IBRD_R = 43; // IBRD=int(80000000/(16*115,200)) = int(43.40278)
  UART0_FBRD_R = 26; // FBRD = round(0.40278 * 64) = 26
  UART0_LCRH_R = 0x0070;              // 8-bit word length, enable FIFO
  UART0_IFLS_R &= ~0x07;              // clear TX interrupt FIFO level field
  UART0_IFLS_R += UART_IFLS_TX1_8;    // interrupt for TX FIFO <= 1/8 full
  UART0_IM_R |= UART_IM_TXIM;         // enable TX FIFO interrupt
  UART0_CTL_R |= UART_CTL_UARTEN;     // enable UART
  GPIO_PORTA_PCTL_R = (GPIO_PORTA_PCTL_R&0xFFFFFF00)+0x00000011; // UART
  GPIO_PORTA_AMSEL_R &= ~0x03;        // disable analog on PA1-0
  GPIO_PORTA_AFSEL_R |= 0x03;         // enable alt funct on PA1-0
  GPIO_PORTA_DEN_R |= 0x03;           // enable digital I/O on PA1-0
  NVIC_PRI0_R = (NVIC_PRI0_R&0xFF00FFFF)|0x00400000; // priority 2
  NVIC_PRI1_R = (NVIC_PRI1_R&0xFFFF00FF)|0x00004000; // priority 2
  NVIC_EN0_R = (NVIC_EN0_INT2|NVIC_EN0_INT5);
  EnableInterrupts();
}
```

```c
// check the status of the printer on PC4
// high means not ready
void checkStatus(void){
  if(GPIO_PORTC_DATA_R&0x10)           // PC4=1 if DTR is -5 to -15
    UART0_CTL_R &= ~UART_CTL_TXE;      // busy, so disable transmitter
  else
    UART0_CTL_R |= UART_CTL_TXE;       // not busy, so enable transmitter
}
// copy from software TX FIFO to hardware TX FIFO
// stop when software TX FIFO is empty or hardware TX FIFO is full
void copySoftwareToHardware(void){
  char letter;
  while(((UART0_FR_R&UART_FR_TXFF) == 0) && (TxFifo_Size() > 0)){
    TxFifo_Get(&letter);
    UART0_DR_R = letter;
  }
}
// output ASCII character to Printer
// spin if TxFifo is full
void Printer_OutChar(char data){
  while(TxFifo_Put(data) == FIFOFAIL){};
  checkStatus();
  UART0_IM_R &= ~UART_IM_TXIM;         // disable TX FIFO interrupt
  copySoftwareToHardware();
  UART0_IM_R |= UART_IM_TXIM;          // enable TX FIFO interrupt
}
// interrupt when hardware TX FIFO goes from 3 to 2 or less items
void UART0_Handler(void){
  if(UART0_RIS_R&UART_RIS_TXRIS){      // hardware TX FIFO <= 2 items
    UART0_ICR_R = UART_ICR_TXIC;       // acknowledge TX FIFO
    // copy from software TX FIFO to hardware TX FIFO
    copySoftwareToHardware();
    if(TxFifo_Size() == 0){            // software TX FIFO is empty
      UART0_IM_R &= ~UART_IM_TXIM;     // disable TX FIFO interrupt
    }
  }
}
// interrupt on any change of the status of the printer on PC4
void GPIOPortC_Handler(void){
  GPIO_PORTC_ICR_R = 0x10;             // acknowledge flag4
  checkStatus();
}
```

Program 7.1. Software implementation of a printer interface with DTR synchronization.

7.3. RS422/USB/RS423/RS485 Balanced Differential Lines

In order to increase the baud rate and maximum distance, the balanced differential line protocols were introduced. The RS422 signal is encoded in a differential signal, A-B. There are many RS422 interface chips, for example, SP301 or MAX13433. A full duplex RS422 channel is implemented in Figure 7.9 with MAX13433 drivers. Often terminating resistors are placed at the ends of long cables. E.g., we can place a 100 Ω resistor at across the A' B' inputs.

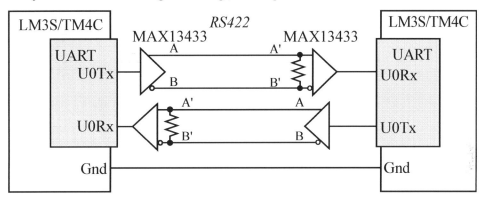

Figure 7.9. RS422 serial channel.

Because each signal requires 2 wires, 5 wires (ground included) are needed to implement a full duplex channel. With RS232 one typically connects one receiver to one transmitter. But with RS422, up to 10 receivers can be connected to one transmitter. Table 7.3 summarizes four common EIA standards.

Specification	RS232D	RS423A	RS422	RS485
Mode of operation	single-ended	single-ended	differential	differential
Drivers on one line	1	1	1	32
Receivers on one line	1	10	10	32
Maximum distance (feet)	50	4,000	4,000	4,000
Maximum data rate	20 kb/s	100 kb/s	10 Mb/s	10 Mb/s
Maximum driver output	±25 V	±6 V	-0.25 to +6V	-7 to +12V
Driver Output (loaded)	±5 V	±3.6 V	±2 V	±1.5 V
Driver Output (unloaded)	±15 V	±6 V	±5 V	±5 V
Driver Load Impedance	3kΩ to 7kΩ	450Ω min	100Ω	54Ω
Receiver input voltage	±15 V	±12 V	±7 V	-7 to +12 V
Receiver input sensitivity	±3 V	±200 mV	±200 mV	±200 mV
Receiver input resistance	3kΩ to 7kΩ	4kΩ min	4kΩ min	12kΩ min

Table 7.3. Specifications for the RS232 RS423A RS422 and RS485 protocols.

The maximum baud rate at 40 feet is 10 Mbps. At 4000 feet, the baud rate can only be as high as 100 Kbps. Table 7.4 shows two implementations of the RS422 protocol.

DB25 Pin	EIA-530 Name	DB37 Pin	RS449 Name	Signal	Description	DTE	DCE
1		1		FG	Frame Ground/Shield		
2	BA (A)	4	SD (A)	TxD	Transmit Data	out	in
14	BA (B)	22	SD (B)	TxD	Transmit Data	out	in
3	BB (A)	6	RD (A)	RxD	Receive Data	in	out
16	BB (B)	24	RD (B)	RxD	Receive Data	in	out
4	CA (A)	7	RS (A)	RTS	Request to Send	out	in
19	CA (B)	25	RS (B)	RTS	Request to Send	out	in
5	CB (A)	9	CS (A)	CTS	Clear to Send	in	out
13	CB (B)	27	CS (B)	CTS	Clear to Send	in	out
6	CC (A)	11	DM (A)	DSR	Data Set Ready	in	out
22	CC (B)	29	DM (B)	DSR	Data Set Ready	in	out
20	CD (A)	12	TR (A)	DTR	Data Terminal Rdy	out	in
23	CD (B)	30	TR (B)	DTR	Data Terminal Rdy	out	in
7	AB	19	SG	SG	Signal Ground		
8	CF (A)	13	RR (A)	DCD	Data Carrier Detect	in	out
10	CF (B)	31	RR (B)	DCD	Data Carrier Detect	in	out
15	DB (A)	5	ST (A)	TxC	Transmit Clk (DCE)	in	out
12	DB (B)	23	ST (B)	TxC	Transmit Clk (DCE)	in	out
17	DD (A)	8	RT (A)	RxC	Receive Clock	in	out
9	DD (B)	26	RT (B)	RxC	Receive Clock	in	out
18	LL	10	LL		Local Loopback	out	in
21	RL	14	RL	RL	Remote Loopback	out	in
24	DA (A)	17	TT (A)	TCK	Speed Selector DCE	out	in
11	DA (B)	35	TT (B)	TCK	Speed Selector DCE	out	in
25	TM	18	TM	TM	Test mode	in	out

Table 7.4. Pin assignments for the EIA-530 and RS449 protocols.

In this section we will introduce the electrical specifications of the **Universal Serial Bus** (USB). Figure 7.10 shows the two types of USB connectors. A single host computer controls the USB, and there can only be one host per bus. The host controls the scheduling of all transactions using a token-based protocol. The USB architecture is a tiered star topology, similar to 10BaseT Ethernet. The host is at the center of the star, devices are attached to the host. The number of nodes on the bus can be extended using USB hubs. Up to 127 devices can be connected to any one USB bus at any one given time. USB plug'n'plug is implemented with dynamically loadable and unloadable drivers. When the user plugs the device into the USB bus, the host will detect the connection, interact with the newly inserted device, and load the appropriate driver. The USB device can be used without explicitly installing drivers or rebooting. When the device is unplugged, the host will automatically unload its driver.

Figure 7.10. USB connectors.

USB uses four shielded wires, +5V power, GND, and a twisted pair differential data signals, as listed in Table 7.5. It uses a NRZI (Non Return to Zero Invert) encoding scheme to send data with a sync field to synchronize the host and receiver clocks. The D+ signal has a 15 kΩ pull-down resistor to ground, and the D- signal has a 1.5 kΩ pull-up resistor to +3.6V. Like the other protocols in this section, the data is encoded as a differential signal between D+ and D-. In general, a differential '1' exists when D+ is greater than D-. More specifically, a differential '1' is transmitted by pulling D+ over 2.8V and D- under 0.3V. The transmitter creates a differential '0' by making D- greater than 2.8V and D+ less than 0.3V. The receiver recognizes the differential '1' when D+ is 0.2V greater than D-. The receiver will consider the input as a differential '0' when D+ 0.2V less than D-. The polarity of the signal is inverted depending on the speed of the bus. Therefore the terms 'J' and 'K' states are used in signifying the logic levels. At low speed, a 'J' state is a differential '0'. At high speed, a 'J' state is a differential '1'. USB interfaces employ both differential and single ended outputs. Certain bus states are indicated by single ended signals on D+, D- or both. For example, a single ended zero (SE0) signifies device reset when held for more than 10mS. More specifically, SE0 is generated by holding both D- and D+ low (< 0.3V). USB can operate at three speeds. The low speed/full speed bus has a characteristic impedance of 90Ω. High Speed mode uses a constant current protocol to reduce noise.

High speed data is clocked at 480Mb/s
Full speed data is clocked at 12Mb/s
Low speed data is clocked at 1.5Mb/s

Pin Number	Color	Function
1	Red	VBUS (5 V)
2	White	D-
3	Green	D+
4	Black	Ground

Pin Number	Function
1	VBUS (5 V)
2	D-
3	D+
4	ID
5	Ground

Table 7.5. USB signals.

7.3.1. RS422 Output Specifications

Table 7.6 shows the output voltage levels for RS422.

	Output Voltage
True or Mark	$-6 \leq A-B \leq -2V$
Transition	$-2 \leq A-B \leq +2V$
False or Space	$+2 \leq A-B \leq +6V$

Table 7.6. Output voltage levels for the RS422 differential line protocol.

A key RS422 specification is that the output impedance of the A and B outputs should be balanced. If the input/output impedances are balanced then added noise in the cable creates a common mode voltage, and the common mode rejection of the input will eliminate it.

$$R_{Aout} = R_{Bout} \leq 100 \, \Omega$$

The time in the transition region must be less than

> 10% for baud rates above 5Mbps, and
> 20ns for baud rates below 5Mbps

7.3.2. RS422 Input Specifications

Let A' and B' be the voltages at the input. Table 7.7 defines the input encoding.

	Input Voltage
True or Mark	A'-B' $\leq$ -0.2V
Transition	-0.2V $\leq$ A'-B' $\leq$ +0.2V
False or Space	+0.2V $\leq$ A'-B'

Table 7.7. Input voltage thresholds for the RS422 differential line protocol.

As mentioned earlier, to provide noise immunity the common mode input impedances must also be balanced

$$4 \text{ k}\Omega \leq R_{A'in} = R_{B'in}$$

The balanced nature of the interface produces good noise immunity. The differential input impedance is specified by Figure 7.11. Any point within the shaded region is allowed.

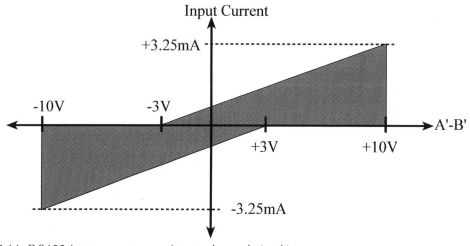

Figure 7.11. RS422 input current versus input voltage relationship.

Even though the ground connection in the RS422 cable is optional, it is assumed the grounds are connected somewhere. In particular, the interface will operate with a common mode voltage up to 7 volts.

$$|A'+B'|/2 \leq +7V$$

Checkpoint 7.5: What is the effect of capacitance on a serial channel?

7.3.3. RS485 Half Duplex channel

RS485 can be either half duplex or full duplex. The RS485 protocol, illustrated in Figure 7.12, implements a half duplex channel using differential voltage signals. The Sipex SP483 or Maxim MAX13431 implements the half duplex RS485 channel. One of the advantages of RS485 is that up to 32 devices can be connected onto a single serial bus. When more than one transmitter can driver the serial bus, the protocol is also called "multi-drop". Normally, we make both DE and RE active on all devices. To transmit the computer sends the serial frame from the TxD output of the UART port. The transmitted frame is echoed into the serial receiver of the SCI RxD line. To receive a frame the computer accepts a serial frame on the RxD line in the usual manner. When careful when selecting the resistances on a half duplex network so that the total driver impedance is about 54Ω.

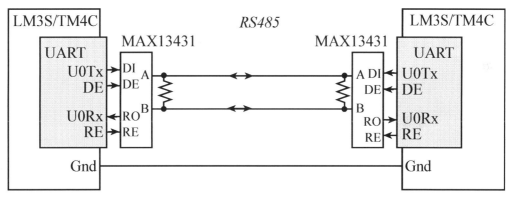

Figure 7.12. A half duplex serial channel is implemented with RS485 logic.

RS422, RS485, Ethernet, and CAN are high-speed communication channels. This means the bandwidth and slew rate on the signals are higher than RS232. There is a correspondence between rise time (τ) of a digital signal and equivalent sinusoidal frequency (f). The derivative of $A \cdot \sin(2\pi f t)$ is $2\pi f \cdot A \cdot \cos(2\pi f t)$. The maximum slew rate of this sinusoid is $2\pi f \cdot A$. Approximating the slew rate as A/τ, we get a correspondence between f and τ

$$f = 1/\tau$$

For example, if the rise time is 5 ns, the equivalent frequency is 200 MHz. Notice that this equivalent frequency is independent of baud rate. So even at 1000 bits/sec, if the rise time is 5 ns, then the signal has a strong 200 MHz frequency component! To deal with this issue, the RS232 protocol limits the slew rate to a maximum of 30V/μs. This means it will take about 1 μs for a signal to rise from -12 to +12 V. Consequently, RS232 signals have frequency components less than 1 MHz. However, to transmit faster than RS232, the protocol must have faster rise times. Electrical signals travel at about 0.6 to 0.9 times the speed of light. This velocity factor (VF) is a property of the cable. For example, VF for RG-6/U coax cable is 0.75, whereas VF is only 0.66 for RG-58/U coax cable. Using the slower 0.66 estimate, the speed is $v = 2 \cdot 10^8$ m/s. According to wave theory, the wavelength is $\lambda = v/f$. Estimating the frequency from rise time, we get

$$\lambda = v * \tau$$

In our example, a rise time of 5 ns is equivalent to a wavelength of about 1 m. As a rule of thumb, we will consider the channel as a *transmission line* if the length of the wire is greater than $\lambda/4$. Another requirement is for the diameter of the wire to be much smaller than the wavelength. In a transmission line, the signals travel down the wires as waves according to the wave equation. Analysis of the wave equation is outside the scope of this book. However, you need to know that when a wave meets a change in impedance, some of the energy will transmit (a good thing) and some of the energy will reflect (a bad thing). Reflections are essentially noise on the signal, and if large enough, they will cause bit errors in transmission. We can reduce the change in impedance by placing terminating resistors on both ends of a long high-speed cable, as shown in Figure 7.12. These resistors reduce reflections; hence they improve signal to noise ratio.

7.4. Logic Level Conversion

There are many 3.3-V devices we wish to interface to a 5-V microcontroller, and there are many 5-V devices we wish to interface to a 3.3-V microcontroller. This section will study various methods to convert one logic level to another. We begin with a 5-V output interfaced to a 3.3-V input. Many 3.3-V inputs are 5-V tolerant, which means no special interface circuits are required. One of the simplest ways to convert 5-V logic into 3.3-V logic is to use a resistor divider as shown in Figure 7.13. A Schottky diode can also be used to convert 5 V into 3.3 V, and convert a 0.4 V into a 0.5 V. The Schottky diode must be fast and have a low voltage drop. The 7407 is another way to convert between logic families. When the 7407 input is 5 V, its output floats, and the 3.3-V pull-up makes a 3.3-V signal. When the 7407 input is low, its output is low.

Many 5 V inputs are 3.3 V tolerant, which means no special interface circuits are required. The 7407 can also be used to interface 3.3-V logic into 5-V logic. The V_{IH} of the 7407 is 2 V, so when the 7407 input is 3.3 V, its output floats, and the 5-V pull-up makes a 5-V signal. When the 7407 input is low, its output is low. A MOSFET, like the BSS138, is a popular method to convert logic levels because it is fast and efficient. SparkFun makes a breakout board with resistor-divider and BSS138 circuits (www.sparkfun.com BOB-12009).

We can produce the same open collector behavior of any I/O port that has a direction register. We initialize the port by writing a zero to the data port. On subsequent accesses to the open collector port, we write the complement to the direction register. I.e., if we want the I/O port bit to drive low, we set the direction register bit to 1, and if we want the I/O port bit to float (open collector), we set the direction register bit to 0.

It is good design practice to read the errata for the microcontrollers in your system. For example, the LM3S811 errata states, "GPIO buffers are not 5-V tolerant when used in open-drain mode. Pulling up the open-drain pin above 4V results in high current draw." Furthermore, "The pins associated with GPIO signals PB6, PC5, and PC6 are not 5-V tolerant. Applying a voltage to any of these pins that is greater than V_{DD} (3.3V) will have undetermined results." These particular mistakes apply specifically to silicon version C0 for the LM3S811. However, in general, one must read the errata for the microcontroller you are using.

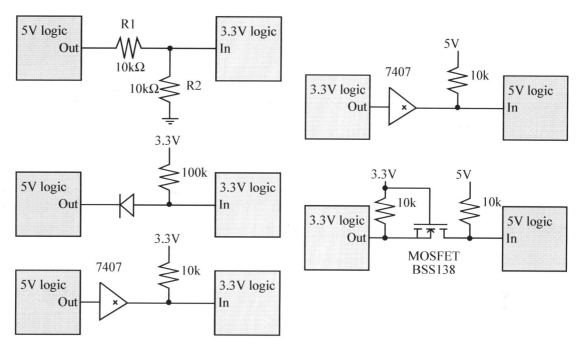

Figure 7.13. Circuits to interface between 5-V logic and 3.3-V logic.

7.5. Synchronous Transmission and Receiving using the SSI

SSI allows microcontrollers to communicate synchronously with peripheral devices and other microcontrollers. The SSI system can operate as a master or as a slave. The channel can have one master and one slave, or it can have one master and multiple slaves. With multiple slaves, the configuration can be a star (centralized master connected to each slave), or a ring (each node has one receiver and one transmitter, where the nodes are connected in a circle.) The master initiates all data communication.

Stellaris and Tiva microcontrollers have 0 to 4 **Synchronous Serial Interface** or SSI modules. Another name for this protocol is **Serial Peripheral Interface** or SPI. The fundamental difference between a UART, which implements an asynchronous protocol, and a SSI, which implements a synchronous protocol, is the manner in which the clock is implemented. Two devices communicating with asynchronous serial interfaces (UART) operate at the same frequency (baud rate) but have two separate clocks. With a UART protocol, the clock signal is not included in the interface cable between devices. Two UART devices can communicate with each other as long as the two clocks have frequencies within ±5% of each other. Two devices communicating with synchronous serial interfaces (SSI) operate from the same clock (synchronized). With a SSI protocol, the clock signal is included in the interface cable between devices. Typically, the master device creates the clock, and the slave device(s) uses the clock to latch the data in and send data out.

The SSI protocol includes four I/O lines. The slave select **SSI0Fss** is an optional negative logic control signal from master to slave signal signifying the channel is active. The second line, **SCK**, is a 50% duty cycle clock generated by the master. The **SSI0Tx** (master out slave in, MOSI) is a data line driven by the master and received by the slave. The **SSI0Rx** (master in slave out, MISO) is a data line driven by the slave and received by the master. In order to work properly, the transmitting device uses one edge of the clock to change its output, and the receiving device uses the other edge to accept the data. Figure 7.14 shows the I/O port locations of the synchronous serial ports for the three microcontrollers discussed in this book. Tables 2.4 and 2.5 describe how to attach I/O pins to the SSI modules on the TM4C123/MSP432E4.

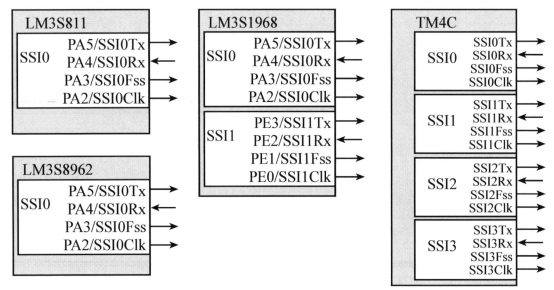

Figure 7.14. Synchronous serial port pins on four LM3S/TM4C/MSP432E401Y microcontrollers.

On the LM3S/TM4C the shift register can be configured from 4 to 16 bits. The shift register in the master and the shift register in the slave are linked to form a distributed register. Figure 7.15 illustrates communication between master and slave. Typically, the microcontroller and the I/O device slave are so physically close we do not use interface logic.

The SSI on the LM3S/TM4C employs two hardware FIFOs. Both FIFOs are 8 elements deep and 4 to 16 bits wide, depending on the selected data width. When performing I/O the software puts into the transmit FIFO by writing to the **SSI0_DR_R** register and gets from the receive FIFO by reading from the **SSI0_DR_R** register.

If there is data in the transmit FIFO, the SSI module will transmit it. With SSI it transmits and receives bits at the same time. When a data transfer operation is performed, this distributed 8- to 32-bit register is serially shifted 4 to 16 bit positions by the SCK clock from the master so the data is effectively exchanged between the master and the slave. Data in the master shift register are transmitted to the slave. Data in the slave shift register are transmitted to the master. Typically, the microcontroller is master and the I/O module is the slave, but one can operate the microcontroller in slave mode. When designing with SSI, you will need to consult the data sheets for your specific microcontroller.

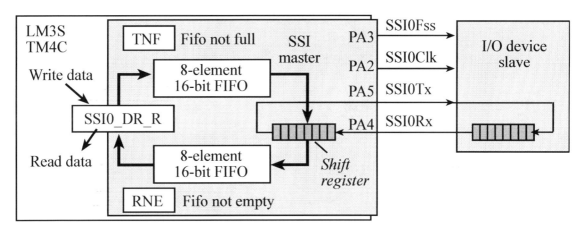

Figure 7.15. A synchronous serial interface between a microcontroller and an I/O device.

Table 7.8 lists the SSI0 registers on the LM3S/TM4C. The LM3S/TM4C can operate in slave mode, but we will focus on master mode. The **PCTL** bits are defined in Tables 2.4 and 2.5.

Address	31-6	3	2	1	0	Name
$400F.E61C		SSI3	SSI2	SSI1	SSI0	SYSCTL_RCGCSSI_R

	31-16	15-8	7	6	5-4	3-0	
$4000.8000		SCR	SPH	SPO	FRF	DSS	SSI0_CR0_R

	31-16	15-0	
$4000.8008		Data	SSI0_DR_R

	7	6	5	4	3	2	1	0	
$4000.8004					SOD	MS	SSE	LBM	SSI0_CR1_R
$4000.800C		BSY			RFF	RNE	TNF	TFE	SSI0_SR_R
$4000.8010				CPSDVSR					SSI0_CPSR_R
$4000.8014					TXIM	RXIM	RTIM	RORIM	SSI0_IM_R
$4000.8018					TXRIS	RXRIS	RTRIS	RORRIS	SSI0_RIS_R
$4000.801C					TXMIS	RXMIS	RTMIS	RORMIS	SSI0_MIS_R
$4000.8020							RTIC	RORIC	SSI0_ICR_R
$4005.8420	SEL	SEL	SEL	SEL	SEL	SEL	SEL	**SEL**	GPIO_PORTA_AFSEL_R
$4005.841C	DEN	DEN	DEN	DEN	**DEN**	**DEN**	DEN	**DEN**	GPIO_PORTA_DEN_R
$4005.8400	DIR	DIR	DIR	DIR	DIR	DIR	DIR	**DIR**	GPIO_PORTA_DIR_R
$400F.E608	GPIOH	GPIOG	GPIOF	GPIOE	GPIOD	GPIOC	GPIOB	**GPIOA**	SYSCTL_RCGCGPIO_R

Table 7.8. The LM3S/TM4C SSI0 registers. Each register is 32 bits wide. Bits 31 – 8 are zero.

The SSI clock frequency is established by the 8-bit field **SCR** field in the `SSI0_CR0_R` register and the 8-bit field **CPSDVSR** field in the `SSI0_CPSR_R` register. **SCR** can be any 8-bit value from 0 to 255. **CPSDVSR** must be an even number from 2 to 254. Let f_{BUS} be the frequency of the bus clock. The frequency of the SSI is

$$f_{SSI} = f_{BUS} / (\textbf{CPSDVSR} * (1 + \textbf{SCR}))$$

Common control features for the SSI module include:
    Baud rate control register, used to select the transmission rate
    Data size 4 to 16 bits, set 4-bit DSS field equal to size-1

Mode bits in the control register to select
>Master versus slave
>Freescale mode with clock polarity and clock phase
>TI synchronous serial mode
>Microwire mode

Interrupt arm bit
Ability to make the outputs open drain (open collector)

Common status bits for the SPI module include:
>BSY, SSI is currently transmitting and/or receiving a frame,
>>or the transmit FIFO is not empty
>RFF, SSI receive FIFO is full
>RNE, SSI receive FIFO is not empty
>TNF, SSI transmit FIFO is not full
>TFE, SSI transmit FIFO is empty

The key to proper transmission is to select one edge of the clock (shown as "T" in Figure 7.16) to be used by the transmitter to change the output, and use the other edge (shown as "R") to latch the data in the receiver. In this way data is latched during the time when it is stable. Data available is the time when the output data is actually valid, and data required is the time when the input data must be valid.

During transmission, the output data will be valid from $S5_{max}$ after the clock edge until $S5_{min}$ after the next clock edge. The maximum S5 time is 1 system bus period (e.g., 20ns) and the minimum is 0. When receiving the setup time (S8) is 1 system bus period and the hold time (S9) is 2 system bus periods. In order for the communication to occur without error, the data available from the device that is driving the data line must overlap (start before and end after) the data required by the other device that is receiving the data. It is this overlap that will determine the maximum frequency at which synchronous serial communication can occur. The concepts of data available and data required were presented previously in Section 4.2.

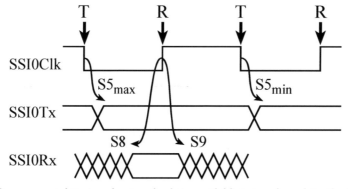

Figure 7.16. Synchronous serial timing showing the data available interval overlaps the data required interval.

Checkpoint 7.6: What are the definitions of setup time and hold time?

Observation: When transmitting multiple frames, the Fss signal goes low and high for each frame. If you need Fss to fall and rise once, you'll have to program it as a GPIO.

The Freescale SPI timing is shown in Figure 7.17 (FRF=00).

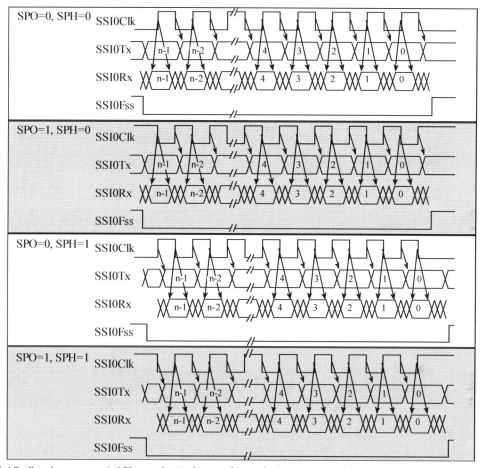

Figure 7.17. Synchronous serial Freescale single transfer mode (n is 4 to 16 bits).

The SPI transmits data at the same time as it receives input. In the Freescale modes, the SPI changes its output on the opposite edge of the clock as it uses to shift data in. There are three mode control bits (**MS**, **SPO**, **SPH**) that affect the transmission protocol. If the device is a master (**MS**=0) it generates the **SCLK**, and data is output on the **SSI0Tx** pin, and input on the **SSI0Rx** pin. The **SPO** control bit specifies the polarity of the SCLK. In particular, the **SPO** bit specifies the logic level of the clock when data is not being transferred. The **SPH** bit affects the timing of the first bit transferred and received. If **SPH** is 0, then the device will shift data in on the first (and 3rd, 5th, 7th, ... etc.) clock edge. If **SPH** is 1, then the device will shift data in on the second (and 4th, 6th, 8th, ... etc.) clock edge. The data is transmitted MSB first.

The TI synchronous serial timing is shown in Figure 7.18 (FRF=01). There is a third protocol called Microwire (FRF=10). Refer to the data sheets for details of these modes.

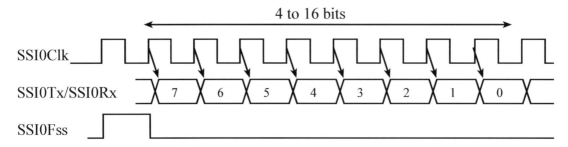

Figure 7.18. *Synchronous serial TI single transfer mode.*

Example 7.2. Interface the Maxim MAX5353 digital to analog converter.

Solution: This first example shows a synchronous serial interface between the microcontroller and a Maxim MAX5353 12-bit digital to analog converter as drawn in Figure 7.19. A digital to analog converter (DAC) accepts a digital input (in our case a number between 0 and 4095) and creates an analog output (in our case a voltage between 0 and V_{REF}*GAIN.) Detailed discussion of converters will be presented later in Chapter 8. Here in this section we will focus on the digital hardware and software aspects of the serial interface.

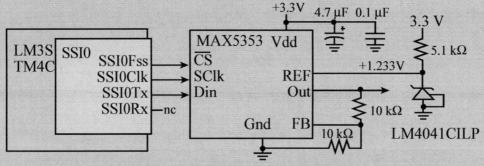

Figure 7.19. *A 12-bit DAC interfaced to the SSI port.*

Table 7.9 and Figure 7.20 describe the protocol. The first 3 bits sent will be zero, then the 12 data bits that specify the analog output, and then one more zero will be sent. The SSI0Fss control signal will be low during the 16-bit transmission. As with any SPI interface, there are basic interfacing issues to consider.

Word size. In this case we need to transmit 16 bits to the DAC. The MAX5353 data sheet specifies that the first three bits are command codes, the next 12 bits are the DAC output (MSB transmitted first), and the last bit is zero. *Bit order.* The MAX5353 requires the most significant bits first.

Clock phase, clock polarity. There are two issues to resolve. Since the MAX5353 samples its serial input data on the rising edge of the clock, the SSI must change the data on the falling edge. SPO=SPH=0 (Figure 7.17) and SPO=SPH=1 both satisfy this requirement. The second issue is which edge comes first the rise or the fall. In this interface it probably doesn't matter.

Bandwidth. We look at the timing specifications of the MAX5353. The minimum clock low width of 40 ns means the shortest SSI period we can use is 100 ns. The commands are:

C2	C1	C0	D11 : D0 MSB LSB	S0	Description
X	0	0	12 bits of data	0	Load input register; DAC register immediately updated.
X	0	1	12 bits of data	0	Load input register; DAC register unchanged.
X	1	0	XXXXXXXXXXXX	X	Update DAC register from input register.
1	1	1	XXXXXXXXXXXX	X	Shutdown
0	1	1	XXXXXXXXXXXX	X	No operation

Table 7.9. MAX5353 protocols

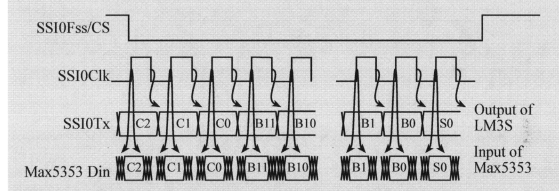

Figure 7.20. MAX5353 DAC serial timing.

The ritual initializes the Freescale SPI master mode, 16-bit data, and 8 MHz bandwidth (Program 7.2). To change the DAC output, one 16-bit transmission is sent (**DAC_Out**). The data returned in this case is not significant because the SSI0Rx pin in Figure 7.19 is left not connected. In this example, the bus clock is 16 MHz, and the SSI clock will be 8 MHz.

```
void DAC_Init(uint16_t data){
  SYSCTL_RCGCSSI_R |= 0x01;       // activate SSI0
  SYSCTL_RCGCGPIO_R |= 0x01;      // activate port A
  while((SYSCTL_PRGPIO_R&0x01) == 0){};// ready?
  GPIO_PORTA_AFSEL_R |= 0x2C;     // enable alt funct on PA2,3,5
  GPIO_PORTA_DEN_R |= 0x2C;       // configure PA2,3,5 as SSI
  GPIO_PORTA_PCTL_R = (GPIO_PORTA_PCTL_R&0xFF0F00FF)+0x00202200;
  GPIO_PORTA_AMSEL_R = 0;         // disable analog functionality on PA
  SSI0_CR1_R = 0x00000000;        // disable SSI, master mode
  SSI0_CPSR_R = 0x02;             // 8 MHz SSIClk
  SSI0_CR0_R &= ~(0x0000FFF0);    // SCR = 0, SPH = 0, SPO = 0 Freescale
  SSI0_CR0_R |= 0x0F;             // DSS = 16-bit data
  SSI0_DR_R = data;               // load 'data' into transmit FIFO
  SSI0_CR1_R |= 0x00000002;       // enable SSI
}
void DAC_Out(uint16_t code){
  while((SSI0_SR_R&0x00000002)==0){};// SSI Transmit FIFO Not Full
  SSI0_DR_R = code; }             // data out, no reply
// send the 16-bit code to the SSI, return a reply
// you will need to enable PA4 as SSI0Rx to use this routine
```

```
uint16_t DAC_Out2(uint16_t code){   uint16_t receive;
  while((SSI0_SR_R&0x00000002)==0){};// SSI Transmit FIFO Not Full
  SSI0_DR_R = code;                  // data out
  while((SSI0_SR_R&0x00000004)==0){};// SSI Receive FIFO Not Empty
  receive = SSI0_DR_R;               // acknowledge response
  return receive;
}
```
Program 7.2. Functions to initialize and to send data to the DAC using the SSI (MAX5353_xxx).

Example 7.3. Interface the ADXL362 3-axis MEMS Accelerometer.

Solution: This second SSI example shows a synchronous serial interface between the computer and an ADXL362, which measures acceleration in three dimensions. Here in this section we will focus on the hardware and software aspects of the serial interface (Figure 7.21). Again, the basic interfacing issues to consider for this interface are:

Word size. Two write a command we need to transmit 24 bits to the ADXL362. To read a result, we first transmit 16 bits and then we receive 8 bits. The software will implement 8-bit transmissions with the SPI module.

Bit order. The ADXL362 requires the most significant bits first.

Clock phase, clock polarity. Since the ADXL362 samples its serial input data on the rising edge of the clock, the SPI must changes the data on the falling edge. SPO=SPH=0 and SPO=SPH=1 both satisfy this requirement. We will use the SPO=SPH=0 mode as suggested in the Analog Devices data sheet, refer back to Figure 7.17.

Bandwidth. We look at the timing specifications of the ADXL362. The maximum SCLK frequency is 8 MHz, and the minimum clock low/high widths is 50 ns, so the shortest SPI period we can use is 125ns.

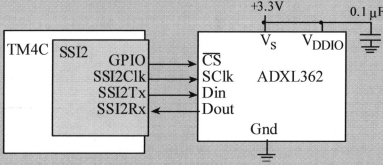

Figure 7.21. A three-axis accelerometer interfaced to the SSI port.

The first 8 bits sent will specify read (0x0B) or write (0x0A) as shown in Figure 7.22. The second 8 bits will specify the register from which to read, or to which to write. On a read operation the last 8 bits will be the value returned from the ADXL362 to the microcontroller. On a write operation, the last 8 bits will be the value written from the microcontroller to the ADXL362. Because we want the CS signal to remain low for the entire 24-bit transfer, we will implement it using the regular I/O pin functions.

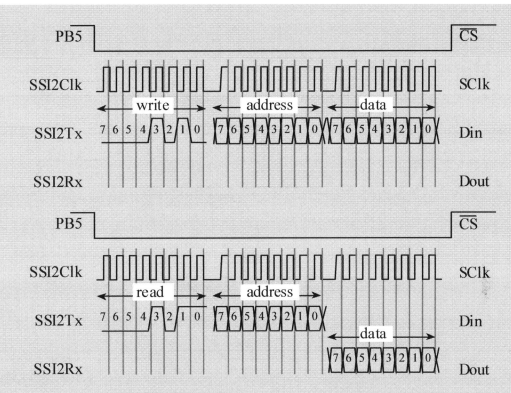

Figure 7.22. ADXL362 serial timing.

Recall that when the software outputs to the SSI data register, the 8-bit register in the SSI is exchanged with the 8-bit register in the ADXL362. To communicate with the ADXL362, three 8-bit transmissions are exchanged.

```
#define PB5     (*((volatile uint32_t *) 0x40005080))
void ADXL362_Init(void){
  SYSCTL_RCGCSSI_R  |= 0x04;            // activate SSI2
  SYSCTL_RCGCGPIO_R |= 0x02;            // activate port B
  while((SYSCTL_PRGPIO_R&0x02) == 0){};// ready?
  GPIO_PORTB_PCTL_R = (GPIO_PORTB_PCTL_R&0x00F0FFFF)+0x22020000;
  GPIO_PORTB_AMSEL_R = 0;               // disable analog functionality on PA
  GPIO_PORTB_AFSEL_R |= 0xD);           // enable alt funct on PB4,6,7
  GPIO_PORTB_DIR_R  |= 0x20;            // make PB5 out (!CS signal)
  GPIO_PORTB_DEN_R  |= 0xF0;            // enable digital I/O on PB4,5,6,7
  PB5 = 0x20;                           // !CS = 1
  SSI2_CR1_R = 0x00000000;              // disable SSI, master mode
  SSI2_CPSR_R = 0x08;                   // 2 MHz SSIClk (16/8)
  SSI2_CR0_R &= ~(0x0000FFF0);          // clear SCR, SPH and SPO
  SSI2_CR0_R |= 0x07;                   // DSS = 8-bit data
  SSI2_CR1_R |= 0x00000002;             // enable SSI
}
```

```c
// send the 8-bit code, wait for reply, return reply
// basically this eliminates the FIFO, doing one at a time
uint8_t sendAfterWaiting(uint8_t code){uint8_t dummy;
  while(SSI2_SR_R&SSI_SR_RNE){   // optional step
    dummy = SSI2_DR_R;   // flush any leftover bytes in receiver
  }
  while((SSI2_SR_R&SSI_SR_TFE)==0){};// wait until FIFO empty
  SSI2_DR_R = code;                    // data out
  while((SSI2_SR_R&SSI_SR_RNE)==0){};// wait until response
  return SSI2_DR_R;                    // acknowledge response
}
void ADXL362_Write(uint8_t reg, uint8_t data){
  PB5 = 0;                      // !CS = 0
  sendAfterWaiting(0x0A);       // send write operation
  sendAfterWaiting(reg);        // send register address
  sendAfterWaiting(data);       // write value
  PB5 = 0x20;                   // !CS = 1
}
uint8_t ADXL362_Read(uint8_t reg){ uint8_t data;
  PB5 = 0;                      // !CS = 0
  sendAfterWaiting(0x0B);       // send read operation
  sendAfterWaiting(reg);        // send register address
  data = sendAfterWaiting(0);   // read value
  PB5 = 0x20;                   // !CS = 1
  return data;                  // right justify
}
```
Program 7.3. Functions to initialize, to send and to receive data from the ADXL362 using the SPI.

Example 7.4. Design an output port expander using a shift register and SSI.

Solution: Sometimes we need more output pins than available on our microcontroller. In general, the proper design approach would be to upgrade to a microcontroller with more pins. However in situations where we do not have the time or money to change microcontrollers, we can interface a 74HC595 shift register to the SSI port for a quick solution providing additional output pins. Basically, three pins of the SSI (SS, MOSI, and SCLK) will be converted to eight digital outputs **Q** on the 74HC595, as shown in Figure 7.23.

Additional shift registers can be chained together (connect the **QH'** outputs of one to the **SER** inputs of the next) to provide additional outputs without requiring more TM4C123 pins. The gate input, **G***, of the 74HC595 is grounded so the eight **Q** outputs will be continuously driven. The SPI clock output is connected to the 74HC595 clock input (**SCK**) and the SPI data output is connected to the 74HC595 data input (**SER**). The Freescale SPI mode (SPO=0, SPH=0) is selected to the TM4C123 changes the output data on the fall of the clock and the 74HC595 shifts data in on the rise. Because the 74HC595 is fast (maximum clock 25 MHz, and setup time of 20 ns), we run the SSI as fast as possible. If the bus clock is 16 MHz and the divider is set to 2, and the SSI clock will be 8 MHz.

After eight bits are transferred from the TM4C123 to the 74HC595, software will create a rising edge of **RCK**, causing the new data to be latched into the 74HC595. If there is just one 74HC595 like Figure 7.23, we can use the automatic SS feature of the SPI to create a rising edge latch on **RCLK**. To enable SPI, we set bits 2,3,5 **AFSEL** for Port A. The **PCTL** code for SSI on Port A is 2 (Table 2.4). In this solution, we perform one SSI transmission to change all 8 bits of the port output (Program 7.4). The SS pulse, see Figure 7.17, occurs automatically and does not require software overhead to produce. However, if we were chaining multiple shift registers, we would not use the automatic Fss feature; rather we would output all the data and then manually latch it in with explicit outputs on **RCK** by using PA3 as a regular GPIO port.

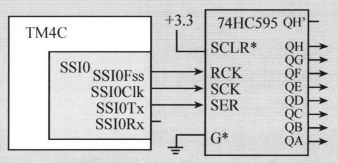

Figure 7.23. Interface between the TM4C123 and a 74HC595 shift register.

Program 7.4 assumes the system clock rate less than or equal to 50 MHz. For bus clocks faster than 50 MHz, you will need to change the clock divider so that the SPI clock will be less than or equal to 25 MHz. At 8 MHz, it takes about 1 µs to output 8 bits to the port.

```
void Port_Init(void){
  SYSCTL_RCGCSSI_R  |= 0x01;          // activate SSI0
  SYSCTL_RCGCGPIO_R |= 0x01;          // activate port A
  while((SYSCTL_PRGPIO_R&0x01) == 0){};// ready?
  GPIO_PORTA_AFSEL_R |= 0x2C;         // enable alt funct on PA2,3,5
  GPIO_PORTA_PCTL_R = (GPIO_PORTA_PCTL_R&0xFF0F00FF)+0x00202200;
  GPIO_PORTA_AMSEL_R = 0;             // disable analog functionality on PA
  GPIO_PORTA_DEN_R |= 0x2C;           // enable digital I/O on PA2,3,5
  SSI0_CR1_R = 0x00000000;            // disable SSI, master mode
  SSI0_CPSR_R = 0x02;                 // 8 MHz SSIClk
  SSI0_CR0_R &= ~(0x0000FFF0);        // SCR = 0, SPH = 0, SPO = 0 Freescale
  SSI0_CR0_R = (SSI0_CR0_R&~0x0F)+0x07; // 8-bit data
  SSI0_CR1_R |= 0x00000002;           // enable SSI
}
void Port_Out(uint8_t code){
  while((SSI0_SR_R&0x02)==0){};       // wait until room in FIFO
  SSI0_DR_R = code;                   // data out
}
```
Program 7.4. Software to control an output parallel port expanded using the SSI (74HC595_xxx).

Checkpoint 7.7: How would you change Program 7.4 to run as fast as possible assuming the bus frequency is 80 MHz?

7.6. Inter-Integrated Circuit (I²C) Interface

7.6.1. The Fundamentals of I²C

Ever since microcontrollers have been developed, there has been a desire to shrink the size of an embedded system, reduce its power requirements, and increase its performance and functionality. Two mechanisms to make systems smaller are to integrate functionality into the microcontroller and to reduce the number of I/O pins. The inter-integrated circuit I²C interface was proposed by Philips in the late 1980s as a means to connect external devices to the microcontroller using just two wires. The SSI interface has been very popular, but it takes 3 wires for simplex and 4 wires for full duplex communication. In 1998, the I²C Version 1 protocol become an industry standard and has been implemented into thousands of devices. The I²C bus is a simple two-wire bi-directional serial communication system that is intended for communication between microcontrollers and their peripherals over short distances. This is typically, but not exclusively, between devices on the same printed circuit board, the limiting factor being the bus capacitance. It also provides flexibility, allowing additional devices to be connected to the bus for further expansion and system development. The interface will operate at baud rates of up to 100 kbps with maximum capacitive bus loading. The module can operate up to a baud rate of 400 kbps provided the I²C bus slew rate is less than 100ns. The maximum interconnect length and the number of devices that can be connected to the bus are limited by a maximum bus capacitance of 400pF in all instances. These parameters support the general trend that communication speed can be increased by reducing capacitance. Version 2.0 supports a high speed mode with a baud rate up to 2.4 MHz (supported by LM4F/TM4C).

Figure 7.24 shows a block diagram of a communication system based on the I²C interface. The master/slave network may consist of multiple masters and multiple slaves. The **Serial Clock Line** (SCL) and the **Serial Data line** (SDA) are both bidirectional. Each line is open drain, meaning a device may drive it low or let it float. A logic high occurs if all devices let the output float, and a logic low occurs when at least one device drives it low. The value of the pull-up resistor depends on the speed of the bus. 4.7 kΩ is recommended for baud rates below 100 kbps, 2.2 kΩ is recommended for standard mode, and 1 kΩ is recommended for fast mode.

Checkpoint 7.8: Why is the recommended pull-up resistor related to the bus speed?

Checkpoint 7.9: What does open drain mean?

The SCL clock is used in a synchronous fashion to communicate on the bus. Even though data transfer is always initiated by a master device, both the master and the slaves have control over the data rate. The master starts a transmission by driving the clock low, but if a slave wishes to slow down the transfer, it too can drive the clock low (called **clock stretching**). In this way, devices on the bus will wait for all devices to finish. Both address (from Master to Slaves) and information (bidirectional) are communicated in serial fashion on SDA.

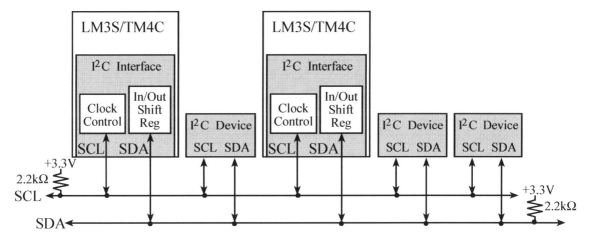

Figure 7.24. Block diagram of an I²C communication network.

The bus is initially idle where both SCL and SDA are both high. This means no device is pulling SCL or SDA low. The communication on the bus, which begins with a START and ends with a STOP, consists of five components:

START (S) is used by the master to initiate a transfer
DATA is sent in 8-bit blocks and consists of
  7-bit address and 1-bit direction from the master
  control code for master to slaves
  information from master to slave
  information from slave to master
ACK (A) is used by slave to respond to the master after each 8-bit data transfer
RESTART (R) is used by the master to initiate additional transfers without releasing the bus
STOP (P) is used by the master to signal the transfer is complete and the bus is free

The basic timings for these components are drawn in Figure 7.25. For now we will discuss basic timing, but we will deal with issues like stretching and arbitration later. A slow slave uses clock stretching to give it more time to react, and masters will use arbitration when two or more masters want the bus at the same time. An idle bus has both SCL and SDA high. A transmission begins when the master pulls SDA low, causing a START (S) component. The timing of a RESTART is the same as a START. After a START or a RESTART, the next 8 bits will be an address (7-bit address plus 1-bit direction). There are 128 possible 7-bit addresses, however, 32 of them are reserved as special commands. The address is used to enable a particular slave. All data transfers are 8 bits long, followed by a 1-bit acknowledge. During a data transfer, the SDA data line must be stable (high or low) whenever the SCL clock line is high. There is one clock pulse on SCL for each data bit, the MSB being transferred first. Next, the selected slave will respond with a positive acknowledge (Ack) or a negative acknowledge (Nack). If the direction bit is 0 (write), then subsequent data transmissions contain information sent from master to slave.

For a write data transfer, the master drives the RDA data line for 8 bits, then the slave drives the acknowledge condition during the 9th clock pulse. If the direction bit is 1 (read), then subsequent data transmissions contain information sent from slave to master. For a read data transfer, the slave drives the RDA data line for 8 bits, then the master drives the acknowledge condition during the 9th clock pulse. The STOP component is created by the master to signify the end of transfer. A STOP begins with SCL and SDA both low, then it makes the SCL clock high, and ends by making SDA high. The rising edge of SDA while SCL is high signifies the STOP condition.

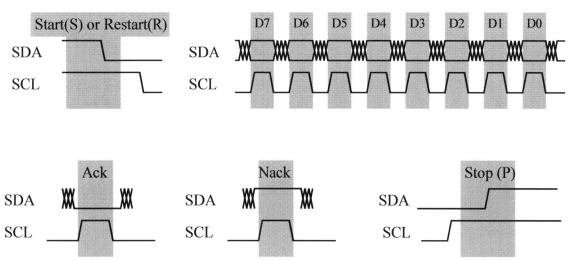

Figure 7.25. Timing diagrams of I²C components.

Checkpoint 7.10: What happens if no device sends an acknowledgement?

Figure 7.26 illustrates the case where the master sends 2 bytes of data to a slave. The shaded regions demark signals driven by the master, and the white areas show those times when the signal is driven by the slave. Regardless of format, all communication begins when the master creates a START component followed by the 7-bit address and 1-bit direction. In this example, the direction is low, signifying a write format. The 1st through 8th SCL pulses are used to shift the address/direction into all the slaves. In order to acknowledge the master, the slave that matches the address will drive the SDA data line low during the 9th SCL pulse. During the 10th through 17th SCL pulses sends the data to the selected slave. The selected slave will acknowledge by driving the SDA data line low during the 18th SCL pulse. A second data byte is transferred from master to slave in the same manner. In this particular example, two data bytes were sent, but this format can be used to send any number of bytes, because once the master captures the bus it can transfer as many bytes as it wishes. If the slave receiver does not acknowledge the master, the SDA line will be left high (Nack). The master can then generate a STOP signal to abort the data transfer or a RESTART signal to commence a new transmission. The master signals the end of transmission by sending a STOP condition.

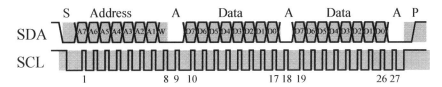

Figure 7.26. I²C transmission of two bytes from master to slave

Figure 7.27 illustrates the case where a slave sends 2 bytes of data the master. Again, the master begins by creating a START component followed by the 7-bit address and 1-bit direction. In this example, the direction is high, signifying a read format. During the 10th through 17th SCL pulses the selected slave sends the data to the master. The selected slave can only change the data line while SCL is low and must be held stable while SCL is high. The master will acknowledge by driving the SDA data line low during the 18th SCL pulse. Only two data bytes are shown in Figure 7.27, but this format can be used to receive as any many bytes the master wishes. Except for the last byte all data are transferred from slave to master in the same manner. After the last data byte, the master does not acknowledge the slave (Nack) signifying 'end of data' to the slave, so the slave releases the SDA line for the master to generate STOP or RESTART signal. The master signals the end of transmission by sending a STOP condition.

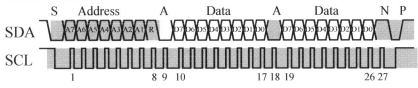

Figure 7.27. I²C transmission of two bytes from slave to master.

Figure 7.28 illustrates the case where the master uses the RESTART command to communicate with two slaves, reading one byte from one slave and writing one byte to the other. As always, the master begins by creating a START component followed by the 7-bit address and 1-bit direction. During the first start, the address selects the first slave and the direction is read. During the 10th through 17th SCL pulses the first slave sends the data to the master. Because this is the last byte to be read from the first slave, the master will not acknowledge letting the SDA data float high during the 18th SCL pulse, so the first slave releases the SDA line. Rather than issuing a STOP at this point, the master issues a repeated start or RESTART. The 7-bit address and 1-bit direction transferred in the 20th through 27th SCL pulses will select the second slave for writing. In this example, the direction is low, signifying a write format. The 28th pulse will be used by the second slave pulls SDA low to acknowledge it has been selected. The 29th through 36th SCL pulses send the data to the second slave. During the 37th pulse the second slave pulls SDA low to acknowledge the data it received. The master signals the end of transmission by sending a STOP condition.

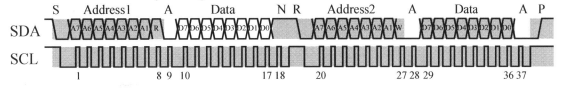

Figure 7.28. I²C transmission of one byte from the first slave and one byte to a second slave.

Checkpoint 7.11: Is the communication in Figure 7.28 full duplex, half duplex, or simplex?

Table 7.10 lists some addresses that have special meaning. A write to address 0 is a general call address, and is used by the master to send commands to all slaves. The 10-bit address mode gives two address bits in the first frame and 8 more address bits in the second frame. The direction bit for 10-bit addressing is in the first frame.

Address	R/W	Description
0000 000	0	General call address
0000 000	1	Start byte
0000 001	x	CBUS address
0000 010	x	Reserved for different bus formats
0000 011	0	Reserved
0000 1xx	x	High speed mode
1111 0xx	x	10-bit address
1111 1xx	X	Reserved

Table 7.10. Special addresses used in the I²C network.

7.6.2. I²C Synchronization

The I²C bus supports multiple masters. If two or more masters try to issue a START command on the bus at the same time, both clock synchronization and arbitration will occur. **Clock synchronization** is procedure that will make the low period equal to the longest clock low period and the high is equal to the shortest one among the masters. Figure 7.29 illustrates clock synchronization, where the top set of traces is generated by the first master, and the second set of traces is generated by the second master. Since the outputs are open drain, the actual signals will be the wired-AND of the two outputs. Each master repeats these steps when it generates a clock pulse. It is during step 3) that the faster device will wait for the slower device

1. Drive its SCL clock low for a fixed amount of time
2. Let its SCL clock float
3. Wait for the SCL to be high
4. Wait for a fixed amount of time, stop waiting if the clock goes low

Because the outputs are open drain, the signal will be pulled to a logic high by the 2 kΩ resistor only if all devices release the line (output a logic high). Conversely, the signal will be a logic low if any device drives it low. When masters create a START, they first drive SDA low, then drive SCL low. If a group of masters are attempting to create START commands at about the same time, then the wire-AND of their SDA lines has its 1 to 0 transition before the wire-AND of their SCL lines has its 1 to 0 transition. Thus, a valid START command will occur causing all the slaves to listen to the upcoming address. In the example shown in Figure 7.29, Master #2 is the first to drive its clock low. In general, the SCL clock will be low from the time the first master drives it low (time 1 in this example), until the time the last master releases its clock (time 2 in this example.) Similarly, the SCL clock will be high from the time the last master releases its clock (time 2 in this example), until the time the first master drives its clock low (time 3 in this example.)

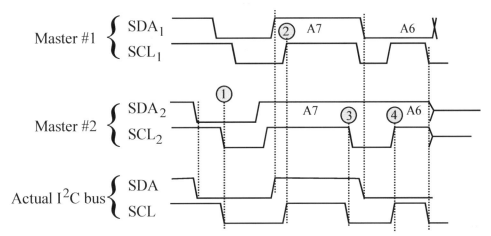

Figure 7.29. *I²C timing illustrating clock synchronization and data arbitration.*

The relative priority of the contending masters is determined by a **data arbitration** procedure. A bus master loses arbitration if it transmits logic "1" while another master transmits logic "0". The losing masters immediately switch over to slave receive mode and stop driving the SCL and SDA outputs. In this case, the transition from master to slave mode does not generate a STOP condition. Meanwhile, a status bit is set by hardware to indicate loss of arbitration. In the example shown in Figure 7.29, master #1 is generating an address with A7=1 and A6=0, while master #2 is generating an address with A7=1 and A6=1. Between times 2 and 3, both masters are attempting to send A7=1, and notice the actual SDA line is high. At time 4, master #2 attempts to make the SDA high (A6=1), but notices the actual SDA line is low. In general, the master sending a message to the lowest address will win arbitration.

Checkpoint 7.12: If Master 1 sends address 0x30 and Master 2 sends address #0x0F, which one wins arbitration?

The third synchronization mechanism occurs between master and slave. If the slave is fast enough to capture data at the maximum rate, the transfer is a simple synchronous serial mechanism. In this case the transfer of each bit from master to slave is illustrated by the following interlocked sequences, see Figure 7.25.

Master sequence
1. Drive its SCL clock low
2. Set the SDA line
3. Wait for a fixed amount of time
4. Let its SCL clock float
5. Wait for the SCL to be high
6. Wait for a fixed amount of time
7. Stop waiting if the clock goes low

Slave sequence (no stretch)

6. Capture SDA data on low to high edge of SCL

If the slave is not fast enough to capture data at the maximum rate, it can perform an operation called **clock stretching**. If the slave is not ready for the rising edge of SCL, it will hold the SCL clock low itself until it is ready. Slaves are not allowed to cause any 1 to 0 transitions on the SCL clock, but rather can only delay the 0 to 1 edge. The transfer of each bit from master to slave with clock stretching is illustrated by the following sequences

Master sequence
1. Drive its SCL clock low
2. Set the SDA line
3. Wait for a fixed amount of time
4. Let its SCL clock float
5. Wait for the SCL clock to be high
6. Wait for a fixed amount of time
7. Stop waiting if the clock goes low

Slave sequence (clock stretching)
1. Wait for the SCL clock to be low
2. Drive SCL clock low
3. Wait until it's ready to capture
4. Let its SCL float
5. Wait for the SCL clock to be high
6. Capture the SDA data

Clock stretching can also be used when transferring a bit from slave to master

Master sequence
1. Drive its SCL clock low
2. Wait for a fixed amount of time

4. Let its SCL clock float
5. Wait for the SCL clock to be high
6. Capture the SDA input
7. Wait for a fixed amount of time,
8. Stop waiting if the clock goes low

Slave sequence (clock stretching)
1. Wait for the SCL clock to be low
2. Drive SCL clock low
3. Wait until next data bit is ready
4. Let its SCL float
5. Wait for the SCL clock to be high

Observation: Clock stretching allows fast and slow devices to exist on the same I^{2}C bus Fast devices will communicate quickly with each other, but slow down when communicating with slower devices.

Checkpoint 7.13: Arbitration continues until one master sends a zero while the other sends a one. What happens if two masters attempt to send data to the same address?

7.6.3. TM4C I2C Details

TM4C microcontrollers have zero to ten I^{2}C modules, see Figure 7.30. The LM3S microcontrollers implement just a subset of the standard. They support master and slave modes, can generate interrupts on start and stop conditions, and allow I^{2}C networks with multiple masters. On the other hand, the LM3S microcontrollers do not support general call, or 10-bit addressing. As shown in Figure 7.24, microcontroller pins SDA and SCL can be connected directly to an I^{2}C network. Because I^{2}C networks are intended to connect devices on the same PCB, no special hardware interface electronics are required. Table 7.11 lists the I^{2}C ports on the TM4C. The TM4C can operate in slave mode, but we will focus on master mode. Tables 2.4 and 2.5 describe how to attach I/O pins to the I²C modules on the TM4C123/MSP432E.

	7	6	5	4	3	2	1	0	Name
$4002.0000				SA				R/S	I2C0_MSA_R
$4002.0004		BSBSY	IDLE	ARBLST	DATACK	ADRACK	ERR	BUSY	I2C0_MCS_R
$4002.0008	DATA	DATA	DATA	DATA	DATA	DATA	DATA	DATA	I2C0_MDR_R
$4002.000C				TPR					I2C0_MTPR_R
$4002.0010								IM	I2C0_MIMR_R
$4002.0014								RIS	I2C0_MRIS_R
$4002.0018								MIS	I2C0_MMIS_R
$4002.001C								IC	I2C0_MICR_R
$4002.0020			SFE	MFE				LPBK	I2C0_MCR_R
$4000.5420	SEL	SEL	SEL	SEL	**SEL**	**SEL**	SEL	SEL	GPIO_PORTB_AFSEL_R
$4000.551C	DEN	DEN	DEN	DEN	**DEN**	**DEN**	DEN	DEN	GPIO_PORTB_DEN_R
$4000.550C	SEL	SEL	SEL	SEL	SEL	SEL	**SEL**	**SEL**	GPIO_PORTB_ODR_R
$400F.E608	GPIOH	GPIOG	GPIOF	GPIOE	GPIOD	GPIOC	**GPIOB**	GPIOA	SYSCTL_RCGCGPIO_R
$400F.E620					I2C3	I2C2	I2C1	I2C0	SYSCTL_RCGCI2C_R

Table 7.11. The TM4C I2C master registers. Each register is 32 bits wide. Bits 31 – 8 are zero.

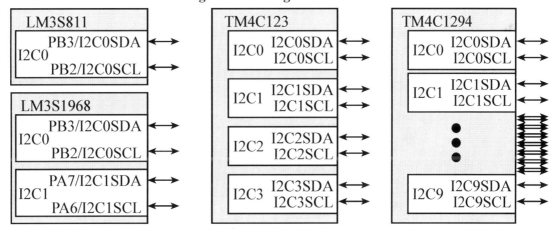

Figure 7.30. I/O port pins for I2C on various LM3S/TM4C/MSP432E4 microcontrollers.

There are seven steps to initialize I2C0 in master mode.

Step 1) we activate the I2C0 clock by setting bit 0 the **SYSCTL_RCGCI2C_R** register.

Step 2) Since I2C0 uses two pins on Port B, we need to activate Port B clock in register **SYSCTL_RCGCGPIO_R**.

Step 3) We connect I2C0 to pins PB3/PB2 by setting the alternative function. I.e., we set bits 2 and 3 of **GPIO_PORTB_AFSEL_R** register.

Step 4) I2C0 uses open drain mode on the data bit, so we set bit 3 of **GPIO_PORTB_ODR_R** register. The port will drive the line low when it wants to output a zero. Conversely, to make the line high, open drain mode causes the output to float and the pull-up resistor in Figure 7.24 causes the line to go high. The SCL pin must not be configured as an open-drain signal, although the port causes it to act as if it were an open drain signal.

Step 5) We enable the digital circuits by setting 2 and 3 of **GPIO_PORTB_DEN_R** register.

Step 6) We enable master mode by setting the **MFE** bit in the `I2C0_MCR_R` register.

Step 7) Lastly, `I2C0_MTPR_R` is the I²C Master Timer Period Register, which determines the baud rate transferred as a master. It is the master's responsibility to generate the I²C clock. The timing of the I²C interface is derived from the bus clock. We set the baud rate by writing to the **TPR** field in the `I2C0_MTPR_R` register. Let f_{BUS} be the frequency of the bus clock and let $t = 1/f_{BUS}$ be the period of the bus clock. The I²C clock period (t_{bit}) will be 20*(**TPR**+1)*t. In standard mode, we set the I²C clock frequency to be about 100 kHz (t_{bit} =10 μs). In fast mode, we set the I²C clock frequency to be about 400 kHz. For example, to set the I²C to standard speed, assuming Δt is given in ns,

$$t_{bit} = 20*(\textbf{TPR}+1)*t = 10000 \text{ ns}$$

or **TPR** = 500/t -1

The **R/S** bit (bit 0) of the Master's Slave Address Register, `I2C0_MSA_R`, specifies whether the next data transfer will be a receive from slave (equals 1) or a write to slave (equals 0). Bits 7 through 1 of the Master's Slave Address Register contain the unique 7-bit address of the slave that the master is addressing. Generally, no two devices should have the same slave address. Many devices can permanently accept a new slave address from the master. `I2C_MCS_ACK` specifies whether or not the master acknowledges data received from the slave. `I2C_MCS_ACK` is only used when the I²C Bus is a receiver, not a transmitter.

When receiving data as a master, this bit determines if an acknowledgement will be sent during the 9th clock bit. 1 means an acknowledgement will be sent (Ack), and 0 means no acknowledgement will be sent (Nack). The master may not acknowledge the final byte sent by a slave. It must issue a STOP or repeated START condition after the negative acknowledge. The I²C module will always acknowledge address matches, provided it is enabled. A START or repeated start (RESTART) will be sent if software writes a 1 to the `I2C_MCS_START` bit, provided this microcontroller is the current bus master. Attempting a repeated start when the bus is owned by another master will result in loss of arbitration. `I2C_MCS_START` is part of the Master's Control/Status Register, which is a special register. Reads from this register return information about the master's status, and writes to this register control the master's next transmission. For example, set the `I2C_MCS_ACK`, `I2C_MCS_STOP`, and/or `I2C_MCS_START` bits and the `I2C_MCS_RUN` bit to begin the next transmission. Afterwards, read the `I2C_MCS_ARBLST` or `I2C_MCS_ERROR` bits to see if the master has lost arbitration of the bus. Many applications may also require reading the `I2C_MCS_ADRACK` and `I2C_MCS_DATACK` bits to ensure that the slave has acknowledged its address and any data. An unresponsive device may represent a serious hardware problem to which the system must react.

The master can be in three modes: idle, transmit and receive. Table 7.12 summarizes the modes and operation of I²C. Combinations not listed in this table are either illegal or no operation.

State	R/S	ACK	STOP	START	Run	Action (and new state)
Idle	0	X	0	1	1	START condition followed by SEND (goes to

State						Action
						Master Transmit)
	0	X	1	1	1	START condition followed by a SEND and STOP condition (remains in Idle).
	1	0	0	1	1	START condition, RECEIVE operation with negative ACK (goes to the Master Receive)
	1	0	1	1	1	START condition followed by RECEIVE and STOP condition (remains in Idle state)
	1	1	0	1	1	START condition followed by RECEIVE (goes to the Master Receive state)
Master Transmit	X	X	0	0	1	SEND operation (remains in Master Transmit state)
	X	X	1	0	0	STOP condition (goes to Idle state)
	X	X	1	0	1	SEND followed by STOP condition (goes to Idle state)
	0	X	0	1	1	Repeated START condition followed by a SEND (remains in Master Transmit state)
	0	X	1	1	1	Repeated START condition followed by SEND and STOP condition (goes to Idle state)
	1	0	0	1	1	Repeated START, then RECEIVE operation with a negative ACK (goes to Master Receive state)
	1	0	1	1	1	Repeated START condition followed by a SEND and STOP condition (goes to Idle state)
	1	1	0	1	1	Repeated START condition followed by RECEIVE (goes to Master Receive state)
Master Receive	X	0	0	0	1	RECEIVE operation with negative ACK (remains in Master Receive state)
	X	X	1	0	0	STOP condition (goes to Idle state)
	X	0	1	0	1	RECEIVE followed by STOP condition (goes to Idle state)
	X	1	0	0	1	RECEIVE operation (remains in Master Receive state)
	1	0	0	1	1	Repeated START, then RECEIVE operation with a negative ACK (remains in Master Receive state).
	1	0	1	1	1	Repeated START, then RECEIVE and STOP condition (goes to Idle state)
	1	1	0	1	1	Repeated START condition followed by RECEIVE (remains in Master Receive state)
	0	X	0	1	1	Repeated START condition followed by SEND (goes to Master Transmit state)
	0	X	1	1	1	Repeated START condition followed by SEND and STOP condition (goes to Idle state)

Table 7.12. I²C commands, actions, and state changes.

Figure 7.31 and Table 7.13 describe the timing of I²C. Let t_{bit} be the I²C period as defined by the bus clock and the **TPR** field in the `I2C0_MTPR_R` register. Because SCL and SDA are open-drain-type outputs, which the controller can only actively drive low. The setup time, t_{setup}, is the time before the rise of SCL that the data on SDA will be valid. The data remains valid throughout the time SCL is high. The hold time, t_{hold}, is the time after the fall of SCL that the data on SDA continues to be valid. The rise time, t_{rise}, of SDA and SCL depends on external signal capacitance and pull-up resistor values (**). The fall time, t_{fall}, is specified up to a maximum 50 pF load.

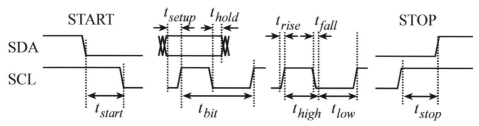

Figure 7.31. *I²C timing intervals.*

Parameter	Minimum	Typical	Maximum
t_{start}	36 *t*		
t_{high}	24 *t*		
t_{low}	36 *t*		
t_{setup}	18 *t*		
t_{hold}	2 *t*		
t_{rise}	**	**	**
t_{fall}		9 ns	10 ns
t_{stop}	24 *t*		

Table 7.13. I²C timing intervals, where *t* is the system bus period (depends on external factors).**

7.6.4. I²C Single Master Example

The objective of this example is to present a low-level device driver for an I²C network where this microcontroller is the only master, as shown in Program 7.5. This simple example will employ busy-wait synchronization. `I2C_Init` first enables the I²C interface, starting out in master mode. Since this is the only master, it does not need to initialize any of the slave mode settings.

```
// Assumes a 50 MHz bus clock, 20*(TPR+1)*20ns = 10us, with TPR=24
#define TPR (500/20 - 1)
void I2C_Init(void){
  SYSCTL_RCGCI2C_R  |= 0x0001;                     // activate I2C0
  SYSCTL_RCGCGPIO_R |= 0x0002;                     // activate port B
  while((SYSCTL_PRGPIO_R&0x0002) == 0){};// ready?
  GPIO_PORTB_AFSEL_R |= 0x0C;                      // enable alt funct on PB2,3
  GPIO_PORTB_ODR_R  |= 0x08;                       // enable open drain on PB3
  GPIO_PORTB_PCTL_R = (GPIO_PORTB_PCTL_R&0xFFFF00FF)+0x00003300; // I2C
  GPIO_PORTB_DEN_R  |= 0x0C;                       // enable digital I/O on PB2,3
```

```
    I2C0_MCR_R = 0x00000010;           // master function enable
    I2C0_MTPR_R = TPR;                  // configure for 100 kbps clock
}
```
Program 7.5. TM4C123 I²C initialization in single master mode (I2C_xxx).

Program 7.6 contains the function **I2C_Send2** that transmits two bytes to a slave, creating a transmission shown in Figure 7.26. In a system with multiple masters this should check to see if the bus is idle first. Because this system has just one master, the bus should be idle. However, the first line of the function makes certain that the I²C hardware is free and there is not already another transaction pending. By setting the **I2C_MCS_START** bit of the I²C Master Control/Status Register, **I2C0_MCS_R**, the microcontroller will create a START condition. In a system with multiple masters, it should check to see if it lost bus arbitration (**I2C_MCS_ARBLST**). The slave address (with bit 0 equal to 0) will be sent. The two data bytes are sent, then the STOP is issued. If there is a possibility the slave doesn't exist, data doesn't get acknowledged, or bus arbitration is lost then this program could have checked **I2C_MCS_ERROR** after each transfer.

```
uint32_t I2C_Send2(uint8_t slave, uint8_t data1, uint8_t data2){
  while(I2C0_MCS_R&0x00000001){};    // wait for I2C ready
  I2C0_MSA_R = (slave<<1)&0xFE;      // MSA[7:1] is slave address
  I2C0_MSA_R &= ~0x01;                // MSA[0] is 0 for send
  I2C0_MDR_R = data1&0xFF;            // prepare first byte
  I2C0_MCS_R = (  I2C_MCS_START       // generate start/restart
               | I2C_MCS_RUN);        // no ack, no stop,master enable
  while(I2C0_MCS_R&0x00000001){};    // wait for transmission done
                                      // check error bits
  if((I2C0_MCS_R&
     (I2C_MCS_DATACK|I2C_MCS_ADRACK|I2C_MCS_ERROR)) != 0){
    I2C0_MCS_R = I2C_MCS_STOP;       // stop, no ack, disable
    return (I2C0_MCS_R&                // return error bits if nonzero
         (I2C_MCS_DATACK|I2C_MCS_ADRACK|I2C_MCS_ERROR));
  }
  I2C0_MDR_R = data2&0xFF;            // prepare second byte
  I2C0_MCS_R = (I2C_MCS_STOP          // no ack, stop, no start
             | I2C_MCS_RUN);          // master enable
  while(I2C0_MCS_R&0x00000001){};    // wait for transmission done
  return (I2C0_MCS_R&                 // return error bits
         (I2C_MCS_DATACK|I2C_MCS_ADRACK|I2C_MCS_ERROR));
}
```
Program 7.6. I²C transmission in single master mode (I2C_xxx).

Program 7.7 contains the function **I2C_Recv2** that receives two bytes from a slave, creating a transmission shown in Figure 7.27. By setting the **I2C_MCS_START** bit, the microcontroller will create a START condition. During the first transfer, the **Tx/Rx** bit is 1, so the slave address (with bit 0 equal to 1) will be sent, and the master goes into receive mode. During the next two transfers, the master is in receive mode, so data flows into the microcontroller. To trigger any data transfer, the software writes a valid value to **I2C0_MCS_R** with the **I2C_MCS_RUN** bit set. During the first data transfer **I2C_MCS_ACK** is 1, creating a positive acknowledgement.

Conversely during the second data transfer `I2C_MCS_ACK` is 0, creating a negative acknowledgement and signaling to the slave that this is the last data to be transferred. The STOP is requested when the `I2C_MCS_STOP` bit of the `I2C0_MCS_R` register is set, and it automatically is issued after the final data transfer.

This is a simple concept that may seem complicated at first. Basically:
    0) Initialize: turn on GPIOB and I2C modules, set up GPIOB pins, set up I2C master
    1) Put slave address into `I2C0_MSA_R`[7:1]
    2) (receive only) Set `I2C0_MSA_R`[0]
    2) (transmit only) Clear `I2C0_MSA_R`[0] and put first byte into `I2C0_MDR_R`[7:0]
    3) Write a valid value to `I2C0_MCS_R`[3:0] to start transfer
    4) Wait for transfer to finish: `while(I2C0_MCS_R&0x00000001){};`
    5) (receive only) Read first byte from `I2C0_MDR_R`[7:0]
    6) repeat Steps 2-5 until done

Timing is complicated and potentially confusing. Fortunately, it is handled by hardware you are already given. There are two complicated things for which you are responsible. The first issue is understanding what codes to send to the I²C slave. The second complicated issue is Step 3. There are 4 bits to write to `I2C0_MCS_R`[3:0], and they depend on the master's current state and next state as described in Table 7.12.

```
#define I2C_MCS_ACK        0x00000008  // Data Acknowledge Enable
#define I2C_MCS_ADRACK     0x00000004  // Acknowledge Address
#define I2C_MCS_STOP       0x00000004  // Generate STOP
#define I2C_MCS_START      0x00000002  // Generate START
#define I2C_MCS_ERROR      0x00000002  // Error
#define I2C_MCS_RUN        0x00000001  // I2C Master Enable
#define MAXRETRIES 5       // number of receive attempts before giving up

uint16_t I2C_Recv2(uint8_t slave){ uint8_t data1,data2;
  int retryCounter = 1;
  do{
    while(I2C0_MCS_R&0x00000001){};   // wait for I2C ready
    I2C0_MSA_R = (slave<<1)&0xFE;     // MSA[7:1] is slave address
    I2C0_MSA_R |= 0x01;               // MSA[0] is 1 for receive
    I2C0_MCS_R = ( I2C_MCS_ACK        // positive data ack
                 | I2C_MCS_START      // no stop, yes start/restart
                 | I2C_MCS_RUN);      // master enable
    while(I2C0_MCS_R&0x00000001){};   // wait for transmission done
    data1 = (I2C0_MDR_R&0xFF);        // MSB data sent first
    I2C0_MCS_R = (I2C_MCS_STOP
                 | I2C_MCS_RUN);      // generate stop, no start
                                      // master enable
    while(I2C0_MCS_R&0x00000001){};   // wait for transmission done
    data2 = (I2C0_MDR_R&0xFF);        // LSB data sent last
    retryCounter = retryCounter + 1;  // increment retry counter
  }                                   // repeat if error
  while(((I2C0_MCS_R&(I2C_MCS_ADRACK|I2C_MCS_ERROR)) != 0)
   && (retryCounter <= MAXRETRIES));
```

```
    return (data1<<8)+data2;        // usually returns 0xFFFF on error
}
```
Program 7.7. I²C reception in single master mode (I2C_xxx).

Figure 7.32 shows a logic analyzer measurement taken with Program 7.7 communicating with a Texas Instruments TMP102 thermometer. The main program calls `I2C_Recv2(0x48);` The first transmission sends the 0x91 (slave address, read) command. It then receives two transmissions, which is the temperature encoded with 0.0625°C resolution.

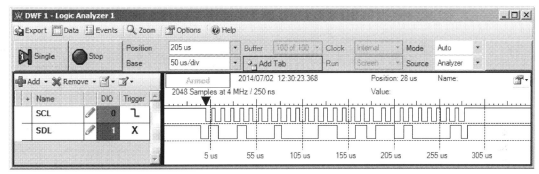

Figure 7.32. Logic analyzer transmission of I2C_Recv2, with one output and two inputs.

I²C can be very difficult to configure. We suggest you observe the SCL and SDL on an oscilloscope. First check the clock rate and second verify the output high and output low voltages are within specification. Look up V_{IH} and V_{IL} of both the microcontroller and the remote sensor. The high voltage measured by the scope should be higher than V_{IH} of both devices. The low voltage measured by the scope should be lower than V_{IL} of both devices.

7.7. Introduction to Universal Serial Bus (USB)

7.7.1. Basic Principles

The Universal Serial Bus (USB) is a host-controlled, token-based high-speed serial network that allows communication between many of devices operating at different speeds. The objective of this section is not to provide all the details required to design a USB interface, but rather it serves as an introduction to the network. There is 650-page document on the USB standard, which you can download from **http://www.usb.org**. In addition, there are quite a few web sites setup to assist USB designers, such as the one titled "USB in a NutShell" at **http://www.beyondlogic.org/usbnutshell/**.

The standard is much more complex than the other networks presented in this chapter. Fortunately, however, there are a number of USB products that facilitate incorporating USB into an embedded system. In addition, the USB controller hardware handles the low-level protocol. USB devices usually exist within the same room, and are typically less than 4 meters from each other. USB 2.0 supports three speeds.

High Speed - 480Mbits/s
Full Speed - 12Mbits/s
Low Speed - 1.5Mbits/s

The original USB version 1.1 supported just full speed mode and a low speed mode. The Universal Serial Bus is host-controlled, which means the host regulates communication on the bus, and there can only be one host per bus. On the other hand, the On-The-Go specification, added in version 2.0, includes a Host Negotiation Protocol that allows two devices negotiate for the role of host. The USB host is responsible for undertaking all transactions and scheduling bandwidth. Data can be sent by various transaction methods using a token-based protocol. USB uses a tiered star topology, using a hub to connect additional devices. A hub is at the center of each star. Each wire segment is a point-to-point connection between the host and a hub or function, or a hub connected to another hub or function, as shown in Figure 7.33. Because the hub provides power, it can monitor power to each device switching off a device drawing too much current without disrupting other devices. The hub can filter out high speed and full speed transactions so lower speed devices do not receive them. Because USB uses a 7-bit address, up to 127 devices can be connected.

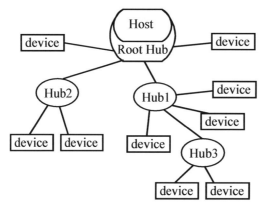

Figure 7.33. USB network topology.

The electrical specification for USB was introduced in Figure 7.10 and Table 7.5, using four shielded wires (+5V power, D+, D- and ground). The D+ and D- are twisted pair differential data signals. It uses **Non Return to Zero Invert** (**NRZI**) encoding to send data with a sync field to synchronize the host and receiver clocks.

USB drivers will dynamically load and unload. When a device plugged into the bus, the host will detect this addition, interrogate the device and load the appropriate driver. Similarly, when the device is unplugged, the host will detect its absence and automatically unload the driver. The USB architecture comprehends four basic types of data transfers:

- Control Transfers: Used to configure a device at attach time and can be used for other device-specific purposes, including control of other pipes on the device.
- Bulk Data Transfers: Generated or consumed in relatively large quantities and have wide dynamic latitude in transmission constraints.
- Interrupt Data Transfers: Used for timely but reliable delivery of data, for example, characters or coordinates with human-perceptible echo or feedback response characteristics.
- Isochronous Data Transfers: Occupy a prenegotiated amount of USB bandwidth with a prenegotiated delivery latency. (Also called streaming real-time transfers).

Isochronous transfer allows a device to reserve a defined about of bandwidth with guaranteed latency. This is appropriate for real-time applications like in audio or video applications. An isochronous pipe is a stream pipe and is, therefore, always unidirectional. An endpoint description identifies whether a given isochronous pipe's communication flow is into or out of the host. If a device requires bidirectional isochronous communication flow, two isochronous pipes must be used, one in each direction.

A USB device indicates its speed by pulling either the D+ or D- line to 3.3 V, as shown in Figure 7.34. A pull-up resistor attached to D+ specifies full speed, and a pull-up resistor attached to D- means low speed. These device-side resistors are also used by the host or hub to detect the presence of a device connected to its port. Without a pull-up resistor, the host or hub assumes there is nothing connected. High speed devices begin as a full speed device (1.5k to 3.3V). Once it has been attached, it will do a high speed chirp during reset and establish a high speed connection if the hub supports it. If the device operates in high speed mode, then the pull-up resistor is removed to balance the line.

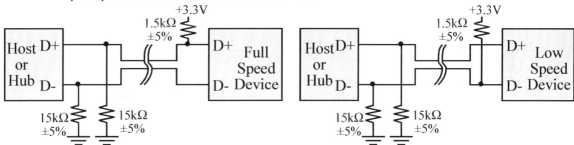

Figure 7.34. Pull-up resistors on USB devices signal specify the speed.

Like most communication systems, USB is made up of several layers of protocols. Like the CAN network presented earlier, the USB controllers will be responsible for establishing the low-level communication. Each USB transaction consists of three packets

- Token Packet (header),
- Optional Data Packet, (information) and
- Status Packet (acknowledge)

The host initiates all communication, beginning with the Token Packet, which describes the type of transaction, the direction, the device address and designated endpoint. The next packet is generally a data packet carrying the information and is followed by a handshaking packet, reporting if the data or token was received successfully, or if the endpoint is stalled or not available to accept data. Data is transmitted least significant bit first. Some USB packets are shown in Figure 7.35. All packets must start with a sync field. The **sync** field is 8 bits long at low and full speed or 32 bits long for high speed and is used to synchronize the clock of the receiver with that of the transmitter. **PID** (Packet ID) is used to identify the type of packet that is being sent, as shown in Table 7.14.

The **address** field specifies which device the packet is designated for. Being 7 bits in length allows for 127 devices to be supported. Address 0 is not valid, as any device which is not yet assigned an address must respond to packets sent to address zero. The **endpoint** field is made up of 4 bits, allowing 16 possible endpoints. Low speed devices, however can only have 2 additional endpoints on top of the default pipe. **Cyclic Redundancy Checks** are performed on the data within the packet payload. All token packets have a 5-bit CRC while data packets have a 16-bit CRC. EOP stands for **End of packet**. **Start of Frame** Packets (SOF) consist of an 11-bit frame number is sent by the host every 1ms ± 500ns on a full speed bus or every 125 μs ± 0.0625 μs on a high speed bus.

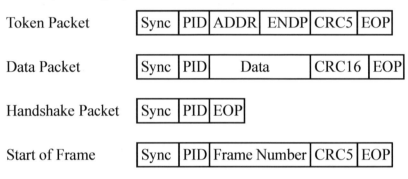

Figure 7.35. *USB packet types.*

Group	PID Value	Packet Identifier
Token	0001	OUT Token, Address + endpoint
	1001	IN Token, Address + endpoint
	0101	SOF Token, Start-of-Frame marker and frame number
	1101	SETUP Token, Address + endpoint
Data	0011	DATA0
	1011	DATA1
	0111	DATA2 (high speed)
	1111	MDATA (high speed)
Handshake	0010	ACK Handshake, Receiver accepts error-free data packet
	1010	NAK Handshake, device cannot accept data or cannot send data
	1110	STALL Handshake, Endpoint is halted or pipe request not supported
	0110	NYET (No Response Yet from receiver)
Special	1100	PREamble, Enables downstream bus traffic to low-speed devices.
	1100	ERR, Split Transaction Error Handshake
	1000	Split, High-speed Split Transaction Token
	0100	Ping, High-speed flow control probe for a bulk/control endpoint

Table 7.14. USB PID numbers.

USB **functions** are USB devices that provide a capability or function such as a Printer, Zip Drive, Scanner, Modem or other peripheral. Most functions will have a series of buffers, typically 8 bytes long. **Endpoints** can be described as sources or sinks of data, shown as `EP0In`, `EP0Out` etc. in Figure 7.36. As the bus is host centric, endpoints occur at the end of the communications channel at the USB function. The host software may send a packet to an endpoint buffer in a peripheral device. If the device wishes to send data to the host, the device cannot simply write to the bus as the bus is controlled by the host. Therefore, it writes data to endpoint buffer specified for input, and the data sits in the buffer until such time when the host sends a IN packet to that endpoint requesting the data. Endpoints can also be seen as the interface between the hardware of the function device and the firmware running on the function device.

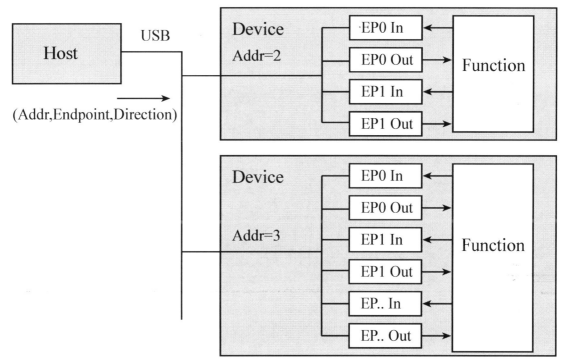

Figure 7.36. USB data flow model.

While the device sends and receives data on a series of endpoints, the client software transfers data through pipes. A **pipe** is a logical connection between the host and endpoint(s). Pipes will also have a set of parameters associated with them such as how much bandwidth is allocated to it, what transfer type (Control, Bulk, Iso or Interrupt) it uses, a direction of data flow and maximum packet/buffer sizes. **Stream Pipes** can be used send unformatted data. Data flows sequentially and has a pre-defined direction, either in or out. Stream pipes will support bulk, isochronous and interrupt transfer types. Stream pipes can either be controlled by the host or device. **Message Pipes** have a defined USB format. They are host-controlled, which are initiated by a request sent from the host. Data is then transferred in the desired direction, dictated by the request. Therefore message pipes allow data to flow in both directions but will only support control transfers.

7.7.2. Modular USB Interface

There are two approaches to implementing a USB interface for an embedded system. In the modular approach, we will employ a USB-to-parallel, or USB-to-serial converter. The modular approach is appropriate for adding USB functionality to an existing system. For about $30, we can buy a converter cable with a USB interface to connect to the personal computer (PC) and a serial interface to connect to the embedded system, as shown in Figure 7.37. The embedded system hardware and software is standard RS232 serial. These systems come with PC device drivers so that the USB-serial-embedded system looks like a standard serial port (COM) to the PC software. The advantage of this approach is that software development on the PC and embedded system is simple. The disadvantage of this approach is none of the power and flexibility of USB is utilized. In particular, the bandwidth is limited by the RS232 line, and the data stream is unformatted. Similar products are available that convert USB to the parallel port. Companies that make these converters include

IOGear Inc.	http://www.iogear.com
Wyse Technology	http://www.wyse.com
D-Link Corporation	http://www.dlink.com
Computer Peripheral Sys, Inc.	http://www.cpscom.com
Jo-Dan International, Inc.	http://www.jditech.com

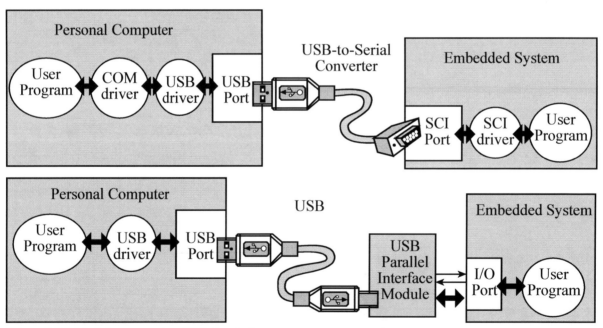

Figure 7.37. Modular approach to USB interfacing.

The second modular approach is to purchase a USB parallel interface module. These devices allow you to send and receive data using parallel handshake protocols similar to the input/output examples in Chapter 3. They typically include a USB-enabled microcontroller and receiver/transmit FIFO buffers. This approach is more flexible than the serial cable method, because both the microcontroller module and the USB drivers can be tailored personalized. In particular, some modules allow you to burn PID and VID numbers into EEPROM. The advantages/disadvantages of this approach are similar to the serial cable, in that the data is unformatted and you will not be able to implement high bandwidth bulk transfers or negotiate for real-time bandwidth available with isochronous data transfers. Companies that make these modules include

Future Tech. Devices Inter. Ltd.	http://www.ftdichip.com/
ActiveWire, Inc.	http://www.activewireinc.com
DLP Design, Inc.	http://www.dlpdesign.com
Elexol Pty Ltd.	http://www.elexol.com

7.7.3. Integrated USB Interface

The second approach to implementing a USB interface for an embedded system is to integrate the USB capability into the microcontroller itself. This method affords the greatest flexibility and performance, but requires careful software design on both the microcontroller and the host. Over the last 15 years USB has been replacing RS232 serial communication as the preferred method for connecting embedded systems to the personal computer. Manufacturers of microcontrollers have introduced versions of the product with USB capability. Every company that produces microcontrollers has members of the family with USB functionality. Examples include the Microchip PIC18F2455, Atmel AT89C5131A, FTDI FT245BM, Freescale MCF51Jx, STMicrosystems STM32F102, Texas Instruments MSP430F5xx, and Texas Instruments TM4C123. Figure 7.38 shows the USB configuration on the EK-TM4C123GXL LaunchPad Evaluation Kit, which is capable of operating as a device, a host or on-the-go (OTG). To use USB populate R25 and R29. The LM4F120 can be only configured as a USB device. The TivaWare® software library has 14 example projects for this evaluation board, including serial port translator, secure digital card, mouse, and keyboard interface.

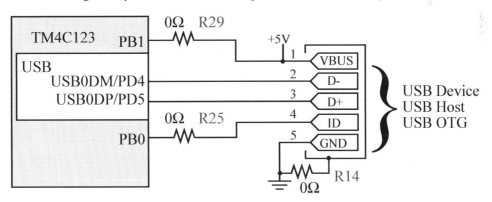

Figure 7.38. The TM4C123 LaunchPad kit supports USB host, device, and OTG modes.

To operate a USB interface at full speed DMA synchronization will be required, so that data is transferred directly from memory to USB output, or from USB input to memory.

7.8. Exercises

7.1 For each term give a definition in 32 words or less.
a) Asynchronous
b) Baud rate
c) Bandwidth
d) Break
e) DCE
f) DTE
g) Even parity
h) Full duplex
i) Frame
j) Framing error
k) Half duplex
l) Mark
m) Negative logic
n) NRZ
o) NRZI
p) Open drain
q) Overrun
r) Positive logic
s) Simplex
t) Space
u) Synchronous

7.2 In 32 words or less, describe the similarities and differences between these pairs of terms
a) Baud rate versus bandwidth
b) Positive logic versus negative logic
c) XON versus XOFF
d) Full duplex versus half duplex
e) DS275 versus MAX232
f) SCI versus SPI
g) NRZ versus NRZI
h) DTE versus DCE

7.3 What fundamental electrical property is used to transfer digital data across a distance?
A) Voltage
D) Frequency
B) Current
E) Phase
C) Energy
F) Wavelength

7.4 Look up in the LM3S/LM4F/TM4C data sheet what the BUSY bit in the UART operates. In particular, when is it set? When is it clear? Can the software clear it? Can it trigger an interrupt?

7.5 In 16 words or less, explain why the UART receiver ISR in Program 5.9 interrupts on both RXRIS and RTRIS. In particular when does RXRIS trigger and when does RTRIS trigger?

7.6 Draw a plot similar to Figure 7.2 for the binary data 00110111.

7.7 Draw a plot similar to Figure 7.2 for the binary data 10011100.

7.8 Consider a serial port operating with a baud rate of 10,000 bits per second. Draw the waveform occurring at the U0Tx output (voltage levels are +3.3 and 0) when the ASCII 'a' (0x61) is transmitted on UART0. The protocol is 1 start, 8 data and 1 stop bit. UART0 is initially idle, and the software writes the 0x61 to **UART0_DR_R** at time=0. Show the U0Tx line before and after the frame, assuming the channel is idle before and after the frame.

7.9 Consider a serial port operating with a baud rate of 1000 bits per second. Draw the waveform occurring at the U1Tx output (voltage levels are +3.3 and 0) when the ASCII 'B' (0x42) is transmitted on UART1. The protocol is 1 start, 8 data and 1 stop bit. UART1 is initially idle, and the software writes the 0x42 to **UART1_DR_R** at time=0. Show the U1Tx line before and after the frame, assuming the channel is idle before and after the frame.

7.10 Assume the UART baud rate is 1000 bits/sec and the protocol is 1 start, 8 data, no parity and 1 stop bit. What is the channel bandwidth? If we used 2 stop bits instead of 1 without changing the baud rate, would the bandwidth be higher, lower or the same?

7.11 The data in Figure 7.39 was measured on a U0Rx serial input, which we think is one frame, but it might be two frames. The serial format is 1-start, 8-bit, and 1-stop bit.
a) What is the baud rate in Case A?

b) What data is being transferred in Case A? Give the number(s) in hexadecimal.
c) What is the baud rate in Case B?
d) What data is being transferred in Case B? Give the number(s) in hexadecimal.

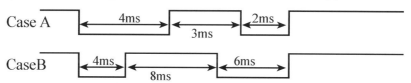

Figure 7.39. Serial transmission for exercise 7.11.

7.12 RXRIS and RTRIS interrupts are armed so that interrupts occur when new data arrives into the LM3S/LM4F/TM4C. Consider the situation like Program 5.9 in which a software FIFO queue is used to buffer data between the ISR and the main program. The ISR reads data from the UART and saves the data by calling `Fifo_Put`. When the main program wants input it calls `UART_InChar`, which in turn calls `Fifo_Get`. Experimental observations show this FIFO is usually empty, and has at most 3 elements. What does it mean? Choose A-F.
  A) The system is CPU bound
  B) Bandwidth could be increased by increasing FIFO size
  C) The system is I/O bound
  D) The FIFO could be replaced by a global variable
  E) The latency is small and bounded
  F) Interrupts are not needed in this system

7.13 A slave device will be interfaced to the master LM3S/LM4F/TM4C using SSI. The timing is shown in Figure 7.40. There are three signals that will be outputs of the LM3S/LM4F/TM4C and inputs to the device (Enable, Clock, and Data). The timing of the external device is shown below. Assuming a Freescale SPI protocol, what SPH, SPO mode should you use?

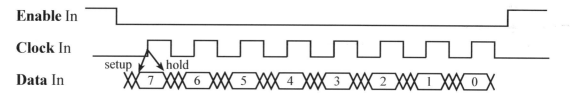

Figure 7.40. Serial transmission for exercise 7.13.

D7.14. Write software that creates the SSI protocol on a regular digital port. More specifically, interface the DAC in Example 7.2 to port pins PD3, PD2, PD1, and PD0. (PD0 is not used).

D7.15. Write software that creates the SSI protocol on a regular digital port. More specifically, interface the ADC in Example 7.3 to port pins PB3, PB2, PB1, and PB0.

D7.16. Write software that creates the SSI protocol on a regular digital port. More specifically, interface the 74HC595 in Example 7.4 to port pins PD3, PD2, PD1, and PD0. (PD0 is not used).

D.7.17 This problem addresses the issue of capacitive loading on a high-speed serial transmission line like SSI. The SSI ports of two microcontrollers are connected with a VERY long cable. We will model this cable as a single resistor in series with a capacitor, as shown in Figure 7.41. For this question, assume an ideal transmitter (output impedance of 0) and an ideal receiver (input impedance of infinity). Let the resistance R be 1 Ω, and the capacitance C be 10 nF. Consider a 3.3-V 100-ns pulse (T = 100 ns) on the output of the transmitter (labeled as V_{in}) as might occur with a 5-Mbps SSI transmission. Derive an equation for V_{out} as a function of time for the first 100 ns. Show your work. Calculate values for V_{out} at time equals 0 and time equals 100 ns. Create a V_{in} versus time sketch similar to the right of Figure 7.41, and add a sketch of V_{out} on this same plot. Show both 0 to time T and time T to time 2T.

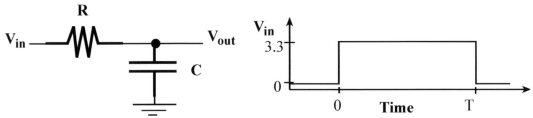

Figure 7.41. Circuit model for exercises D7.17 and D7.18.

D.7.18 Solve exercise D7.17 with R = 10 Ω, C = 100 nF and T = 10 ns. Justify whether or not this system will work?

D7.19 The output of one shift register is connected to the input of a second shift register, as shown in Figure 7.42. The two registers are controlled by the same 50% duty cycle Clock. The period of the clock is t_1. The data is shifted out on the falling edge of Clock. The time t_2 = [10ns min, 200ns max] is the delay from the falling edge of Clock until when the output is valid. The data is shifted into the second register on the rising edge of Clock. The time t_3 = 50ns is the time before the rising edge of Clock that the data must be valid. The time t_4 = 20ns is the time after that same rising edge of Clock that the data must continue to be valid. What is the smallest t_1 clock period that reliably transverses data from one shift register to the other?

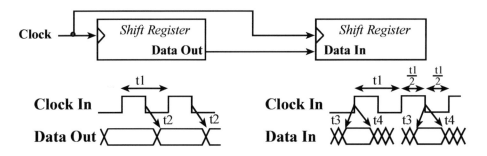

Figure 7.42. Circuit and timing for exercise D7.19.

7.9. Lab Assignments

Lab 7.1 The overall objective is to redesign the Lab 6.2 or 6.3 data acquisition system to employ two microcontrollers. The slave microcontroller will perform the data acquisition (position or temperature), and the master microcontroller will interface to the PC. The two microcontrollers will be linked using their SSI ports. The master microcontroller will fetch data from the slave, and transmit it to the PC, using UART interrupt synchronization. The user will interact with a PC running PuTTY.

Lab 7.2 The overall goal is to develop an interrupting UART device driver that implements fixed-point input/output. The fixed-point constant is 0.001. The full-scale range is from 0 to 65.534. The **Fix_InDec** function should provide for flexible operation. For example "50.5" is returned as 50500, and "1.4595" is returned as 1460. The operator creates the input by typing on the keyboard, and the output is a number passed back as the return parameter of the function. Notice that **Fix_InDec** rounds the input to the closest fixed-point result (e.g., 1.4595 rounds to 1.460 and 1.4604 also rounds to 1.460). Some numbers like 1.2345678 might be considered illegal because they cause overflow of intermediate results. In the comments of your software, please discuss why you chose your particular implementation method over the other available choices. Please handle the backspace character, allowing the operator to erase characters. In particular, you are free to use iterative or recursive algorithms. You are free to modify the prototypes as well as handle illegal inputs in any way you feel is appropriate. You must detect illegal input, but you have a choice as to how your system responds to the illegal input. One possibility for handling an illegal number would be to return 65535, which you could define as an illegal number. A second possibility for when an illegal number is typed is to output an error message, and require the operator to enter the number again. For example, 50 is output as "0.050" and 12345 is output as "12.345".

Lab 7.3. The overall goal is to interface a DS1620 temperature controller to the microcontroller. Create the functions to initialize, set high temperature, set low temperature and read current temperature. When reading the temperature you will have to start a conversion and wait for the conversion to be complete. Perform experiments to determine the accuracy of the temperature measurement. Interface three LEDs to TH TL TCOM outputs, and perform experiments to verify the temperature controller in the DS1620 is operating properly.

Lab 7.4. The overall objective of this lab is to design, implement and test an output port expander. You will use three I/O pins of the microcontroller, and four 74HC595 shift registers. You will design hardware and software that supports four 8-bit output ports. The output ports do not need to be readable. Measure how long it takes for the microcontroller to perform outputs to all 32 bits.

Lab 7.5 The overall objective of this lab is to design, implement and test an input port expander. You will use thee I/O pins of the microcontroller, and four 74HC165 shift registers. You will design hardware and software that supports four 8-bit input ports. The input ports do not need to be latched by an external signal. Measure how long it takes for the microcontroller to perform inputs from all 32 bits.

Lab 7.6 The objective of this lab is to design a digital clock using a DS1307 external clock chip. The first step is to interface the clock chip to the microcontroller using an I^{2}C network. The second step is to design low-level drivers to allow the microcontroller to send and receive data from the DS1307. The next software layer includes functions like `SetTime` `FormatTime` and `ReadTime`. The highest level is a main program that implements a digital clock using an LED or LCD display. 2, 3, or 4 momentary switches will be used to control the operation of the digital clock.

Lab 7.7 The objective of this lab is to design a digital thermometer using a DS1631A external thermometer chip. The first step is to interface the thermometer chip to the microcontroller using an I^{2}C network. The second step is to design low-level drivers to allow the microcontroller to send and receive data from the DS1631A. The next software layer includes functions like `SetMode` and `ReadTemperature`. The highest level is a main program that implements a digital thermometer using an LED or LCD display. 2 or 3 momentary switches will be used to control the operation of the digital thermometer.

8. Analog Interfacing

Chapter 8 objectives are to:

- Design analog amplifiers and filters
- Study building blocks for data acquisition including DAC and ADC
- Discuss the functionality of the ADC on the TM4C

Most embedded systems include components that measure and/or control real world parameters. These real world parameters, like position, speed, temperature, and voltage, usually exist in a continuous or analog form. Therefore, the design of an embedded system involving these parameters rarely uses only binary or digital logic. Rather, we will often need to amplify, filter and eventually convert to digital form these signals. In this chapter we will develop the analog circuit building blocks used in the design of data acquisition systems and control systems. A computer engineer interested in the field of embedded systems will find more job opportunities if he or she can develop microcontroller skills along with analog circuit design skills.

8.1. Resistors and Capacitors

8.1.1. Resistors

As engineers, we use resistors and capacitors for many purposes. The resistor or capacitor type is defined by the manufacturing process, the materials used, and the testing performed. The performance and cost of these devices vary significantly. For example, a 5% 1/4 watt carbon resistor costs less than 1¢, while a 0.01% thin film or wirewound resistor may cost $6. It is important to understand both our circuit requirements and the resistor parameters so that we match the correct resistor type to each application yielding an acceptable cost-performance balance. We must specify in our technical drawings the device type and tolerance (e.g., 1% metal film), so that your prototype can be effectively manufactured. The characteristics of various resistor types are shown in Table 8.1.

Type	Range	Tolerance	Temperature coef	Max Power
Carbon composition	1 Ω to 22 MΩ	5 to 10 %	200 to 700 ppm/°C	1 W
Carbon film	1 Ω to 22 MΩ	1 to 10 %	200 to 1500 ppm/°C	2 W
Metal film	0.01 Ω to 68 MΩ	>0.05 %	2 to 300 ppm/°C	1 W
Thick and thin film	0.001 Ω to 100 GΩ	0.1 to 20 %	5 to 1000 ppm/°C	100 W
Wirewound	0.005 Ω to 167 kΩ	>0.0005 %	10 to 900 ppm/°C	30 W

Table 8.1. General specification of various types of resistor components.

Carbon composition resistors are manufactured with hot-pressed carbon granules. Various amounts of filler are added to achieve a wide range of resistance values. Film resistors are made by depositing pure metals, such as nickel, or an oxide film, such as tin-oxide, onto an insulating ceramic rod or substrate. A wirewound resistor, is made by winding a thin metal alloy wire such as Nichrome onto an insulating ceramic in the form of a spiral helix.

The least expensive type of a through-hole resistor is carbon film. We add them to digital circuits as +3.3-V pull-ups. In order to improve the accuracy and stability of our precision analog circuits, we will use resistors with a lower tolerance and smaller temperature coefficient. For most applications 1% thick film or metal film resistors will be sufficient to build our analog amplifier circuits. For surface mount construction we use either thick film or thin film resistors, which come in a wide range of sizes and tolerances. Wirewound resistors are manufactured by twisting a very long very thin wire like a spring. The wire is coiled up and down a shaft in such a way to try and cancel the inductance. Since some inductance remains, wirewound resistors should not be used for high frequency (above 1 MHz) applications.

Observation: All resistors produce white (thermal) noise.

Observation: Wirewound resistors do not generate 1/f noise, whereas the other types do.

8.1.2. Capacitors

Similarly, capacitors come in a wide variety of sizes and tolerances. Polarized capacitors operate best when only positive voltages are applied. Nonpolarized or bipolar capacitors operate for both positive and negative voltages. We select a capacitor based on the following parameters: capacitance value, polarized/nonpolarized, maximum voltage level, tolerance, leakage current (resistance), temperature coefficient, useful frequency response, and temperature range. The maximum voltage rating is important for high voltage circuits, but is of lesser importance for embedded microcontroller systems. Table 8.2 compares various capacitor types we could use in our circuit. Ceramic capacitors are labelled with three digits **xyz**; the capacitance is $xy*10^z$ pF. For example, 473 means $47*10^3$ pF = 47 nF.

Type	Range	Tolerance	Temp coef	Applications
C0G/NP0 Ceramic	0.1pF to 0.47µF	±1%	Excellent	Analog signal processing
X7R/Y5V Ceramic	10 pF to 47µF	-20 to +80%	Good	Decoupling
Polypropylene	100pF to 10µF	Excellent	Good	Analog signal processing
Tantalum	0.1µF to 220µF	±10%	Poor	Bulk decoupling
Electrolytic	0.47µF to 0.01F	±20%	Ghastly	Bulk decoupling

Table 8.2. General specification of various types of capacitor components (Analog Engineer's Pocket Reference, Art Kay and Tim Green editors, Texas Instruments, 2014).

We will use capacitors for two purposes in our microcontroller-based embedded systems. First, we will place them on the DC power lines to filter the supply voltage to our circuits. A voltage supply typically will include ripple, which is added noise on top of the DC voltage level. There are two physical locations to place the supply filters. The first location is at the entry point of the supply voltage onto the circuit board. There are two approaches to this board-level filter. If the supply noise is mostly voltage ripple, then two capacitors in parallel can be used. The large amplitude polarized capacitor (e.g., 1 to 47 µF electrolytic or tantalum) will remove low frequency large amplitude voltage noise, and the nonpolarized capacitor (e.g., 0.01 to 0.47 µF ceramic) will remove high frequency voltage noise. Two different types of capacitors are used because they are effective (i.e., behave like a capacitor) at different frequencies. The ∏ filter (CLC) is very effective in removing current spike noise. The CLC parameters depend on the amplitude of the current ripple. The inductor in Figure 8.1 can be a ferrite bead. At DC the bead is essentially a short current. The ferrite bead increases both its real and reactive impedance at high frequencies. The bead should be selected to have large impedance at the digital clock frequency. The Panasonic EXC-ELDR25C has a DC resistance of 0.08 Ω, can conduct 7A DC, but has an 80-Ω impedance at 24 MHz. The inductance prevents noise ripples at the bus clock frequency from affecting analog circuits also powered with the same +3.3V supply.

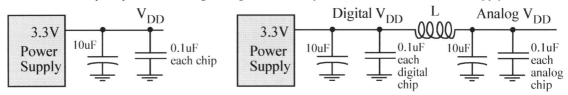

Figure 8.1. DC supply filters.

In addition to the board-level supply filter, we will add bypass capacitors at the supply pins of each chip. It will be important to place these capacitors as close to the pin as physically possible as illustrated in Figure 8.2. A nonpolarized capacitor (e.g., 0.01 to 0.1 µF ceramic) will smooth the supply voltage as seen by the chip. Placing the capacitor close to the chip prevents current surges from one chip from affecting the voltage supply of another.

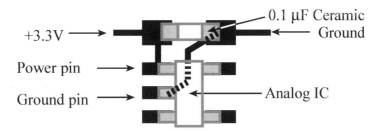

Figure 8.2. PCB layout positioning the bypass capacitor close to the chip.

The second application of capacitors in our embedded systems will be in the linear analog circuits of the low pass filter, the high pass filter, the derivative circuit, and the integrator circuit. For these applications we will select a nonpolarized capacitor even if the signal amplitude is always positive. In addition, we usually want a low-tolerance, low-leakage capacitor to improve the accuracy of the linear analog circuit. Ceramic capacitors are a low-cost medium-quality choice for analog circuit design. They come in three tolerances, see Table 8.2. The best ceramic is **C0G**, which has a 1 to 10% tolerance and a temperature coefficient of 30ppm/°C or ±0.3% over -55 to 125°C. The middle grade is **X7R** ceramic, which has a 5 to 20% tolerance and a temperature coefficient of ±15% over -55 to 125°C. The lowest cost ceramic is **Z5U**, which has a 20% tolerance and a temperature coefficient of 22 to -56% over 10 to 86°C. We can use Z5U for bypass capacitors on power lines, but we should use either **C0G** or **X7R** for analog filters.

Part number	Type	Tolerance	Cost
06035A102FAT2A	C0G	±1%	$0.467
06035C102JAT2A	X7R	±5%	$0.129
06035C102KAT2A	X7R	±10%	$0.019

Table 8.3. Cost of AVX ceramic capacitors: 1000 pF, 50 V, surface mount 0603 package (2017 prices on www.digikey.com for quantity 10).

We can model a capacitor as a series combination of capacitance, resistance, and inductance. A 10-μF ceramic capacitor has an equivalent series resistance (ESR) of 4 mΩ. The tantalum capacitor has an ESR of 0.1Ω and the electrolytic cap has an ESR if 1Ω. In this analysis all three have an equivalent series inductance (ESL) of 200pH. Figure 8.3 shows the effect of ESL and ESR on capacitor frequency response. The solid line is an ideal capacitance with impedance of $1/(2\pi \cdot f \cdot C)$. Above 5 MHz the inductance is seen in the ceramic capacitor.

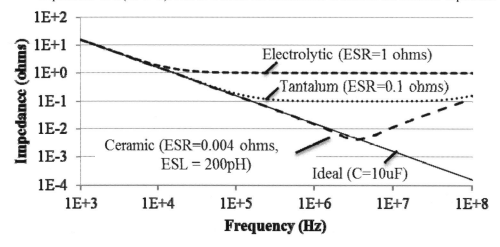

Figure 8.3. Impedance of a 10 uF capacitor as a function of frequency $(R + 1/(2\pi f C_j) + 2\pi f L_j)$.

8.2. Op Amps

While the design of analog electronics is not an explicit objective of this book, we will include a brief discussion of analog circuit design issues often related to embedded systems. For example, low cost, small size, single voltage supply, and low power are four characteristics typical of embedded systems. Other factors to consider are reliability, noise, frequency response, availability, and temperature range.

Over a dozen manufacturers produce thousands of analog integrated circuits. Manufacturers include: Analog Devices, Avago Technologies, Cirrus Logic, Fairchild, Honeywell, Intersil, Linear Technology, Maxim IC, Microchip, NXP Semiconductors, ON Semiconductor, Silicon Laboratories, Texas Instruments, and Toshiba. The fact that there are so many op amps available makes the choice confusing. The manufacturers do publish a selection guide of their products to assist in finding an appropriate part from the ones they produce.

Table 8.4 lists performance parameters of four op amps, and the choice of these particular devices typify the kinds of op amps used in an embedded system, but the selection of these particular four is not meant as a recommendation. Many op amps come in a variety of package sizes and are available in 1, 2, or 4 op amps per package. We will use rail-to-rail op amps, like the ones in Table 8.4 to design analog circuits that run on a single +3.3 V supply, sorted from lowest power to highest speed. The OPA330 and MAX494 are low-power devices.

The op amp **rails** are its two power supplies, $-V_s$ and $+V_s$. A typical op amp powered with ±12-V rails operates properly when its inputs and outputs are between -10 to +10 V. A **rail-to-rail** op amp operates in a linear fashion for output voltages all the way from the minus rail to the plus rail. Some op amps operate "rail-to-rail" for both the inputs and output.

Single op amp	OPA330			OPA350
Double op amp	OPA2330	MAX492	TLC2272	OPA2350
Quad op amp	OPA4330	MAX494	TLC2274	OPA4350
Description	Low power	Low power	Medium speed	High speed
K, Open loop gain	100 dB	108 dB	104 dB	122 dB
R_{cm}, Input impedance	4 pF		$10^{12}\,\Omega \parallel 8pF$	$10^{13}\,\Omega \parallel 6.5pF$
R_{diff}, Input impedance	2 pF	2 MΩ	$10^{12}\,\Omega \parallel 8pF$	$10^{13}\,\Omega \parallel 2.5pF$
V_{os}, Offset voltage	50 μV	0.5 mV	3 mV	0.5 mV
I_{os}, Offset current	500 pA	6 nA	100 pA	10 pA
I_b, Bias current	1 nA	60 nA	100 pA	10 pA
e_n, Noise density	55 nV/√Hz	25 nV/√Hz	50 nV/√Hz	5 nV/√Hz
f_1, Gain*bandwidth product	350 kHz	500 kHz	2.18 MHz	38 MHz
dV/dt, Slew rate	0.16 V/μs	0.2 V/μs	3.6 V/μs	22 V/μs
$+V_s$, Voltage supply	1.5 to 5.5 V	2.7 to 6 V	0 to 5 or ±5 V	2.7 to 5.5 V
I_s, Supply current	35 μA	170 μA	3 mA	7.5 mA
Cost	$2	$6	$2	$4

Table 8.4. Parameters for various rail-to-rail CMOS op amps used in this chapter (with $-V_s$ grounded).

Checkpoint 8.1: What is the relationship between bandwidth and supply current?

Observation: The op amp should be much faster than the signal you are trying to process.

8.2.1. Ideal Op Amp Model

We will begin our discussion of op amps with the **ideal op amp**. For simple analog circuits using high quality devices, the ideal model will be sufficient. With an ideal op amp, the output voltage, V_{out}, is linearly related to the difference between the input voltages,

$$V_{out} = K \cdot (V_y - V_x)$$

where the gain, K, is a very large number, as shown in Figure 8.4.

Voltage ranges of the inputs and outputs are bounded by the supply voltages, $+V_s$ and $-V_s$. Op amp circuits found in many traditional analog design textbooks are powered with ±12-V supplies. On the other hand, it will reduce system cost to power our embedded systems with a single voltage. More specifically, if the microcontroller operates with +3.3-V supply, we will run the analog circuits also on +3.3 V. In particular, we will use rail-to-rail op amps and set $+V_s$ to +3.3 V and $-V_s$ to ground.

No matter how we power the analog circuits, we assume the input and output voltages will exist between $-V_s$ and $+V_s$. The input currents into the op amp are very small. Because the input impedance of the op amp is large, the ideal model assumes I_x and I_y are zero.

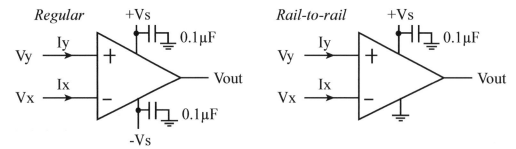

Figure 8.4. *Regular op amp and a single supply rail-to-rail op amp.*

If a **feedback resistor** is placed between the output and the negative terminal of the op amp, then this **negative feedback** will select an output such that V_x is very close to V_y. In the ideal model, we let $V_x=V_y$. One way to justify this behavior is to recall that $V_{out}=K\bullet(V_y-V_x)$. Since K is very large, the only way for V_{out} to be between $-V_s$ and $+V_s$ is for V_x to be very close to V_y.

We can design a **threshold detector** using positive or no feedback. A threshold detector has a binary output (true or false) depending on whether or not an input signal is greater than a threshold value. Positive feedback or no feedback drives V_{out} to equal $-V_s$ or $+V_s$. If a feedback resistor is placed between the output and the positive terminal of the op amp, then this feedback will saturate the output to either the positive or negative supply. With no feedback, the output will also saturate. In both cases the output will saturate to the positive supply if $V_y>V_x$, and to the negative supply if $V_x>V_y$. Positive feedback can be used to create hysteresis.

8.2.2. Realistic Op Amp Model

Although the input impedance of an op amp is large, it is not infinite and some current enters the input terminals. Figure 8.5 illustrates the definition of input impedance. We define the **common-mode input impedance**, R_{cm}, as the common mode voltage divided by the common mode current. We define the **differential input impedance**, R_{diff} as the differential voltage divided by the differential current. These parameters vary considerably from op amp to op amp. The CMOS and FET devices have very large input impedances.

The next realistic parameter we will define is **open-loop output impedance**. When the op amp is used without feedback, the op amp output impedance is defined as the open circuit voltage divided by the short circuit current. The output impedance is a measure of how much current the op amp can source or sink. The open loop output impedance of the TLC2274 is 140Ω.

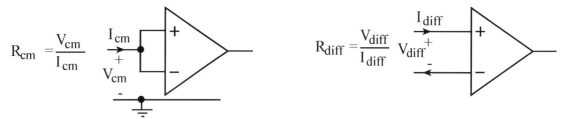

Figure 8.5. Definition of op amp input impedance.

> **Checkpoint 8.2:** The MAX494 has an output short-circuit current of 30 mA. Assuming an output of 3.3 V, what is its output impedance?

As illustrated in Figure 8.6, the op amp **offset voltage** V_{os} is defined as the voltage difference between V_y and V_x which yields an output of zero. The **offset current** I_{os} is defined as the current difference between inputs. There will be an output error equal to the offset voltage times the gain of amplifier. Similarly, the offset current creates an offset voltage through the resistors of the circuit. Some op amps provide the capability to add an external potentiometer to nullify the two offset errors. The use of a null offset pot increases manufacturing costs and incurs a labor cost to adjust it periodically. Therefore, the overall system cost may be reduced by using more expensive op amps that do not require a null offset pot. Alternatively, if the gain and offset are small enough not to saturate the output, then the offset error can be corrected in software by adding/subtracting an appropriate calibration constant.

The op amp **bias current** I_b is defined as the common current coming out of both V_y and V_x. We can reduce the effect of bias current by selecting resistors in our circuit in order to equalize the effective impedance to ground from the two input terminals.

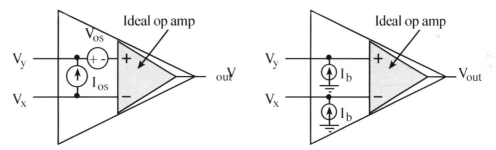

Figure 8.6. Definition of op amp offset voltage, offset current, and bias current.

> **Checkpoint 8.3:** Consider the situation where a MAX494 is used to create an analog amplifier with gain 100. What will be the output error due to offset voltage?

> **Checkpoint 8.4:** Why can't a TLC2274 is be used to create an amplifier with gain 1000?

The **input voltage noise**, V_n, arises from the thermal noise generated in the resistive components within the op amp. Due to the white-noise process, the magnitude of the noise is a function of the bandwidth (BW) of the analog circuit. This parameter varies quite a bit from op amp to op amp. To calculate the RMS amplitude of the input voltage noise, we need to calculate, $V_n = e_n \cdot \sqrt{(BW)}$. To reduce the effect of noise we can limit the bandwidth of the analog system using an analog low pass filter or buy a better op amp. The **output voltage noise** will be the input voltage noise multiplied by the gain of the circuit.

There are two approaches to defining the transient response of our analog circuits. In the frequency domain we can specify the frequency and phase response. In the time domain, we can specify the step response. For most simple analog circuits designed with op amps, the frequency response depends on the op amp performance and the analog circuit gain. If the unity-gain op amp frequency response is f_1, then the frequency response at gain, G, will be f_1/G. The **bandwidth**, BW, is defined as the frequency at which the gain (V_{out}/V_{in}) drops to 0.707 of the original. The voltage gain in decibels G_{db} is related to the voltage gain in V/V. When V_{out}/V_{in} equals 0.707 ($\sqrt{\frac{1}{2}}$), the $G_{db} = -3$ dB.

$$G_{db} = 20 \cdot \log_{10} \frac{V_{out}}{V_{in}}$$

The output **slew rate** is the maximum slope that the output can generate. Slew rate is important if the circuit must response quickly to changes in input (e.g., a sensor detecting discrete events). Alternatively, bandwidth is important if the circuit is responding to a continuously changing input (e.g., audio and video).

> **Checkpoint 8.5:** Consider the situation where a MAX494 is used to create an analog amplifier with a gain of 100. What will be the bandwidth of this circuit? Given this bandwidth, what will be the RMS output voltage noise?

When we consider the performance of a linear amplifier normally we specify the voltage gain, input impedance and output impedance. These three parameters can be lumped into a single parameter, A_{db}, called the **power gain**. Let V_{in}, R_{in} be the inputs and V_{out}, R_{out} be the outputs of our amplifier. The input and output powers are $P_{in} = V^2_{in}/R_{in}$ and $P_{out} = V^2_{out}/R_{out}$ respectively. Then the power gain in decibels has voltage gain and impedance components.

$$A_{db} = 10 \cdot \log_{10} \frac{P_{out}}{P_{in}} = 20 \cdot \log_{10} \frac{V_{out}}{V_{in}} + 10 \cdot \log_{10} \frac{R_{in}}{R_{out}}$$

8.2.3. Op Amp Circuit Design

The open loop or saturated mode performance of a OPA2350 op amp can be studied by looking at the simple circuit in Figure 8.7. We use saturated mode to create a **threshold detector**. When the input, V_1, is above the reference, V_t, then the output, V_2, saturates to +3.3 V. Similarly, when V_1 is below V_t then V_2 saturates to 0 V.

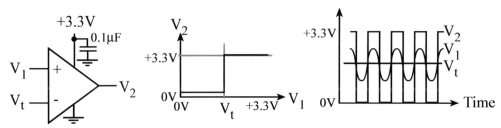

Figure 8.7. *A positive logic voltage comparator using a OPA2350 op amp.*

The OPA2350 will operate on a supply range of 2.7 to 5.5 V. Notice that it will create a digital output when $+V_s$ is 3.3V and $-V_s$ is ground. The input voltages on input pins of the OPA2350 must be between -0.3 and +3.6V. The short circuit output current of the OPA2350 is 80 mA. The OPA2350 slew rate is 22V/µsec, with a settling time of 500ns.

A wide range of analog circuits can be designed by following these simple design rules.

1. Choose quality components.

It is important to use op amps with good enough parameters. Similarly we should use low tolerance resistors and capacitors. On the other hand, once the preliminary prototype has been built and tested, then we could create alternative designs with less expensive components. Because a working prototype exists, we can explore the cost/performance tradeoff.

2. Negative feedback is required to create a linear mode circuit.

As mentioned earlier, the negative feedback will produce a linear input/output response. In particular, we place a resistor between the negative input terminal and the output (Figure 8.8). A corollary to this is we almost never place a resistor between the positive input terminal and the output.

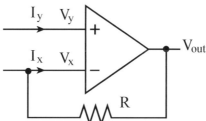

Figure 8.8. Negative feedback is created by placing a resistor between the - input and output.

3. Assume no current flows into the op amp inputs.

Since the input impedance of the op amp is large compared to the other resistances in the circuit, we can assume that $I_x=I_y=0$.

4. Assume negative feedback equalizes the op amp input voltages.

If the analog circuit is in linear mode with negative feedback, then we can assume $V_x=V_y$.

5. Choose resistor values in the 1 kΩ to 1 MΩ range.

In order to have the resistors in the circuit be much larger than the output impedance of the op amp and much smaller than the input impedance of the op amp, we choose resistors in the 1 kΩ to 1 MΩ range. If we can, it is better to restrict values to the 10 kΩ to 100 kΩ range. If we choose resistors below 1 kΩ, then currents will increase. If the currents get too large the batteries will drain faster and the op amp may not be able to source or sink enough current. As the resistors go above 1 MΩ, the white noise increases, the current errors (I_{os}, I_b, I_n) become more significant. In addition, low tolerance resistors are expensive in sizes above 2 MΩ.

6. The analog circuit bandwidth depends on the gain and the op amp performance.

Let the unity-gain op amp frequency response be f_1, and let the analog circuit gain be G. The frequency response or bandwidth of the analog circuit, BW, will be f_1/G. Design the circuit bandwidth 10 times faster than the signal you are trying to process.

7. Equalize the effective resistance to ground at the two op amp input terminals.

To study the bias currents, consider all other voltage sources as shorts to ground, and all other current sources as open circuits. Adjust the resistance values in the circuit so that the impedance from the + terminal to ground is the same as the impedance from the - terminal to ground. In this way, the bias currents will create a common mode voltage, which will not appear at the op amp output because of the common mode rejection of the op amp.

8. The impedance is the voltage divided by the current.

If the analog circuit has a single input voltage, then the input impedance, Z_{in}, is simply the input voltage divided by the input current as shown in the top of Figure 8.9. The output impedance, Z_{out}, is the open circuit output voltage divided by the short circuit output current.

If the input stage of the analog circuit is a differential amplifier with two input voltages, then we can specify the common mode input impedance, Z_{cm}, and the differential mode input impedance, Z_{diff}. See bottom of Figure 8.9.

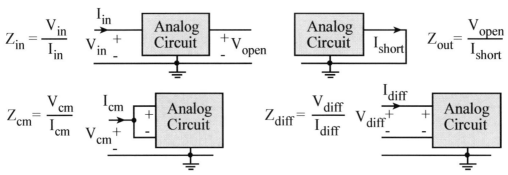

Figure 8.9. *Definition of input and output impedance for an analog circuit with a single input and input impedance for an analog circuit with two inputs.*

> **Observation:** In most cases, the differential mode input impedance of the analog circuit will be the differential mode input impedance of the op amp.

9. Match input impedances to improve CMRR.

If the input stage of the analog circuit is a differential amplifier with two input voltages, a very important performance parameter is called the **common mode rejection ratio** (CMRR). It is assumed that the signal of interest is the differential voltage, whereas common mode voltages are considered noise. I.e., V_{diff} will be the desired signal and V_{cm} is the added noise. The CMRR is defined to be the ratio of the differential gain divided by the common mode gain. See Figure 8.10. Therefore a differential amplifier with a large CMRR will pass the signal and reject the noise. In decibels, it is calculated as

$$CMRR = 20 \cdot \log_{10}(G_{diff}/G_{cm})$$

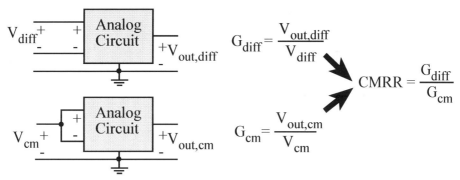

Figure 8.10. Definition of common mode rejection ratio (CMRR).

Remember, V_{diff} is the signal and V_{cm} is noise. There are two sets of impedances we must match to achieve a good CMRR. Each amplifier input has a separate input impedance to ground, shown as Z_{in1} and Z_{in2} in Figure 8.11. To improve CMRR, we make Z_{in1} equal to Z_{in2}. Similarly, the signal source (V_{diff}) has a separate output impedance, Z_{out1} and Z_{out2}. Again, we try to make Z_{out1} equal to Z_{out2}. Unfortunately, if Z_{in1} does not equal Z_{in2} or if Z_{out1} does not equal Z_{out2} then a common mode signal (e.g., added noise in the cable) will appear as a differential signal to the analog circuit and thus be present in the output.

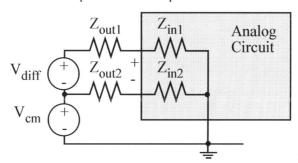

Figure 8.11. Circuit model for improving common mode rejection ratio (CMRR).

10. Rail-to-rail considerations.
When designing with rail-to-rail op amps, we must guarantee that the voltages at all input and output pins of the op amps never go outside the range of $-V_s$ to $+V_s$.

We will begin with an inverting amplifier (Figure 8.12), which is a simple linear mode analog circuit. The gain is the R_2/R_1 ratio. Notice that the gain response is independent of R_3. Thus, we can choose R_3 to be the parallel combination of $R_1 \| R_2$ so that the effect of the bias currents is reduced. Because I_y is zero, V_y is also zero. Because of negative feedback, V_x equals V_y. Thus, V_x equals zero too. Because V_x is zero, I_{in} is V_{in}/R_1 and I_2 is $-V_{out}/R_2$. Because I_x is zero, I_{in} equals I_2. Setting I_{in} equal to I_2 yields

$$V_{out} = -(R_2/R_1) \cdot V_{in}$$

The input impedance (Z_{in}) of this circuit (defined as V_{in}/I_{in}) is R_1. If the circuit were built with an OPA227, which has a gain bandwidth product of 8 MHz, the bandwidth of this circuit will be 8 MHz divided by the gain, R_2/R_1.

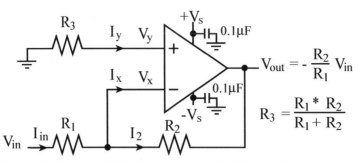

Figure 8.12. Inverting amplifier, built with OP227, powered with +12 and -12V.

Common error: This low input impedance of Z_{in} may cause loading on the previous analog stage.

Observation: The inverting amplifier input impedance is independent of the op amp input impedance.

The negative feedback will reduce the output impedance of the amplifier, Z_{out}, to a value much less than the output impedance of the op amp itself, R_{out}. To calculate Z_{out}, we first determine the open circuit voltage

$$V_{open} = -(R_2/R_1) \cdot V_{in}$$

We next determine the short circuit current, I_{short}. This means we consider what would happen if the output were shorted to ground. If the output is shorted, the circuit is no longer in feedback mode, and V_x will not equal V_y. In fact, V_x will be a simple voltage divider from V_{in} through R_1 and R_2 to ground,

$$V_x = V_{in} \cdot R_2/(R_1+R_2)$$

Because of the large open-loop gain, the ideal output will attempt to become

$$V_o = K \cdot (V_y - V_x) = -K \cdot V_{in} \cdot R_2/(R_1+R_2)$$

The short circuit current will be a function of the ideal output voltage, and the output resistance of the op amp,

$$I_{short} = V_o/R_{out}$$

The output impedance of the circuit is defined to be the open circuit voltage divided by the short circuit current, which for this inverting amplifier is

$$Z_{out} = V_{open}/I_{short} = R_{out} \cdot (R_2+R_1)/(K \cdot R_1)$$

Observation: The output impedance of analog circuits using op amps with negative feedback is typically in the mΩs.

A **mixed-signal design** includes both analog and digital components. The classic approach to combining analog and digital circuits is to power the analog system with a low noise ±12 V power supply, maintain separate analog and digital grounds, and connect the analog ground to the digital ground only at the ADC. One of the limitations of the ADC built into a microcontroller is that analog signals extending beyond the 0 to +3.3 V range will permanently damage the microcontroller. One approach to allowing signed analog voltages, while still using a single voltage supply and protecting the microcontroller is to create an analog ground that is at a different potential from the digital ground. For our microcontroller systems, which are powered with a +3.3 V supply, we will create an analog ground that is at 1.5 V relative to the digital ground. Thus, analog signals ranging from -1.5 to +1.5 V are actually 0 to +3 V relative to the microcontroller digital ground. The first step to implementing this approach is to use an analog reference chip, like the ones shown in Table 8.5, to create a low-noise +1.50 V signal (the analog ground). The second step is to connect power to the analog circuits with $-V_s$ set to digital ground, and the $+V_s$ set to the +3.3 V supply. We will use rail-to-rail op amps that operate on 3.3 V power. The last step is to replace all connections to analog ground with the low-noise +1.50 V reference voltage. Figure 8.13 shows the inverting amplifier, redesigned from Figure 8.12 to operate on a single +3.3 V supply. The signals V_{in} and V_{out} are allowed to vary from 0 to +3 V relative to digital ground, but relative to the analog ground, these signals will vary from -1.5 to +1.5 V. An adjustable shunt voltage references like the LM4041 also can be used to create constant analog voltages.

Part	Voltage (V)	±Accuracy (mV)
AD1580, AD589, REF1004, MAX6120, LT1034, LM385	1.2	1 to 15
MAX6101, REF3312, ADR1581	1.25	2
MAX6108	1.6	3
ADR420, ADR520, REF191, MAX6191, LT1790, LM4120	2.048	1 to 10
AD580, REF03, REF1004, MAX6192, MAX6225, LT1389, LM336	2.5	1 to 75
AD1583, ADR530, ADR423, REF193, MAX6163, LT1461, LM4120	3	1.5 to 10
ADR366, REF196, MAX6331, LT1461, LM3411, LM4120	3.3	4 to 10
AD1584, ADR540, ADR292, REF198, MAX6241, LT1790, LM4040	4.096	2 to 8

Table 8.5. Parameters of various precision reference voltage chips.

Common Error: Precision reference chips do not provide much output current and should not be used to power other chips.

We can analyze the response of the circuit in Figure 8.13 by assuming an ideal op amp. Since there are no currents into the inputs, the voltage at the positive input will be 1.50 V. Because of negative feedback, the voltage at the negative input will also be 1.50 V. The current through R_1 will be $(V_{in}-1.5)/R_1$. The current through R_2 will be $(1.5-V_{out})/R_2$. Since there are no currents into the inputs of the op amp, these two currents are equal,

$$(V_{in} - 1.5)/R_1 = (1.5 - V_{out})/R_2$$

or $V_{out} = 1.5 - (V_{in} - 1.5) R_2/R_1$

Define the analog signals relative to 1.5V, i.e., $V'_{in} \equiv (V_{in}-1.5)$ and $V'_{out} \equiv (V_{out}-1.5)$. The circuit in Figure 8.13 implements a negative gain inverter.

$$V'_{out} = -(R_2/R_1) V'_{in}$$

Observation: It is important in this scheme to separate the digital and analog grounds avoiding direct connections between the two grounds.

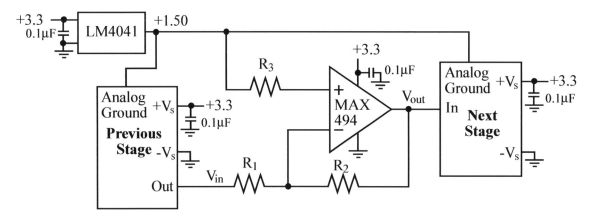

Figure 8.13. Inverting amplifier with an effective -1.5 V to +1.5 V analog signal range.

The second linear mode circuit we will study is the noninverting amplifier, as shown in Figure 8.14. The gain is $1+R_2/R_1$. The noninverting amplifier cannot have a gain less than 1. Just like the inverting amp, the gain response is independent of R_3. So, we choose R_3 to be the parallel combination $R_1 \| R_2$ so that the effect of the bias currents is reduced.

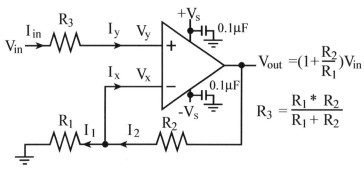

Figure 8.14. Noninverting amplifier.

Because I_y is zero, V_y equals V_{in}. Because of negative feedback V_x equals V_y. Thus, V_x equals V_{in} too. Calculating currents we get, I_1 is V_{in}/R_1 and I_2 is $(V_{out}-V_{in})/R_2$. Because I_x is zero, I_1 equals I_2. Setting I_1 equal to I_2 yields

$$V_{out} = (1 + R_2/R_1) \cdot V_{in}$$

Using the simple op amp rules I_y is zero, so the input impedance (Z_{in}) of this circuit (defined as V_{in}/I_{in}) would be infinite. In this situation we specify the amplifier input impedance to be the op amp input impedance. If the circuit were built with an OPA350, which has a gain bandwidth product of 38 MHz, the bandwidth of this circuit will be 38 MHz divided by the gain.

Checkpoint 8.6: If the noninverting amplifier in Figure 8.14 were built with an OPA350, what would be the input impedance of the amplifier?

The calculation of the output impedance of this amp follows the same approach as the inverting amp,

$$Z_{out} = R_{out} \cdot (R_2 + R_1)/(K \cdot R_1)$$

The following design process can be used to build any analog circuit in the form of

$$V_{out} = A_1 V_1 + A_2 V_2 + \ldots + A_n V_n + B$$

where $A_1 A_2 \ldots A_n$ B are constants and $V_1 V_2 \ldots V_n$ are input voltages. The circuit will be designed with one op amp beginning with the boiler plate shown in Figure 8.15.

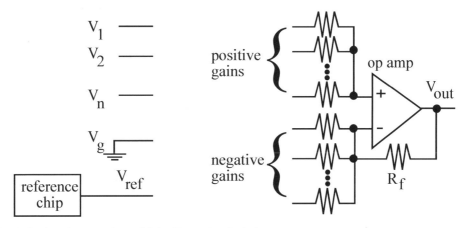

Figure 8.15. Boiler plate circuit model for linear circuit design.

The **first step** is to choose a reference voltage from available reference voltage chips, like ones shown in the Table 8.5. Some of the manufacturers that produce voltage references are Analog Devices, Texas Instruments, Linear Technology, and Maxim. The parameters to consider when choosing a voltage reference are voltage, package configuration, accuracy, temperature coefficient, and power. In particular, let V_{ref} be this reference voltage.

> **Common Error:** If you use resistor divider from the power supply to create a voltage constant, then the power supply ripple will be added directly to your analog signal.

The **second step** is to rewrite the design equation in terms of the reference voltage, V_{ref}. In particular, we make $A_{ref} = B/V_{ref}$.

$$V_{out} = A_1 V_1 + A_2 V_2 + \ldots + A_n V_n + A_{ref} V_{ref}$$

where $A_1 A_2 \ldots A_n$ A_{ref} are constants and $V_1 V_2 \ldots V_n$ are input voltages. The **third step** is to add a ground input to the equation. Ground is zero volts ($V_g = 0$), but it is necessary to add this ground so that the sum of all the gains is equal to one.

$$V_{out} = A_1 V_1 + A_2 V_2 + \ldots + A_n V_n + A_{ref} V_{ref} + A_g V_g$$

Choose A_g such that $A_1 + A_2 + \ldots + A_n + A_{ref} + A_g = 1$

In other words, let $A_g = 1 - (A_1 + A_2 + \ldots + A_n + A_{ref})$

The **fourth step** is to choose a feedback resistor, R_f, in the range of 10 kΩ to 1 MΩ. The larger the gains, the larger the value of R_f must be. Then calculate input resistors to create the desired gains. In particular,

$$|A_1| = R_f/R_1 \quad \text{so } R_1 = R_f/|A_1|$$
$$|A_2| = R_f/R_2 \quad \text{so } R_2 = R_f/|A_2|$$
$$|A_n| = R_f/R_n \quad \text{so } R_n = R_f/|A_n|$$
$$|A_{ref}| = R_f/R_{ref} \quad \text{so } R_{ref} = R_f/|A_{ref}|$$
$$|A_g| = R_f/R_g \quad \text{so } R_g = R_f/|A_g|$$

Observation: We will get a low-cost solution if we choose standard resistor values, as shown in Tables 9.1, 9.2, 9.3, or 9.4.

The **last step** is to build the circuit. If the gain is positive, then the input resistor is connected to the positive terminal of the op amp. Conversely, if the gain is negative, then the input resistor is connected to the negative terminal of the op amp. The feedback resistor, R_f, will always be connected from the negative input to the output.

Example 8.1. Design an analog circuit with two inputs and one output. The first input varies from 1 to 1.5 V and the second input varies from 0.75 to 1 V. The output will be equally sensitive to the two inputs, so the gain of the second input will be twice the gain of the first input. The output should range from 0 to 3 V. In particular, build a circuit with a transfer function of $V_{out} = 3V_1 + 6V_2 - 7.5$.

Solution: The **first step** is to choose a reference voltage. The REF3312 +1.25-V voltage reference will be used. The **second step** is to rewrite the design equation in terms of the reference voltage. Notice that $A_{ref} = -7.5/1.25 = -6$.

$$V_{out} = 3V_1 + 6V_2 - 6V_{ref}$$

The **third step** is to add a ground input to the equation so that the sum of all the gains is equal to one. Notice, $A_g = 1 - (3 + 6 - 6) = -2$

$$V_{out} = 3V_1 + 6V_2 - 6V_{ref} - 2V_g$$

The **fourth step** is to choose a feedback resistor, $R_f = 60$ kΩ. This value is a common multiple of the gains: 6,3,2. Then calculate input resistors to create the desired gains.

$$R_1 = R_f/3 = 20 \text{ kΩ}$$
$$R_2 = R_f/6 = 10 \text{ kΩ}$$
$$R_{ref} = R_f/6 = 10 \text{ kΩ}$$
$$R_g = R_f/2 = 30 \text{ kΩ}$$

The **last step** is to build the circuit, as shown in Figure 8.16. The positive gain inputs are connected to the plus input of the op amp and the negative gain inputs are connected to the minus input of the op amp input.

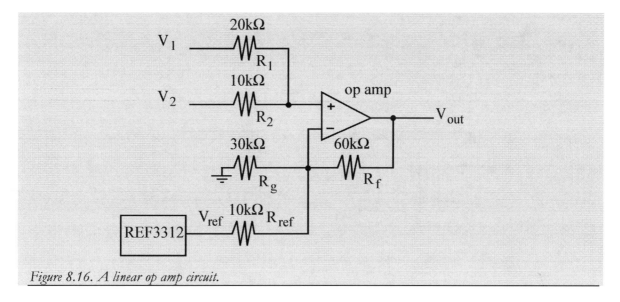

Figure 8.16. *A linear op amp circuit.*

Tables 9.1, 9.2, 9.3 and 9.4 list standard resistance values. We designing analog circuits we need to select from these specific values to reduce cost and simplify manufacturing.

The **instrumentation amp** will amplify a differential voltage, $V_{out} = G \cdot (V_2 - V_1)$. We use instrumentation amps in applications that require a large gain (above 100), a high input impedance, and a good common mode rejection ratio. An instrumentation amp can be built using three high quality op amps as shown in Figure 8.17.

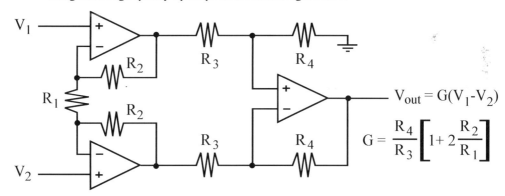

Figure 8.17. *Instrumentation amplifier made with three op amps.*

Observation: In order to achieve quality performance with a 3-op-amp instrumentation amp circuit, we must use precision resistors and quality op amps.

Common Error: If you use a potentiometer in place of one the gain resistors in the above circuit, then fluctuations in the potentiometer resistance that can occur with temperature, vibration, and time will have a strong effect on the amplifier gain.

Because of the wide range of applications that require instrumentation amplifiers, chip manufacturers have developed integrated solutions. In many cases we can achieve higher performance at reduced cost by utilizing an integrated circuit. The gain is selected by external jumpers or external resistors. The MAX4460, AD627, and INA122 are single-supply rail-to-rail instrumentation amps. The AD627 and INA122 have the same pins and the same formula for calculating gain, as shown in Figure 8.18.

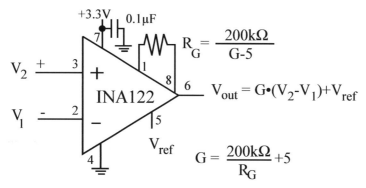

Figure 8.18. Integrated instrumentation amplifier made with INA122 or AD627.

Common Error: If you use a potentiometer as the R_G gain resistors in the above circuit, then fluctuations in the potentiometer resistance that can occur with temperature, vibration, and time will have a strong effect on the amplifier gain.

We can use a voltage comparator to detect events in an analog waveform. The input voltage range is determined by the analog supply voltages of the comparator. The output is takes on two values, shown an V_h and V_l in Figure 8.19. A comparator with **hysteresis** has two thresholds, V_{t+} and V_{t-}. In both the positive and negative logic cases the threshold (V_{t+} or V_{t-}) depends on the present value of the output.

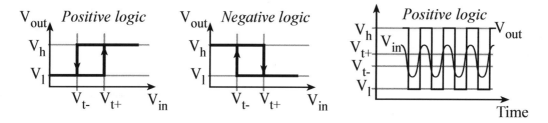

Figure 8.19. Input/output response of voltage converters with hysteresis.

Hysteresis prevents small noise spikes from creating a false trigger.

Performance Tip: In order to eliminate false triggering, we select a hysteresis level ($V_{t+} - V_{t-}$) greater than the noise level in the signal.

In Figure 8.20, a rail-to-rail op amp is used to design a voltage comparator. Since the output swings from 0 to 3.3 V, it can be connected directly to an input pin of the microcontroller. On the other hand, since +3.3 and 0 are used to power the op amp, the analog input must remain in the 0 to +3.3 V range. The hysteresis level is determined by the amplitude of the output and the $R_1/(R_1+R_2)$ ratio. If the output is at 0V, the voltage at the +terminal is $V_{in}*R_2/(R_1+R_2)$. The output switches when the voltage at the +terminal goes above 1.65. By solving for $V_{in}*200k/(10k+200k)=1.65$, we see V_{in} must go above +1.73 for the output to switch. Similarly, if the output is at +3.3 V, the voltage at the +terminal can be calculated as $V_{in}+(3.3-V_{in})*R_1/(R_1+R_2)$. The output switches back when the voltage at the +terminal goes below 1.65. By solving for $V_{in}+(3.3-V_{in})*R_1/(R_1+R_2)=1.65$, we see V_{in} go below +1.57 before the +terminal of the op amp falls below 1.65 V. In linear mode circuits we should not use the supply voltage to create voltage references, but in a saturated mode circuit, power supply ripple will have little effect on the response.

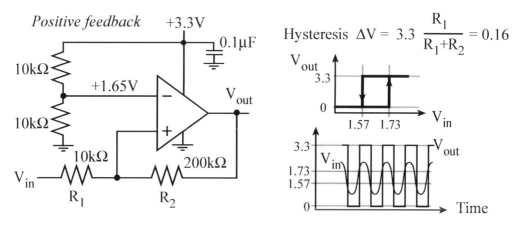

Figure 8.20. *A voltage comparator with hysteresis using a rail to rail op amp.*

In some medical and industrial applications we need to design analog instrumentation that is isolated from earth ground. In an industrial setting, isolation is one way to reduce noise pickup from large EM fields produced by heavy machinery. In medical applications we need to protect the patient from potentially dangerous microshocks, Thus, the medical instrument must be isolated. There are three approaches to isolation, as shown in Figure 8.21. In the first approach, shown at the top of Figure 8.21, an analog isolation barrier is created between the preamp and amp. This was the original approach used in analog instruments before the advent of mixed analog-digital systems. It is expensive, bulky and introduces a very large transfer error. It is not appropriate for embedded applications that use a microcontroller. In the second approach, we use digital isolation. The 6N139 optical isolator is an effective low-cost mechanism to implement digital isolation. This is the most common approach used for new designs when a hard connection between the data acquisition system and building is required. It is fast, small, cheap, and will not introduce errors. The third approach runs the entire system with batteries. This is a very attractive approach due to the availability of high-quality low-power LED/LCD displays and wireless networks, such as Bluetooth, ZigBee, and 802.11b.

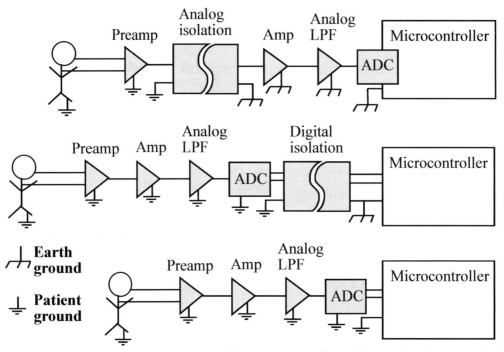

Figure 8.21. Analog isolation, digital isolation, and battery-powered all provide protection from microshocks.

8.3. Analog Filters

8.3.1. Simple Analog Filters

We will use a low-pass filter to remove unwanted high frequency signals. We can add a capacitor in parallel with the feedback resistor in the inverting amplifier to create a simple one pole low-pass filter. The impedance of a resistor, R_2 in parallel the capacitor, C is a function of frequency, see Figure 8.22.

$$Z = \frac{R_2}{1 + j\omega R_2 C}$$

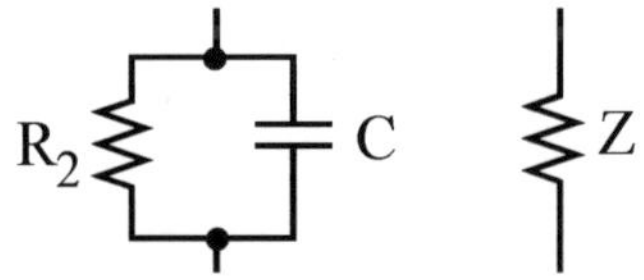

Figure 8.22. A parallel combination of a resistor and capacitor make a frequency-dependent impedance.

In Figure 8.23, the feedback path has both a resistor and capacitor. Therefore the gain of the circuit is $-Z/R_1$, which exhibits low-pass behavior. The cutoff frequency is defined to be the frequency at which the gain drops to 0.707 of its original value. In this simple low-pass filter, the cutoff frequency, f_c, is $1/(2\pi R_2 C)$. The gain drops off at high frequencies.

$$V_{out}/V_{in} = (-R_2/R_1)*(1/(1+j\omega R_2 C))$$

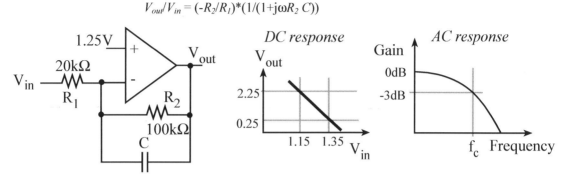

Figure 8.23. One pole low-pass analog filter.

We classify this low-pass filter as one pole, because the transfer function has only one pair of poles in the s-plane. One pole low-pass filters have a gain versus frequency response of

$$\left|\frac{V_{out}}{V_{in}}\right| = \frac{-R_2}{R_1}\sqrt{\frac{1}{1+(f/f_c)^2}}$$

We will use a high-pass filter (HPF) to remove unwanted low frequency signals. Sound and biopotentials are examples of signals where the low frequency components are not wanted. If we place a capacitor in series it will block low frequencies. One of the difficulties in high-pass filters for embedded systems is the fact that we often power the analog electronics with a single supply. Because we cannot process negative voltages, we must use an analog ground different from the digital ground when building high-pass filters, as illustrated in Figure 8.24. The filter on the left is a simple approach. The middle one is low cost and not appropriate for high fidelity systems. The one on the right is appropriate for differential signals. Each of these filters has a one-pole HPF shape, with $2\pi f = \omega$, and $1/f_c = 2\pi RC$.

$$H(\omega) = \frac{j\omega RC}{1 + j\omega RC} \qquad \text{or} \qquad |H(\omega)| = \frac{\omega RC}{\sqrt{1+\omega^2 R^2 C^2}} = \frac{f/f_c}{\sqrt{1+(f/f_c)^2}}$$

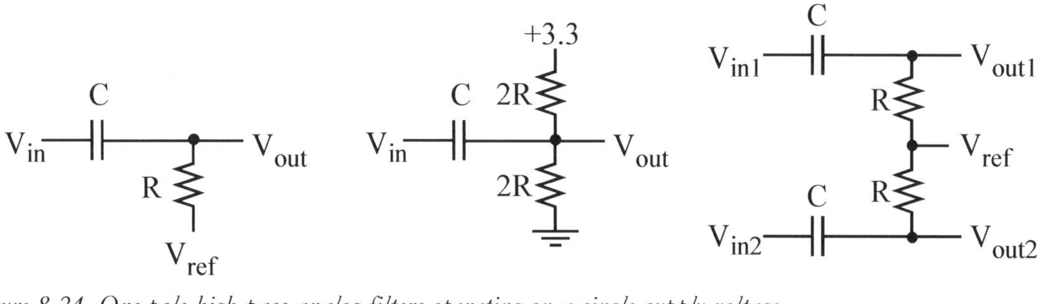

Figure 8.24. One pole high-pass analog filters operating on a single supply voltage.

8.3.2. Butterworth Filters

Higher order analog filters can be designed using multiple capacitors. One of the advantages of the two-pole Butterworth analog filter is that as long as the capacitors maintain the 2/1 ratio, the analog circuit will be a Butterworth filter. Figure 8.25 is a template for a Salen-Key form of the filter. The design steps for the two-pole Butterworth low-pass filter are as follows:

1) Select the cutoff frequency, f_c
2) Divide the two capacitors by $2\pi f_c$ (let C_{1A}, C_{2A} be the new capacitor values)
   $C_{1A} = 141.4\mu F/2\pi f_c$
   $C_{2A} = 70.7\mu F/2\pi f_c$
3) Locate two standard value capacitors (with the 2/1 ratio) with the same order of magnitude as the desired values. We can create capacitors with a 2/1 ratio using three capacitors of the same value. Let C_{1B}, C_{2B} be these standard value capacitors, let x be this convenience factor
   $C_{1B} = C_{1A}/x$
   $C_{2B} = C_{2A}/x$
4) Adjust the resistors to maintain the cutoff frequency
   $R = 10k\Omega \cdot x$
5) V_{ref} can be set to analog ground, e.g., 1.50V or 0V.

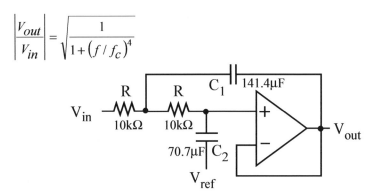

Figure 8.25. Two-pole Butterworth low-pass analog filter (see the file LPF.xls).

The analog filters in this section all require low leakage, high accuracy and low temperature coefficient capacitors like C0G ceramic.

Performance Tip: If you choose standard value resistors near the desired values, you will save money and the circuit will still be a Butterworth filter. The only difference is that the cutoff frequency will be slightly off from the original specification.

We can use a similar approach to design a two-pole Butterworth high-pass filter. Figure 8.26 is a template for a Salen-Key form of the HPF. The design steps for the two-pole Butterworth high-pass filter are as follows:

1) Select the cutoff frequency, f_c
2) Divide the two capacitors by $2\pi f_c$ (let C_A be the new capacitor values)
   $C_A = 10\mu F/2\pi f_c$
3) Locate a standard value capacitor with the same order of magnitude as the desired value. Let C_B, be this standard value, let x be this convenience factor

$C_B = C_A/x$

4) Adjust the two resistors to maintain the cutoff frequency
   $R_1 = 70.7\text{k}\Omega \cdot x$ and $R_2 = 141.4\text{k}\Omega \cdot x$
5) V_{ref} should be set to analog ground, e.g., 1.50V.

$$\left|\frac{V_{out}}{V_{in}}\right| = \sqrt{\frac{(f/f_c)^4}{1+(f/f_c)^4}}$$

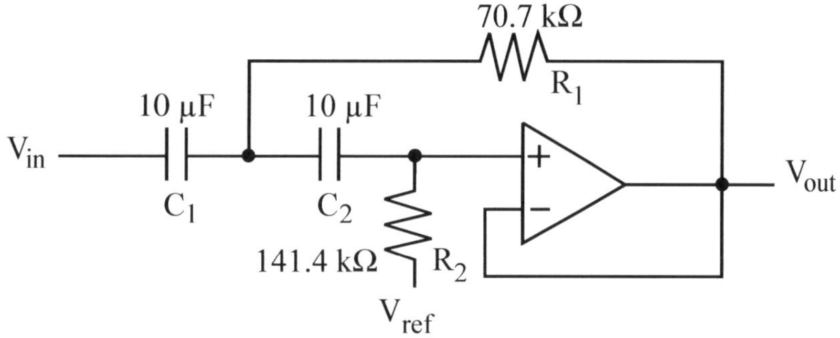

Figure 8.26. Two-pole Butterworth high-pass analog filter.

Many analog IC manufacturers provide design tools. **FilterPro** is a free design tool from Texas Instruments you can use to design analog filters (www.ti.com). Using these design tools you will be able to create filters much better than the ones presented in this book.

8.4. Digital to Analog Converters

An **analog signal** is one that is continuous in both amplitude and time. Neglecting quantum physics, most signals in the world exist as continuous functions of time in an analog fashion (e.g., voltage, current, position, angle, speed, force, pressure, temperature, and flow etc.) In other words, the signal has an amplitude that can vary over time, but the value cannot instantaneously change. To represent a signal in the digital domain we must approximate it in two ways: amplitude quantizing and time quantizing. From an amplitude perspective, we will first place limits on the signal restricting it to exist between a minimum and maximum value (e.g., 0 to +3V), and second, we will divide this amplitude range into a finite set of discrete values. The **range** of the system is the maximum minus the minimum value. The **precision** of the system defines the number of values from which the amplitude of the digital signal is selected. Precision can be given in number of alternatives, binary bits, or decimal digits. The **resolution** is the smallest change in value that is significant. Figure 8.27 shows a temperature waveform (solid line), with a corresponding digital representation sampled at 1 Hz and stored as a 5-bit integer number with a range of 0 to 31 °C. Because it is digitized in both amplitude and time, the digital samples (individual dots) in Figure 8.27 must exist at an intersection of grey lines. Because it is a time-varying signal (mathematically, this is called a function), we have one amplitude for each time, but it is possible for there to be 0, 1, or more times for each amplitude.

The second approximation occurs in the time domain. Time quantizing is caused by the finite sampling interval. For example, the data are sampled every 1 second in Figure 8.27. In practice we will use a periodic timer to trigger an analog to digital converter (ADC) to digitize information, converting from the analog to the digital domain. Similarly, if we are converting from the digital to the analog domain, we use the periodic timer to output new data to a digital to analog converter (DAC). The **Nyquist Theorem** states that if the signal is sampled with a frequency of f_s, then the digital samples only contain frequency components from 0 to ½ f_s. Conversely, if the analog signal does contain frequency components larger than ½ f_s, then there will be an **aliasing** error during the sampling process. Aliasing is when the digital signal appears to have a different frequency than the original analog signal.

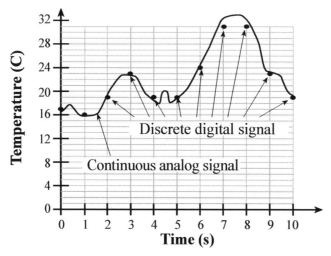

Figure 8.27. An analog signal is represented in the digital domain as 5-bit discrete samples.

8.4.1. DAC Operation and Performance Measures

A DAC converts digital signals into analog form as illustrated in Figure 8.28. Although one can interface a DAC to a regular output port, most DACs are interfaced using high-speed synchronous protocols. The DAC output can be current or voltage. Additional analog processing may be required to filter, amplify or modulate the signal. We can also use DACs to design variable gain or variable offset analog circuits.

The DAC **precision** is the number of distinguishable DAC outputs (e.g., 4096 alternatives, 12 bits). The DAC **range** is the maximum and minimum DAC output. The DAC resolution is the smallest distinguishable change in output. The **resolution** is the change in output that occurs when the digital input changes by 1. The units of range and resolution are in volts or amps depending on whether the output is voltage or current.

Range(volts) = Precision(alternatives) • Resolution(volts)

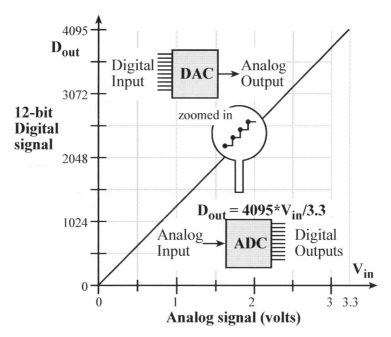

Figure 8.28. A 12-bit DAC provides analog output. A 12-bit ADC provides analog input.

The DAC **accuracy** is (Actual - Ideal) / Ideal where Ideal is referred to the National Institute of Standards and Technology (NIST). One can choose the full scale **range** of the DAC to simplify the use of fixed-point math. For example, if an 8-bit DAC had a full scale range of 0 to 2.55 volts, then the resolution would be exactly 10 mV. This means that if the DAC digital input were 123 (decimal), then the DAC output voltage would be 1.23 volts.

A DAC **gain error** is a shift in the slope of the V_{out} versus digital input static response. A DAC **offset error** is a shift in the V_{out} versus digital input static response. The DAC transient response has three components: delay phase, slewing phase, ringing phase. During the delay phase, the input has changed but the output has not yet begun to change. During the slewing phase, the output changes rapidly. During the ringing phase, the output oscillates while it stabilizes. For purposes of **linearity**, let m,n be digital inputs, and let $f(n)$ be the analog output of the DAC, see Figure 8.29. One quantitative measure of linearity is the correlation coefficient of a linear regression fit of the $f(n)$ responses. If Δ is the DAC resolution, it is linear if

$f(n+1)-f(n) = f(m+1)-f(m) = \Delta$ for all n, m

The DAC is **monotonic** if

$\text{sign}(f(n+1)-f(n)) = \text{sign}(f(m+1)-f(m))$ for all n, m

Conversely, the DAC is **nonlinear** if

$f(n+1)-f(n) \neq f(m+1)-f(m)$ for some n, m

Practically speaking all DACs are nonlinear, but the worst nonlinearity is nonmonotonicity. The DAC is **nonmonotonic** if

$\text{sign}(f(n+1)-f(n)) \neq \text{sign}(f(m+1)-f(m))$ for some n, m

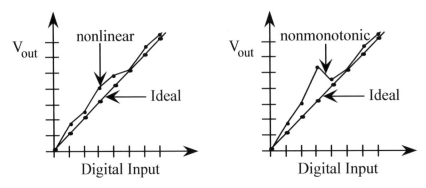

Figure 8.29. *Nonlinear and nonmonotonic DACs.*

Example 8.2. Design a 2-bit binary-weighted DAC with a range of 0 to +3.3V using resistors.

Solution: We begin by specifying the desired input/output relationship of the 2-bit DAC. There are two possible solutions depending upon whether we want a resolution of 0.825 V or 1.1 V, as shown as V_1 and V_2 in Table 8.6. Both solutions are presented in Figure 8.30.

N	Q_1	Q_0	V_1 (V)	V_2 (V)
0	0	0	0.000	0.0
1	0	3.3	0.825	1.1
2	3.3	0	1.650	2.2
3	3.3	3.3	2.475	3.3

Table 8.6. Specifications of the 2-bit DAC.

Assume the output high voltage (V_{OH}) of the microcontroller is 3.3 V, and its output low voltage (V_{OL}) is 0. With a binary-weighted DAC, we choose the resistor ratio to be 2/1 so Q_1 bit is twice as significant as the Q_0 bit, as shown in Figure 8.30. Considering the circuit on the right, if both Q_1 and Q_0 are 0, the output V_2 is zero. If Q_1 is 0 and Q_0 is +3.3V, the output V_2 is determine by the resistor divider network

$$Q_0 \;\; 3.3V \;\; 20k\Omega \;\; V_2 \;\; 10k\Omega \;\; Q_1$$

which is 1.1V. If Q_1 is +3.3V and Q_0 is 0, the output V_2 is determine by the network

$$Q_1 \;\; 3.3V \;\; 10k\Omega \;\; V_2 \;\; 20k\Omega \;\; Q_0$$

which is 2.2V. If both Q_1 and Q_0 are +3.3V, the output V_2 is +3.3V. The output impedance of this DAC is approximately 20 kΩ, which means it cannot source or sink much current.

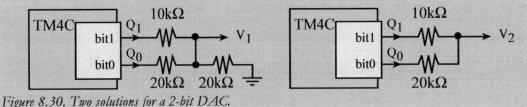

Figure 8.30. *Two solutions for a 2-bit DAC.*

Example 8.3. Design a 6-bit DAC with a range of 0 to +3.3V using E96 1% standard resistors.

Solution: 1% error is of course 1 part out of 100, while 6-bit precision will be only 1 part out of 64. So we expect it will be possible to build a 6-bit DAC with 1% parts. We begin the design by specifying the desired input/output relationship of the 6-bit DAC. All zeros will map to $V_{out}=0$, and all ones will map to $V_{out}=3.3\text{V}$. The exponential weighting of the bits corresponds to the basis elements in a 6-bit number. Let $b_5, b_4, b_3, b_2, b_1, b_0$ be the 6-bit DAC input, the desired performance is

$$V_{out} = 3.3\text{V}*(32*b_5 + 16*b_4 + 8*b_3 + 4*b_2 + 2*b_1 + b_0)/63$$

One set of E96 standard resistors with values approximating the desired exponential weighting is shown on the left of Figure 8.31. The performance of the DAC is shown on the left. This DAC remains monotonic even if the resistors are varied by ±1%.

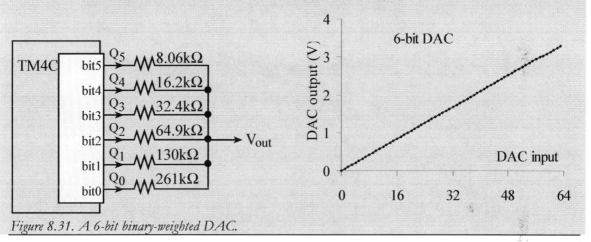

Figure 8.31. A 6-bit binary-weighted DAC.

It is not feasible to construct a DAC with more than 8 bits using the binary weighted technique for three reasons. First, if one chooses the resistor values from the practical 10kΩ to 1MΩ range, then the maximum precision would be 1MΩ/10kΩ, which equals 100 or about 7 bits. The second problem is that it would be difficult to avoid nonmonotonicity because a small percentage change in the small resistor (e.g., the one causing the largest gain) would overwhelm the effects of the large resistor (e.g., the one causing the smallest gain.) For example, if you tried to add a 7th bit to the DAC in Figure 8.31 using a 523 kΩ 1% resistor, the DAC could become nonmonotonic. Third, the summing DAC includes the errors, uncertainty and noise in the digital power supply. To address all three of these limitations, the R-2R ladder is used. It is practical to build resistor networks such that all the resistances are equal. To create a 2R component, we use two resistors in series. Resistance errors will change all resistors equally. This type of error affects the slope, V_{fs}, but not the linearity or the monotonicity. In Figure 8.32 each of the three digital inputs (bit2, bit1, bit0) controls a current switch. When the digital signal is true, the reference voltage is applied to the ladder. When the digital signal is false, that position on the ladder is a virtual ground (0V). Using a reference voltage instead of the V_{OH} of the microcontroller greatly reduces the noise in the output.

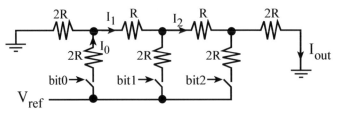

Figure 8.32. 3-bit unsigned R-2R DAC.

To analyze this circuit we will consider the three basis elements (1, 2, and 4). If these three cases are demonstrated, then the law of superposition guarantees the other five will work. When one of the digital inputs is true then V_{ref} is connected to the R-2R ladder, and when the digital input is false, then the connection is grounded. See Figure 8.33.

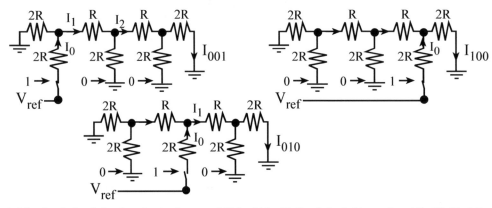

Figure 8.33. Analysis of the three basis elements {100, 010, 001} of the 3-bit unsigned R-2R DAC.

In each of the three test cases, the current across the active switch is $I_0 = V_{ref} / (3R)$. This current is divided by 2 at each branch point. I.e., $I_1 = I_0/2$, and $I_2 = I_1/2$. Current injected from the lower bits will be divided more times. Since each stage divides by two, the exponential behavior is produced. An actual DAC is implemented with a current switch rather than a voltage switch. Nevertheless, this simply circuit illustrates the operation of the R-2R ladder function. When the input is 001, V_{ref} is presented to the left. The effective impedance to ground is $3R$, so the current injected into the R-2R ladder is $I_0 = V_{ref} / (3R)$. The current is divided in half three times, and $I_{001} = V_{ref} / (24R)$.

When the input is 010, V_{ref} is presented in the middle. The effective impedance to ground is still $3R$, so the current injected into the R-2R ladder is $I_0 = V_{ref} / (3R)$. The current is divided in half twice, and $I_{010} = V_{ref} / (12R)$.

When the input is 100, V_{ref} is presented on the right. The effective impedance to ground is once again $3R$, so the current injected into the R-2R ladder is $I_0 = V_{ref} / (3R)$. The current is divided in half once, and $I_{100} = V_{ref} / (6R)$.

Using the Law of Superposition, the output voltage is a linear combination of the three digital inputs, $I_{out} = (4b_2 + 2b_1 + b_0)V_{ref} / (3R)$. A current to voltage circuit is used to create a voltage output. To increase the precision one simply adds more stages to the R-2R ladder.

Many manufacturers, like Analog Devices, Texas Instruments, Sipex and Maxim produce DACs. These DACs have a wide range of performance parameters and come in many configurations. The following paragraphs discuss the various issues to consider when selecting a DAC. Although we assume the DAC is used to generate an analog waveform, these considerations will generally apply to most DAC applications.

Precision/range/resolution. These three parameters affect the quality of the signal that can be generated by the system. The more bits in the DAC the finer the control the system has over the waveform it creates. As important as this parameter is, it is one of the more difficult specifications to establish a priori. A simple experimental procedure to address this question is to design a prototype system with a very high precision (e.g., 12, 14, 16, or 20 bits.) The software can be modified to use only some of the available precision. For example, the 12-bit Max5353 hardware developed in Example 7.2, can be reduced to 4, 8, or 10 bits using the following functions. The bottom bits are set to zero, instead of shifting so the rest of the system will operate without change.

```
void DAC_Out4(uint16_t code){
    DAC_Out(code&0xFF00);}    // ignore bottom 8 bits

void DAC_Out8(uint16_t code){
    DAC_Out(code&0xFFF0);}    // ignore bottom 4 bits

void DAC_Out10(unsigned int code){
    DAC_Out(code&0xFFFC);}    // ignore bottom 2 bits
```
Program 8.1. Software used to test how many bits are really needed.

Multiple versions of the software (e.g., 4-bit, 8-bit, 10-bit, and 12-bit DAC) are used to see experimentally the effect of DAC precision on the overall system performance. Figure 8.34 illustrates how DAC precision affects the quality of the generated waveform.

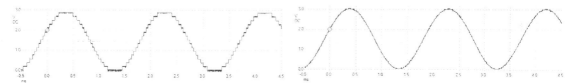

Figure 8.34. The waveform on the left uses a 4-bit DAC, while on one on the right uses a 12-bit DAC.

Channels. Even though multiple channels could be implemented using multiple DAC chips, it is usually more efficient to design a multiple channel system using a multiple channel DAC. Some advantages of using a DAC with more channels than originally conceived are future expansion, automated calibration, and automated testing. A multiple channel DAC allows you to update all channels at the same time.

Configuration. DACs can have voltage or current outputs. Current output DACs can be used in a wide spectrum of applications (e.g., adding gain and filtering), but do require external components. DACs can have internal or external references. An internal reference DAC is easier to use for standard digital input/analog output applications, but the external reference DAC can often be used in variable gain applications (multiplying DAC). Sometimes the DAC generates a unipolar output, while other times the DAC produces bipolar outputs.

Speed. There are a couple of parameters manufacturers use to specify the dynamic behavior of the DAC. The most common is settling time, another is maximum output rate. When operating the DAC in variable gain mode, we are also interested in the gain/bandwidth product of the analog amplifier. When comparing specifications reported by different manufacturers it is important to consider the exact situation used to collect the parameter. In other words, one manufacturer may define settling time as the time to reach 0.1% of the final output after a full scale change in input given a certain load on the output, while another manufacturer may define settling time as the time to reach 1% of the final output after a 1 volt change in input under a different load. The speed of the DAC together with the speed of the computer/software will determine the effective frequency components in the generated waveforms. Both the software (rate at which the software outputs new values to the DAC) and the DAC speed must be fast enough for the given application. In other words, if the software outputs new values to the DAC at a rate faster than the DAC can respond, then errors will occur. Figure 8.35 illustrates the effect of DAC output rate on the quality of the generated waveform. According to the Nyquist Theorem states the digital data rate must be greater than twice the maximum frequency component of the desired analog waveform. However, both waveforms in Figure 8.35 satisfy the Nyquist Theorem, but increasing the output rate by eight improves the signal to noise ratio by eight. 31 dB is a ratio of about 35 to 1, and 49 dB is a ratio of about 281 to 1. If the goal is to create a sine wave at a fixed frequency, we could improve the SNR greatly by using an analog low pass filter.

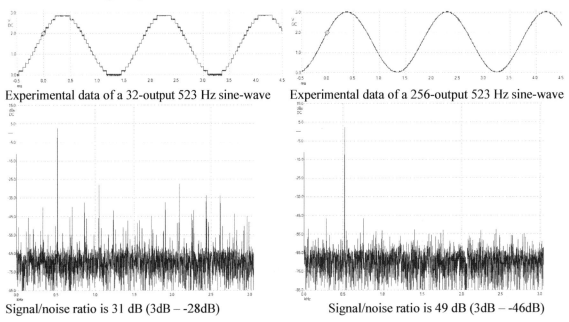

Figure 8.35. The waveform on the right was created by a 12-bit DAC with eight times the output rate than the left. Voltage versus time data on top and the Fourier Transform (frequency spectrum dB versus kHz) of the data on the bottom. There is a point in the spectrum at 0, which is the DC component. However, the signal is the 523 Hz bump with a magnitude of 3dB, representing the sine wave. The noises are all the other points not at 0 or 523 Hz. The largest noise on the left is -28 dB. The largest noise on the right is -46 dB.

Power. There are three power issues to consider. The first consideration is the type of power required. Older devices require three power voltages (e.g., +5 and -5 V), while most devices will operate on a single voltage supply (e.g., +2.7, +3.3, or +5 V.) If a single supply can be used to power all the digital and analog components then the overall system costs will be reduced. The second consideration is the amount of power required. Some devices can operate on less than 0.1 mW and are appropriate for battery-operated systems or for systems where excess heat is a problem. The last consideration is the need for a low-power sleep mode. Some battery operated systems need the DAC only intermittently. In these applications, we wish to give a shutdown command to the DAC, so that it draws less current when not needed.

Interface. Three approaches exist for interfacing the DAC to the computer. In a digital logic or parallel interface, the individual data bits are connected to a dedicated computer output port. For example, a 12-bit DAC requires a 12-bit output port bits to interface. The software simply writes to the parallel port(s) to change the DAC output. The second approach is called μP-bus or microprocessor-compatible. These devices are intended to be interfaced onto the address/data bus of an expanded mode microcontroller. The third approach is a high-speed serial interface like I2C or SPI. The SSI/MAX5353 interface is an example of a high-speed serial interface. This approach requires the fewest number of I/O pins. Even if the microcontroller does not support the SPI interface directly, these devices can be interfaced to regular I/O pins via the bit-banging software approach.

Package. DIP packages are convenient for creating and testing an original prototype. On the other hand surface mount packages require less board space. Because surface mount packages do not require holes in the PC board, circuits with these devices are easier/cheaper to produce.

Cost. Cost is always a factor in engineering design. Beside the direct costs of the individual components in the DAC interface, other considerations that affect cost include: 1) power supply requirements; 2) manufacturing costs; 3) the labor involved in individual calibration if required; and 4) software development costs.

8.4.2. DAC Waveform Generation

One application that requires a DAC is waveform generation. In this section, we will discuss various software methods for creating analog waveforms with a DAC. In each case, we will be using the MAX5353 hardware/software interface introduced in Example 7.2. In addition, we will use an output capture interrupt for the timing, so that the waveform generation occurs in the background. The rituals for initializing the periodic interrupt are shown in Chapter 6. In order to get a fair comparison between the various methods, each implementation will generate 32 interrupts per waveform. In the first approach, we assume there exists a time to voltage function, called **Wave()**, which we can call to determine the next DAC value to output. The waveform on the left of Figure 8.36 could be generated by Programs 8.2 and 8.3.

```
const float A=2048.0, B=1000.0, C=2*pi*31.25, D=-500.0, E=2*pi*125.0;
uint16_t Wave(float time){ float result;
   result = A + B*cos(C*time) + D*sin(E*time);
return (uint16_t) result;
}
```

Program 8.2. Waveform is defined by a mathematical function.

The simplest solution generates an output compare interrupt at a regular rate. The advantage of this approach is that complex waveforms can be encoded with a small amount of data. In this particular example, the entire waveform can be stored as 5 data points (**2048.0 1000.0 31.25 -500.0 125.0**). The disadvantage of this technique is that not all waveforms have a simple function, and this software will run slower as compared to the other techniques. If you were to implement this approach, performance would be improved by replacing the floating-point math with fixed-point.

```
float Time; // incremented every 1ms
void Timer0A_Handler(void){
  TIMER0_ICR_R = 0x01;   // acknowledge
  Time = Time+0.001;
  DACout(Wave(Time));
}
```
Program 8.3. Periodic interrupt used to create the analog output waveform.

In the second approach, we put the waveform information in a large statically allocated global array, see Program 8.4. Every interrupt we fetch a new value out of the data structure and output it to the DAC. In this case the output compare interrupt also occurs at a regular rate. Assume the ritual initializes **Time=0**. Running at 80 MHz, the ISR in Program 8.4 executes in less than 1 μs.

```
uint16_t Time;   // incremented every 1ms, 0-31
const uint16_t Wave[32]= { 3048,2675,2472,2526,2755,2957,
  2931,2597,2048,1499,1165,1139,1341,1570,1624,1421,1048,714,624,863,
  1341,1846,2165,2206,2048,1890,1931,2250,2755,3233,3472,3382};

uint16_t Time; // every 1ms
void Timer0A_Handler(void){
  TIMER0_ICR_R = 0x01; // acknowledge
  Time = (Time+1)&0x1F;
  DACout(Wave[Time]));
}
```
Program 8.4. Periodic interrupt used to create the analog output waveform.

Since the output rate is equal and fixed these first two methods have the same performance as illustrated in the left of Figure 8.36. The solid line is the desired waveform and the dotted line is the actual generated curve.

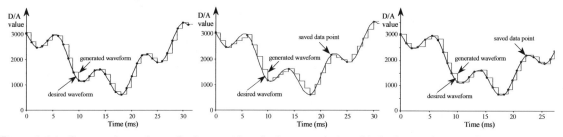

Figure 8.36. Generated waveforms. Left uses either the function or the table lookup technique. Center uses a small table and interpolation. Right uses a variable output rate.

If the size of the table gets large it is possible to store a smaller table in memory and use linear interpolation to recover the data points in between the stored samples. The center of Figure 8.36 shows the generated waveform derived from only 9 of the original 32 data points. To simplify the software, the first data point is repeated as the last data point. For each point we will need to save both the DAC value and time length of the current line segment. For the 9 saved data points we simply output the data, but for the other points, we must perform a linear interpolation to get the value to output to the DAC, see Program 8.5. Assume the ritual initializes I=J=0. Signed 16-bit numbers are used so the subtractions operate properly. In other words, some of the intermediate calculations can be negative.

```
int16_t I;   // incremented every 1ms
int16_t J;   // index into these two tables
const int16_t Time[10] = {0,2,6,10,14,18,22,25,30,32};   // time in msec
const int16_t Wave[10] =
   {3048,2472,2931,1165,1624,624,2165,1890,3472,3048};   // last=first
uint16_t Time; // every 1ms
void Timer0A_Handler(void){
  TIMER0_ICR_R = 0x01; // acknowledge
  if((++I)==32) {I=0; J=0;}
  if(I==Time[J])
    DACout(Wave[J]);
  else if (I==Time[J+1]){
    J++;
    DACout(Wave[J]);}
  else
    DACout(Wave[J]+((Wave[J+1]-Wave[J])*(I-t[J]))/(Time[J+1]-Time[J]));
}
```

Program 8.5. Periodic interrupt used to create the analog output waveform.

The software in the previous techniques changes the DAC at a fixed rate. While most of the time this is adequate, there are some waveforms for which uneven times between outputs seem appropriate. In our test signal, there are places in the wave where the signal varies slowly in addition to places in the wave with rapidly changing values. Notice the data points in this figure are placed at uneven time intervals to match the various phases of this signal. This generated waveform is still created with 32 points, but placing the points closer together during phases with large slopes improves the overall accuracy, see Program 8.6. The table data structure will encode both the voltage (as a DAC value) and the time. The time parameter is stored as a Δt in Timer0 cycles to simplify servicing the output compare interrupt.

```
uint16_t Time;   // incremented every sample, 0 to 31
const uint16_t Wave[32] = {
   3048,2675,2472,2526,2817,2981,2800,2337,1901,1499,1165,1341,1570,1597,
   1337,952,662, 654, 863,1210,1605,1950,2202,2141,1955,1876,2057,
   2366,2755,3129,3442,3382};
const uint16_t dt[32] = { // time increment in Timer0 cycles
   2000,2000,2000,2500,2500,2000,2000,1500,1500,2000,4000,2000,2500,
   2000,2000,2000, 2000,1500,1500,1500,1500,2000,2500,2000,2000,2000,
   1500,1500,1500,2000,2500,2000};
```

```c
void Timer0A_Handler(void){
  Time = (Time+1)&0x1F;
  DACout(Wave[Time]);         // this output amplitude
  TIMER0_TAILR_R = dt[Time];  // this time duration
  TIMER0_ICR_R = 0x01;        // acknowledge
}
```
Program 8.6. Periodic interrupt at unequal rates used to create the analog output waveform.

8.4.3. PWM DAC

We can use the PWM module to create a DAC. The PWM digital signal is fed through an analog low-pass filter to create a DC voltage linearly related to the duty cycle of the PWM wave. The precision of the DAC will be the precision of the counter used to create the PWM. In particular, the precision will be the number we put into the **PWM_0_LOAD_R** register of the TM4C. For example, if we call Program 6.8 with **PWM0_Init(1024,512)**, then the DAC precision will be 1024 alternatives or 10 bits. The frequency of the DAC is the rate at which the DAC output can be changed and depends on the frequency of the PWM output. In particular, we can change the PWM duty cycle only once per cycle. If the bus clock is 8 MHz, and we initialize with **PWM0_Init(4000,2000)**, the PWM period will be 1ms. This means we can update the 12-bit DAC 1000 times a second. However, if the bus clock is 50 MHz, and we initialize with **PWM0_Init(250,125)**, the PWM period will be 10µs. This means we can update the 8-bit DAC 100,000 times a second. It is usually best for the PWM signal frequency to be 10 to 100 times higher than the desired bandwidth of analog signals to be produced. Generally, the higher the PWM frequency, the lower the order of filter required, and the easier it is to build a suitable filter. Let f_{max} be the highest frequency component we wish to create in the analog output, and let f_{PWM} be the PWM frequency. We want the analog LPF to reject f_{PWM} and pass f_{max}.

Example 8.4. Design a system that outputs a 1-kHz sine wave using a PWM DAC.

Solution: This solution will use the periodic timer shown in Program 6.6 and the PWM driver shown in Program 6.8. We will run the PWM output wave at 100 kHz, which is much faster than the desired sine wave. We can create an 8-bit PWM DAC using a bus clock of 50 MHz and initializing with **PWM0_Init(250,125)**. 50 MHz/100kHz is 250 alternatives, which about 8 bits. To change the amplitude of the DAC we adjust the duty cycle of the PWM by calling **PWM0_SetDuty()**. We will output 32 points to the 8-bit PWM DAC for each cycle in the sine wave, see Program 8.7. The LPF must pass 1 kHz and reject 100 kHz. We choose the LPF cutoff at 3 kHz, which is between f_{PWM} and f_{max}. A passive LPF can be made with a resistor and capacitor, $f_c = 1/(2\pi RC)$. An R equal to 22 kΩ and C equal to 2.2nF will create a LPF at 3.3 kHz, as shown in Figure 8.37. If we need low output impedance, we could replace the passive filter with an active filter like Figure 8.25.

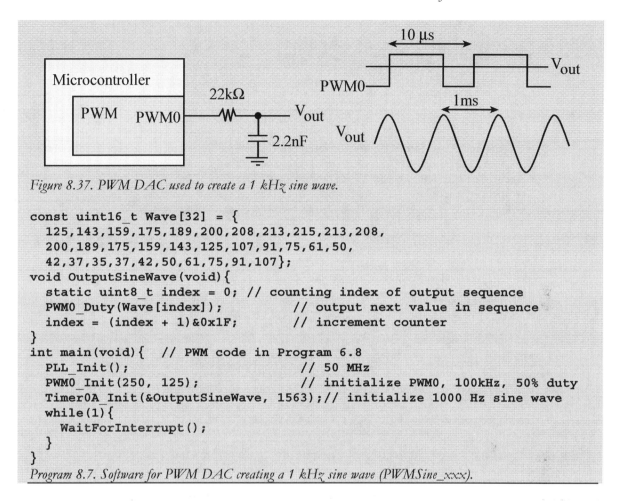

Figure 8.37. PWM DAC used to create a 1 kHz sine wave.

```
const uint16_t Wave[32] = {
  125,143,159,175,189,200,208,213,215,213,208,
  200,189,175,159,143,125,107,91,75,61,50,
  42,37,35,37,42,50,61,75,91,107};
void OutputSineWave(void){
  static uint8_t index = 0;  // counting index of output sequence
  PWM0_Duty(Wave[index]);    // output next value in sequence
  index = (index + 1)&0x1F;  // increment counter
}
int main(void){   // PWM code in Program 6.8
  PLL_Init();                           // 50 MHz
  PWM0_Init(250, 125);                  // initialize PWM0, 100kHz, 50% duty
  Timer0A_Init(&OutputSineWave, 1563);// initialize 1000 Hz sine wave
  while(1){
    WaitForInterrupt();
  }
}
```

Program 8.7. Software for PWM DAC creating a 1 kHz sine wave (PWMSine_xxx).

8.5. Analog to Digital Converters

8.5.1. ADC Parameters

An ADC converts an analog signal into digital form. The input signal is usually an analog voltage (V_{in}), and the output is a binary number. The ADC **precision** is the number of distinguishable ADC inputs (e.g., 4096 alternatives, 12 bits). The ADC **range** is the maximum and minimum ADC input (volts, amps). The ADC **resolution** is the smallest distinguishable change in input (volts, amps). The resolution is the change in input that causes the digital output to change by 1.

Range(volts) = Precision(alternatives) • Resolution(volts)

Normally we don't specify accuracy for just the ADC, but rather we give the accuracy of the entire system (including transducer, analog circuit, ADC and software). Therefore, accuracy will be described later in Chapter 10 as part of the systems approach to data acquisition systems. An ADC is **monotonic** if it has no missing codes. This means if the analog signal is a slow rising voltage, then the digital output will hit all values sequentially. The ADC is linear if the resolution is constant through the range. Let $f(x)$ be the input/output ADC transfer function. One quantitative measure of **linearity** is the correlation coefficient of a linear regression fit of the $f(x)$ responses. The ADC **speed** is the time to convert, called t_c. The ADC **cost** is a function of the number and price of internal components. There are four common encoding schemes for an ADC. Table 8.7 shows two encoding schemes for a 12-bit unipolar ADC, and Table 8.8 shows two encoding schemes for a 12-bit bipolar ADC.

Unipolar Codes	Straight Binary	Complementary Binary
$+V_{max}$	1111,1111,1111	0000,0000,0000
$+V_{max}/2$	1000,0000,0000	0001,1111,1111
$+V_{max}/1024$	0000,0000,0001	1111,1111,1110
$+0.00$	0000,0000,0000	1111,1111,1111

Table 8.7. Unipolar codes for a 12-bit ADC with a range of 0 to $+V_{max}$.

Bipolar Codes	Offset Binary	Two's Complement Binary
$+V_{max}$	1111,1111,1111	0111,1111,1111
$+V_{max}/2$	1000,0000,0000	0100,0000,0000
$+V_{max}/512$	1000,0000,0001	0000,0000,0001
$+0.00$	1000,0000,0000	0000,0000,0000
$-V_{max}/512$	0111,1111,1111	1111,1111,1111
$-V_{max}/2$	0100,0000,0000	1100,0000,0000
$-V_{max}$	0000,0000,0000	1000,0000,0000

Table 8.8. Bipolar codes for a 12-bit ADC with a range of $-V_{max}$ to $+V_{max}$.

The TM4C uses straight binary, has a precision of 12 bits, and has a range of 0 to 3.3 V. The LM3S has a 10-bit ADC with a range of 0 to 3V. To convert between straight binary and complementary binary we simply complement (change 0 to 1, change 1 to 0) all the bits. To convert between offset binary and 2's complement, we complement just the most significant bit. The exclusive-or operation can be used to complement bits.

Just like the DAC, one can choose the full scale range to simplify the use of fixed-point math. For example, if a 12-bit ADC had a full scale range of 0 to 4.095 volts, then the resolution would be exactly 1 mV. This means that if the ADC input voltage were 1.234 volts, then the result would be 1234 (decimal).

The **total harmonic distortion** (THD) of a signal is a measure of the harmonic distortion present and is defined as the ratio of the sum of the powers of all harmonic components to the power of the fundamental frequency. Basically, it is a measure of all the noise processes in an ADC and usually is given in dB full scale. A similar parameter is **signal-to-noise and distortion ratio** (SINAD), which is measured by placing a pure sine wave at the input of the ADC (signal) and measuring the ADC output (signal plus noise). We can compare precision in bits to signal-to-noise ratio in dB using the relation dB = $20 \log_{10}(2^n)$. For example, the 12-bit MAX1247 ADC has a SINAD of 73 dB. Notice that $20 \log_{10}(2^{12})$ is 72 dB.

Dynamic range, expressed in dB, is defined as the range between the noise floor of a device and its specified maximum output level. The dynamic range is the range of signal amplitudes which the ADC can resolve. If an ADC can resolve signals from 1 mV to 1 V, it has a dynamic range of 20*log(1V/1mV) = 60dB. Dynamic range is important in communication applications, where signal strengths vary dramatically. If the signal is too large, it saturates the ADC input. If the signal is too small, it gets lost in the quantization noise.

The **effective number of bits** (ENOB) specifies the dynamic performance of an ADC at a specific input frequency and sampling rate. In an ideal situation, ADC error consists only of quantization noise (resolution = range/precision). As the input frequency increases, the overall noise (particularly in the distortion components) also increases, thereby reducing the ENOB and SINAD.

8.5.2. ADC Conversion Techniques

The most pervasive method for ADC conversion is the **successive approximation** technique, as illustrated in Figure 8.38. A 12-bit successive approximation ADC is clocked 12 times. At each clock another bit is determined, starting with the most significant bit. For each clock, the successive approximation hardware issues a new "guess" on V_{dac} by setting the bit under test to a "1". If V_{dac} is now higher than the unknown input, V_{in}, then the bit under test is cleared. If V_{dac} is less than V_{in}, then the bit under test is remains 1. In this description, *bit* is an unsigned integer that specifies the bit under test. For a 12-bit ADC, *bit* goes 2048, 1024, 512, 256, 128, 64,...,1. D_{out} is the ADC digital output, and Z is the binary input that is true if V_{dac} is greater than V_{in}.

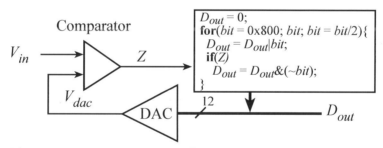

Figure 8.38. A 12-bit successive approximation ADC.

> **Observation:** The speed of a successive approximation ADC relates linearly with its precision in bits.

The **sigma delta** analog to digital converter is used in many audio applications. It is a cost effective approach to 16-bit 44 kHz sampling (CD quality audio). Sigma delta converters have a DAC, a comparator and digital processing similar to the successive approximation technique. While successive approximation converters have DACs with the same precision as the ADC, sigma delta converters use DACs with a much smaller precision than the ADC. A 1-bit DAC is simply a digital signal itself. The digital signal processing will run at a clock frequency faster than the overall ADC system, called **oversampling**. It uses complex signal processing to drive the output voltage V_0 to equal the unknown input V_{in} in a time-averaged sense. The "delta" part of the sigma delta converter is the subtractor, where $V_1=V_0-V_{in}$. Next comes the "sigma" part that implements an analog integration. If V_0 to equal the unknown input V_{in} in a time-averaged sense, then V_2 will be zero. The comparator tests the V_2 signal. If V_2 is positive then V_0 is made smaller. If V_2 is negative then V_0 is made larger. This DAC-subtractor-integrator-comparator-digital loop is executed at a rate much faster than the eventual digital output rate.

A very simple algorithm, shown in Figure 8.39 is run continuously. For every time through the outer while loop there is one ADC output. This algorithm is much too simple to be appropriate in an actual converter, but it does illustrate the sigma delta approach. For a 10-bit conversion, the DAC output rate is 1024 times the ADC conversion rate. We assume the input voltage, V_{in}, is between 0 and +3 V. *DAC* is an output of the sigma-delta processing that sets the 1-bit DAC. *Z* is the comparator output, which is an input to the signal processing.

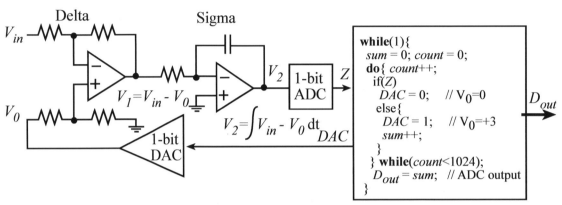

Figure 8.39. Block diagram of the sigma-delta ADC conversion technique.

In this very simple solution, the *DAC* is set to 1 (V_0=+3) if *Z* is 0 (V_2<0). Conversely, the *DAC* is set to 0 (V_0=0) if *Z* is 1 (V_2>0). Each time the DAC is set to 1, *sum* is incremented. At the end of 1024 passes, the value *sum* is recorded as the ADC sample. Since there are 1024 passes through the loop, *sum* will vary from 0 to 1023. For example, if the V_{in} is 1.5 V, then half of the *DAC* outputs will be 1 and the other half 0. This will make V_0 oscillate between 0 and 3V, with a 50% duty cycle, V_1 will oscillate between -1.5 and +1.5 with a 50 % duty cycle, and the time-averaged V_2 will be zero.

A second example is illustrated in Figure 8.40. The input V_{in} is 2.25 V, so the output should be 2.25/3*1024 or 768. The sigma delta will adjust the *DAC* output so that $V_1=V_0-V_{in}$ is equal to 0 in a time-averaged sense. V_1 is 2.25 V for 25% of the time and -0.75 V for 75% of the time. Three out of every four DAC outputs are high, so three out of every four time, V_2 will be above 0. Therefore after 1024 times through the loop, 768 of them will increment *sum*, yielding the correct ADC result. If the V_{in} input rises, a higher percentage of *DAC* outputs will be high, increasing *sum*. If the V_{in} input falls, a lower percentage of *DAC* outputs will be high, decreasing *sum*.

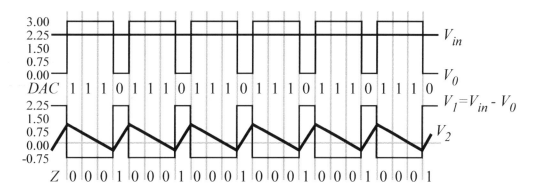

Figure 8.40. Example operation of a sigma-delta conversion.

In a real sigma delta the overclock rate is typically 8 to 1 or 16 to 1. Multiple bits are obtained each time through the output-input cycle using DSP algorithms.

Another ADC technique is called **flash**. Flash converters are very expensive and very fast. Figure 8.41 shows a 3-bit flash. The MAX104 is a ±5V, 1Gsps, 8-Bit ADC with on-chip 2.2GHz sample/hold.

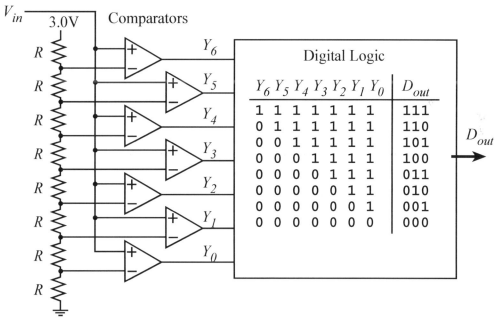

Figure 8.41. Block diagram of a 3-bit flash ADC.

8.5.3. Sample/Hold

A **sample and hold** (S/H) is an analog latch, illustrated in Figure 8.42. An alternative name for this analog component is track and hold. The purpose of the S/H is to hold the ADC analog input constant during the conversion. There is S/H at the input of most ADC, including the ones on the Stellaris and Tiva microcontrollers. The first phase of most analog to digital conversions is the sampling phase, where the input voltage V_{in}, is recorded as a charge on the capacitor C.

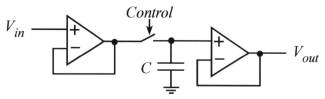

Figure 8.42. The sample and hold has a digital input, an analog input and an analog output.

The digital input, *Control*, determines the S/H mode. The S/H is in **sample mode**, where V_{out} equals V_{in}, with the switch is closed. The S/H is in **hold mode**, V_{out} is fixed because the switch is open. The **acquisition time** is the time for the output to equal the input after the control is switched from hold to sample. This is the time to charge the capacitor C. The **aperture time** is the time for the output to stabilize after the control is switched from sample to hold. This is the time to open the switch, which is usually quite fast. The **droop rate** is the output voltage slope (dV_{out}/dt) when *Control* equals hold. Normally the gain, K, should be one and the offset, V_{off}, should be zero. The gain and offset error specify how close is the V_{out} to the desired V_{in} when *Control* equals sample,

$V_{out} = K V_{in} + V_{off}$

To choose the capacitor, C:
1) One should use a high quality capacitor with high
   insulation resistance and low dielectric absorption.
2) A larger value of C will decrease (improve) the droop rate.
   If the droop current is I_{DR}, then the droop rate will be $dV_{out}/dt = I_{DR}/C$
3) A smaller C will decrease (improve) the acquisition time.

The system will require a sample and hold if the input signal could change during an ADC conversion. There will be a time during which the ADC samples the input voltage, t_{samp}. Let the maximum slew rate of the input signal be dV_{in}/dt. If the slew rate times the sampling time is larger than the ADC resolution, we should add a sample and hold module to keep the analog input stable during conversion. The LM3S/TM4C ADC modules have a built-in S/H.

8.5.4. Internal ADC

Table 8.9 shows the ADC register bits required to perform periodic sampling. We will show timer triggering on one and on two channels. For more complex configurations refer to the specific data sheet. The value in **ADC0_PC_R** specifies the maximum sampling rate, see Table 8.10. The TM4C123 and MSP432E401Y can sample up to 1 million samples per second, but the maximum sampling rate on many LM3S microcontrollers is only 500K. Refer to the data sheet of your specific microcontroller for maximum possible sampling rate, the number of channels, and the number of bits. The ADC has four sequencers, but we will present code using sequencers 2 and 3. We set the **ADC0_SSPRI_R** register to 0x3210 to make sequencer 0 the highest and sequencer 3 the lowest priority. We need to make sure each sequencer has a unique priority. We set bits 15–12 (**EM3**) in the **ADC0_EMUX_R** register to specify how sequencer 3 will be triggered. Similarly, we set bits 11–8 (**EM2**) to specify how sequencer 2 will be triggered. Table 8.11 shows the various ways to trigger an ADC conversion. In this section, we will use timer triggering (**EM3**=0x5).

Address	31-2						1	0	Name
$400F.E638							ADC1	ADC0	SYSCTL_RCGCADC_R

	31-14	13-12	11-10	9-8	7-6	5-4	3-2	1-0	
$4003.8020		SS3		SS2		SS1		SS0	ADC0_SSPRI_R

	31-16	15-12	11-8	7-4	3-0	
$4003.8014		EM3	EM2	EM1	EM0	ADC0_EMUX_R

	31-4	3	2	1	0	Name
$4003.8000		ASEN3	ASEN2	ASEN1	ASEN0	ADC0_ACTSS_R
$4003.8028		SS3	SS2	SS1	SS0	ADC0_PSSI_R
$4003.8004		INR3	INR2	INR1	INR0	ADC0_RIS_R
$4003.8008		MASK3	MASK2	MASK1	MASK0	ADC0_IM_R
$4003.8FC4		Speed				ADC0_PC_R
$4003.800C		IN3	IN2	IN1	IN0	ADC0_ISC_R

	31-28	27-24	23-20	19-16	15-12	11-8	7-4	3-0	
$4003.8040	MUX7	MUX6	MUX5	MUX4	MUX3	MUX2	MUX1	MUX0	ADC0_SSMUX0_R

	31-16	15-12	11-8	7-4	3-0	
$4003.8060		MUX3	MUX2	MUX1	MUX0	ADC0_SSMUX1_R
$4003.8080		MUX3	MUX2	MUX1	MUX0	ADC0_SSMUX2_R
$4003.80A0					MUX0	ADC0_SSMUX3_R

	31	30	29	28	27	26	...	8	7	6	5	4	3	2	1	0	
$4003.8044	TS7	IE7	END7	D7	TS6	IE6	...	D2	TS1	IE1	END1	D1	TS0	IE0	END0	D0	ADC0_SSCTL0_R

	15	14	13	12	11	10	9	8	7	6	5	4	3	2	1	0	
$4003.8064	TS3	IE3	END3	D3	TS2	IE2	END2	D2	TS1	IE1	END1	D1	TS0	IE0	END0	D0	ADC0_SSCTL1_R
$4003.8084	TS3	IE3	END3	D3	TS2	IE2	END2	D2	TS1	IE1	END1	D1	TS0	IE0	END0	D0	ADC0_SSCTL2_R
$4003.80A4													TS0	IE0	END0	D0	ADC0_SSCTL3_R

	31-10	11-0	
$4003.8048		DATA	ADC0_SSFIFO0_R
$4003.8068		DATA	ADC0_SSFIFO1_R
$4003.8088		DATA	ADC0_SSFIFO2_R
$4003.80A8		DATA	ADC0_SSFIFO3_R

Table 8.9. Some of the ADC registers. Each register is 32 bits wide. LM3S has 10-bit data

Value	Description
0x7	1M samples/second
0x5	500K samples/second
0x3	250K samples/second
0x1	125K samples/second

Table 8.10. The Speed bits in the ADC0_PC_R register.

When using sequencer 0, we can specify up to eight channels to convert on a single command. Sequencers 1 and 2 allow up to four channels to be converted. The control bits for sequencer 2 are shown in Table 8.9. The **ADC0_SSMUX2_R** register specifies which channels to convert and the **ADC0_SSCTL2_R** register specifies how many samples to take and if/when to interrupt. For example, if we wish to convert channels 2, 3, and 6, we set **MUX0**, **MUX1**, and **MUX2** fields to 2,3,6 respectively in **ADC0_SSMUX2_R**. The **ADC0_SSCTL2_R** bit **END2** is set to specify three conversions, assuming END0=0 and END1=0. The **IE2** bit can be set to request an interrupt after the third conversion. 10-bit results from sequencer 2 are read from the **ADC0_SSFIFO2** register. All other bits in the **ADC0_SSCTL2_R** register will be clear for this example (no temperature or differential measurements).

The **ADC0_RIS_R** register has flags that are set when the conversion is complete, assuming the **IE0** bit is set. The **ADC_IM_R** register has interrupt arm bits. The **ADC0_ISC_R** register has interrupt trigger bits. The **IN3** bit is set when both **INR3** and **MASK3** are set. We clear the **INR3** and **IN3** bits by writing an 8 to the **ADC0_ISC_R** register. The interrupt vector for ADC sequencer 3 is at 0x00000084.

Value	Event
0x0	Software start
0x1	Analog Comparator 0
0x2	Analog Comparator 1
0x3, 0x9-0x0E	Reserved
0x4	External (GPIO PB4)
0x5	Timer
0x6	PWM0
0x7	PWM1
0x8	PWM2
0xF	Always (continuously sample)

Table 8.11. The ADC EM3, EM2, EM1, and EM0 bits in the ADC0_EMUX_R register.

There are 13 steps to configure the ADC to sample a single channel at a periodic rate. The most accurate sampling method is timer-triggered sampling (**EM3**=0x5). On the TM4C123, the MUX fields are 4 bits wide, allowing us to specify channels 0 to 11. On the MSP432E401Y, the channel ranges from 0 to 19 (see Table 3.2).

Step 1. We enable the ADC clock in the **SYSCTL_RCGCADC_R** register.

Step 2. Bits 3 – 0 of the **ADC0_PC_R** register specify the maximum sampling rate of the ADC. In this example, we will sample slower than 125 kHz, so the maximum sampling rate is set at 125 kHz. This will require less power and produce a longer sampling time as described the S/H section, creating a more accurate conversion.

Step 3. We will set the priority of each of the four sequencers. In this case, we are using just one sequencer, so the priorities are irrelevant, except for the fact that no two sequencers should have the same priority. The default configuration has Sample Sequencer 0 with the highest priority, and Sample Sequencer 3 as the lowest priority.

Step 4. Next, we need to configure the timer to run at the desired sampling frequency. We enable the Timer0 clock by setting bit 0 of the **SYSCTL_RCGCTIMER_R** register. This initialization is similar to Program 6.6 with two changes. First we set bit 5 of the **TIMER0_CTL_R** register to activate **TAOTE**, which is the Timer A output trigger enable.

Secondly, we do not arm any Timer0 interrupts. The rate at which the timer rolls over determines the sampling frequency. Let *prescale* be the value loaded into **TIMER0_TAPR_R**, and let *period* be the value loaded into **TIMER0_TAILR_R**. If the period of the bus clock frequency is Δt, then the ADC sampling period will be

$$\Delta t * (prescale + 1) * (period + 1)$$

The fastest sampling rate is determined by the speed of the processor handling the ADC interrupts and by the speed of the main program consuming the data from the FIFO. If the bus clock is 80 MHz, the slowest possible sampling rate for this example is $80MHz/2^{32}$, which is about 0.018 Hz, which is every 53 seconds.

Step 5. Before configuring the sequencer, we need to disable it. To disable sequencer 3, we write a 0 to bit 3 (**ASEN3**) in the **ADC0_ACTSS_R** register. Disabling the sequencer during programming prevents erroneous execution if a trigger event were to occur during the configuration process.

Step 6. We configure the trigger event for the sample sequencer in the **ADC0_EMUX_R** register. For this example, we write a 0101 to bits 15–12 (**EM3**) specifying timer trigger mode for sequencer 3.

Step 7. For each sample in the sample sequence, configure the corresponding input source in the **ADC0_SSMUXn** register. In this example, we write the channel number (0, 1, 2, or 3) to bits 3–0 in the **ADC0_SSMUX3_R** register.

Step 8. For each sample in the sample sequence, we configure the sample control bits in the corresponding nibble in the **ADC0_SSCTLn** register. When programming the last nibble, ensure that the END bit is set. Failure to set the END bit causes unpredictable behavior. Sequencer 3 has only one sample, so we write a 0110 to the **ADC0_SSCTL3_R** register. Bit 3 is the **TS0** bit, which we clear because we are not measuring temperature. Bit 2 is the **IE0** bit, which we set because we want to request an interrupt when the sample is complete. Bit 1 is the **END0** bit, which is set because this is the last (and only) sample in the sequence. Bit 0 is the **D0** bit, which we clear because we do not wish to use differential mode.

Step 9. If interrupts are to be used, write a 1 to the corresponding mask bit in the **ADC0_IM_R** register. We want an interrupt to occur when the conversion is complete (set bit 3, **MASK3**).

Step 10. We enable the sample sequencer logic by writing a 1 to the corresponding **ASENn**. To enable sequencer 3, we write a 1 to bit 3 (**ASEN3**) in the **ADC0_ACTSS_R** register.

Step 11. The priority of the ADC0 sequencer 3 interrupts are in bits 13–15 of the **NVIC_PRI4_R** register.

Step 12. Since we are requesting interrupts, we need to enable interrupts in the NVIC. ADC sequencer 3 interrupts are enabled by setting bit 17 in the **NVIC_EN0_R** register.

Step 13. Lastly, we must enable interrupts in the **PRIMASK** register.

The timer starts the conversion at a regular rate. Bit 3 (**INR3**) in the **ADC0_RIS_R** register will be set when the conversion is done. This bit is armed and enabled for interrupting, so conversion complete will trigger an interrupt. The **IN3** bit in the **ADC0_ISC_R** register triggers the interrupt. The ISR acknowledges the interrupt by writing a 1 to bit 3 (**IN3**). The 12-bit result is read from the **ADC0_SSFIFO3_R** register. The book web site for has example code. In order to reduce latency of other interrupt requests in the system, this ISR simply stores the 12-bit conversion in a FIFO, to be processed later in the main program. Program 8.8 shows the initialization and interrupt service routine to affect the periodic sampling. For the port pin, we disable its DEN, clear its DIR, set its AFSEL and enable its AMSEL bit.

```c
void ADC0_InitTimer0ATriggerSeq3PD3(uint32_t period){
  volatile uint32_t delay;
  SYSCTL_RCGCADC_R |= 0x01;          // 1) activate ADC0
  SYSCTL_RCGCGPIO_R |= 0x08;         // Port D clock
  delay = SYSCTL_RCGCGPIO_R;         // allow time for clock to stabilize
  GPIO_PORTD_DIR_R &= ~0x08;         // make PD3 input
  GPIO_PORTD_AFSEL_R |= 0x08;        // enable alternate function on PD3
  GPIO_PORTD_DEN_R &= ~0x08;         // disable digital I/O on PD3
  GPIO_PORTD_AMSEL_R |= 0x08;        // enable analog functionality on PD3
  ADC0_PC_R = 0x01;                  // 2) configure for 125K samples/sec
  ADC0_SSPRI_R = 0x3210;             // 3) seq 0 is highest, seq 3 is lowest
  SYSCTL_RCGCTIMER_R |= 0x01;        // 4) activate timer0
  delay = SYSCTL_RCGCGPIO_R;
  TIMER0_CTL_R = 0x00000000;         // disable timer0A during setup
  TIMER0_CTL_R |= 0x00000020;        // enable timer0A trigger to ADC
  TIMER0_CFG_R = 0;                  // configure for 32-bit timer mode
  TIMER0_TAMR_R = 0x00000002;        // configure for periodic mode
  TIMER0_TAPR_R = 0;                 // prescale value for trigger
  TIMER0_TAILR_R = period-1;         // start value for trigger
  TIMER0_IMR_R = 0x00000000;         // disable all interrupts
  TIMER0_CTL_R |= 0x00000001;        // enable timer0A 32-b, periodic
  ADC0_ACTSS_R &= ~0x08;             // 5) disable sample sequencer 3
  ADC0_EMUX_R = (ADC0_EMUX_R&0xFFFF0FFF)+0x5000; // 6) timer trigger
  ADC0_SSMUX3_R = 4;                 // 7) PD3 is analog channel 4
  ADC0_SSCTL3_R = 0x06;              // 8) set flag and end after first sample
  ADC0_IM_R |= 0x08;                 // 9) enable SS3 interrupts
  ADC0_ACTSS_R |= 0x08;              // 10) enable sample sequencer 3
  NVIC_PRI4_R = (NVIC_PRI4_R&0xFFFF00FF)|0x00004000; // 11)priority 2
  NVIC_EN0_R = 1<<17;                // 12) enable interrupt 17 in NVIC
  EnableInterrupts();                // 13) enable interrupts
}
void ADC0Seq3_Handler(void){
  ADC0_ISC_R = 0x08;         // acknowledge ADC sequence 3 completion
  Fifo_Put(ADC0_SSFIFO3_R);  // pass to foreground
}
```
Program 8.8. Software to sample data using the ADC (ADCT0ATrigger_xxx).

Checkpoint 8.7: If the input voltage is 1.0V, what value, in 10-bit unsigned right-justified mode, will the LM3S ADC return? What will a TM4C with a 12-bit ADC return?

Sequencer 3 can only sample one analog input. When we wish to sample multiple channels with one trigger, we need to use sequencers 0, 1, or 2. The next example shows how to sample two channels using sequencer 2. The TM4C123 has pins that could be 12 analog input and the MSP432E401Y has 20 possible analog pins. Both of these microcontrollers have two ADC modules and each module has 4 sequencers.

Example 8.5. Write software to sample ADC channels 4 and 5 at 1 kHz. Channel 4 on the TM4C123 is PD3 and channel 5 is PD2.

Solution: This solution will use the periodic timer to establish the 1000 Hz sampling rate, similar to Program 8.8. We specify the channels to sample in the **ADC0_SSMUX2_R** register. 0x0054 means first sample channel 4 then sample channel 5. The **ADC0_SSCTL2_R** bit **END1** is set to specify two conversions, assuming END0=0. The **IE1** bit is set to request an interrupt after the second conversion. We read the two 12-bit results from the **ADC0_SSFIFO2** register. All other bits in the **ADC0_SSCTL2_R** register will be clear for this example (no temperature or differential measurements).

```
void ADC_Init(void){     // assumes a 80 MHz bus clock
  SYSCTL_RCGCADC_R |= 0x01;       // 1) activate ADC0
  SYSCTL_RCGCGPIO_R |= 0x08;      // Port D clock
  SYSCTL_RCGCTIMER_R |= 0x01;     // 4) activate timer0
  Ch1Fifo_Init();                 // initialize FIFOs
  Ch3Fifo_Init();                 // wait for clocks to stabilize
  GPIO_PORTD_DIR_R &= ~0x0C;      // make PD3-2 input
  GPIO_PORTD_AFSEL_R |= 0x0C;     // enable alternate function on PD3-2
  GPIO_PORTD_DEN_R &= ~0x0C;      // disable digital I/O on PD3-2
  GPIO_PORTD_AMSEL_R |= 0x0C;     // enable analog functionality on PD3-2
  ADC0_PC_R = 0x01;               // 2) configure for 125K samples/sec
  ADC0_SSPRI_R = 0x3210;          // 3) Priority of sequencers
  TIMER0_CTL_R = 0x00000000;      // disable timer0A during setup
  TIMER0_CTL_R |= 0x00000020;     // enable timer0A trigger to ADC
  TIMER0_CFG_R = 0;               // configure for 32-bit timer mode
  TIMER0_TAMR_R = 0x00000002;     // configure for periodic mode
  TIMER0_TAPR_R = 0;              // prescale value for trigger
  TIMER0_TAILR_R = 79999;         // 80000 cycles is 1ms
  TIMER0_IMR_R = 0x00000000;      // disable all interrupts
  TIMER0_CTL_R |= 0x00000001;     // enable timer0A 32-b, periodic
  ADC0_ACTSS_R &= ~0x04;          // 5) disable sample sequencer 2
  ADC0_EMUX_R = (ADC0_EMUX_R&0xFFFFF0FF)+0x0500;  // 6) timer trigger
  ADC0_SSMUX2_R = 0x0054;         // 7) PD3-2 are channel 4,5
  ADC0_SSCTL2_R = 0x0060;         // 8) set flag and end after second
  ADC0_IM_R |= 0x04;              // 9) enable SS2 interrupts
  ADC0_ACTSS_R |= 0x04;           // 10) enable sample sequencer 2
  NVIC_PRI4_R = (NVIC_PRI4_R&0xFFFFFF00)|0x00000040; // ADC2 priority 2
  NVIC_EN0_R = 1<<16;             // 12)enable interrupt 16
  EnableInterrupts();             // 13)enable all interrupts
}
void ADC0Seq2_Handler (void){
  ADC0_ISC_R = 0x04;              // acknowledge ADC sequence 2 completion
  Ch1Fifo_Put(ADC0_SSFIFO2_R);    // PD3, Channel 4 first
  Ch3Fifo_Put(ADC0_SSFIFO2_R);    // PD3, Channel 5 second
}
```

Program 8.9. Software to sample channels 4 and 5 at 1 kHz (ADCSWTriggerTwoChan_xxx).

8.5.5. Multiple Access Circular Queue

A multiple access circular queue (**MACQ**) is used for data acquisition and control systems. A MACQ is a fixed length order preserving data structure, see Figure 8.43. The source process (ADC sampling software) places information into the MACQ. Once initialized, the MACQ is always full. The oldest data is discarded when the newest data is **Put** into a MACQ. The sink process can read any of the data from the MACQ. The **Read** function is non-destructive. This means that the MACQ is not changed by the **Read** operation.

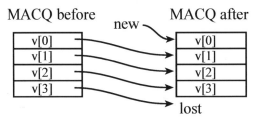

Figure 8.43. *When data is put into a multiple access circular queue, the oldest data is lost.*

For example consider the problem of weather forecasting as shown in Figure 8.44.

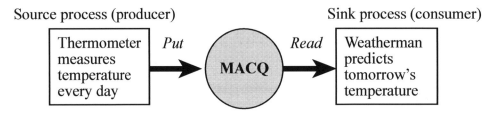

Figure 8.44. *Application of the multiple access circular queue.*

The weatherman measures the temperature every day at 12 noon, and puts the temperature into the MACQ. To predict tomorrow's temperature, she looks at the trend over the last 3 days. Let $T(0)$ be today's temperature, $T(1)$ be yesterday's temperature, etc. We could predict tomorrow's by executing this digital equation

$$U = (170 \cdot T(0) + 60 \cdot T(1) + 36 \cdot T(2))/256$$

The MACQ is useful for implementing digital filters and linear control systems. One common application of the MACQ is the real-time calculation of derivative. Also assume the ADC sampling is triggered every 1 ms. $x(n)$ will refer to the current sample, and $x(n-1)$ will be the sample 1 ms ago. There are a couple of ways to implement the discrete time derivative. The simple approach is

$$d(n) = (x(n) - x(n-1))/\Delta t$$

In practice, this first order equation is quite susceptible to noise. An approach generating less noise calculates the derivative using a higher order equation like

$$d(n) = (x(n) + 3x(n-1) - 3x(n-2) - x(n-3))/\Delta t$$

The C implementation of this discrete derivative uses a MACQ. Since Δ*t* is 1 ms, we simply consider the derivative to have units mV/ms and not actually execute the divide by Δ*t* operation. Signed arithmetic is used because the slope may be negative.

```c
int32_t x[4]; // MACQ (mV)
int32_t d;    // derivative(V/s)
void ADC3_Handler(void){
  ADC0_ISC_R = 0x08;       // acknowledge ADC sequence 3 completion
  x[3] = x[2];  // shift data
  x[2] = x[1];  // units of mV
  x[1] = x[0];
  x[0] = (3300*ADC_SSFIFO3_R)/4096; // in mV
  d = x[0]+3*x[1]-3*x[2]-x[3];      // in V/s
  Fifo_Put(d);  // pass to foreground
}
```

Program 8.9. Software implementation of first derivative using a multiple access circular queue.

When the MACQ holds many data points it can be implemented using a pointer or index to the newest data. In this way, the data need not be shifted each time a new sample is added. The disadvantage of this approach is that address calculation is required during the **Read** access. For example, we could implement a 16-element averaging filter. More specifically, we will calculate the average of the last 16 samples.

```c
uint32_t Data[32];   // two copies
uint32_t *Pt;        // pointer to current
uint32_t Sum;        // sum of the last 16 samples
void LPF_Init(void){
  Pt = &Data[0]; Sum = 0;
}
// calculate one filter output
// called at sampling rate
// Input: new ADC data
// Output: filter output, DAC data
uint32_t LPF_Calc(uint32_t newdata){
  Sum = Sum-*(Pt+16); // subtract the one 16 samples ago
  if(Pt == &Data[0]){
    Pt = &Data[16];   // wrap
  } else{
    Pt--;             // make room for data
  }
  *Pt = *(Pt+16) = newdata; // two copies of the new data
  return Sum/16;
}
```

Program 8.10. Digital low pass filter implemented by averaging the previous 16 samples (cutoff = $f_s/32$).

See Volume 3, Embedded Systems: Real-Time Operating Systems for ARM Cortex-M Microcontrollers for more details on the design and implementation of digital filters.

8.6. Exercises

8.1 For each term give a definition in 32 words or less.
a) Thermal noise
b) CLC or π filter
c) Voltage comparator
d) Differential amplifier
e) Low pass filter
f) Rail to rail
g) Common mode rejection ratio
h) Output impedance
i) Shunt reference
j) Hysteresis
k) Analog isolation
l) Complementary binary
m) Analog multiplexor
n) Sample and hold

8.2 For each op amp parameter give a definition in 32 words or less.
a) Open loop gain
b) Gain bandwidth product
c) Offset voltage
d) Offset current
e) Bias current
f) Common mode input impedance
g) Differential mode input impedance
h) Slew rate
i) Noise density
j) Supply current

8.3 For each DAC parameter give a definition in 32 words or less.
a) Precision
b) Range
c) Resolution
d) Monotonic
e) Linearity
f) Settling time
g) Supply current
h) Slew rate

8.4 For each ADC parameter give a definition in 32 words or less.
a) Precision
b) Range
c) Resolution
d) Total harmonic distortion
e) No missing codes
f) Conversion time
g) Supply current
h) Bandwidth

8.5 For each pair of terms explain the similarities and differences in 32 words or less
a) Carbon versus metal film resistor
b) Ceramic versus tantalum capacitor
c) Linear versus buck-boost regulator
d) Regular versus rail-to-rail op amp
e) Op amp versus instrumentation amp
f) One-pole versus two-pole filter
g) Li-ion versus NiMH battery
h) Negative feedback versus positive feedback

8.6 Describe each ADC type in 32 words or less.
a) Flash
b) Sigma delta
c) Successive approximation
d) Pipeline

8.7 A microcontroller with an **8-bit** *DAC* is used to create a software-controlled analog output. The 8-bit signed DAC uses **offset binary** format is connected to a parallel port of the microcontroller and has a -2V to +2V range. N is the port output (DAC input) and V is the DAC output.
a) Derive the linear relationship between N and V. Show both the equation that expresses V in terms of N, and the equation that expresses N in terms of V.
b) What is the **DAC precision**? State the units.
c) What is the **DAC resolution**? State the units.

8.8 A microcontroller with a **10-bit** *DAC* is used to create a software-controlled analog output. The 10-bit signed DAC uses **two's complement** format is connected to a parallel port of the microcontroller and has a -1V to +1V range. N is the port output (DAC input) and V is the DAC output.
a) Derive the linear relationship between N and V. Show both the equation that expresses V in terms of N, and the equation that expresses N in terms of V.
b) What is the **DAC precision**? State the units.
c) What is the **DAC resolution**? State the units.

8.9 If a 10-bit ADC has a range of 0 to +10 volts, what is its resolution?

8.10 Give three equivalent ways to specify the precision of a 12-bit ADC.

8.11 If an 8-bit ADC has a range of 0 to 3 volts, what will be output for an input voltage of 1 volt assuming **straight binary** encoding?

8.12 If an 8-bit ADC takes inputs ranging from -2.5 V to +2.5 V, what will the output corresponding to +1 V assuming **offset binary** encoding?

D8.13. Design 2-bit analog to digital converter. The range is 0 to 3.2 V. For input voltages between 0 and 0.8V make the output 00, for inputs between 0.8 and 1.6V make the output 01, for inputs between 1.6 and 2.4V make the output 10, and for inputs between 2.4 and 3.2V make the output 11. Use rail-to-rail op amps powered at 3.3V. Specify resistor values and capacitor values.

D8.14. Design a variable gain analog amplifier. The analog input is V_{in}, the 3-bit digital inputs (connected to a microcontroller output digital output) are B_2, B_1, B_0 and the analog output is V_{out}. Use an analog switch, such as a MAX4783. Exactly one of the digital inputs will be one, and the gain should be

B2,B1,B0	Gain (V_{out}/V_{in})
001	1
010	10
100	100

D8.15. Design an analog circuit with the following specifications
  two single-ended inputs (not differential)
  any input impedance is OK
  transfer function $\quad V_{out} = 5 \cdot V_1 - 3 \cdot V_2 + 5$
You are limited to one OPA227 op amp and one reference chip (you choose it). Give chip numbers but not pin numbers. Specify all resistor values. You will use +12 and –12V analog supply voltages.

D8.16. Design an instrumentation amp, using an AD627, with the following transfer function.
  $V_{out} = 500 \cdot (V_2 - V_1)$

D8.17. Design an analog circuit with the following transfer function $V_{out} = 3 - 2 \cdot V_{in}$. The input is a single voltage (not differential). The input range is 0 to 1.5V and the output range is 0 to 3V. Use an analog reference and one rail-to-rail op amp. Show your work and label all chip numbers and resistor values. You do not have to show pin numbers.

D8.18 The input, V_{in}, is differential, not single-ended. Design an analog circuit with a transfer function of $V_{out} = 30*V_{in}+1.5$ powered by a single +3.3 V supply. You may use any of the analog chips in this chapter. The input range is -0.05 V to +0.05 V, and the output range is 0 to +3 V. Label all chips, resistors and capacitors as needed.

D8.19. The input, V_{in}, is single-ended, not differential. Design an analog circuit with a transfer function of $V_{out} = 30*V_{in}-3$ using one rail-to-rail op amp powered by a single +3.3 V supply. The input range is 0.1 V to 0.2 V, and the output range is 0 to +3 V.

D8.20. Design a two-pole Butterworth analog high pass filter with a cutoff of 10 Hz.

D8.21. Design a two-pole Butterworth analog low pass filter with a cutoff of 250 Hz.

D8.22. Design a two-pole Butterworth analog low pass filter with a cutoff of 10000 Hz.

D8.23. Design an analog band pass filter with a cutoffs of 1 and 1000 Hz. First, design a two-pole Butterworth analog high pass filter, and then follow it with a two-pole Butterworth analog low pass filter.

8.7. Lab Assignments

Lab 8.1. The overall objective of this lab is to design a variable gain amplifier, where the gain is controlled by a digital output of the microcontroller. The gain settings are 10, 20, 50, and 100. As preparation, design two separate circuits using different approaches. Evaluate the bandwidth, noise, error due to offset voltage, and cost. Build and experimentally measure bandwidth, noise, and offset error.

Lab 8.2. Most digital music devices rely on high-speed DACs to create the analog waveforms required to produce high-quality sound. In this lab you will create a very simple sound generation system that illustrates this application of the DAC. Your goal is to play your favorite song. For the first step, you will build a DAC. You are free to design your DAC with a precision anywhere from 5 to 8 bits. You will convert the binary bits (digital) to an analog output current using a simple resistor network. The third step is to convert the DAC analog output to speaker current using an audio amplifier. The fourth step is to design a low-level device driver for the DAC. The fifth step is to design a data structure to store the sound waveform. The sixth step is to organize the music software into a device driver. The last step is to write a main program that inputs from three binary switches and performs the three public functions.

Lab 8.3. The overall objective of this lab is design the music player described in Lab 8.2 with a two channel DAC chip, and design two audio amplifiers. The system will implement two-channel stereo sound.

9. System-Level Design

Chapter 9 objectives are to:

- Discuss issues associated with manufacturability and testability
- Describe power sources
- Design methods to charge batteries
- Present approaches for low-power design
- Introduce PCB design

Chapters 1 to 8 of this book have presented embedded systems from an interfacing or component level. The remaining chapters will focus on systems level design. The chapter begins with a discussion of selecting resistor and capacitor components. Next, it will describe power sources, including batteries and battery chargers. One of the important aspects of embedded systems is low-power operation. The chapter will conclude with an introduction to PCB design.

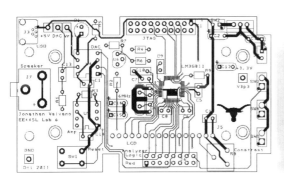

9.1. Design for Manufacturability

Using standard values for resistors and capacitors makes finding parts quicker. Standard values for 1% resistors range from 10 Ω to 2.2 MΩ. We can multiply a number in Tables 9.1, 9.2, 9.3, and 9.4 by powers of 10 to select a standard value resistor. For example, if we need a 5 kΩ 1% resistor, the closest number is 49.9*100, or 4.99 kΩ.

Sometimes we need a pair of resistors with a specific ratio. There are 19 pairs of resistors with a 2 to 1 ratio (e.g., 20/10). There is only one pair with a 3 to 1 ratio, 102/34. Similarly, there is only one pair with a 4 to 1 ratio, 102/25.5. There are 19 pairs of resistors with a 5 to 1 ratio (e.g., 100/20). There are 5 pairs of resistors with a 7 to 1 ratio (e.g., 93.1/13.3, 105/15, 140/20, 147/21, 196/28). There are no pairs with ratios of 6, 8, or 9.

Using standard values can greatly reduce manufacturing costs because parts are less expensive, and parts for one project can be used in other projects. Ceramic capacitors can be readily purchased as E6, E12, or E24 standards. Filters scale over a fairly wide range. If a resistor is increased by a factor of x and the capacitor is reduced by a factor of x, the filter response will remain unchanged. For example, the response of a filter that uses 100 kΩ and 0.1 µF will be the same as a filter with 20 kΩ and 0.5 µF. Resistors that are too low will increase power consumption in the circuit, and resistor values that are too high will increase noise. 1% resistors below 100 Ω and above 10 MΩ are hard to obtain. Precision capacitors below 10 pF and above 1 µF are hard to obtain. High-speed applications use lower values of resistors in the 100 Ω to 1 kΩ range, precision equipment operates best with resistors in the 100 kΩ to 1 MΩ range, while portable equipment uses higher values in the 100 kΩ to 10 MΩ range.

E12 standard values for 10% resistors range from 10 Ω to 22 MΩ. We can multiply a number in Table 9.1 by powers of 10 to select a standard value 10% resistor or capacitor. The E6 series is every other value and typically available in 20% tolerances.

10	12	15	18	22	27	33	39	47	56	68	82

Table 9.1. E12 Standard resistor and capacitor values for 10% tolerance.

E24 standard values for 5% resistors range from 10 Ω to 22 MΩ. We can multiply a number in Table 9.2 by powers of 10 to select a standard value 5% resistor. For example, if we need a 25 kΩ 5% resistor, the closest number is 24*1000, or 24 kΩ. Capacitors range from 10 pF to 10 µF, although ceramic capacitors above 1 µF can be quite large. The physical dimensions of a capacitor also depend on the rated voltage. You can also get 1% resistors and 1% capacitors in the E24 series. For example, if you need a 0.05 µF capacitor, you can choose an 0.047µF E12, or a 0.051µF E24 capacitor.

10	11	12	13	15	16	18	20	22	24	27	30
33	36	39	43	47	51	56	62	68	75	82	91

Table 9.2. E24 Standard resistor and capacitor values for 5% tolerance.

Table 9.3 shows the E96 standard resistance values for 1 % resistors. Table 9.4 shows E192 standard resistance values for 0.5, 0.25, 0.1% tolerances. Tables 9.1 and 9.2 refer to both resistors and capacitors, but the E96 and E192 standards refer only to resistors.

10.0	10.2	10.5	10.7	11.0	11.3	11.5	11.8	12.1	12.4	12.7	13.0
13.3	13.7	14.0	14.3	14.7	15.0	15.4	15.8	16.2	16.5	16.9	17.4
17.8	18.2	18.7	19.1	19.6	20.0	20.5	21.0	21.5	22.1	22.6	23.2
23.7	24.3	24.9	25.5	26.1	26.7	27.4	28.0	28.7	29.4	30.1	30.9
31.6	32.4	33.2	34.0	34.8	35.7	36.5	37.4	38.3	39.2	40.2	41.2
42.2	43.2	44.2	45.3	46.4	47.5	48.7	49.9	51.1	52.3	53.6	54.9
56.2	57.6	59.0	60.4	61.9	63.4	64.9	66.5	68.1	69.8	71.5	73.2
75.0	76.8	78.7	80.6	82.5	84.5	86.6	88.7	90.9	93.1	95.3	97.6

Table 9.3. E96 Standard resistor values for 1% tolerance(resistors only, not for capacitors).

Checkpoint 9.1: Let R = 100 kΩ. Find an E24 capacitor such that $1/(2\pi RC)$ is as close to 1000 Hz as possible.

Checkpoint 9.2: Rather than using an E96 resistor, find two E24 resistors such that the series combination is as close to 127 kΩ as possible.

10.0	10.1	10.2	10.4	10.5	10.6	10.7	10.9	11.0	11.1	11.3	11.4
11.5	11.7	11.8	12.0	12.1	12.3	12.4	12.6	12.7	12.9	13.0	13.2
13.3	13.5	13.7	13.8	14.0	14.2	14.3	14.5	14.7	14.9	15.0	15.2
15.4	15.6	15.8	16.0	16.2	16.4	16.5	16.7	16.9	17.2	17.4	17.6
17.8	18.0	18.2	18.4	18.7	18.9	19.1	19.3	19.6	19.8	20.0	20.3
20.5	20.8	21.0	21.3	21.5	21.8	22.1	22.3	22.6	22.9	23.2	23.4
23.7	24.0	24.3	24.6	24.9	25.2	25.5	25.8	26.1	26.4	26.7	27.1
27.4	27.7	28.0	28.4	28.7	29.1	29.4	29.8	30.1	30.5	30.9	31.2
31.6	32.0	32.4	32.8	33.2	33.6	34.0	34.4	34.8	35.2	35.7	36.1
36.5	37.0	37.4	37.9	38.3	38.8	39.2	39.7	40.2	40.7	41.2	41.7
42.2	42.7	43.2	43.7	44.2	44.8	45.3	45.9	46.4	47.0	47.5	48.1
48.7	49.3	49.9	50.5	51.1	51.7	52.3	53.0	53.6	54.2	54.9	55.6
56.2	56.9	57.6	58.3	59.0	59.7	60.4	61.2	61.9	62.6	63.4	64.2
64.9	65.7	66.5	67.3	68.1	69.0	69.8	70.6	71.5	72.3	73.2	74.1
75.0	75.9	76.8	77.7	78.7	79.6	80.6	81.6	82.5	83.5	84.5	85.6
86.6	87.6	88.7	89.8	90.9	92.0	93.1	94.2	95.3	96.5	97.6	98.8

Table 9.4. E192 Standard resistor values for tolerances better than 1% (resistors only, not for capacitors).

9.2. Power

9.2.1. Regulators

Every embedded system needs power to operate. The source of power could be

- 120 VAC 60 Hz
- DC power, like +5V on USB or +12V in an automobile
- Battery
- Energy harvesting like solar or EM field pickup

Many embedded systems are powered with an AC adapter. Other names for this adapter that takes 120 VAC in and outputs an unregulated DC voltage include power adapter, power block, wall wart, wall cube, and power brick. A 9-V, 500-mA unregulated AC adapter means the voltage is above 9 V for all currents less than 500 mA. However, the actual voltage can vary considerably. For example, the voltage might range from 13 V at no current down to 9.5V at 500 mA. Similarly, battery voltage is not constant, decreasing with age and use. Therefore, a **regulator** will be used to provide a constant voltage to power the system. In this section we will introduce two types of regulators: linear and switching. There are many considerations when choosing a regulator, and we will discuss some of these considerations.

Linear regulators take an input voltage, V_{in}, and create a constant output voltage, V_{out}. Capacitors on both the input and output are needed to stabilize the linear regulator (Figure 9.1).

The band gap reference includes a diode to create a constant reference voltage, V_{ref}, in a similar way we used the LM4041 in Chapter 8. The two resistors will sense the output voltage. The sensed output voltage is compared to V_{ref}. The output of the error amp will control the **pass element**, which is a transistor device whose resistance can be adjusted. It is classified as a linear regulator because the pass element operates in its linear region. The regulator uses feedback to maintain a constant V_{out} regardless of load current, I_{out} and input voltage V_{in}. Some regulators, like the 78x05 series use a bipolar transistor for the pass element. While other regulators, like LP2981, use a MOSFET for the pass element. In either case, current can only flow from input to output across the pass element, meaning V_{in} must be greater than V_{out}. Furthermore, because the pass element is resistive, power will be lost as voltage drops across the regulator. The quiescent current of a regulator, I_q, is the current needed to operate internal functions of the regulator, like the band gap reference.

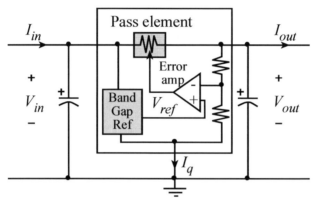

Figure 9.1. Architecture of a linear regulator.

Linear regulators need an input voltage, V_{in}, larger than the output voltage, V_{out}. If the current is I_{out}, then power is lost in the regulator and dissipated as heat in the amount of $P_d = (V_{in}-V_{out})*I_{out}$. Linear regulators have a **voltage dropout**, V_{do}, which specifies how much higher V_{in} needs to be above V_{out} in order for it to work. The dropout voltage for the 5-V 78L05 regulator is 2 V, meaning V_{in} must be larger than +7 V for the output to be +5V. Conversely, the dropout voltage for the LP2981 is only 0.2 V. For example, a 3.7-V Li-ion battery with a LP2981-3.3 regulator could be used to power a 3.3 V system. Regulators with a low dropout voltage are classified as LDO regulators. Figure 9.2 shows some circuits based on linear regulators. Because of the resistive nature of the pass element, linear regulators are not very efficient. The efficiency is defined as the output power divided by the input power:

$$\eta = (V_{out} * I_{out})/(V_{in} * I_{in}) = (P_{out})/(P_{out} + P_d)$$

Regulator	I_{out}	V_{do}	I_q
LM2937-3.3	500 mA	1.45 V	10 mA
LM1117-3.3	800 mA	1.2 V	5 mA
LP2950-3.3	100 mA	0.38 V	75 µA
LP2981-3.3	100 mA	0.2 V	0.6 mA
TPS78233	150 mA	0.13 V	500 nA

Table 9.5. Typical parameters for some 3.3-V linear regulators (see respective data sheets).

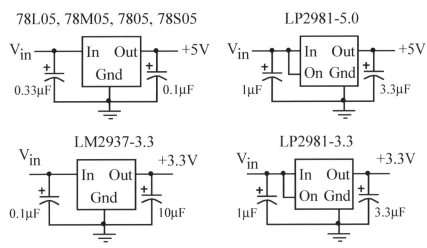

Figure 9.2. Linear regulator circuits.

In order to improve efficiency and provide for flexible input voltage, there is class of regulators that use an inductor and a switching circuit to create the constant output voltage. Three types of switching regulators are shown in Figure 9.3. A **buck** regulator uses switch at the input to dump power into an LC circuit. Like the linear regulator, a buck regulator requires the input voltage to be larger than the output voltage. A **boost** regulator rearranges the switch and inductor, but it also uses the switch to dump power into an LC circuit. When the switch is open power flows into the capacitor; when the switch is closed power is stored on the inductor. A boost regulator can only be used in situations where the input voltage is less than the output voltage. A **buck-boost** regulator uses a method called flyback so the output voltage can be regulated regardless of whether the input voltage is larger or smaller than the desired output. In each case, the duty cycle of the switch is controlled so that the output voltage remains constant. Assuming ideal components these switching regulators will be 100% efficient.

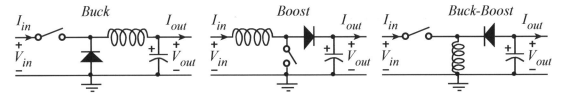

Figure 9.3. Architecture of a switching regulator.

The TPS63002 buck-boost regulator will create a +5-V output as long as the input is between 1.8 and 5.5 V, making it suitable for battery-powered applications. Buck-boost regulators are rated for their efficiency and have a sizeable high frequency switching noise in the power line. The frequency of the noise depends on how the buck-boost regulator is built. Since linear regulators have lower noise, they are more suitable for precision analog electronics. Noise on V_{out} can come from the input, like the 60 Hz ripple from a 120 VAC transformer, or noise can come from within the regulator, like the switching regulator. Let V_{inrms} be the rms AC amplitude on the input, and V_{outrms} be the rms AC amplitude at the output. **Power supply rejection ratio** (*PSRR*) is the amount of ripple on the input that is transferred to the output.

$$PSRR = 20 * \log_{10}(V_{inrms}/V_{outrms})$$

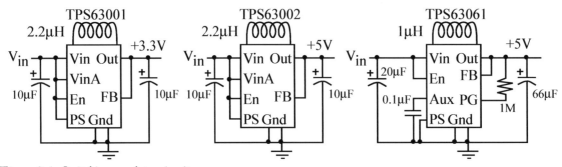

Figure 9.4. Switching regulator circuits.

Checkpoint 9.3: Which regulator in Figures 9.2 and 9.4 can create a 5V supply from a 3.7V battery?

Regulators can be **fixed output** (e.g., 3.3 or 5 V) or **adjustable**. Adjustable regulators use two resistors to set the output voltage level. Many regulator families have both fixed output and adjustable versions such as LM1086, LM1117, LT1761, TPS63000, and TPS63060. Adjustable regulators are convenient if the system needs to have a user-controlled switch to decide whether the system is powered at 3.3 or 5 V. We choose a regulator that has a **maximum current** larger than the current requirements of the system. For example, the maximum currents for the 78L05, 78M05, 7805, and 78S05 are 100 mA, 500 mA, 1 A, and 2 A, respectively. The **line regulation** parameter specifies how the output voltage will vary as a function of input voltage. The **load regulation** parameter specifies how the output voltage will vary as a function of the current to the system. Table 9.6 characterizes the strengths of a linear versus switching regulator.

Observation: Regulators are very different from each other, so it is essential to read the data sheet and follow all the design recommendations concerning external components, heat sinks, and layout.

Linear Regulator	Switching Regulator
Small size	High efficiency
Low noise, low output ripple	Wide range of input voltage
Simple design	V_{out} less than or greater than V_{in}
Fast transient response	Low power dissipation (small heat sink)
Low quiescent current	Can have multiple output voltages
Low cost	With a transformer, could be isolated
Good PSRR	Best for large currents
No EMI emissions	

Table 9.6. Advantages of regulator types.

The TM4C microcontrollers need a regulated 3.3-V supply voltage. However, many microcontrollers will operate on a wide range of supply voltages

MSP430L092 will run on supply voltage from 0.9 to 1.65V
MSP430F20x2 will run on supply voltage from 1.8 to 3.6V
MSP432P401 will run on supply voltage from 1.62 to 3.7V

For some applications it may be possible to run the system directly off a battery without a regulator. Without regulated power, these systems will have more variability in bus clock frequency and analog voltage. However, without a regulator the system can be smaller, lower cost, and by design it will be 100% battery efficient.

Observation: I_q is important when putting a microcontroller into deep sleep for a long time.

Many analog IC manufacturers provide design tools. **WebBench** is a free design tool from Texas Instruments you can use to design power supply circuits (webench.ti.com/). Using these design tools you will be able to create supply circuits much better than the ones presented in this book.

9.2.2. Low-Power Design

To save energy, our parents taught us to "turn off the light when you leave the room." We can use this same approach to conserve energy in our embedded system. There are many ways to place analog circuits in low-power mode. Some analog circuits have a low-power mode that the software can select. For example, the MAX5353 12-bit DAC requires 280 µA for normal operation, but the software can place it into shut-down mode, reducing the supply current to 2 µA. Some analog circuits have a digital input that the microcontroller can control placing the circuit in active or low-power mode. For example, the TPA731 audio amplifier has a **CD** pin, see Figure 9.5. When this pin is low, the amplifier operates normally with a supply current of 3 mA. However, when **CD** pin is above 2 V, the supply current drops down to 65 µA. So, when the software wishes to output sound, it sends a command to the MAX5353 to turn on and makes PB0 equal to 0. Conversely, when the software wishes to save power, it sends a shutdown command to the MAX5353 and makes PB0 high.

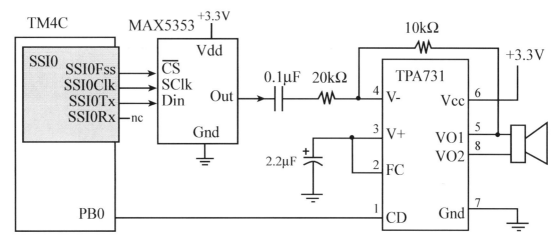

Figure 9.5. Audio amplifier that can be placed into low-power mode.

The most effective way to place analog circuit in a low-power state is to actually shut off its power. Some regulators have a digital signal the microcontroller can control to apply or remove power to the analog circuit. For example, when the OFF pin of the MAX604 regulator is high, the voltage output is regulated to +3.3 V, as shown in Figure 9.6. Conversely, when the **OFF** pin is low, the regulator goes into shut-down mode, and no current is supplied to the analog circuit. When the software wishes to turn off power to the analog circuit, it makes PB0 equal to 0. Conversely, when the software wishes enable the analog circuit, it makes PB0 high. The microcontroller itself always will be powered. However, most microcontrollers can put themselves into a low-power state.

We can save power by designing with low-power components. Many analog circuits require a small amount of current, even when active. The MAX494 requires only 200 µA per amplifier. If there are ten op amps in the circuit, the total supply current will be 2 mA. For currents less than 8 mA, we can use the output port itself to power the analog circuit, as shown in Figure 9.7. To activate the analog circuit, the microcontroller makes the PB0 high. To turn the power to the analog circuit off, the microcontroller makes PB0 low.

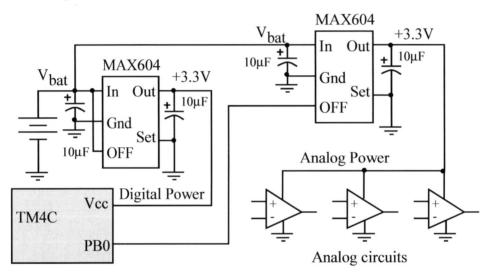

Figure 9.6. Power to analog circuits can be controlled by switching on/off the regulator.

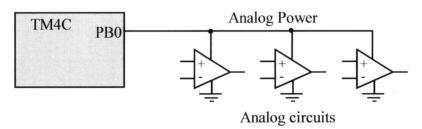

Figure 9.7. Power to analog circuits can be delivered from a port output pin.

As we mentioned in Chapter 1, considerable power can be saved by reducing the supply voltage. A microcontroller operating at 3.3 V requires less than half the power for an equivalent +5 V system. Power can be saved by turning off modules (like the timer, ADC, UART, and SSI) when not in use. Whenever possible, slowing down the bus clock with the PLL will save power. Many microcontrollers can put themselves into a low-power sleep mode to be awakened by a timer or external event.

There are a number of factors that affect the supply current to a microcontroller. The first is bus frequency; power increases linearly with bus frequency. A second factor is turning on/off I/O ports. A third factor is activating a sleep mode. The TM4C has a **sleep mode**, a deep sleep mode, and a hibernate mode. The deep sleep mode on a LM3S811 requires about 1 mA of supply current. Hibernate mode on the TM4C123 requires 5 µA. Conversely, the deep sleep mode on a MSP430F2012 microcontroller is less than 1 µA.

9.2.3. Battery Power

A **battery** is a source of energy that can be used in an embedded system to make the system portable. Another application of batteries is to supply power to a mission-critical system when the regular AC power is lost temporarily. Typically, a battery has three parts. The anode is the negative terminal of the battery, the cathode is the positive terminal, and the electrolyte is a liquid solution that accepts stores and releases energy. These three components can be constructed from many different materials and configured in an almost endless array of sizes and shapes. The type, size, and shape of the materials play a major role in determining the battery performance. A **primary battery** is used once and discarded, and a **secondary battery** can be recharged and reused.

There are many parameters to consider when selecting a battery. **Nominal voltage** is the typical voltage of the battery when fully charged. Some batteries maintain a fairly constant voltage output while energy is being discharged. However, other batteries will drop the voltage steadily during usage. Physical parameters of the battery (such as **volume**, **weight**, and **shape**) often play a significant role in the overall appeal of an embedded system. **Peak current** is the maximum current the battery can deliver. Shelf-life, operating temperature, and storage temperature are other parameters to consider when choosing a battery. **Memory effect** is an observable condition in some rechargeable batteries that causes them to hold less charge over time. The **energy storage** of a battery is typically defined in amp-hours, because the voltage is assumed constant. The standard units of energy are watt-hours (1 W-hr is 3600 J). One can estimate the operation time of a battery-powered embedded system by dividing the energy storage by the average current required to run the system. The **power budget** embodies this concept. Let E be the battery specification in amp-hours and t_{life} be the desired lifetime of the product; then we can estimate the average current our system is allowed to draw:

$$Average\ Current \leq E/t_{life}$$

Checkpoint 9.4: A medical device implanted inside the body has a 500 mA-hour battery that must last 5 years. What is your power budget?

Heavy-duty batteries, were first made with zinc-carbon in the mid-1800s, but now are made with zinc chloride. They are a low-cost, low-performance battery but are not appropriate for most embedded applications. An **alkaline** battery is made with alkaline manganese. Alkaline batteries are appropriate for situations that require long shelf life, but size and weight are not important. There are two kinds of **lead acid** batteries. Flooded lead-acid vent inflammable gasses and require additional water to maintain the proper specific gravity of the acid. Valve-regulated lead-acid (VRLA also called sealed lead battery) have about a two-to-one advantage over the flooded type battery in specific energy and energy density. In the VRLA cell, the vent for the gas space incorporates a pressure relief valve to minimize the gas loss and to prevent direct contact between the headspace and outside air. Lead acid batteries can be used for backing up power on systems that require large currents. Lead acid batteries have a maximum storage time of six months at temperatures between 20 and 30°C, after which they require a freshening charge. Zinc chloride, alkaline, and lead-acid batteries all have voltages that drop as energy is drained from them. In these systems, the voltage can be monitored as a measure of the energy left in the battery. However, embedded systems that use these types of battery will require a voltage regulator to maintain a constant voltage for the electronics. For example, the MSP430 microcontroller will operate with a power supply voltage anywhere from 1.8 to 3.6 V. The TM4C microcontrollers will run with a supply voltage from 3.0 to 3.6 V.

Nickel-cadmium (NiCad) and **nickel-metal hydride** (NiMH) are low-cost rechargeable batteries that used to be popular for embedded systems. NiMH batteries have about twice the storage capacity as NiCad. Certain NiCad batteries gradually lose their maximum energy capacity if they are repeatedly recharged after being only partially discharged. Most NiMH batteries do not suffer from a memory-effect. The NiMH batteries operate between 10 to 55°C, and have a projected life of seven and a half years at 30 °C. You should cycle new NiMH batteries three to five times to achieve peak performance. Cycling or conditioning a NiMH battery is performed by completely discharging it then completely recharging it. At room temperature, NiMH batteries will self-discharge in 30 to 60 days without usage, depending on environmental conditions. In general, you can expect NiMH batteries to last up to 500 recharges.

The search for a lighter battery that uses metallic lithium as its anode was driven by the fact that lithium is the lightest and the most electropositive of metals. The specific energy of lithium metal (1727Ah/lb) is greater than lead (118Ah/lb) and cadmium (218Ah/lb). There are a whole range of batteries based on Lithium, both single use (used in cameras) and rechargeable. The most common rechargeable type is called **Lithium-ion** (Li-ion). When energy is being discharged, the lithium ion moves from the anode to the cathode. During charging, the lithium ion moves from the cathode to the anode. Because of their excellent energy-to-weight and energy-to-size ratios, Lithium-ion rechargeable batteries are commonly employed in portable embedded systems. Table 9.7 shows energy storage for typical AA-sized batteries (50 mm tall by 14 mm diameter). Table 9.8 lists some Li-Ion rechargeable batteries.

Battery	Voltage (V)	Energy (mAh)	Type
Alkaline	1.5	2000	Primary
Lithium	1.5	3000	Primary
NiCad	1.2	1200	Secondary
NiMH	1.2	1800	Secondary
Li-ion	3.6	1900	Secondary

Table 9.7. Energy storage for different AA-sized battery types.

Battery	Shape	Energy (mAh)	Size (mm)	Weight (g)
GMB power GM041124	Thin Pack	60	4.0 × 11 × 24	1.2
GMB power GM041429	Thin Pack	120	4.0 × 14 × 29	2.8
GMB power GM041842	Thin Pack	250	4.0 × 18 × 42	5.5
TrustFire 10440	AAA	600	10.25 × 46.25	9.5
UltraFire 14500	AA	900	14 × 50	31
Tenergy 30027	Cylinder	2200	69 × 19	54
Tenergy 31000	Pack	4400	65 × 37 × 18	170
Tenergy 31002	Pack	6600	69 × 54 × 18	255

Table 9.8. Example 3.7 V Li-Ion batteries (http://www.powerstream.com/ http://www.gmbattery.com/ http://www.batteryjunction.com/).

Checkpoint 9.5: A medical device implanted inside the body has an average current 20 µA and must last for 1 year. Which battery in Table 9.8 would you choose?

To make the system more convenient, battery-powered embedded systems could include a built-in recharger. The system can run and charge while plugged in. However, it will also run off the battery when AC power is not available. There are many battery charging circuits available. Figure 9.8 shows a charging circuit a NiMH battery. *R1* is depends on the smallest V_{in} from the wall cube. If V_{in} is larger than 10 V, the set $R1 = (10-5V)/5mA = 1 kΩ$. *R3* and *R4* are part of a thermal shut-down safety circuit, which is not implemented in this simple version. The *Rsense* resistor sets the fast programming current: $I_{fast} = 2.5V/Rsense$, which is 1 A for this circuit. The PGM0, PGM1, PGM2, and PGM3 pins specify the charging protocol and the number of NiMH cells, which is 7 cells or 8.4 V for this circuit. Figure 9.9 shows a charging circuit for a Li-ion battery. You should refer to the specific datasheets to work through the resistor and capacitor values depending on the voltage and current of the battery.

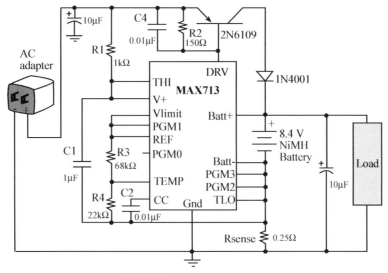

Figure 9.8. Charging circuit for an 8.4-V NiMH battery with a 1-A charging current.

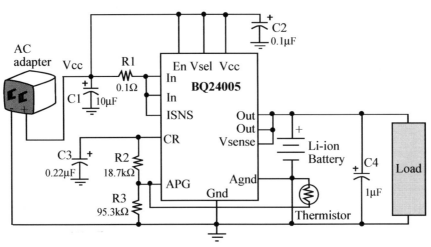

Figure 9.9. Charging circuit for a 7.2-V Li-ion with a 1.2-A charging current.

9.2.4. Measurement of current

In order to evaluate the performance of a low power system we need a method to measure supply current on the actual system. A simple method would be to charge up a battery with known energy storage and measure how long the battery runs the system. There are a couple of circuits that could be used to measure current, allowing us to visualize supply current versus time. If a +5V supply is available we could use an instrumentation amplifier as shown in Figure 9.10. In this application, we expect the current to be less than 30 mA, so a shunt resistance of 2Ω is selected, because the 3.3V supply will only drop to 3.24V and not affect system operation. Since the instrumentation amp is powered with 5V and ground, the voltages at the two inputs (pins 2 and 3) will be around 3.3V operating the amp in its linear region. If the supply current is 30 mA, the output voltage will be 2.68V. When using this circuit, I will calibrate the measurement by using three or four precision resistors in place of the system.

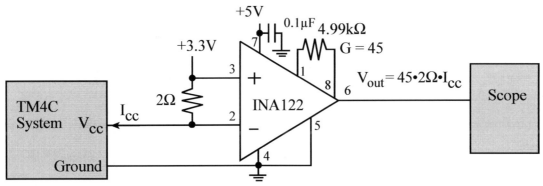

Figure 9.10. Circuit used to measure supply current, I_{cc}.

To illustrate the use of Figure 9.10, two TM4C123 systems were designed. Both systems toggled an output pin every 5 seconds. Both systems ran on the internal oscillator at 16 MHz. The first system used SysTick and the second one used hibernation to enter deep sleep while waiting the 5 seconds. Program 9.1 shows the low-level software to hibernate. The real time clock (RTC) is programmed to wake up the processor in a specified number of seconds. Since the hibernation module runs on the 32768Hz external crystal, the software must wait for the **WRC** bit in the **CTL** register after every access. Initialization turns on the hibernation module and connects to the 32768Hz clock. The **VDD3ON** bit is set so the digital outputs remain powered while sleeping. To start hibernation we set the match time in **RTCM0**, clear the RTC counter, and issue a hibernation request. It takes some time to activate, so the dummy for-loop is added to the execution never returns from hibernation.

```
void static HIBready(void){
  while((HIB_CTL_R&0x80000000)==0){};   // wait for WRC, Write Complete
}
void HIB_Init(void){
  SYSCTL_RCGCHIB_R |= 0x01;
  while((SYSCTL_PRHIB_R&0x01)==0){}; // wait for clock
  HIBready(); HIB_CTL_R = 0x40;    // CLK32EN Hibernation module clock
  HIBready(); HIB_CTL_R = 0x140;   // VDD3ON internal switches
}
void HIB_Hiberate(uint32_t seconds){ int i;
  HIBready(); HIB_RTCM0_R = seconds; // time to sleep
  HIBready(); HIB_RTCSS_R = 0;       // no subseconds time
  HIBready(); HIB_RTCLD_R = 0;       // clear RTC counter
  HIBready(); HIB_CTL_R = 0x014B;    // go to sleep
  for(i=1000;i;i--){};               // this should not finish
}
void HIB_SetData(uint32_t data){
  HIBready(); HIB_DATA_R = data;     // save important information
}
uint32_t HIB_GetData(void){
  HIBready();
  return HIB_DATA_R;    // recover data
}
```
Program 9.1. Software to put the microcontroller into deep sleep (HIB_4C123).

Program 9.2 shows two high-level programs that toggle an output pin every 5 seconds. The one on the left uses SysTick and requires a continuous 7.1 mA to operate. The program on the right uses hibernation. When the software wakes up from hibernation, it begins execution at the reset vector. However, there are 16 words (64 bytes) of backup RAM that can be used to remember data after hibernation. The **SetData** and **GetData** functions allow the user to save and restore information in the first of these 16 words (**HIB_DATA_R**). In this example, the number of times the system has been awakened is stored in this backup RAM. If the count is odd the output is set high, and if the count is even the output is cleared low. It takes about 10.4 ms to execute this simple program, during which time the processor used about 5 mA of current. However in deep sleep for 4.99sec, the system only draws 0.1 mA. Figure 9.11 plots the instantaneous current during the wakeup. Notice the current surge occurring during the initial wakeup.

```
int main(void){                            int main(void){ uint32_t count;
  SysTick_Init();  // 16 MHz                 HIB_Init();
  SYSCTL_RCGCGPIO_R |= 0x00000020;           SYSCTL_RCGCGPIO_R |= 0x00000020;
  while((SYSCTL_PRGPIO_R&0x20)==0);          while((SYSCTL_PRGPIO_R&0x20)==0);
  GPIO_PORTF_DIR_R = 0x0E;                   GPIO_PORTF_DIR_R = 0x0E;
  GPIO_PORTF_DEN_R = 0x0E;                   GPIO_PORTF_DEN_R = 0x0E;
  while(1){                                  count = HIB_GetData();
    SysTick_Wait10ms(500); // 5 sec          count++;       // number of times
    PF2 ^= 0x04;    // toggle                HIB_SetData(count);
  }                                          PF2 = (count&0x01)<<2; // toggle
}                                            HIB_Hiberate(6);  // 5 second
                                             while(1){ // should not be here
                                             }
                                           }
```

Program 9.2. System that toggles an output every 5 seconds (HIB_4C123).

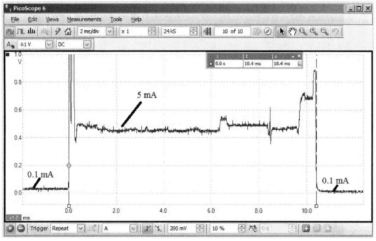

Figure 9.11. Instantaneous current measured versus time measured with the circuit in Figure 9.10. This measurement was taken on a TM4C123 LaunchPad. $I_{cc} = V/90$.

Observation: The hibernation system executes about 50 lines of C code each time it runs. The processor is running at 16 MHz. However, because the HIB module runs on the 32kHz clock, these 50 lines require 10.4 ms to execute.

Figure 9.10 was also used to study the effect of the PLL on supply current. In each system Program 5.10 on a TM4C123 LaunchPad was used to create a periodic interrupt that toggled an output pin. Running the system at 16 MHz with the internal oscillator required an average current of 3.5 mA, while running the system at 16 MHz with the PLL required 8 mA.

Next I ran the periodic interrupt at a range of bus frequencies. I then ran the same frequencies a second time, but this time I executed a **WaitForInterrupt** (**WFI**), which powers down some of the processor while waiting for the interrupt. Figure 9.12 shows the results of the study.

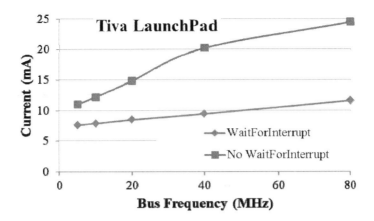

Figure 9.12. System current depends on bus frequency and one can save power by executing WFI when waiting for an interrupt. The measurements were performed on a TM4C123 LaunchPad with the PLL active.

Sometimes one wishes to measure current but there is not extra supply voltage like the +5V supply used in Figure 9.10. For this situation we can use a **current sense amplifier** that is specifically designed to collect power off the same supply with which we are measuring current. The circuit in Figure 9.13 will measure supply current up to 5V yielding a linear output from 0 to 100 mV. Again it uses a shunt resistor to create a voltage linear with current. For large currents we should use a Kelvin 4-terminal shunt resistor to improve accuracy. However this circuit is specially designed to operate linearly enough though the input voltages are similar to the supply. Gain is determined by the resistor ratio, 4999/100 = 50.

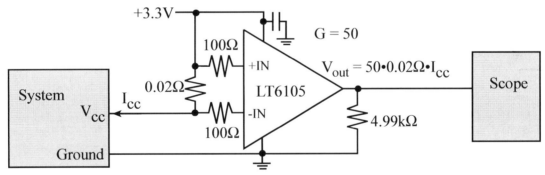

Figure 9.13. Circuit used to measure supply current, I_{cc}.

9.3 Tolerance

The first step in effective communication is establishing a clear agreement on the definition of terms. Broadly put, **tolerance** means putting up with error. However, the dictionary definitions of tolerance and error are not adequate for engineering design as listed in Table 9.9. From an engineering perspective, tolerance is the quantitative difference between the desired parameter and the actual value. For example, a ±1% 1000Ω resistor may have an actual value from 990Ω to 1010Ω. Sometimes a parameter is listed as the minimum and maximum values. For example, the offset voltage of an OPA2350 op amp will be between -500 and +500 µV. Specifying tolerance will have a profound impact on both price and performance. If we over specify (use 1% resistors when 5% would have been OK), then the system costs increase. However, if we under specify (use 5% resistors for a system needing 1%), then some of the devices we manufacture will not operate properly.

Tolerance	Error
1) a fair, objective, and permissive attitude toward those whose opinions, practices, race, religion, nationality, etc., differ from one's own; freedom from bigotry.	1) a deviation from accuracy or correctness; a mistake, as in action or speech: His speech contained several factual errors.
2) a fair, objective, and permissive attitude toward opinions and practices that differ from one's own.	2) belief in something untrue; the holding of mistaken opinions.
3) interest in and concern for ideas, opinions, practices, etc., foreign to one's own; a liberal, undogmatic viewpoint.	3) the condition of believing what is not true: in error about the date.
4) the act or capacity of enduring; endurance: My tolerance of noise is limited.	4) a moral offense; wrongdoing; sin.
5) Medicine/Medical, Immunology. a) the power of enduring or resisting the action of a drug, poison, etc.: a tolerance to antibiotics. b) the lack of or low levels of immune response to transplanted tissue or other foreign substance that is normally immunogenic.	5) Baseball. a misplay that enables a base runner to reach base safely or advance a base, or a batter to have a turn at bat prolonged, as the dropping of a ball batted in the air, the fumbling of a batted or thrown ball, or the throwing of a wild ball, but not including a passed ball or wild pitch.

Table 9.9. Dictionary definitions of tolerance and error http://dictionary.reference.com

Another factor related to tolerance is temperature coefficient. Properties of most devices will change with temperature. For example, the Temperature Coefficient of Resistance (TCR) is defined as

$$\text{TCR (ppm/°C)} = \frac{R - R_{25}}{R_{25}(T - 25°)} * 10^6$$

where R_{25} is the temperature at 25°C, and R is the resistance at temperature T, as T varies from -55°C to +125°C.

We will do a simple example illustrating the effect of tolerance on performance. We will revisit the LED interface initially presented in Figure 1.23. Back in Chapter 1, we assumed the V_{CE} of the PN2222 was 0.3V, the desired LED operating point was 1.9V 10mA, and we calculated the resistor value to be (3.3-1.9-0.3V)/10mA = 110Ω.

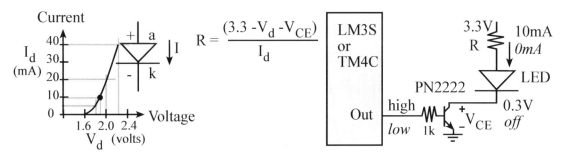

Figure 1.23. Open collector used to interface a light emitting diode.

The LED data sheet specifies the light intensity is linear with current. So we will determine the effect of tolerance on LED current. In particular we will investigate an E24 5%-resistor with nominal value of 110 Ω. This resistor can vary from 104.5 to 115.5 Ω, causing the LED current to vary from 9.5 to 10.5 mA. In this case the ±5% resistance tolerance converted simply to a ±5% variability in light intensity. To consider the effect of tolerance in the PN2222 transistor, we need to know the possible range for V_{CE}. At saturation, the V_{CE} has a maximum of 0.3 V. Assuming a worst case of V_{CE} equal to 0, the LED current could increase to 12.7 mA. Putting these two effects together is presented in Table 9.10. Realistically, the V_{CE} probably varies only from 0.2 to 0.3 V, so the LED current could vary from 9.5 to 11.5 mA, or -5% to +15% from desired.

	$V_{CE} = 0$	$V_{CE} = 0.1$	$V_{CE} = 0.2$	$V_{CE} = 0.3$
R=104.5	13.4	12.4	11.5	10.5
R=110	12.7	11.8	10.9	10.0
R=115.5	12.1	11.3	10.4	9.5

Table 9.10. LED current varies with R and V_{CE}.

Table 9.10 is a simple two-dimensional analysis. If we added the effect of temperature, then the problem becomes three dimensional. As the number of parameters increases so does the complexity of the problem. It is important to determine which parameters have the largest effect on output. These parameters will then require lower tolerance. For example, assume the circuit has three resistors (R_1, R_2, and R_3) and the output is the voltage V. Each resistor has an error due to tolerance and/or temperature drift ($\pm\delta_1$, $\pm\delta_2$, and $\pm\delta_3$). We could use differential calculus to determine the relative sensitivity to each parameter (S_1, S_2, and S_3). We should focus our money and attention on the parameters having the largest effect on performance. We can calculate the error in V (δ_V) due to errors in R. Let x be the measurand and let Δx be the desired resolution. Using the sensitivity of V to x, we can determine a design metric.

$$S_1 = \partial V/\partial R_1 \qquad S_2 = \partial V/\partial R_2 \qquad S_3 = \partial V/\partial R_3$$

$$\delta_V = \pm(\partial V/\partial R_1)\cdot\delta_1 \pm (\partial V/\partial R_2)\cdot\delta_2 \pm (\partial V/\partial R_3)\cdot\delta_3$$

$$\delta_x = (\pm(\partial V/\partial R_1)\cdot\delta_1 \pm (\partial V/\partial R_2)\cdot\delta_2 \pm (\partial V/\partial R_3)\cdot\delta_3)/(\partial V/\partial x) \leq \Delta x$$

9.4. Design for Testability

Design for test (DFT) is important and should be incorporated at every stage of a design. We can greatly increase reliability by testing the device before it is used, while it is being used, and after it has been used. Having a plan for testing should be a part of every project. In software we can add verbose output, event logging, assertions, diagnostics, resource monitoring, test points, and fault injection hooks. The approach to hardware testing mimics the two important factors for software testing: visibility and control. **Visibility** allows the test engineer to observe system behavior, and **control** allows the test procedure to manipulate inputs to modules within the system. We need to observe the outputs, side effects and internal states. An important parameter is the intrusiveness of the observation. In other words, we need to be careful to not introduce delays that affect the dynamic behavior. Storing data into an array (software dump) at run time can usually be performed with delay overheads of less than 1 µs.

We can add hardware features to our system to facilitate testing. The first is the ability to quickly and reliable attach hardware test equipment to the device. For low production products we can add test points to the circuit to make it easy to attach oscilloscopes and voltmeters. For higher production projects we can add test connectors that directly connect to test equipment like logic analyzers or signal generators. Basically, we need to create known inputs, and record responses at strategic points in the system.

The **verbose** modes found in some compilers add debugging output along with data output. This approach may cause significant slowdown in dynamic behavior, so it will be important to measure timing behavior with and without verbose mode. For aspects of the system that are not real time, verbose is an excellent testing feature. When you add the "-v" switch with a command in Unix, the mail program will display its communication with the local mail server.

Logging or a software **dump** is an excellent tool for real-time systems. If you place the following dump at strategic places in your system, after a short while, the `Buffer` contains the last 256 times the debugging instrument was invoked. Subtracting "3" compensates for the time it takes to run the instrument up to the point at which it reads the SysTick timer. The entire instrument is 12 assembly instructions and runs in about 16 bus cycles. In most cases this will be minimally intrusive.

```
uint32_t Buffer[256];
uint8_t N=0;
__inline void RecordTime(void){
  Buffer[N] = NVIC_ST_CURRENT_R-3;
  N++;   // 8-bit variable goes 0 to 255
}
```

Fault injection is the ability to introduce error conditions to see if the system properly handles them. A **monitor** is a debugging mechanism to observe internal behavior. A good example of a monitor is configuring unused pins as outputs and writing strategic information to these pins during execution. If we connect the pins to a logic analyzer, we have a minimally intrusive debugging instrument. Other examples of monitors include streaming data out unused ports such as a UART, CAN, or SPI. We could log debug data onto flash drive. A **testing interface** is a hardware or software connection between the system and test procedure.

Checkpoint 9.6: Identify three design for test features on the PCB shown on page 451.

9.5. Printed Circuit Board Layout and Enclosures

Begin with choosing a PCB size you need to hold all the components. Next you choose a box with which to enclose the system. The enclosure is an important aspect of the design. It determines the look and feel of the device. Many manufacturers give suggested PCB sizes for their enclosures. One way to make sure everything will fit is to create a mockup. One collects all the electrical and mechanical components of the system. The PCB design is printed on paper and glued to cardboard or wood matching the thickness of the PCB. Mounting holes are drilled and components are attached to the PCB mockup. All components are assembled and mounted into the enclosure to verify proper fit. The left side of Figure 9.14 shows a mockup of an LM3S811-based recorder with an LCD touch screen. The right side shows the final system. The hardware, software, and PCB layout for this system can be downloaded from the book web site as ezLCD_811.

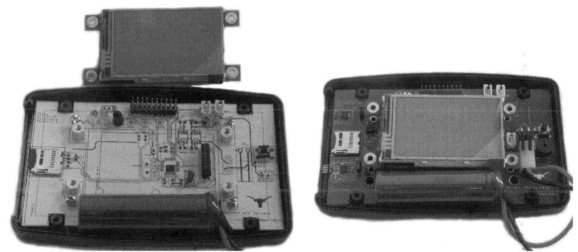

Figure 9.14. Mockup and PCB for an LM3S811 system using a PacTec enclosure.

Laying out a PCB is an art that is best learned by practice under the watchful eye of a master craftsman. So, show your PCB art to others and solicit feedback. For many years, it has been common PCB design practice to avoid 90° corners in PCB traces. Many claim that a 90° turn produces a ¼ turn inductor, which will affect the frequency response of the signals. The standard practice is to use a **mitered corner** replacing a 90° turn with two 45° turns. The reality is that the signal-integrity benefits of avoiding 90° angles are insignificant at the frequencies/edge-rates seen in microcontroller circuits (even up to and past 1 GHz/100ps). [Johnson, H and Graham, M, High-Speed Digital Design: a Handbook of Black Magic, Prentice Hall: New Jersey, 1993.] However, there are a few simple reasons to continue to avoid 90° angles. There is a higher possibility of an acid-trap forming during etching on the inside of the angle (especially in acute angles). An acid trap causes over-etching that can be a yield issue in PC boards with small trace widths. Routing at 45° typically reduces overall trace length and uses less board area. Most PCB designers think it looks better. Looking good is an important factor for anyone who appreciates the art of PCB layout.

This paragraph contains suggestions to consider when laying out a PCB. Make sure the "Snap to Grid" mode is active. You should experiment with different settings of the snap. A net defines which pins and pads must be connected. Just like software variables, nets should have descriptive names. You should visualize the unconnected nets while placing components on the board. Place all components initially on the board to facilitate input/output connectors. You also want to minimize the lengths and the crossings of the nets. Extra time spent placing nets will be recovered many times over while the drawing the traces. Think about how it will be soldered, by placing though-hole components on the top side so soldering will occur on the bottom. Surface mount components can go on either side. Add top and bottom silk labeling to assist in construction, debugging, and device operation. For example, all components need labels (e.g., U1 R1 C1 J1 etc.), shown both on the board and the circuit diagram. Add test points at strategic points to assist in debugging. One inexpensive test point can be made by placing two holes 0.1 inch apart then soldering a U wire into it. You can also purchase test point hardware such as Keystone Electronics 5000 through 5004, which uses a 0.090 in pad with 0.043 in hole. One way to connect all traces is to go left-right on one side and up-down on the other side. Table 9.11 shows the trace width is a function of the length and current in the trace. $1 oz/ft^2$ is a typical trace thickness.

Length	Temperature Rise	Current	Thickness	Resistance	Trace Width
5"	1 C	100 mA	$1\ oz/ft^2$	1 Ω	2 mil
5"	1 C	200 mA	$1\ oz/ft^2$	0.47 Ω	5 mil
5"	1 C	500 mA	$1\ oz/ft^2$	0.13 Ω	20 mil
5"	1 C	1 A	$1\ oz/ft^2$	0.05 Ω	50 mil
5"	1 C	2 A	$1\ oz/ft^2$	0.02 Ω	120 mil

Table 9.11. Minimum trace width for various current levels

In order to verify all components will fit, you must analyze dimensions in 3-D. You can use a design tool like AutoCAD, or you can fabricate a mockup with actual parts. You can cut cardboard in the shape of the PCB, drill holes in the cardboard, and glue the parts onto the cardboard. Figure 9.15 shows an embedded system using a LM3S8962, placed in an enclosure.

Figure 9.15. An EEG system designed with a LM3S8962 that fits into a case (designed by Katy Loeffler).

It is important to read data sheets and application notes for each component of the system. The manufacturer will suggest tips and reference designs for their parts. Figures 9.16 and 9.17 show the suggested layouts for Texas Instruments microcontrollers. In Figure 9.16, the layouts show connections to power and ground on the microcontroller. It is recommended to place a capacitor as close to the chip as possible. The circles in Figure 9.17 represent vias to the power and ground planes. A) is the best practice because there is minimal inductance between capacitor, pins and power planes. B) is not recommended because the distance from pins to vias increases inductance in power rails. C) is acceptable because the distance from pin to via is short.

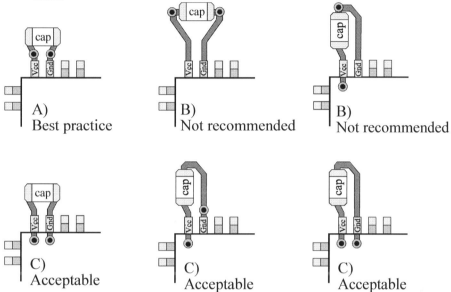

Figure 9.17. Capacitor layout suggestions (System Design Guidelines for Stellaris® Microcontrollers, AN01283).

Figure 9.17 shows a suggested crystal layout, where an important objective is to minimize the effective capacitance from OSC0 to ground and from OSC1 to ground inherent in the board layout itself (called C_{board}). When calculating the capacitor values needed for the oscillator one must account for this board capacitance. The values of these capacitors also depend on the crystal load capacitance, C_L. A typical equation is $C1 = C2 = 2\times(C_L - C_{board})$. This means one should not place power or ground planes under the crystal or the OSC0 and OSC1 signals.

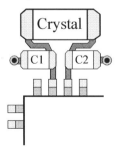

Figure 9.17. Crystal layout suggestion (System Design Guidelines for Stellaris® Microcontrollers, AN01283).

9.6. Exercises

9.1 Find a combination of an E12 capacitor and E24 resistor with a frequency response, $1/(2\pi RC)$ of 1000 Hz. Select the resistor as close to 100 kΩ as possible.

9.2 Find a combination of an E12 capacitor and E24 resistor with a frequency response, $1/(2\pi RC)$ of 12 kHz. Select the resistor as close to 100 kΩ as possible.

9.3 Find two E24 resistors that can be combined to create a resistance as close to 5.12 kΩ as possible.

9.4 Find two E24 resistors that can be combined to create a resistance as close to 25.6 kΩ as possible.

9.5 Assume the system requires 50 mA at 3.3V. What is the dropout voltage of an LM317LZ? Use this regulator to design a 3.3V regulated power supply. What is the range of input voltages that can be used for your circuit?

9.6 Assume the system requires 25 mA at 2.5V. What is the dropout voltage of an LM317LZ? Use this regulator to design a 2.5V regulated power supply. What is the range of input voltages that can be used for your circuit?

9.7 A low-power system requires an average current of 50 µA at 3.7V. The system must run for 6 months. Which battery in Table 9.8 would you use?

9.8 A low-power system requires an average current of 25 µA at 3.7V. The system must run for 3 months. Which battery in Table 9.8 would you use?

9.9 If your ARM microcontroller requires 50 mA to run, how do you build the system with an average current of less than 10 mA?

9.10 If your crystal has a load capacitance of 18 pF, and the PCB has a board capacitance of 6pF, what capacitor values would you use to build the oscillator?

D9.11 Write a C program that knows the E24 capacitors values and E192 resistors values. The input to the program is the desired frequency response, $1/(2\pi RC)$, and the output is an E192 resistor value and an E24 capacitor value that combines to create the desired frequency.

9.7. Lab Assignments

Lab 9.1 Design, build and test a Li-Ion battery charger. Assume the battery voltage is 3.7 V.

Lab 9.2 Design, build and test a NiMH battery charger. Assume the battery voltage is 4.2 V.

Lab 9.3 Design, build and test complete battery-powered embedded system. Begin with a requirements document. Design and layout a PCB. Place the components in an enclosure like the one shown in Figure 9.8. Include design for test features.

10. Data Acquisition Systems

Chapter 10 objectives are to:

- Define performance criteria to evaluate our overall data acquisition system
- Introduce specifications necessary to select the proper transducer
- Develop a methodology for designing data acquisition systems
- Analyze the sources of noise and suggest methods to reduce their effect
- Illustrate concepts of this chapter with case studies

Embedded systems are different from general-purpose computers in a sense that embedded systems have a dedicated purpose. As part of this purpose, many embedded systems are required to collect information about the environment. A system that collects information is called a data acquisition system. In this chapter, we will use the two terms, data acquisition system and instrument interchangeably. Previous chapters presented the basic building blocks to acquire data into the computer, and in this chapter we will combine these blocks into data acquisition systems. Sometimes the acquisition of data is fundamental purpose of the system, such as with a voltmeter, a thermometer, a tachometer, an accelerometer, an altimeter, a manometer, a barometer, an anemometer, an audio recorder, or a camera. At other times, the acquisition of data is an integral part of a larger system such as a control system or communication system. Control systems will be presented in Volume 3, and communication systems will be discussed in Chapter 11.

10.1. Introduction

Figure 10.1 illustrates the integrated approach to instrument design. In this section, we begin with the clear understanding of the problem. We can use the definitions in this section to clarify the design parameters as well as to report the performance specifications. Next in Section 10.2, we will define the parameters and discuss the physics in order to select a suitable transducer. Performing differentiation and integration is described in Section 10.3. In Section 10.4, we put together the analog and digital components, introduced in Chapter 8, to build data acquisition systems. The use of period/pulse/frequency as a means of collecting information was developed in Chapter 6. Noise can never be eliminated, but we will study techniques in Section 10.5 to reduce its effect on our system. The integrated approach to design will be illustrated using the case studies in Section 10.6.

The **measurand** is the physical quantity, property, or condition that the instrument measures. See Figure 10.2. The measurand can be inherent to the object (like position, mass, or color), located on the surface of the object (like the human EKG, or surface temperature), located within the object (e.g., fluid pressure, or internal temperature), or separated from the object (like emitted radiation.)

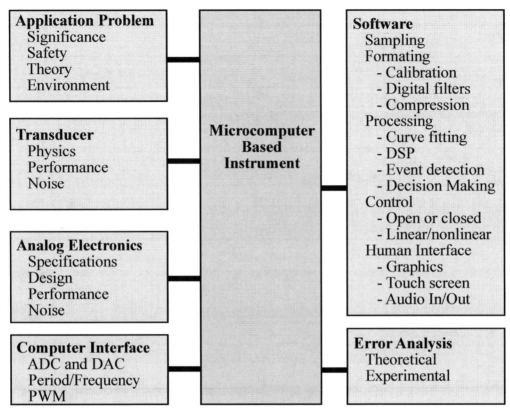

Figure 10.1. *Individual components are integrated into a data acquisition system.*

In general, a **transducer** converts one energy type into another. In the context of this book, the transducer converts the measurand into an electrical signal that can be processed by the microcontroller-based instrument. Typically, a transducer has a primary sensing element and a variable conversion element. The primary sensing element interfaces directly to the object and converts the measurand into a more convenient energy form. The output of the variable conversion element is an electrical signal that depends on the measurand. For example, the primary sensing element of a pressure transducer is the diaphragm, which converts pressure into a displacement of a plunger. The variable conversion element is a strain gauge that converts the plunger displacement into a change in electrical resistance. If the strain gauge is placed in a bridge circuit, the voltage output is directly proportional to the pressure. Some transducers perform a direct conversion without having a separate primary sensing element and variable conversion element. The instrumentation contains **signal processing**, which manipulates the transducer signal output to select, enhance, or translate the signal to perform the desired function, usually in the presence of disturbing factors. The signal processing can be divided into stages. The **analog signal processing** consists of instrumentation electronics, isolation amplifiers, amplifiers, analog filters, and analog calculations. The first analog processing involves calibration signals and preamplification. Calibration is necessary to produce accurate results. An example of a calibration signal is the reference junction of a thermocouple. The second stage of the analog signal processing includes filtering and range conversion. The analog signal range should match the ADC analog input range. Examples of analog calculations include: RMS calculation, integration, differentiation, peak detection, threshold detection, phase lock loops, AM FM modulation/demodulation, and the arithmetic calculations of addition, subtraction, multiplication, division, and square root. When period, pulse width, or frequency measurement is used, we typically use an analog comparator to create a digital logic signal to measure (see Figures 6.5, 8.7 and 8.20). Whereas the Figure 10.1 outlined design components, Figure 10.2 shows the data flow graph for a data acquisition system or control system. The **control system** uses an actuator to drive a parameter in the real world to a desired value while the data acquisition system has no actuator because it simply measures the parameter in a nonintrusive manner.

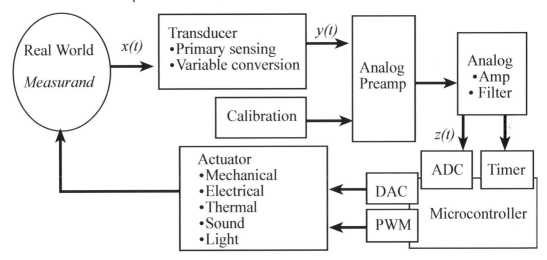

Figure 10.2. Signal paths a data acquisition system.

The **data conversion element** performs the conversion between the analog and digital domains. This part of the instrument includes: hardware and software computer interfaces, ADC, DAC, S/H, analog multiplexer, and calibration references. The **ADC** converts the analog signal into a digital number. In Chapter 6, we saw the period, pulse width, and frequency measurement approach provides a low-cost high-precision alternative to the traditional ADC. The **digital signal processing** includes: data acquisition (sampling the signal at a fixed rate), data formatting (scaling, calibration), data processing (filtering, curve fitting, FFT, event detection, decision making, analysis), control algorithms (open or closed loop). The **human interface** includes the input and output which is available to the human operator. The advantage of computer-based instrumentation is that, devices that are sophisticated but easy to use and understand are possible. The **inputs** to the instrument can be audio (voice), visual (light pens, cameras), or tactile (keyboards, touch screens, buttons, switches, joysticks, roller balls). The **outputs** from the instrument can be numeric displays, CRT screens, graphs, buzzers, bells, lights, and voice. If the system can deliver energy to the real world then it is classified as a control system. Control systems will be developed in the next book (Volume 3). In this chapter, we focus on data acquisition.

Whenever reporting specifications of our instrument, it is important to give the definitions of each parameter, the magnitudes of each parameter, and the experimental conditions under which the parameter was measured. This is because engineers and scientists apply a wide range of interpretations for these terms.

10.1.1. Accuracy

The **instrument accuracy** is the absolute error referenced to the National Institute of Standards and Technology (NIST) of the entire system including transducer, electronics, and software. Let x_{mi} be the values as measured by the instrument, and let x_{ti} be the true values from NIST references. In some applications, the signal of interest is a relative quantity (like temperature or distance between objects). For relative signals, accuracy can be appropriately defined many ways:

$$\text{Average accuracy (with units of x)} = \frac{1}{n}\sum_{i=0}^{n}|x_{ti} - x_{mi}|$$

$$\text{Maximum error (with units of x)} = \max|x_{ti} - x_{mi}|$$

$$\text{Standard error (with units of x)} = \sqrt{\frac{1}{n}\sum_{i=0}^{n}(x_{ti} - x_{mi})^2}$$

In other applications, the signal of interest is an absolute quantity. For these situations, we can specify errors as a percentage of reading or as a percentage of full scale:

$$\text{Average accuracy of reading (\%)} = \frac{100}{n}\sum_{i=0}^{n}\frac{|x_{ti} - x_{mi}|}{x_{ti}}$$

$$\text{Average accuracy of full scale (\%)} = \frac{100}{n}\sum_{i=0}^{n}\frac{|x_{ti} - x_{mi}|}{x_{t\max}}$$

$$\text{Maximum accuracy of reading (\%)} = 100\max\frac{|x_{ti} - x_{mi}|}{x_{ti}}$$

$$\text{Maximum accuracy or full scale (\%)} = 100\max\frac{|x_{ti} - x_{mi}|}{x_{t\max}}$$

Observation: The definitions of the terms accuracy, resolution, and precision vary considerably in the technical literature. It is good practice to include both the definitions of your terms as well as their values in your technical communication.

Since the Celsius and Fahrenheit temperature scales have arbitrary zeroes (e.g., 0°C is the freezing point of water), it is inappropriate to specify temperature error as a percentage of reading or as a percentage of full scale when Celsius and Fahrenheit scales are used. When specifying temperature error, we should use average accuracy, maximum error, or standard error. These errors have units of °C or °F.

Typically, we calibrate a quantitative data acquisition system by determining a transfer function that relates the measured variable, x, to raw measurements such as the ADC sample. Accuracy is limited by two factors: resolution and calibration drift. **Calibration drift** is the change in the transfer function occurring over time used to calculate the measured variable from the raw measurements.

10.1.2. Resolution

The **instrument resolution** is the smallest input signal difference, Δ_x that can be detected by the entire system including transducer, electronics, and software. The resolution of the system is sometimes limited by noise processes in the transducer itself (e.g., thermal imaging) and sometimes limited by noise processes in the electronics (e.g., thermistors, RTDs, and thermocouples). The **coefficient of variation** is the standard deviation divided by the mean, CV is σ/μ. $1/CV$ is μ/σ is a simple estimate of the signal to noise ratio (SNR).

The **spatial resolution** (or **spatial frequency response**) of the transducer is the smallest distance between two independent measurements. The size and mechanical properties of the transducer determine its spatial resolution. When measuring temperature, a metal probe will disturb the existing medium temperature field more than a glass probe. Hence, a glass probe has a smaller spatial resolution than a metal probe of the same size. Noninvasive imaging systems exhibit excellent spatial resolution because the instrument does not disturb the medium, which is being measured. The spatial resolution of an imaging system is the medium surface area from which the radiation originates that is eventually focused onto the detector during the imaging of a single pixel, the so-called instantaneous field of view, IFOV. When measuring force, pressure, or flow, the spatial resolution is the effective area over which the measurement is obtained. Another way to illustrate spatial resolution is to attempt to collect a 2-D or 3-D image of the measurand. The spatial resolution is the distance between points.

10.1.3. Precision

Precision is the number of distinguishable alternatives, n_x, from which the given result is selected. Precision can be expressed in alternatives, bits or decimal digits. Consider a thermometer instrument with a temperature range of 0 to 100 °C. The system displays the output using 3 digits (e.g., 12.3 °C). In addition, the system can resolve each temperature T from the temperature $T+0.1$°C. This system has 1001 distinguishable outputs, and hence has a precision of 1001 alternatives or about 10 bits. For a linear system, there is a simple relationship between range (r_x), resolution (Δ_x) and precision (n_x). Range is equal to resolution times precision

$$r_x\ (100°C) = \Delta_x\ (0.1°C) \bullet n_x\ (1001 \text{ alternatives})$$

where "range" is the maximum minus minimum temperature, and precision is specified in terms of number of alternatives. Table 10.1 illustrates the relationship between alternatives and decimal digits.

Alternatives	Decimal digits
1000	3
2000	3 ½
4000	3 ¾
10000	4

Table 10.1. Definition of decimal digits.

Observation: A good rule of thumb to remember is $2^{10 \bullet n} \approx 10^{3 \bullet n}$.

10.1.4. Reproducibility and Repeatability

Reproducibility (or repeatability) is a parameter that specifies whether the instrument has equal outputs given identical inputs over some period of time. This parameter can be expressed as the full range or standard deviation of output results given a fixed input, where the number of samples and time interval between samples are specified. One of the largest sources of this type of error comes from transducer drift. **Statistical control** is a similar parameter based on a probabilistic model that also defines the errors due to noise. The parameter includes the noise model (e.g., normal, chi-squared, uniform, salt and pepper) and the parameters of the model (e.g., average, standard deviation).

10.2. Transducers

In this section, we will start with quantitative performance measures for the transducer. Next, specific transducers will be introduced. Rather than give an exhaustive list of all available transducers, the intent in this section is to illustrate the range of possibilities, and to provide specific devices to use in the design sections later in the chapter.

10.2.1. Static Transducer Specifications

The input or measurand is x. The output is y. A transducer converts x into y. In this subsection, we assume the input parameter, x, is constant or static.

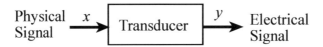

Figure 10.3. Transducers in this book convert a physical signal into an electrical signal.

The input x and the output y can be either absolute or differential. An absolute signal represents a parameter that exists in a single place at a single time. A differential signal is derived from the difference between two signals that exist at different places or at different times. Voltage is indeed defined as a potential difference, but when the voltage is referred to ground, we consider it an absolute quantity. On the other hand, if the signal is represented by the voltage difference between two points neither of which are ground, then we consider the signal as differential. Table 10.2 illustrates four types of transducers.

Type	Input->output	Example
Absolute → absolute	$x \rightarrow y$	Thermistor converts absolute temperature to a resistance
Relative → absolute	$\Delta x \rightarrow y$	Mass balance converts mass difference to an angle
Absolute → relative	$x \rightarrow \Delta y$	Strain gauge converts displacement to a resistance difference
Relative → relative	$\Delta x \rightarrow \Delta y$	Thermocouple converts temperature difference to voltage difference

Table 10.2. Four types of transducers.

The **static sensitivity** is the slope, m, of the straight line through the static calibration curve that gives minimal mean squared error. Let x_i, y_i be the input/output signals of the transducer as shown in the figure below. The **linearity** is a measure of the straightness of the static calibration curve. Let $y_i = f(x_i)$ be the transfer function of the transducer. A linear transducer satisfies:

$$f(ax_1 + bx_2) = af(x_1) + bf(x_2)$$

for any arbitrary choice of the constants a and b. Let $y_i = mx_i + b$ be the best fit line through the transducer data. Linearity (or deviation from it) as a figure-of-merit can be expressed as percentage of reading or percentage of full scale. Let y_{max} be the largest transducer output.

$$\text{Average linearity of reading (percent)} = \frac{100}{n} \sum_{i=1}^{n} \frac{|y_i - mx_i - b|}{y_i}$$

$$\text{Average linearity of full scale (percent)} = \frac{100}{n} \sum_{i=1}^{n} \frac{|y_i - mx_i - b|}{y_{max}}$$

Two definitions for sensitivity are used for temperature transducers. The static sensitivity is:

$$m = \frac{\partial y}{\partial x}$$

If the transducer is linear then the static sensitivity is the slope, m, of the straight line through the static calibration curve which gives the minimum mean squared error. If x_i and y_i represent measured input/output responses of the transducer, then the least squares fit to $y_i = mx_i + b$ is

$$m = \frac{n \sum_{i=1}^{n} x_i y_i - \sum_{i=1}^{n} x_i \sum_{i=1}^{n} y_i}{n \sum_{i=1}^{n} x_i^2 - \left(\sum_{i=1}^{n} x_i\right)^2} \quad \text{and} \quad b = \frac{\sum_{i=1}^{n} y_i \sum_{i=1}^{n} x_i^2 - \sum_{i=1}^{n} x_i y_i \sum_{i=1}^{n} x_i}{n \sum_{i=1}^{n} x_i^2 - \left(\sum_{i=1}^{n} x_i\right)^2}$$

Thermistors can be manufactured to have a resistance value at 25 °C ranging from 4 Ω to 20 MΩ. Because the interface electronics can just as easily convert any resistance into a voltage, a 20 MΩ thermistor is not more sensitive than a 30 Ω thermistor. In this situation, it makes more sense to define **fractional sensitivity** as:

$$\alpha = \frac{1}{R} \frac{\partial R}{\partial T} \text{ (units 1/°C)} \qquad \alpha = \frac{1}{y} \frac{\partial y}{\partial x}$$

Unfortunately, transducers are often sensitive to factors other than the signal of interest. Environmental issues involve how the transducer interacts with its external surroundings (e.g., temperature, humidity, pressure, motion, acceleration, vibration, shock, radiation fields, electric fields and magnetic fields.) **Specificity** is a measure of relative sensitivity of the transducer to the desired signal compared to the sensitivity of the transducers to these other unwanted influences. A transducer with a good specificity will respond only to the signal of interest and be independent of these disturbing factors. On the other hand, a transducer with a poor specificity will respond to the signal of interest as well as to some of these disturbing factors. If all these disturbing factors are grouped together as noise, then the **signal-to-noise ratio (S/N)** is a quantitative measure of the specificity of the transducer.

The **input range** is the allowed range of input, x. The **input impedance** is the phasor equivalent of the steady state sinusoidal effort (voltage, force, pressure) input variable divided by the phasor equivalent of steady state flow (current, velocity, flow) input variable. The output signal strength of the transducer can be specified by the output resistance, R_{out}, and output capacitance, C_{out}.

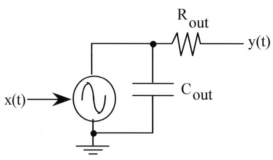

Figure 10.4. *Output model of a transducer.*

The input impedance of a thermal sensor is a measure of the thermal perturbation that occurs due to the presence of the probe itself in the medium. For example, a thermocouple needle inserted into a laser-irradiated medium will affect the medium temperature because heat will conduct down the stainless steel shaft. A thermocouple has a low input impedance (which is bad) because the transducer itself loads (reduces) the medium temperature. On the other hand, an infrared detector measures surface medium temperature without physical contact. Infrared detectors therefore have a very high input impedance (which is good) because the presence of the transducer has no effect on the temperature to be measured. In the case of temperature sensors, the driving force for heat transfer is the temperature difference, ΔT. The resulting heat flow, q, can be expressed using Fourier's law of thermal conduction:

$$q = kA \frac{\partial T}{\partial x} \approx k 4\pi a^2 \frac{\Delta T}{a}$$

where k is the probe thermal conductivity, and A is the probe surface area and a is the radius of a spherical transducer. The steady state input impedance of a spherical temperature probe can thus be approximated by:

$$Z \equiv \frac{\Delta T}{a} \approx \frac{1}{4\pi a k}$$

Again, the approximation in the above equation assumes a spherical transducer. Similar discussions can be constructed for the sinusoidal input impedance. Since most thermal events can be classified as step events rather than sinusoidally varying events, most researchers prefer the use of time constant to describe the transient behavior of temperature transducers.

Some transducers are completely passive (e.g., thermocouple, EKG electrode), and others are active requiring external power (e.g. ultrasonic crystals, strain gauge, microphone, and thermistors.) **Electrical isolation** is a critical factor in medical instrumentation. Some transducers are inherently isolated (e.g., thermistors, thermocouples, microphones), while others are not isolated (e.g., EKG electrodes, pacemakers, blood pressure catheters.) Minimization of errors is important for all instruments. The sensitivity to disturbing factors (electric fields, magnetic fields, radiation, vibration, shock, acceleration, temperature, humidity) must be determined before a device can be used.

The **zero drift** is the change in the static sensitivity curve intercept, b, as a function of time or other factor (see the Figure 10.5). The **sensitivity drift** is the change in the static sensitivity curve slope, m, as a function of time or some other factor. These drift factors determine how often the transducer must be calibrated. For example, thermistors have a drift much larger than that of RTDs or thermocouples. Transducers may be aged at high temperatures for long periods of time to improve their reproducibility.

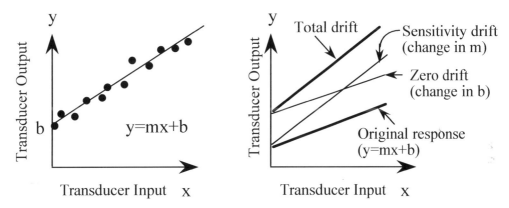

Figure 10.5. The two types of transducer drift: sensitivity drift and zero drift.

The transducer is often a critical device involving both the cost and performance of the entire system. A quality transducer may produce better signals but at an increased cost. An important manufacturing issue is the availability of components. The availability of a device may be enhanced by having a second source (more than one manufacturer produces the device). The use of standard cables and connectors will simplify the construction of your system. The power requirements, size and weight of the device are important in some systems, and thus should be considered when selecting a transducer.

10.2.2. Dynamic Transducer Specifications

The **transient response** is the combination of the **delay** phase, the **slewing** phase, and the **ringing** phase as shown in Figure 10.6. The total transient response is the time for the output, $y(t)$, to reach 99% of its final value after a step change in input, $x(t)=u_0(t)$.

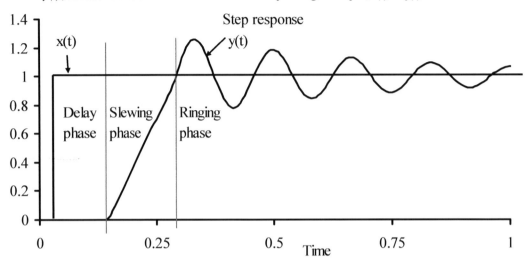

Figure 10.6. *The step response often has delay, slewing and ringing phases.*

The transient response of a temperature transducer to a sudden change in signal input can sometimes be approximated by an exponential equation (assuming first-order response):

$$y(t) = y_f + (y_0 - y_f) e^{-t/\tau}$$

where y_0 and y_f are the initial and final transducer outputs respectively. The **time constant**, τ, of a transducer is the time to reach 63.2% of the final output after the input is instantaneously increased. This time is dependent on both the transducer and the experimental setup. Manufacturers often specify the time constant of thermistors and thermocouples in well-stirred oil (fastest) or still air (slowest). In your applications, one must consider the situation. If the transducer is placed in a high flow liquid like an artery or a water pipe, it may be reasonable to use the stirred oil time constant. If the transducer is in air or embedded in a solid, then thermal conduction in the medium will determine the time constant almost independently of the transducer.

The **frequency response** is a standard technique to describe the dynamic behavior of linear systems. Let $y(t)$ be the system response to $x(t)$. Let

$$x(t) = A\sin(\omega t) \quad y(t) = B\sin(\omega t + \phi) \qquad \omega = 2\pi f$$

The magnitude B/A and the phase ϕ responses are both dependent on frequency. Differential equations can be used to model linear transducers. Let $x(t)$ be the time domain input signal. Let $X(j\omega)$ be the frequency domain input signal. Let $y(t)$ be the time domain output signal. Let $Y(j\omega)$ be the frequency domain output signal.

Classification	differential equation	gain response	phase response
ZERO ORDER	$y(t) = m\,x(t)$	$Y/X = m$ = static sensitivity	
FIRST ORDER	$y'(t) + a\,y(t) = b\,x(t)$	$Y/X = b/\sqrt{a^2+\omega^2}$	$\phi = \arctan(-\omega/a)$
SECOND ORDER	$y''(t) + a\,y'(t) + b\,y(t) = c\,x(t)$		
TIME DELAY	$y(t) = x(t-T)$	$Y/X = \exp(-j\omega T)$	

Table 10.3. Classifications of simple linear systems.

10.2.3. Nonlinear Transducers

Nonlinear characteristics include hysteresis, saturation, bang-bang, breakdown, and dead zone. **Hysteresis** is created when the transducer has memory. We can see in the Figure 10.7 that when the input was previously high it falls along the higher curve, and when the input was previously low it follows along the lower curve. Hysteresis will cause a measurement error, because for any given sensor output, y, there may be two possible measurand inputs. **Saturation** occurs when the input signal exceeds the useful range of the transducer. With saturation, the sensor does not respond to changes in input value when the input is either too high or too low. **Breakdown** describes a second possible result that may occur when in the input exceeds the useful range of the transducer. With breakdown, the sensor output changes rapidly, usually the result of permanent damage to the transducer. Hysteresis, bang bang and dead zone all occur within the useful range of the transducer. **Bang bang** is a sudden large change in the output for a small change in the input. If the bang bang occurs predictably, then it can be corrected for in software. A **dead zone** is a condition where a large change in the input causes little or no change in the output. Dead zones cannot be corrected for in software, thus if present will cause measurement errors.

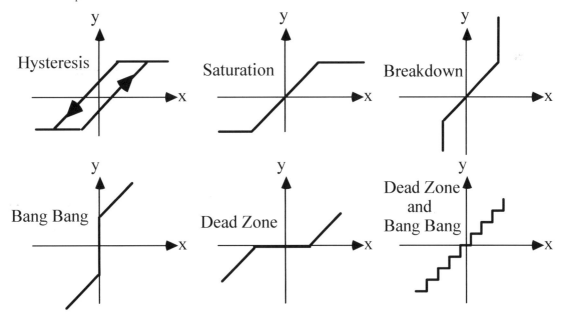

Figure 10.7. Nonlinear transducer responses.

There are many ways to model nonlinear transducers. A nonlinear transducer can be described as a piecewise linear system. The first step is to divide the range of x into a finite subregions, assuming the system is linear in each subregion. The second step is to solve the coupled linear systems so that the solution is continuous. Another method to model a nonlinear system is to use empirically determined nonlinear equations. The first step in this approach is to observe the transducer response experimentally. Given a table of x and y values, the second step is to fit the response to a nonlinear equation. Engineers call these empirical fits **performance maps**.

A third approach to model a nonlinear transducer uses a lookup table located in memory. This method is convenient and flexible. Let x be the measurand and y be the transducer output. The table contains x values and the measured y value is used to index into the table. Sometimes a small table coupled with linear interpolation achieves equivalent results to a large table. The spreadsheet **Therm12.xls**, which can be found on the web site, is an example of this approach.

A **nonmonotonic** response is an input/output function that does not have a mathematical inverse. For example, if two or more input values yield the same output value, then the transducer is nonmonotonic. Software will have a difficult time correcting a nonmonotonic transducer. For example, the Sharp GP2Y0A21YK IR distance sensor has a transfer function as shown in Figure 10.8. If you read a transducer voltage of 2 V, you cannot tell if the object is 3 cm away or 12 cm away.

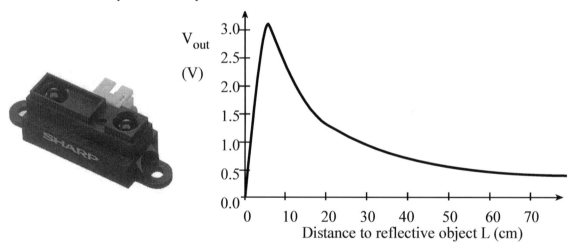

Figure 10.8. The Sharp IR distance sensor exhibits nonmonotonic behavior.

10.2.4. Position Transducers

One of the simplest methods to convert position into an electrical signal uses a position sensitive potentiometer. These devices are inexpensive to build and are sensitive to small displacements. The transducer is constructed from a potentiometer. The fixed part of the potentiometer is called the frame, and the movable part is the armature. The armature is free to move up and down along the measurement axis, see Figure 10.9. The frame is fixed and the armature is attached to the object being measured. The total electrical resistance of the transducer is fixed, but the resistance to the slide arm varies with distance, d.

$$R_{out} = R_{max} * d/d_{max}$$

where d_{max} is distance at full scale and R_{max} is the resistance at full scale. If the material in the potentiometer has uniform resistance, then R_{out} will be linearly related to displacement, d. The disadvantages of this transducer are its low frequency response, its high mechanical input impedance, and it degenerates with time. Nevertheless, this type of transducer is adequate for many applications. This transducer will be interfaced two ways later in Example 10.3.

Figure 10.9. Potentiometer-based position sensor.

The transducer in Figure 10.8 uses IR light to measure distance to a reflecting object. These sensors require a nonuniform power, so placing a 10 µF near the power line of the sensor reduces noise on other components. If the object is more than 6 cm away, the output voltage is inversely related to voltage. If N is the ADC sample, then distance can be calculated as

$$d = c/N \qquad \text{where } c \text{ is a calibration constant}$$

Another method to measure the distance between two objects is to transmit a ultrasonic wave from one object at the other and listen for the reflection (Figure 10.10). The instrument must be able to generate the sound pulse, hear the echo and measure the time, t_{in}, between pulse and echo. If the speed of sound, c, is known, then the distance, d, can be calculated. Our microcontrollers also have mechanisms to measure the pulse width t_{in}.

$$d = c\, t_{in} / 2$$

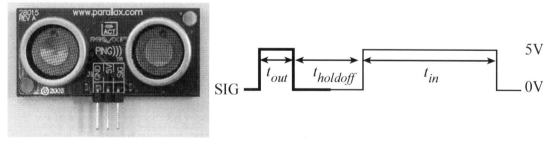

Figure 10.10. An ultrasonic pulse-echo transducer measures the distance to an object, Ping))).

10.2.5. Sound Transducers

A **microphone** is a type of displacement transducer. Sound waves, which are pressure waves travelling in air, cause a diaphragm to vibrate, and the diaphragm motion causes the distance between capacitor plates to change. This variable capacitance creates a voltage, which can be amplified and recorded. The **electret condenser microphone** (ECM) is an inexpensive choice for converting sound to analog voltage. Electret microphones are used in consumer and communication audio devices because of their low cost and small size. For applications requiring high sensitivity, low noise, and linear response, we could use the dynamic microphone, like the ones used in high-fidelity audio recording equipment. The ECM capsule acts as an acoustic resonator for the capacitive electret sensor shown in Figure 10.11. The ECM has a **Junction Field Effect Transistor** (JFET) inside the transducer providing some amplification. This JFET requires power as supplied by the R1 resistor. This local amplification allows the ECM to function with a smaller capsule than typically found with other microphones. ECM devices are cylindrically shaped, have a diameter ranging from 3 to 10 mm, and have a thickness ranging from 1 to 5 mm.

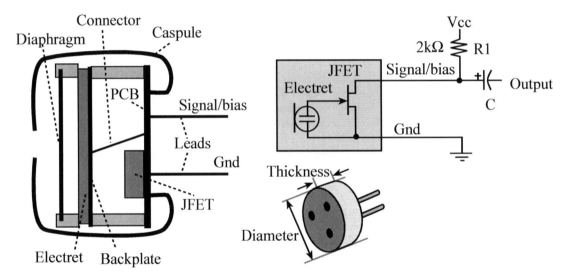

Figure 10.11. Physical and electrical view of an ECM with JFET buffer (Vcc depends on microphone)

An ECM consists of a pre-charged, non-conductive membrane between two plates that form a capacitor. The backplate is fixed, and the other plate moves with sound pressure. Movement of the plate results in a capacitance change, which in turn results in a change in voltage due to the non-conductive, pre-charged membrane. An electrical representation of such an acoustic sensor consists of a signal voltage source in series with a source capacitor. The most common method of interfacing this sensor is a high-impedance buffer/amplifier. A single JFET with its gate connected to the sensor plate and biased as shown in Figure 10.11 provides buffering and amplification. The capacitor C provides high-pass filtering, so the voltage at the output will be less than ±100 mV for normal voice. Audio microphones need additional amplification and band-pass filtering. Typical audio signals exist from 100 Hz to 10 kHz. The presence of the R1 resistor is called "phantom biasing". The electret has two connections: Gnd and Signal/bias. Typically, the metallic capsule is connected to Gnd.

10.2.6. Force and Pressure Transducers

A common device to measure force and pressure is the strain gauge. As a wire is stretched its length increases and its cross-sectional area decreases. These geometric changes cause an increase in the electrical resistance of the wire, R. The transducer is constructed with four sets of wires mounted between a stationary member (frame) and a moving member (armature.) As the armature moves relative to the frame, two wires are stretched (increase in R_1 R_4), and two wires are compressed (decrease in R_2 R_3), as shown in Figure 10.13. The strain gauge is a displacement transducer, such that a change in the relative position between the armature and frame, Δx, causes a change in resistance, ΔR. The sensitivity of a strain gauge is called its **gauge factor**.

$$G = \frac{\Delta R / R}{\Delta x / x}$$

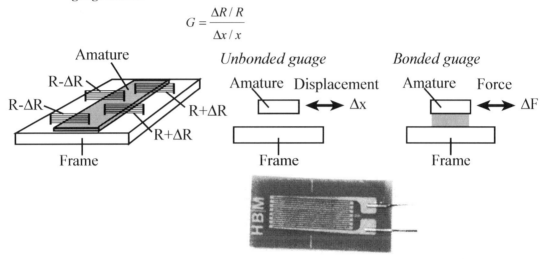

Figure 10.12. Strain gauges used for force or pressure measurement.

The gauge factor for an Advance strain gauge is 2.1. The typical resistance, R, is 120 Ω. If the gauge is bonded onto a material with a spring characteristic:

$$F = -k\,x$$

then the transducer can be used to measure force. The wires each have a significant temperature drift. When the four wires are placed into a bridge configuration, the temperature dependence cancels. A high gain, high input impedance, high CMRR differential amplifier is required.

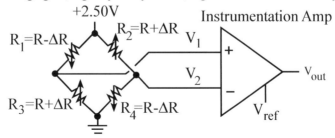

Figure 10.13. Four strain gauges are placed in a bridge configuration.

Force and pressure sensors can also be made from semiconductor materials. These silicon sensors can be made much smaller than strain gauges, but tend to much more nonlinear. A **force sensing resistor** (FSR) is made with a resistive film and converts force into resistance. These sensors are low cost and easy to interface, but tend to be quite nonlinear.

10.2.7. Temperature Transducers

Thermistors are a popular temperature transducer made from a ceramic-like semiconductor. A NTC (negative temperature coefficient) thermistor is made from combinations of metal oxides of manganese, nickel, cobalt, copper, iron, and titanium. A mixture of milled semiconductor oxide powders and a binder is shaped into the desired geometry. The mixture is dried and sintered (under pressure) at an elevated temperature. The wire leads are attached and the combination is coated with glass or epoxy. By varying the mixture of oxides, a range of resistance values from 30 Ω to 20 MΩ (at 25 °C) is possible. Table 10.4 lists the tradeoffs between thermistors and thermocouples.

Thermistors	Thermocouples
More sensitive	More sturdy
Better temperature resolution	Faster response
Less susceptible to noise	Inert, interchangeable V vs. T curves
Less thermal perturbation	Requires less frequent calibration
Does not require a reference	More linear

Table 10.4. **Tradeoffs between thermistors and thermocouples.**

A precision thermometer, an ohmmeter, and a water bath are required to calibrate thermistor probes. The following empirical equation yields an accurate fit over a narrow range of temperature:

$$T = \frac{1}{H_0 + H_1 \ln R} \quad \text{or} \quad R = R_0 \exp\frac{\beta}{T}$$

where T is the temperature in Kelvin, and R is the thermistor resistance in ohms. 0 degrees Celsius is 273.15 degrees Kelvin. It is preferable to use the ohmmeter function of the eventual instrument for calibration purposes so that influences of the resistance measurement hardware and software are incorporated into the calibration process.

The first step in the calibration process is to collect temperature (measured by a precision thermometer) and resistance data (measured by the ohmmeter process of the instrument). The thermistor(s) to be calibrated should be placed as close to the sensing element of the precision thermometer as possible. The water bath creates a stable yet controllable environment in which to perform the calibration. See the spreadsheet **Therm12.xls** on the book web site for an example thermistor calibration.

A thermocouple is constructed by spot welding two different metal wires together. Probe transducers include a protective casing which surrounds the thermocouple junction. Probes come in many shapes including round tips, conical needles and hypodermic needles. Bare thermocouple junctions provide faster response but are more susceptible to damage and noise pickup. Ungrounded probes allow electrical isolation but are not as responsive as grounded probes. Commercial thermocouples have been constructed in 16 to 30 gauge hypodermic needles — a 30 gauge needle has an outside diameter of above 0.03 cm. Bare thermocouples can be made from 30-μm wire producing a tip with an 80 μm diameter. A spot weld is produced by passing a large current through the metal junction that fuses the two metals together. If the wires form a loop, and the junctions are at different temperatures, then a current will flow in the loop. This thermal to electrical energy conversion is called the Seebeck effect (see Figure 10.14). If the temperature difference is small, then the current, I, is linearly proportional to the temperature difference T_1-T_2.

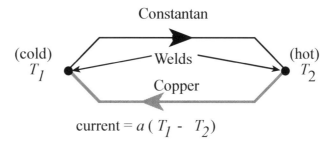

Figure 10.14. When the two thermocouple junctions are at different temperatures current will flow.

If the loop is broken, and an electrical voltage is applied, then heat will be absorbed at one junction and released at the other. This electrical to thermal energy conversion is called the Peltier effect, see Figure 10.15. If the voltage is small, then the heat transferred is linearly proportional to the voltage, V.

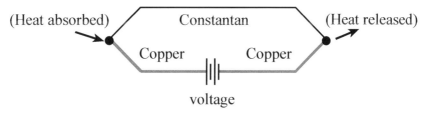

Figure 10.15. When voltage is applied to two thermocouple junctions heat will flow.

If the loop is broken, and the junctions are at different temperatures, then a voltage will develop because of the Seebeck effect. If the temperature difference is small, then the voltage, V, is nearly linearly proportional to the temperature difference T_1-T_2. An amplifier is required to match the transducer output range to the ADC range, see Figure 10.16. Under most conditions we will add an analog low pass filter.

Type - Thermocouple	μV/°C at 20°C	Useful range °C	Comments
T - Copper/Constantan	45	-150 to +350	Moist environment
J - Iron/Constantan	53	-150 to +1000	Reducing environment
K - Chromel/alumel	40	-200 to +1200	Oxidizing environment
E - Chromel/constantan	80	0 to +500	Most sensitive
R S - Platinum/platinum-rhodium	6.5	0 to +1500	Corrosive environment
C – Tungsten/rhenium	12	0 to 2000	High temperature

Table 10.5. Temperature sensitivity and range of various thermocouples.

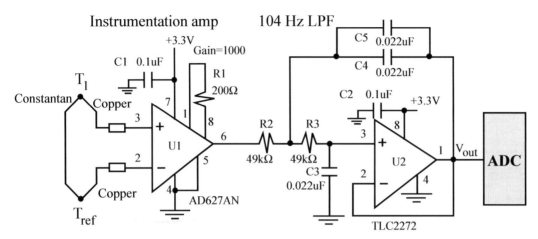

Figure 10.16. A thermocouple converts temperature difference to voltage.

Thermocouples are characterized by: 1) low impedance (resistance of the wires), 2) low temperature sensitivity (45 µV/°C for copper/constantan), 3) low-power dissipation, 4) fast response (because of the metal), 5) high stability (because of the purity of the metals), and 6) interchangeability (again because of the physics and the purity of the metals).

If the temperature range is less than 25°C, then the linear approximation can be used to measure temperature. Let N be the digital sample from the ADC for the unknown medium temperature, T_1. A calibration is performed under the conditions of a constant reference temperature: typically, one uses the extremes of the temperature range (T_{min} and T_{max}). A precision thermometer system is used to measure the "truth" temperatures. Let N_{min} and N_{max} be the digital samples at T_{min} and T_{max} respectively. Then the following equation can be used to calculate the unknown medium temperature from the measured digital sample.

$$T_1 = T_{min} + (N-N_{min}) \cdot \frac{T_{max} - T_{min}}{N_{max} - N_{min}}$$

Because the thermocouple response is not exactly linear, the errors in the above linear equation will increase as the temperature range increases. For instruments with a larger temperature range, a quadratic equation can be used,

$$T_1 = H_0 + H_1 \cdot N + H_2 \cdot N^2$$

where H_0 H_1 and H_2 are determined by calibration of the instrument over the range of interest.

10.3. Discrete Calculus

Given a method to measure position, we can use calculus to determine velocity and acceleration.

$$v(t) = \frac{dx(t)}{dt} \qquad a(t) = \frac{dv(t)}{dt}$$

The continuous derivative can be expressed as a limit.

$$v(t) = \lim \frac{x(t) - x(t - \Delta t)}{\Delta t} \text{ as } \Delta t \to 0$$

The microcontroller can approximate the velocity be calculating the discrete derivative.

$$v(t) = \frac{x(t) - x(t - \Delta t)}{\Delta t}$$

where Δt is the time between velocity samples. At this point we introduce a short hand to describe digital samples. $x(n)$ is the current sample, $x(n-1)$ is the previous sample, and $x(n-2)$ is the sample two times ago. When describing calculations performed on sampled data it is customary to use this notation. The use of n as the time variable emphasizes the signal exists only at discrete times.

$$v(n) = \frac{x(n) - x(n-1)}{\Delta t}$$

This calculation of derivative is very sensitive to errors in the measurement of $x(n)$. A more stable calculation averages two or more derivative terms taken over different time windows. In general we can define such a robust calculation as

$$v(t) = \frac{a}{a+b} \frac{x(t) - x(t - p\Delta t)}{p\Delta t} + \frac{b}{a+b} \frac{x(t - c\Delta t) - x(t - (c+m)\Delta t)}{m\Delta t}$$

If the integers p, m, and c are all positive, this calculation can be performed in real-time. The first term is the derivative over the large time window of $p\Delta t$. The second window term has a smaller size of $m\Delta t$. It normally fits entire inside the first with $c>0$ and $c-m<p$. The coefficients a and b create the weight for combining the short and long intervals. With $a=b=1$, $p=3$, $m=1$, and $c=1$, we get:

$$v(n) = \frac{x(n) + 3x(n-1) - 3x(n-2) - x(n-3)}{6\Delta t}$$

The acceleration can also be approximated by a discrete derivative.

$$a(n) = \frac{x(n) - 2x(n-1) + x(n-2)}{\Delta t^2}$$

Observation: In the above calculations of derivative, a single error in one of the $x(t)$ input terms will propagate to only a finite number of the output calculations.

Observation: Although the central difference calculation of $v(t)=(x(t+\Delta t)-x(t-\Delta t))/2\Delta t$ is theoretically valid, we cannot use it for real-time applications, because it requires knowledge about the future, $x(t+\Delta t)$, which is unavailable at the time $v(t)$ is being calculated.

Similarly, we can perform integration of velocity to determine position.

$$x(t) = \int_0^t v(\tau)d\tau$$

The microcontroller can perform a discrete integration by summation.

$$x(n) = x(0) + \sum_0^n v(m)\Delta t$$

There are two problems with this approach. The first difficulty is determining *x(0)*. The second problem is the accumulation of errors. If one is calculating velocity from position and an error occurs in the measurement of *x(n)*, then that error affects only two calculations of *v(n)*. Unfortunately, if one is calculating position from velocity and an error occurs in the measurement of *v(n)*, then that error will affect all subsequent calculations of *x(n)*.

Observation: In the above calculation of integration, a single error in one of the x(t) input terms will propagate into all remaining output calculations.

The following function, which is quite similar to the derivative, is actually a low pass digital filter.

$$y(n) = \frac{x(n) + x(n-1)}{2}$$

Furthermore, if you sample an input at a regular rate, f_s, then the following is a digital low pass filter with a cutoff of about $f_s/2k$. This filter calculates the average of the last *k* samples (see Program 8.10).

$$y(n) = \frac{1}{k}\sum_{i=0}^{k-1} x(n-i)$$

Checkpoint 10.1: If you sample at 1000 Hz, how do you create a digital low pass filter with a cutoff of 10 Hz?

10.4. Data Acquisition System Design

10.4.1. Introduction and Definitions

Before designing a data acquisition system (DAS) we must have a clear understanding of the system goals. We can classify system as a **Quantitative DAS**, if the specifications can be defined explicitly in terms of desired range (r_x), resolution (Δx), precision (n_x), and frequencies of interest (f_{min} to f_{max}). If the specifications are more loosely defined, we classify it as a **Qualitative DAS**. Examples of qualitative systems include those which mimic the human senses where the specifications are defined using terms like "sounds good", "looks pretty", and "feels right." Other qualitative systems involve the detection of events. We will consider two examples, a burglar detector, and an instrument to diagnose cancer. For binary detection systems like the presence/absence of a burglar or the presence/absence of cancer, we define a true positive (TP) when the condition exists (there is a burglar) and the system properly detects it (the alarm rings.) We define a false positive (FP) when the condition does not exist (there is no burglar) but the system thinks there is (the alarm rings.) A false negative (FN) occurs when the condition exists (there is a burglar) but the system does not think there is (the alarm is silent.) A true negative (TN) occurs when the condition does not exist (the patient does not have cancer) and the system properly detects it (the instrument says the patient is normal.) **Prevalence** is the probability the condition exists, sometimes called pre-test probability. In the case of diagnosing the disease, prevalence tells us what percentage of the population has the disease. **Sensitivity** is the fraction of properly detected events (a burglar comes and the alarm rings) over the total number of events (number of robberies.) It is a measure of how well our system can detect an event. For the burglar detector, a sensitivity of 1 means when a burglar breaks in the alarm will go off. For the diagnostic instrument, a sensitivity of 1 means every sick patient will get treatment. **Specificity** is the fraction of properly handled non-events (a patient doesn't have cancer and the instrument claims the patient is normal) over the total number of non-events (the number of normal patients.) A specificity of 1 means no people will be treated for a cancer they don't have. The **positive predictive value** of a system (PPV) is the probability that the condition exists when restricted to those cases where the instrument says it exists. It is a measure of how much we believe the system is correct when it says it has detected an event. A PPV of 1 means when the alarm rings, the police will come and arrest a burglar. Similarly, a PPV of 1 means if our instrument says a patient has the disease, then that patient is sick. The **negative predictive value** of a system (NPV) is the probability that the condition does not exists when restricted to those cases where the instrument says it doesn't exist. A NPV of 1 means if our instrument says a patient doesn't have cancer, then that patient is not sick. Sometimes the true negative condition doesn't really exist (how many times a day does a burglar not show up at your house?) If there are no true negatives, only sensitivity and PPV are relevant.

Prevalence = (TP + FN) / (TP + TN + FP + FN)
Sensitivity = TP / (TP + FN)
Specificity = TN / (TN + FP)
PPV = TP / (TP + FP)
NPV = TN / (TN + FN)

The **transducer** converts the physical signal into an electrical signal. The **amplifier** converts the weak transducer electrical signal into the range of the ADC (e.g., 0 to +3 V). The **analog filter** removes unwanted frequency components within the signal. The analog filter is required to remove aliasing error caused by the ADC sampling. The **analog multiplexer** is used to select one signal from many sources. The **sample and hold** is an analog latch used to keep the ADC input voltage constant during the ADC conversion. The **clock** is used to control the sampling process. Inherent in digital signal processing is the requirement that the ADC be sampled on a fixed time basis. The **computer** is used to save and process the digital data. A **digital filter** may be used to amplify or reject certain frequency components of the digitized signal. The multiple access circular queue is a convenient data structure to use with the digital filter.

10.4.2. Using Nyquist Theory to Determine Sampling Rate

There are two errors introduced by the sampling process. **Voltage quantizing** is caused by the finite word size of the ADC. The **precision** is determined by the number of bits in the ADC. If the ADC has n bits, then the number of distinguishable alternatives is

$$n_z = 2^n$$

Time quantizing is caused by the finite discrete sampling interval. The **Nyquist Theorem** states that if the signal is sampled at f_s, then the digital samples only contain frequency components from 0 to 0.5 f_s. Conversely, if the analog signal does contain frequency components larger than $\frac{1}{2} f_s$, then there will be an **aliasing** error. Aliasing is when the digital signal appears to have a different frequency than the original analog signal. Simply put, if one samples a sine wave at a sampling rate of f_s,

$$V(t) = A \sin(2\pi f t + \phi)$$

is it possible to determine A f and ϕ from the digital samples? Nyquist Theory says that if f_s is strictly greater than twice f, then one can determine A f and ϕ from the digital samples. In other words, the entire analog signal can be reconstructed from the digital samples. But if f_s less than or equal to f, then one cannot determine A f and ϕ. In this case, the apparent frequency, as predicted by analyzing the digital samples, will be shifted to a frequency between 0 and $\frac{1}{2} f_s$.

In this example, the sampling rate is fixed at 1600 Hz and the signal frequency is varied. When sampling rate is exactly twice the input frequency, the original signal may or may not be properly reconstructed. In this specific case, it is frequency shifted (aliased) to DC and lost.

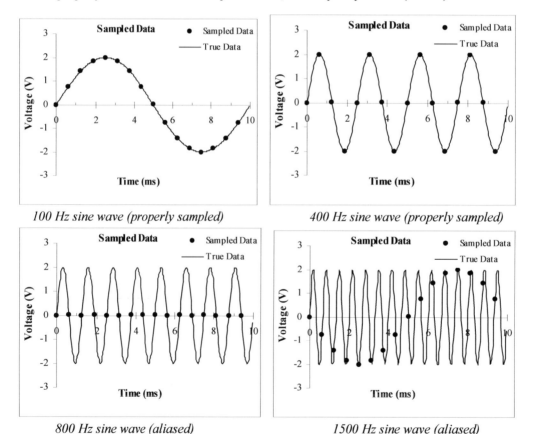

Figure 10.17. Aliasing does not occur when the sampling rate is more than twice the signal frequency.

When sampling rate is slower than twice the input frequency, the original signal cannot be properly reconstructed. It is frequency shifted (aliased) to a frequency between 0 and ½ f_s. In this case the 1500 Hz wave was aliased to 100 Hz.

The choice of **sampling rate**, f_s, is determined by the maximum useful frequency contained in the signal. One must sample at least twice this maximum useful frequency. Faster sampling rates may be required to implement a digital filter and other digital signal processing.

$$f_s > 2 f_{max}$$

Even though the largest signal frequency of interest is f_{max}, there may be significant signal magnitudes at frequencies above f_{max}. These signals may arise from the input x, from added noise in the transducer or from added noise in the analog processing. Once the sampling rate is chosen at f_s, then a low pass analog filter may be required to remove frequency components above ½f_s. A digital filter cannot be used to remove aliasing.

An interesting question arises: how do we determine the maximum frequency component in our input? If we know enough about our system we might be able to derive an equation to determine the maximum frequency. For example, if a mechanical system consists of a mass, friction and a spring, then we can write a differential equation relating the applied force to the position of the object. The second way to find the maximum frequency component in our signal is to measure it with a spectrum analyzer.

> **Valvano Postulate**: If f_{max} is the largest frequency component of the analog signal, then you must sample more than ten times f_{max} in order for the reconstructed digital samples to look like the original signal when plotted on a voltage versus time graph.
>
> **Central Limit Theorem**: As independent random variables are added, their sum tends toward a Normal distribution. Assume the input is constant, as more and more data are added together (averaging), the pmf of the sum (or average) will tend towards a Gaussian shape.
>
> **Oversampling**: We will sample the ADC faster than we need, and average multiple samples to get one reading. This averaging, called oversampling, will improve the signal to noise ratio.

10.4.3. How Many Bits Does One Need for the ADC?

The choice of the **ADC precision** is a compromise of various factors. The desired resolution of the data acquisition system will dictate the number of ADC bits required. If the transducer is nonlinear, then the ADC precision must be larger than the precision specified in the problem statement. For example, let y be the transducer output, and let x be the real world signal. Assume for now, that the transducer output is connected to the ADC input. Let the range of x be r_x. Let the range of y be r_y. Let the required precision of x be n_x. The resolutions of x and y are Δ_x and Δ_y respectively. Let the following describe the nonlinear transducer, $y = f(x)$. The required ADC precision, n_y, (in alternatives) can be calculated by:

$$\Delta_x = r_x / n_x \quad \Rightarrow \quad \Delta_y = \min \{f(x+\Delta_x)-f(x)\} \text{ for all } x \text{ in } r_x \quad \Rightarrow \quad n_y = r_y / \Delta_y$$

For example, consider the nonlinear transducer $y = x^2$. The range of x is $0 \leq x \leq 1$. Thus, the range of y is also $0 \leq y \leq 1$. Let the desired resolution be $\Delta_x = 0.01$. $n_x = r_x/\Delta_x = 100$ alternatives or about 7 bits. From the above equation, $\Delta_y = \min\{(x+0.01)^2 - x^2\} = \min\{0.02x + 0.0001\} = 0.0001$. Thus, $n_y = r_y/\Delta_y = 10000$ alternatives or almost 15 bits.

Checkpoint 10.2: What is the relationship between n_x and n_y if the transducer is linear?

10.4.4. Specifications for the Analog Signal Processing

In general we wish the analog signal processing to map the full scale range of the transducer into the full scale range of the ADC. If the ADC precision is $N=2^n$ in alternatives, and the output impedance of the transducer is R_{out}, then we need an input impedance larger than $N*R_{out}$ to avoid loading the signal. We need the analog circuit to pass the frequencies of interest. When considering noise, we determine the signal equivalent noise. For example, consider the thermistor circuit in Figure 10.16. If we wish to consider noise on signal V_{out}, we calculate the relationship between input temperature T and the signal V_{out}. Next, we determine the sensitivity of the signal to temperature, dV_{out}/dT. If the noise is V_n, then the temperature equivalent noise is $T_n = V_n/(dV_{out}/dT)$. In general, we wish all equivalent noises to be less than the system resolution.

An **analog low pass filter** may be required to remove aliasing. The cutoff of this analog filter should be less than $\tfrac{1}{2} f_s$. Some transducers automatically remove these unwanted frequency components. For example, a thermistor is inherently a low pass device. Other types of filters (analog and digital) may be used to solve the data acquisition system objective. One useful filter is a 60 Hz bandreject filter.

In order to prevent aliasing, one must know the frequency spectrum of the ADC input voltage. This information can be measured with a spectrum analyzer. Typically, a spectrum analyzer samples the analog signal at a very high rate (>1 MHz), performs a Discrete Fourier Transform (DFT), and displays the signal magnitude versus frequency. As defined in Figure 10.2, $z(t)$ is the input to the ADC. Let $|Z(f)|$ be the magnitude of the ADC input voltage as a function of frequency. There are 3 regions in the magnitude versus frequency graph shown in Figure 10.18.

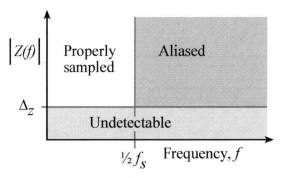

Figure 10.18. To prevent aliasing there should be no measurable signal above $\tfrac{1}{2} f_s$.

We will classify any signal with an amplitude less than the ADC resolution, Δ_z, to be undetectable. This region is labeled "Undetectable". Undetectable signals cannot cause aliasing regardless of their frequency.

We will classify any signal with an amplitude larger than the ADC resolution at frequencies less than $\tfrac{1}{2}f_s$ to be properly sampled. This region is labeled "Properly sampled". It is information in this region that is available to the software for digital processing.

The last region includes signals with amplitude above the ADC resolution at frequencies greater than or equal to $\tfrac{1}{2}f_s$. Signals in this region will be aliased, which means their apparent frequencies will be shifted into the 0 to $\tfrac{1}{2}f_s$ range.

Most spectrum analyzers give the output in decibels full scale (dB_{FS}). For an ADC system with a range of 0 to 3V, the full scale peak-to-peak amplitude for any AC signal is 3 V. If V is the DFT output magnitude in volts

$dB_{FS} = 20\ \log_{10}(V/3)$

Table 10.6 calculates the ADC resolution in dB_{FS}. For a real ADC, the resolution will be a function of other factors other than bits. For example, the MAX1246 12-bit ADC has a minimum Signal-to-Noise+Distortion Ratio (SINAD) of 70 dB, meaning it is not quite 12 bits. The typical SINAD is 73 dB, which is slightly better than 12 bits.

Bits	dB_{FS}
8	-48.2
9	-54.2
10	-60.2
11	-66.2
12	-72.2
13	-78.3
14	-84.3

Table 10.6. ADC resolution in dBFS, assuming full scale is defined as peak-to-peak voltage.

Aliasing will occur if $|Z|$ is larger than the ADC resolution for any frequency larger than or equal to $\tfrac{1}{2}f_s$. In order to prevent aliasing, $|Z|$ must be less than the ADC resolution. Our design constraint will include a safety factor of $\alpha \leq 1$. Thus, to prevent aliasing we will make:

$|Z| < \alpha \Delta_z$ for all frequencies larger than or equal to $\tfrac{1}{2}f_s$

This condition usually be can be satisfied by increasing the sampling rate or increasing the number of poles in the analog low pass filter. We cannot remove aliasing with a digital low pass filter, because once the high frequency signals are shifted into the 0 to $\tfrac{1}{2}f_s$ range, we will be unable to separate the aliased signals from the regular ones. To determine α, the sum of all errors (e.g., ADC, aliasing, and noise) must be less than the desired resolution.

10.5. Analysis of Noise

The consideration of noise is critical for all instrumentation systems. The success of an instrument does depend on careful transducer design, precision analog electronics, and clever software algorithms. But any system will fail if the signal is overwhelmed by noise. Fundamental noise is defined as an inherent and nonremovable error. It exists because of fundamental physical or statistical uncertainties. We will consider three types of fundamental noise:

> Thermal noise, White noise or Johnson noise
> Shot noise
> 1/f noise

Although fundamental noise cannot be eliminated, there are ways to reduce its effect on the measurement objective. In general, added noise includes the many disturbing external factors that interfere with or are added to the signal. We will consider three types of added noise:

> Galvanic noise
> Motion artifact
> Electromagnetic field induction

10.5.1. Thermal Noise

Thermal fluctuations occur in all materials at temperatures above absolute zero. Brownian motion, the random vibration of particles, is a function of absolute temperature. As the particles vibrate, there is an uncertainty as to the position and velocity of the particles. This uncertainty is related to the thermal energy

> The absolute temperature, T (K)
> Boltzmann's constant, $k = 1.67 \cdot 10^{-23}$ joules/K
> Uncertainty in thermal energy $\approx 1/2 \, kT$

Because the electrical power of a resistor is dissipated as thermal power, the uncertainty in thermal energy produces an uncertainty in electrical energy. The electrical energy of a resistor depends on

> Resistance, R (Ω)
> voltage, V (volts)
> Time, (sec).
> Electrical power = V^2/R (watts)
> Electrical energy = $V^2 \ast \text{time}/R$ (watt-sec)

By equating these two energies we can derive (hand wave) an equation for voltage noise similar to the empirical findings of J. B. Johnson. In 1928, he found that the open circuit root-mean-square (RMS) voltage noise of a resistor is given by:

$V_f^2 = 4 \, k \, T \, R \, \Delta \gamma$ where $\Delta \gamma = f_{max} - f_{min}$

where $f_{max}-f_{min}$ is the frequency interval, or bandwidth over which the measurement was taken. For instance, if the system bandwidth is DC to 1000 Hz then $\Delta\gamma$ is 1000 cycles/sec. Similarly, if the system is a bandpass from 10 kHz to 11 kHz, then $\Delta\gamma$ is also 1000 cycles/sec. The term "white noise" comes from the fact that thermal noise contains the superposition of all frequencies and is independent of frequency. It is analogous to optics where "white light" is the superposition of all wavelengths.

	1 Hz	10 Hz	100 Hz	1 kHz	10 kHz	100 kHz	1 MHz
10 kΩ	14 nV	45 nV	142 nV	448 nV	1.4 μV	4.5 μV	14 μV
100 kΩ	45 nV	142 nV	448 nV	1.4 μV	4.5 μV	14 μV	45 μV
1 MΩ	142 nV	448 nV	1.4 μV	4.5 μV	14 μV	45 μV	142 μV

Table 10.7. White noise for resistors at 300K=27°C.

Interestingly, only resistive but not capacitive or inductive electrical devices exhibit thermal noise. Thus a transducer which dissipates electrical energy will have thermal noise, and a transducer which simply stores electrical energy will not.

Figure 10.19 defines root-mean-squared as the square root of the time average of the voltage squared. RMS noise is proportional to noise power. The **crest factor** is the ratio of peak value divided by RMS. The peak value is 1/2 of the peak to peak amplitude, and can be easily measured from recorded data. From Table 10.8, we see that the crest factor is about 4. The crest factor can be defined for other types of noise.

Percent of the time the peak is exceeded	Crest factor (peak/RMS)
1.0	2.6
0.1	3.3
0.01	3.9
0.001	4.4
0.0001	4.9

Table 10.8. Crest factor for thermal noise.

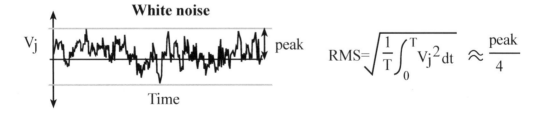

Figure 10.19. Root-mean-squared (RMS) is a time average of the voltage squared.

Observation: When measured with a spectrum analyzer, the response is uniform amplitude at all frequencies, which is why it is classified as white noise.

10.5.2. Shot Noise

Shot noise arises from the statistical uncertainty of counting discrete events. Thermal cameras, radioactive detectors, photomultipliers, and O_2 electrodes count individual photons, gamma rays, electrons, and O_2 particles respectively as they impinge upon the transducer. Let dn/dt be the count rate of the transducer. Let t be the measurement interval or count time. The average count is

$$n = dn/dt \cdot t$$

On the other hand, the statistical uncertainty of counting random events is $\sqrt{n}$. Thus the shot noise is

$$\text{shot noise} = \sqrt{dn/dt \cdot t}$$

The signal to noise ratio (S/N) is

$$S/N = \frac{n}{\sqrt{n}} = \sqrt{dn/dt \cdot t}$$

The solutions are to maximize the count rate (by moving closer or increasing the source) and to increase the count time. There is a clear tradeoff between accuracy and measurement time.

10.5.3. 1/f, Flicker, or Pink Noise

Pink noise is somewhat mysterious. The origin of 1/f noise lacks rigorous theory. It is present in devices that have connections between conductors. Garrett describes it as a fluctuating conductivity. It is of particular interest to low bandwidth applications due to the 1/f behavior. Wire wound resistors do not have 1/f noise, but semiconductors (like MOSFETs) and carbon resistors do. One of the confusing aspects of 1/f noise is its behavior as the frequency approaches 0 Hz. The noise at DC is not infinite because although 1/f is infinite at DC, $\Delta\gamma$ is zero. Garrett gives an equation to calculate the 1/f noise of a carbon resistor.

$$V_c = 10^{-6} \sqrt{1/f} \; R \; I \; \sqrt{\Delta\gamma}$$

where
- V_c is the 1/f voltage noise in volts
- f is the frequency in Hz
- R is the resistance in Ω
- I is the average DC current in amps
- $\Delta\gamma$ is the system bandwidth in Hz

f (Hz)	V_c (μV)
1	316
10	100
100	32
1000	10

Table 10.9. V_c versus frequency for R=10 kΩ, I=1 mA, $\Delta\gamma$ = 1 kHz.

10.5.4. Galvanic Noise

The contact between dissimilar metals will induce a voltage, due to the electrochemistry at the metal-metal interface. Voltages will also develop when a conductive liquid contacts a metal. This problem usually arises as a metal surface within a connector oxidizes (corrosion due to moisture). The materials least susceptible to corrosion are silver, graphite, gold and platinum. For this reason, we use gold-plated connector pins and sockets.

10.5.5. Motion Artifact

Motion can introduce errors in many ways. According to Faraday's Law, a conducting wire that moves in a magnetic field will induce an EMF. This voltage error is proportional to the strength of the magnetic field, the length of the wire that is moving, the velocity of the motion, and the angle between the velocity and the field. Another problem occurring with moving cables is the connector impedance may change or disconnect. Acceleration of the transducer will induce forces inside the device often affecting its response.

10.5.6. Electromagnetic field induction

Usually, the largest source of noise is caused by electromagnetic field induction. According to Faraday's Law, changing magnetic fields can induce a voltage into our circuits. The changing magnetic field must pass through a wire loop, drawn as the shaded area in Figure 10.20. This voltage noise (V_m) is proportional to, the strength of the magnetic field, B (wb/m^2), the area of the loop S(m^2), and a geometric factor, K (volts/wb.) The drawing on the left of Figure 10.20 illustrates the physical situation causing magnetic field pickup. A typical situation causing magnetic field noise occurs when AC power being delivered to a low-impedance load, such as a motor. The voltage V_l is the 120 VAC 60 Hz power line, and V_s is a signal in our instrument. The alternating current (I_l) will create a magnetic field, B. This magnetic field (B) also alternates as it flows through the loop area (S) formed by the wires in our circuit (such as the lead wires between the transducer and the instrument box.) This will induce a current (I_m) along the wire, causing a voltage error (V_m). We can test for the presence of magnetic field pickup by deliberately changing the loop area and observing the magnitude of the noise as a function of the loop area. The drawing on the right of Figure 10.20 illustrates an equivalent circuit we can use to model magnetic field pickup. Basically, we can model magnetic field induced noise as a mutual inductance between an AC current flow and our electronics.

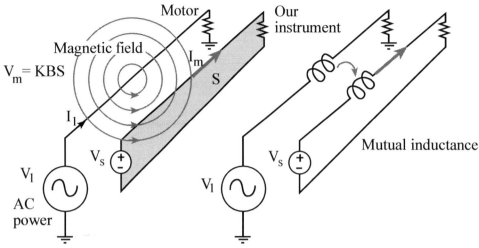

Figure 10.20. Magnetic field noise pickup can be modeled as a transformer.

The second way EM fields can couple into our circuits is via the electric field. Changing electric fields will capacitively couple into the lead wires. The drawing on the left of Figure 10.21 illustrates the physical situation causing electric field pickup. The alternating voltage (V_1) will create an electric field. This electric field also traverses near the wires in our circuit (such as the lead wires between the transducer and the instrument box.) This will induce a displacement current (I_d) along the wire. We can test for the presence of electric field pickup by placing a shield separating our electronics from the source of the field and observing the magnitude of the noise. The drawing on the right of Figure 10.21 illustrates an equivalent circuit we can use to model electric field pickup. Basically, we can model electric field induced noise as a stray capacitance between an AC voltage and our electronics.

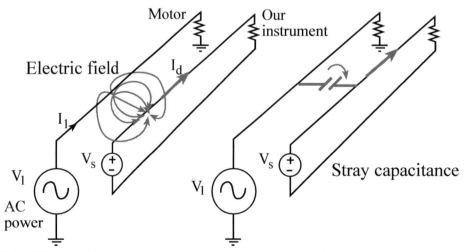

Figure 10.21. Electric field noise pickup can be modeled as a stray capacitance.

Observation: Sometimes the EM fields originate from inside the instrument box, such as high frequency digital clocks and switching power supplies.

10.5.7. Techniques to Measure Noise

There are two objectives when measuring noise. The first objective is to classify the type of noise. In particular, we wish to know if the noise is broadband (i.e., all frequencies, like white noise) or does the noise certain specific frequencies (e.g., 60, 120, 180 Hz,… like 60 Hz EM field pickup). The type of noise is of great importance when determining where the noise is coming from. Classifying the noise type is essential in developing a strategy for reducing the effect of the noise. The second objective is to quantify the magnitude of the noise. Quantifying the noise is helpful in determining if a change to the system design has increased or decreased the noise. The measurement resolution of many data acquisition systems is limited by noise rather than by ADC precision and software algorithms. For these systems, quantitative noise measurements are an important performance parameter of the instrument.

Digital voltmeter (DVM) in AC mode. Root-mean-squared (RMS) is defined the square root of the time-average of the voltage squared (Figure 10.19). If you remove the input signal, the output of the system contains just noise. Because the resistance load is usually constant, squaring the voltage results in a signal proportional to noise power. The averaging calculation gives a measure related to average power, and the square root produces a result with units in volts. RMS noise of a signal can be measured with a DVM using AC mode. Most DVMs in AC mode perform a direct measurement of RMS; hence this method is the most precise. For example, a 3½ digit DVM has a precision of about 11 bits. A calibrated voltmeter in AC mode will be the most accurate quantitative method to measure noise.

Analog oscilloscope. The second method is to connect the signal to an oscilloscope and measure the peak-to-peak noise amplitude, as illustrated in Figure 10.22. The crest factor is the ratio of peak value divided by RMS. The peak value is ½ of the peak-to-peak amplitude, and can be estimated from the scope tracing. From Table 10.8, we see for white noise that the crest factor is about 4, so we can approximate the RMS noise amplitude by dividing the peak-to-peak noise by 8. Because the quantitative assessment of noise with a scope requires visual observation, this method can only be used to approximate the quantitative level of noise. One the other hand, oscilloscope have very high bandwidth, and therefore they are good for classifying high frequency noise. White noise and 1/f noise look random, like left graph in Figure 10.22. For white and 1/f noise, the scope trigger will not be able to capture a repeating waveform. Noise from EM fields on the other hand are repeating and can be triggered by the scope. In fact the **line-trigger** setting on the scope can be used specifically to see if the noise is correlated to the 60 Hz 120VAC power line. In particular, 60 Hz noise will trigger when using the line-trigger setting of the scope. The shape of the noise varies depending on the relative strengths of the fundamental and harmonic frequencies. The graph on the right is a periodic wave with a fundamental plus a 50% strength first harmonic.

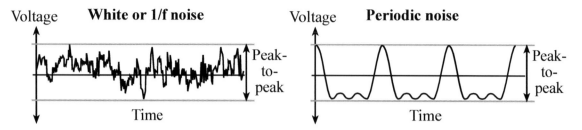

Figure 10.22. Quantifying noise my measuring peak-to-peak amplitude.

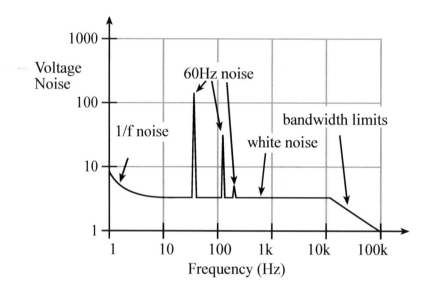

Figure 10.23. Classifying noise my measuring the amplitude versus frequency with a spectrum analyzer.

Spectrum analyzer. The third method uses a spectrum analyzer, which combines a computer, high-speed ADC sampling and the Fast-Fourier-Transform (FFT), as illustrated in Figure 10.23. Being able to see the noise in the frequency domain is particularly useful to classify the type of noise. Both 1/f and white noise components exist in a typical analog amplifier, but the amplitude is usually small compared to EM field noise. The 1/f and white noise levels occurring in electronics can be reduced by dropping the temperature or spending more money buying a better components. Typically, we see the fundamental and multiple harmonics for EM field noise. For example, 60 Hz noise also includes components at 120, 180, 240 Hz etc.

10.5.8. Techniques to Reduce Noise

It is much simpler to reduce noise early in the design process. Conversely, it can be quite expensive to eliminate noise after an instrument has been built. Therefore, it is important to consider noise at every stage of the development cycle. We can divide noise reduction techniques into three categories. The first category involves reducing noise from the source. You can enclose noisy sources in a grounded metal box. If a cable contains a high frequency noise signal, then that signal could be filtered. Magnetic and electric field strength depends on dI/dt and dV/dt. So whenever possible, you should limit the rise/fall times of noisy signals, as shown in Figure 10.24. For example, the square wave on the left will radiate more noise than the smooth signal on the right. When operating a GPIO output pin in 8mA drive mode, we can set the corresponding bit in the SLR register to limit the slew rate on the output.

Figure 10.24. Limiting rise/fall times can reduce radiated noise.

Motors have coils to create electromagnets, and noise can be reduced by limiting the dI/dt in the coil. Cables with noisy signals should be twisted together, so the radiated magnetic fields will cancel. These cables should also be shielded to reduce electric field radiation, and this shield should be grounded on both sides.

The second category of noise reduction involves limiting the coupling between the noise source and your instrument. Whenever possible maximize the distance between the noisy source and the delicate electronics. All transducer cables should use twisted wire, as shown in Figure 10.25.

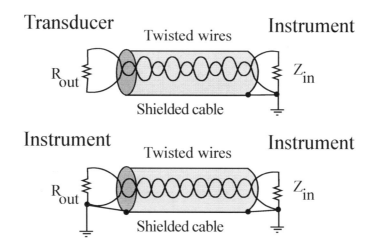

Figure 10.25. Proper cabling can reduce noise when connecting a remote transducer or when connecting two instruments.

For situations where a remote sensor is attached to an instrument, the ground shield should be connected only on the instrument side. If the cable is connected between two instruments, then connect the ground shield on both instruments. For high frequency signals coaxial cable is required. The shield should be electrically insulated to eliminate direct electrical connection to other devices. If noisy signals must exist in the same cable as low-level signals, then separate the two with a ground wire in between. Whenever possible reduce the length of a cable. Similarly minimize the length of leads that extend beyond the ground shield. Place the delicate electronics in a grounded case. You can use optical or transformer isolation circuits to separate the noisy ground from the ground of the delicate electronics.

The last category involves techniques that reduce noise at the receiver. The bandwidth of the system should be as small as possible. In particular, you can use an analog low pass filter to reduce the bandwidth, which will also reduce the noise. You can add frequency-reject digital filters to reduce specific noise frequencies like 60, 120, 180 Hz. You should use power supply decoupling capacitors on each chip to reduce the noise coupling from the power supply to the electronics. Figure 10.26 illustrates how EM field noise will affect our instrument. If the cable has twisted wires then I_{d1} should equal I_{d2}. The input impedance of the amplifier is usually much larger than the source impedance of the signal. Thus, $V_1 - V_2 = R_{s1} I_{d1} - R_{s2} I_{d2}$.

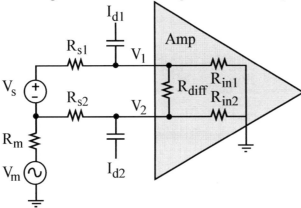

Figure 10.26. *Capacitively coupled displacement currents.*

Some capacitors have a foil wrap surrounding its cylindrical shape. This foil should be grounded. For more information about noise refer to Henry Ott, Noise Reduction Techniques in Electronic Systems, Wiley, 1988 or Ralph Morrison, Grounding and Shielding Techniques, Wiley, 1998.

10.6. Data Acquisition Case Studies

We introduced the design process back in Chapter 1. In this section, we will present the designs of four data acquisition systems. The first example is a quantitative measurement of temperature, the second example is a qualitative measurement of electrocardiogram signals, and the third example compares measurements made with the ADC versus timing measurements using input capture. The last example inputs from a microphone and outputs to a speaker. In the **analysis phase**, we determine the requirements and constraints for our proposed system. In the **high-level design** phase, we define our input/output, break the system in modules, and show the interconnection using data flow graphs. In the **engineering design** phase we design the hardware/software subcomponents using techniques such as simulation, mechanical mockups, and call graphs. We define specific I/O signals, analog circuits, power sources, noise filters, software algorithms, data structures, and testing procedures. During the **implementation** and **testing** phases we build and test the modules. Modularity allows for concurrent development.

Example 10.1. Design an instrument that measures temperature with a range of 30 to 40°C, and a resolution is 0.01°C. The frequency range of interest is 0 to 20 Hz.

Solution: The first decision to make is the choice of transducer. The RTD has a linear resistance versus temperature response. RTDs are expensive, but a good choice for ease of calibration, interchangeability, and accuracy. Thermocouples are inexpensive and a good choice for large temperature ranges, harsh measurement conditions, fast response time, interchangeability, and large temperature ranges. Interchangeability means we can buy multiple transducers and they will all have similar temperature curves. Thermistors will be used in this design because they are inexpensive and have better sensitivity than RTDs and thermocouples. A thermometer built using a thermistor will be harder to mass-produce because each transducer must be separately calibrated in order to create an accurate measurement. From a first glance, we might expect a 10-bit ADC will generate a temperature resolution of 0.01 °C. Recall that range equals resolution times precision. On the other hand, because the thermistor is nonlinear, we will need to verify the resolution specification is met. Figure 10.27 shows the block diagram of the instrument, which also illustrates the data flow in our system.

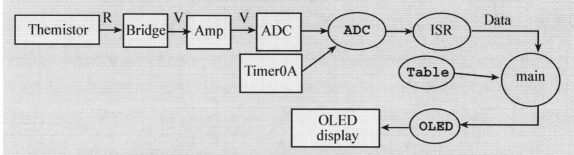

Figure 10.27. Data flow graph of a temperature measurement system using a thermistor.

The resistance bridge is a classic means to convert resistance to voltage. Table 10.10 is used during the design phase to show the signal values as they transverse the electronics. A +2.50 V reference drives the bridge. The value of resistor R_1 is chosen to eliminate errors due to self-heating the thermistor (100 kΩ). The first two columns of Table 10.10 give the thermistor calibration. Since we will be using rail-to-rail electronics, we need to have all voltages between the 0 to +3.3 V range. We can make the bridge output (V_1-V_2) positive by selecting the value of resistor R_2 less than the thermistor resistance at the maximum temperature (18.3 kΩ). For this thermistor, R_2 is selected at 18 kΩ because it is a standard resistor less than 18.3 kΩ. Next, we choose the gain of the amplifier to map the minimum temperature into the +3V limit of the ADC (17.5). Using an INA122 or AD627, a gain resistor of 16 kΩ creates the desired gain of 17.5. Since the thermistor is nonlinear, we will tabulate explicit values to determine the ADC precision required (Table 10.10). There are two possible approaches to the design of the amplifier. If we use an instrumentation amp, the input impedance will be large enough not to affect the bridge. If we use a single op amp differential amp, then the amplifier will load the bridge and affect the bridge response. In this system, the instrumentation amp will be used. The first two columns of Table 10.10 show the resistance temperature calibration of the thermistor. The third column, V_1, is the voltage across the thermistor. The fourth column is the output of the bridge, V_1-V_2. V_{out}, the output of the instrumentation amp, is shown in the fifth column. The ADC value gives the digital output of a 10-bit converter, and the last column will be calculated by our software as a decimal fixed-point with resolution 0.01 °C. We use this table in two ways. Initially, we use theoretical values to design the electronics and software. During the implementation phase, we substitute resistors with standard values to bring down the cost. During the testing phase, we measure actual values to verify proper operation. Measured values for the last two columns will be stored in software as a calibration table. To measure temperature, the software measures the ADC value, and then uses a table look-up and linear interpolation to get the decimal fixed-point temperature (last column). The fixed-point number is output to an OLED display.

Observation: There is an Excel worksheet named Therm10.xls that was used to create this design. You can find it on the book web site.

T (°C)	R_t (kΩ)	V_1 (V)	V_1-V_2 (V)	V_{out} (V)	ADC	T (0.01°C)
30.0	28.241	0.551	0.169	2.961	1009	3000
32.0	25.837	0.513	0.132	2.309	787	3200
34.0	23.665	0.478	0.097	1.698	579	3400
36.0	21.700	0.446	0.064	1.127	384	3600
38.0	19.921	0.415	0.034	0.594	202	3800
40.0	18.307	0.387	0.005	0.096	32	4000

Table 10.10. Signals as they pass through the temperature data acquisition system.

In order to prevent noise in the ADC samples, the noise must be less than the resolution. The resolution of V_1-V_2 is its range (0.169V) divided by its precision (1024), which is about 0.1 mV. Again a safety factor of 1/2 is included. Thus in the frequency range 0 to 20 Hz, the maximum allowable noise referred to the input of the differential amp should be

$$\text{amplifier noise} \leq \frac{\text{resolution}}{2} = 50 \, \mu V$$

A two-pole low pass analog filter (Figure 8.25) is used to pass the temperature signal having frequencies from 0 to 20 Hz, reject noise having frequencies above $\frac{1}{2}f_s$. In order to prevent aliasing, Z must be less than the ADC resolution for all frequencies larger than or equal to $\frac{1}{2}f_s$. As an extra measure of safety, we make the amplitude less than 0.5 Δ_Z for frequencies above $\frac{1}{2}f_s$. Thus,

$$|Z| < 0.5 \, \Delta_Z = 3/2048 \approx 1 \, mV$$

The effective output impedance of the bridge is 18 kΩ. The input impedance of the differential amp must be high enough to not affect the ADC conversion.

$$Z_{in} > 18 \text{ k}\Omega \cdot 2^{n+1} = 36 \text{ M}\Omega$$

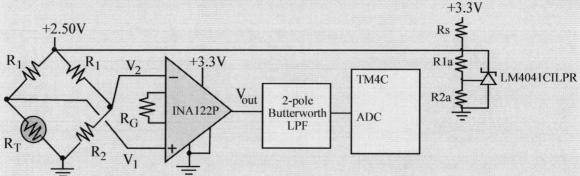

Figure 10.28. *Amplifier and low pass filter.*

To determine the resolution we work backwards, as illustrated in Table 10.11. The basic approach to verifying the temperature resolution is to work backwards through the circuit, showing that a change in ADC value of 1 corresponds to a temperature change of 0.01 °C.

ADC	V_3 (V)	V_1-V_2 (V)	V_1 (V)	R_T (kΩ)	T (°C)	ΔT (°C)
1023	2.997	0.1713	0.5526	28.3774	29.892	
1022	2.994	0.1711	0.5524	28.3664	29.901	0.0087
512	1.500	0.0857	0.4671	22.9752	34.679	
511	1.497	0.0855	0.4669	22.9651	34.690	0.0102
1	0.003	0.0002	0.3815	18.0093	40.391	
0	0.000	0.0000	0.3814	18.0000	40.404	0.0124

Table 10.11. **Equations calculated in reverse to show that the resolution meets the design specification.**

There are three possible approaches to converting ADC sample to temperature (the last two columns of Table 10.10). First, we could fit the transfer function to a polynomial equation, and save the coefficients of that equation as the calibration file. This approach performs well for simple situations. A plot of this data is shown as Figure 10.29. Second, we could calculate the temperature output for each possible ADC and save it in a 1024-entry lookup table. This conversion is fast because we just need to use the ADC data to index into the big table. This method is fast, but requires a lot of ROM. The third approach, shown in Program 10.1, uses a table of paired (ADC,T) data. These points are determined from experimental calibration. To find the corresponding temperature for a given ADC value, the program first searches the table for a pair of ADC-values that surround the input. Extra entries were added at the beginning and end of the table to guarantee the search step will always be successful. Then, it uses linear interpolation to calculate the temperature, given the 53 entries in the table and the ADC input. The output result is a fixed-point number with a resolution of 0.01 °C.

Typically, the resolution of this thermometer will depend on analog noise rather than ADC resolution. The accuracy will depend on both resolution and calibration drift. Drift is defined as a change over time of the data in Figure 10.29 and the numbers in Program 10.1.

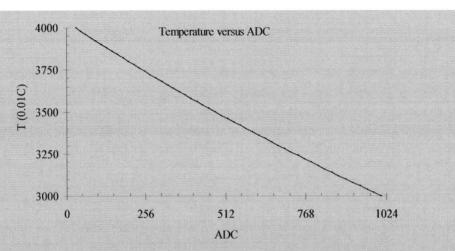

Figure 10.29. *Transfer function between sampled 10-bit ADC and fixed-point temperature.*

The calibration data in **ADCdata** and **Tdata** are stored in EEPROM. In general, we perform time-critical tasks like ADC sampling in the background, and noncritical functions like conversion to temperature and OLED output in the foreground. Therefore, the interrupt service routine passes the measured temperature to the foreground through a FIFO queue, and the main program has the responsibility of outputting the result to the OLED. The completion of this system is left as a laboratory exercise.

```
// table of multiple unsigned (x,y), piece-wise linear function
uint16_t const ADCdata[53]={0,32,49,65,82,99,116,133,150,167,184,
    202,220,237,255,273,291,310,328,346,365,
    384,403,422,441,460,480,499,519,539,559,
    579,599,619,640,660,681,702,723,744,765,
    787,808,830,852,874,896,918,941,964,986,1009,1024
};
uint16_t const Tdata[53]={4000,4000,3980,3960,3940,3920,3900,3880,
    3860,3840,3820,3800,3780,3760,3740,3720,3700,3680,3660,3640,3620,
    3600,3580,3560,3540,3520,3500,3480,3460,3440,3420,
    3400,3380,3360,3340,3320,3300,3280,3260,3240,3220,
    3200,3180,3160,3140,3120,3100,3080,3060,3040,3020,3000,3000
};
```
Program 10.1. *Calibration data for thermistor.*

Checkpoint 10.3: What would be the temperature resolution if the ADC precision were decreased from 10 to 8 bits?

To measure temperature resolution, we use the student's t-test to determine if the system is able to detect the change. To use the student's t test we need to make the following assumptions:
    1) Errors in one data set are independent, not correlated to errors in the other data set;
    2) Errors in each data sample are independent, not correlated to errors within that set;
    3) Errors are normally distributed;
    4) Variance is unknown;
    5) Variances in the two sets are equal.

If a random variable, X, is normally distributed with a mean is μ and a standard deviation of σ, then the probability that it falls between $\pm 1\ \sigma$ is 68 %. I.e.,
$$P(\mu-\sigma < X < \mu+\sigma) = 0.68$$
Similarly,
$$P(\mu-1.96\sigma < X < \mu+1.96\sigma) = 0.95$$
$$P(\mu-2\sigma < X < \mu+2\sigma) = 0.954$$
$$P(\mu-2.58\sigma < X < \mu+2.58\sigma) = 0.99$$
$$P(\mu-3\sigma < X < \mu+3\sigma) = 0.9997$$

The square of the standard deviation is called variance, σ^2. In most situations, we do not know the mean and standard deviation, so we collect data and estimate them. In particular, we take multiple measurements assuming the temperature is constant. Let X_i be repeated measurements under the same conditions, and N is the number of measurements (e.g., $N = 10$).

$$\overline{X} = \frac{1}{N}\sum_i X_i \qquad S^2 = \frac{1}{N-1}\sum_i (X_i - \overline{X})^2$$

The $N-1$ term is used in the calculation of S because there are $N-1$ degrees of freedom. These expressions are unbiased estimates of μ and σ, meaning as the sample size increases the estimates approach truth. Formally, we say the expected value of $\overline{X}$ is μ, or $E(\overline{X}) = \mu$. Similarly, the expected value of S^2 is σ^2, or $E(S^2) = \sigma^2$. For example, we collect two sets of data (e.g., 10 measurements in each set, $N = 10$), and we want to know if the means of two sample sets are different. Consider the measurements in the two data sets as the sum of the true value plus an error:
$$X_{0i} = \mu_0 + e_{0i}$$
$$X_{i1} = \mu_1 + e_{1i}$$
Assumption 1 states that e_{0i} are not correlated to e_{1i}. Assumption 2 states that e_{0i} are not correlated to e_{0j} and e_{1i} are not correlated to e_{1j}. Thermal noise will satisfy these assumptions. We employ a test statistic to test the hypothesis H_0: $\mu_0=\mu_1$. First, we estimate the means and variances of the data (assuming equal sized samples)

$$\overline{X}_0 = \frac{1}{N}\sum_i X_{0i} \qquad S_0^2 = \frac{1}{N-1}\sum_i (X_{0i} - \overline{X}_0)^2$$

$$\overline{X}_1 = \frac{1}{N}\sum_i X_{1i} \qquad S_1^2 = \frac{1}{N-1}\sum_i (X_{1i} - \overline{X}_1)^2$$

From these, we calculate the test statistic t:

$$t = \frac{\overline{X}_1 - \overline{X}_0}{\sqrt{S_0^2/N + S_1^2/N}}$$

The two sets of data, together, have $2N-2$ degrees of freedom. The student's t table, shown as Table 10.12, has two dimensions. In the vertical direction, we specify the degrees of freedom, *df*. For example, if there are 10 data points in each data set, then *df* equals 18. In the horizontal direction we select the probability of being correct. For example, if we wish to be 99% sure of the test, then we select the 99% column. Selecting the row and the column allows us to pick a number threshold. For example, the number in the *df*=18 row, **confidence**=99% column is 2.878.

This means if H_0 is true, then the probability of $t < -2.878 = 0.005$ and the probability of $t > 2.878 = 0.005$.

Therefore, the probability of $-2.878 < t < 2.878 = 0.99$ (confidence interval of 99%)

If we collect data and calculate t such that the test statistic t is greater than 2.878 or less than -2.878, then we claim "we reject the hypothesis H_0". If the test statistic t is between -2.878 and 2.878 we do not claim the hypothesis to be true. In other words we have not proven the means to be equal. Rather, we say "we do not reject the hypothesis H_0".

confidence	80%	90%	98%	99%	99.8%	99.9%
df p=	0.10	0.05	0.01	0.005	0.001	0.0005
8	1.397	1.860	2.896	3.355	4.501	5.041
9	1.383	1.833	2.821	3.250	4.297	4.781
10	1.372	1.812	2.764	3.169	4.144	4.587
11	1.363	1.796	2.718	3.106	4.025	4.437
12	1.356	1.782	2.681	3.055	3.930	4.318
13	1.350	1.771	2.650	3.012	3.852	4.221
14	1.345	1.761	2.624	2.977	3.787	4.140
15	1.341	1.753	2.602	2.947	3.733	4.073
16	1.337	1.746	2.583	2.921	3.686	4.015
17	1.333	1.740	2.567	2.898	3.646	3.965
18	1.330	1.734	2.552	2.878	3.610	3.922
19	1.328	1.729	2.539	2.861	3.579	3.883
20	1.325	1.725	2.528	2.845	3.552	3.850
21	1.323	1.721	2.518	2.831	3.527	3.819
22	1.321	1.717	2.508	2.819	3.505	3.792
23	1.319	1.714	2.500	2.807	3.485	3.767
24	1.318	1.711	2.492	2.797	3.467	3.745
25	1.316	1.708	2.485	2.787	3.450	3.725
26	1.315	1.706	2.479	2.779	3.435	3.707
27	1.314	1.703	2.473	2.771	3.421	3.690
28	1.313	1.701	2.467	2.763	3.408	3.674
29	1.311	1.699	2.462	2.756	3.396	3.659
30	1.310	1.697	2.457	2.750	3.385	3.646
40	1.303	1.684	2.423	2.704	3.307	3.551
50	1.299	1.676	2.403	2.678	3.261	3.496
60	1.296	1.671	2.390	2.660	3.232	3.460
80	1.292	1.664	2.374	2.639	3.195	3.416
100	1.290	1.660	2.364	2.626	3.174	3.390
120	1.289	1.658	2.358	2.617	3.160	3.373
∞	1.282	1.645	2.326	2.576	3.090	3.291

Table 10.12. Student's t distribution table.

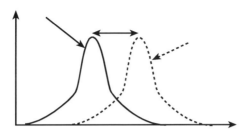

Figure 10.30. Resolution means if the temperature increases by ΔT, the system will probably notice.

Example 10.2. Design a system to measure the electrical activity in the heart. In particular, we wish to measure heart rate.

Solution: Biopotentials are important measurements in many research and clinical situations. **Biopotentials** are electric voltages produced by individual cells and can be measured on the skin surface. The status of the heart, brain, muscles, and nerves can be studied by measuring biopotentials. Electrodes, which are attached to the skin, interface the machine to the body. Electronic instrumentation amplifies and filters the signal. For example, Figure 10.31 shows a normal Lead II **electrocardiogram**, or **EKG**, which is measured with the positive terminal attached to the left arm, the negative terminal attached to the right arm, and ground connected to the right leg. Each wave represents one heartbeat, and the shape and rhythm of this wave contains a lot of information about the health and status of the heart.

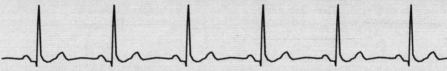

Figure 10.31. Normal II-lead electrocardiogram.

There are two types of electrodes used to record biopotentials. **Nonpolarizable** electrodes like silver/silver chloride involve the following chemical reaction in the electrode at the electrode/tissue interface:

$$AgCl + e^- \leftrightarrow Ag + Cl^-$$

A **nonpolarizable electrode** has a low electrical impedance because electrons can freely pass the electrode/tissue interface. In the electrode, current flows by moving electrons, but in the tissue current flows by physical motion of charged ions (e.g., Na^+, K^+ and Cl^-). A silver/silver chloride electrode does include a half-cell potential of 0.223 V, but since biopotentials are always measured with two electrodes, these half-cell potentials cancel. On the other hand, if you tried to use these electrodes to measure DC voltages, then the above chemical reaction would saturate and fail. Fortunately, biopotentials measured produced by muscles and nerves are AC only and have no DC component.

Polarizable electrodes, made from metals like platinum gold or silver, have a high electrical impedance because electrons cannot freely pass the electrode/tissue interface. Charge can develop at the electrode/tissue interface effectively creating a capacitive barrier. Displacement current can flow across the capacitor, allowing the AC biopotentials to be measured by the electronics. The metallic electrodes also include a half-cell potential, but again, these potentials will cancel.

$$Ag^+ + e^- \leftrightarrow Ag$$

The graphical display of EKG versus time is an example of a qualitative data acquisition system. The measurement of heart rate is quantitative. The parameters of an EKG amplifier include: high input impedance (larger than 1 MΩ), high gain, 0.05 to 100 Hz bandpass filter and good common mode rejection ratio. The EKG signal is about ±1 mV, so an overall gain of about 2000 will produce a range of 0 to +3.3V on V_3. The data flow graph of this system is similar to Figure 10.27. This EKG amp (Figure 10.32) begins with a preamp stage having a good CMRR, high input impedance, and a gain of 10. If the system is battery operated, then it does not need a third or ground electrode. Pin 5 of the AD627 is the analog ground, which in this circuit is the 1.233 V reference voltage. The AD627 is rail-to-rail. A 0.05 Hz passive high pass filter is created by R_4 and C_4. Low-leakage capacitors for C_1, C_2, and C_4 are critical for elimination of DC offset drift. A polypropylene or polystyrene would be a good choice, but a C0G ceramic is acceptable. The remaining gain is performed with a noninverting amplifier (U2a). The LPF is implemented as a 2-pole Butterworth LPF. The 153 Hz cutoff was chosen because it is greater than 100Hz and can be implemented with standard components. If the signal V_1 saturates, you can reduce the gain of the preamp and increase the gain of the amp.

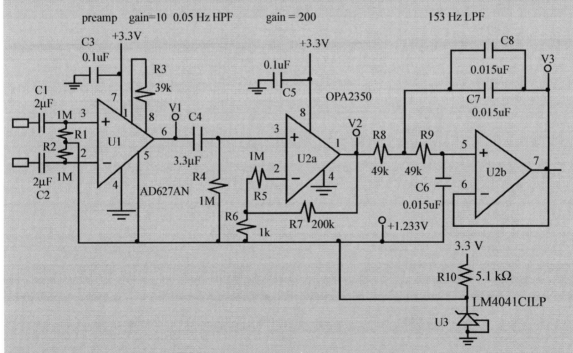

Figure 10.32. *A battery-power EKG instrument.*

Program 10.2 shows the real-time data acquisition and 60 Hz digital notch filter. The design of digital filters will be presented in the third volume of this book series. The ADC sampling occurs in the background and the data are passed to the foreground using a FIFO queue. The large pulse in the EKG, originating from the contraction of the ventricles, is called the R-wave and it occurs once a heartbeat. Program 10.3 shows the foreground process, where there are four calculation steps performed on the EKG data. A low pass filter followed by a high pass filter capture a narrow band of information around 8 Hz. The square function calculates power and the 200 ms wide moving average gives an output very specific for the R-wave. Hysteresis is implemented with two thresholds. A heartbeat is counted (**RCount++**) when the moving average goes below the **LOW** threshold, then above the **HIGH** threshold. This software uses a combined frequency-period method to calculate heart rate. The algorithm to measure heart rate searches for R-waves in a 5-second interval. **Rfirst** is the time (in 1/120 sec units) of the first R-wave and **Rlast** is the time (also in 1/120 sec units) of the last R-wave. (**RCount-1**) is the number of beat-to-beat intervals between **Rfirst** and **Rlast**. The number 7200 is the conversion between the sample period (1/120 sec) and one minute. For example at 72 BPM, there will be 6 R-waves detected in the 5-second interval, making (**Rcount-1**) equal to 5 and the difference **Rlast-Rfirst** will be 500. For more information on EKG systems, see Webster's book *Medical Instrumentation*, published by Wiley 1997, or Pan and Tompkins, "A Real-Time QRS Detection Algorithm," IEEE Transactions on Biomedical Engineering, pp. 230-236, March 1985.

> **Warning:** If you are going to build an EKG, please have a trained engineer verify the safety of your hardware and software before you attach people to your machine.

```
void ADC3_Handler(void){ int16_t data;
  ADC_ISC_R = 0x08;             // acknowledge ADC sequence 3
  data = ADC_SSFIFO3_R -2048;   // twos complement
  Fifo_Put(data);               // pass to foreground
}
```
Program 10.2. Real-time sampling of EKG (see program 8.8), sampled at 120 Hz.

```
int16_t Data;      // ADC sample, -2048 to +2047, 12-bit signed ADC sample
int16_t x[50];     // sampled EKG, 120Hz
int16_t y[50];     // low pass filter, 120Hz
int16_t z[50];     // high pass filter, 120Hz
int16_t w[50];     // squared result, R-wave power, 120Hz
int16_t Rwav;      // moving average of R-wave power, energy
uint16_t n=25;     // 25,26, ..., 49
uint16_t Trigger;
#define HIGH 100   // trigger when over this
#define LOW 20     // reset when under this
uint16_t Rcount;      // number of R-waves
uint16_t Rfirst;      // time of first R-wave
uint16_t Rlast;       // time of last R-wave
uint16_t HeartRate;   // units bpm
void main(void) { uint16_t time;               // units 1/120sec
int16_t lpfSum=0,hpfSum=0,RwavSum=0;           // uses Program 8.8
  ADC_InitTimer0ATriggerSeq3(0,49,8332);       // 120 Hz sampling
```

```
Fifo_Init();
Trigger =0;        // looking for HIGH
for(;;) {
  Rcount = 0;
  Rlast = 0;
  for(time=0;time<600;time++){   // 120 Hz, every 5 second
    while(Fifo_Get(&Data)){};    // Get data from background thread
    Plot(Data);                  // draw voltage versus time plot
    n++; if(n==50) n=25;
    x[n] = x[n-25] = Data;       // new data
    lpfSum = lpfSum+x[n]-x[n-4];
    y[n] = y[n-25] = lpfSum/4;   // Low Pass Filter
    hpfSum = hpfSum+y[n]-y[n-10];
    z[n] = z[n-25] = y[n]-hpfSum/10; // High Pass Filter
    w[n] = w[n-25] = (z[n]*z[n])/10; // Power calculation
    RwavSum = RwavSum+w[n]-w[n-24];  // 200ms wide moving average
    Rwav = RwavSum/24;
    if(Trigger){
      if(Rpow<LOW){
        Trigger = 0;      // found low
      }
    } else{
      if((Rpow>HIGH)&&((time-Rlast)>30)){ // max HR= 240bpm
        Trigger = 1;      // found high
        if(Rcount){
          Rlast = time;   // mark time of last R-wave, units 1/120sec
        } else{
          Rfirst = time;  // mark time of first R-wave
        }
        Rcount++;
      }
    }
  }
  if(Rcount>=2){
    HeartRate = (7200*(int32_t)(Rcount-1))/(int32_t)(Rlast-Rfirst);
  } else{
    HeartRate = 0;
  }
  Output(HeartRate);   // display results
 }
}
```
Program 10.3. Measurement of heart rate.

Checkpoint 10.4: What is the theoretical heart rate resolution of this approach when the HR is 60 BPM?

Common error: There are two reasons for EKG circuits to fail. The first poor contact between the skin and electrode causing a reduction in CMRR, and second is resistive leakage in capacitor C_4. Clean the skin well, use new electrodes and select a high quality C_4.

Example 10.3. Design a system to measure position. The range is 0 to 2 cm and the resolution at least 0.01 cm.

Solution: A Bourns SSHA20B20300 potentiometer can be used to convert position into resistance, see Figure 10.33. In this particular transducer the resistance between pins 1 and 3 is fixed at 20 kΩ, while the resistance between pins 2 and 3 (R_{out}) varies from 0 to 20 kΩ.

$$R_{out} = 5 \cdot x$$

where the units of R_{out} and x are in kΩ and cm respectively. To interface this transducer to the microcontroller we drive the potentiometer with a stable DC voltage using a precision voltage reference. If we were to drive the circuit with the +3.3V power, then any noise ripple in the power line would couple directly into the measurement. Rather, we implement the 3.00V with a precision reference. The ADC produces a digital output dependent on its analog input. The position resolution is 2cm/1024, which is about 0.002cm.

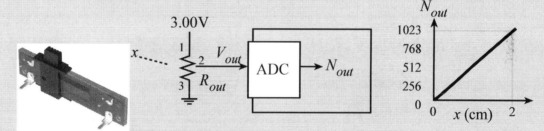

Figure 10.33. Potentiometer interface using an ADC.

Another approach to interface this transducer to the microcontroller would be to use astable multivibrator. The period of a 555 timer is $0.693 \cdot C_T \cdot (R_A + 2R_B)$. In our circuit, R_A is R_{out}, R_B is 1kΩ, and C_T is 2.2µF. Given a fixed R_B, C_T, the period of the square wave, P_{out}, is a linear function of R_{out}. Our microcontrollers have a rich set of mechanisms to measure frequency, pulse width or period. To change the slope and offset of the conversion between R_{out} and P_{out}, the values of R_B and C_T can be adjusted. Even though the period does not include zero, the precision of this measurement is over 3000 alternatives or more than 11 bits. The precision can be improved by increasing the capacitor, C_T, or decreasing the period on the measurement clock. The position resolution is 2cm/3049, which is about 0.001cm. The actual resolution will probably be dominated by noise and not period measurement resolution.

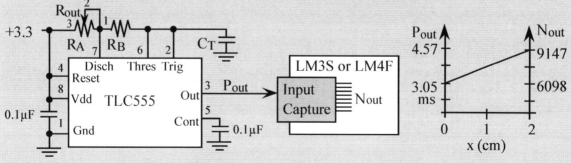

Figure 10.34. Potentiometer interface using input capture.

Example 10.4. Design a system that can input and output sound.

Solution: The electret microphone was described previously in Figure 10.11 as a low-cost, small-size transducer capable of converting sound into voltage. Many electret data sheets suggest an $R1$ of 2 kΩ, but signal-to-noise ratio can be improved by using a 10 kΩ resistor. Because the output of a HPF would normally include positive and negative voltages, we will need a way to offset the circuit so all voltages exist from 0 to +3.3 V, allowing the use of a single supply and rail-to-rail op amps. $R2$ and $R3$ provide an offset for the HPF, so the signal $V2$ will be the sound signal plus 1.65 V. The effective impedance from $V2$ to ground is 11 kΩ, so the HPF cutoff is $1/(2\pi*1\mu F*11k\Omega) = 14$ Hz. The gain of the system is $1+R6/R5$, which will be 101. The capacitor $C2$ will make the signal $V3$ be the amplified sound plus 1.65 V. The gain is selected so the $V3$ signal is 1.65 ±1 V for the sounds we wish to record. The capacitor $C3$ provides a little low pass filtering, causing the amplifier gain to drop to one for frequencies above $1/(2\pi*220pF*100k\Omega) = 7.2$ kHz. A better LPF would be to add an active LPF. The LPF would need a 1.65 V offset like the one in Figure 10.35. If we wish to process sound with frequency components from 100 to 5 kHz, then we should sample at or above 10 kHz. If we sampled sound with a 12-bit ADC, we should select a 12-bit DAC to output the sound. We could improve signal to noise by replacing the +3.3 V connected to $R1$ and $R2$ in Figure 10.35 with a LM4041 adjustable reference and create a low noise 3.0V voltage.

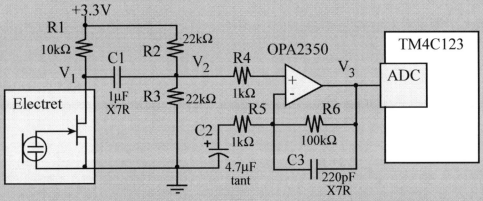

Figure 10.35. An electret microphone can be used to record sound (PD3 on TM4C123).

The LM4041CILP is a shunt reference used to make the analog reference required by the MAX5353 12-bit DAC. This DAC was previously interfaced in Example 7.2. The TPA731 audio amp, first shown in Figure 9.5, can be used to amplify the DAC output providing the current needed to drive a typical 8 Ω speaker. The gain of the audio amplifier is $2*R11/R10$, which for this circuit will be one. This means a 2-V peak-to-peak signal out of the DAC will translate to a 2-V peak-to-peak signal on the speaker. The maximum power that the TPA731 can deliver to the speaker is 700 mW, so the software should limit the sound signal below 2.3 V_{rms} when driving an 8 Ω speaker. The quality of sound can be increased by selecting a better speaker and placing the speaker into an enclosure. For more information on how to design a speaker box, perform a web search on "speaker enclosure".

Software in Program 7.2 can be used to interface the MAX5353 12-bit DAC. Program 10.4 performs the sound input and output. The sampling rate is 10 kHz. The ADC code was previously shown as Program 8.8.

```
void ADC0Seq3_Handler(void){ int32_t data;
  ADC0_ISC_R = 0x08;                // acknowledge ADC sequence 3
  data = ADC0_SSFIFO3_R - 2048;     // sound, -2048 to +2047
                                    // process, filter, record etc.
  DAC_Out(data);                    // output, Program 7.2
}
void main(void){ PLL_Init();                     // 80 MHz TM4C123
  ADC0_InitTimer0ATriggerSeq3PD3(7999); // sample at 10 kHz
  DAC_Init(2048);                                // Program 7.2
  while(1){
    WaitForInterrupt();
  }
}
```
Program 10.4. Real-time sound output input/output.

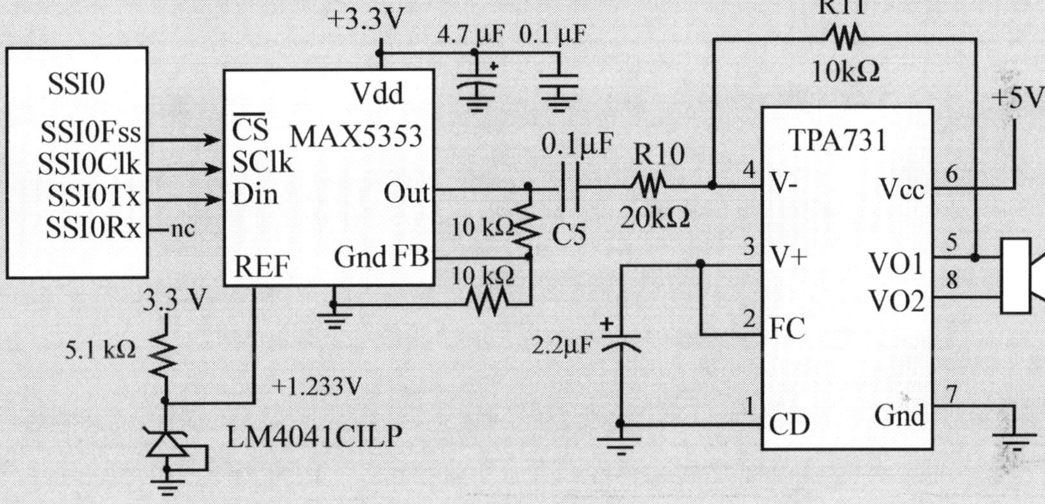

Figure 10.36. A 12-bit DAC and an audio amplifier allow the microcontroller to output sound.

10.7. Exercises

10.1 For each term give a definition in 16 words or less.
a) Measurand   b) Transducer   c) Hysteresis
d) Saturation   e) Breakdown   f) Bang bang
g) Deadzone   h) Phantom bias   i) Prevalence
j) Triple point   k) Positive predictive value   l) Negative predictive value
m) Impedance loading   n) Crest factor

10.2 For each transducer parameter give a definition in 16 words or less.
a) Accuracy b) Linearity   c) Sensitivity   d) Specificity   e) Input impedance
f) Drift   g) First order h) Slew rate   i) Second order   j) Frequency response

10.3 For each instrument parameter give a definition in 16 words or less.
a) Accuracy
b) Maximum error
c) Resolution
d) Precision
e) Reproducibility
f) Signal to noise ratio

10.4 For each transducer give its measurand.
a) Thermistor
b) LVDT
c) Electret
d) Strain gauge
e) Thermocouple
f) Ultrasonic crystal
g) Shaft encoder
h) RTD
i) ADXL202
j) Ag-AgCl electrode

10.5 For each concept give a definition in 16 words or less.
a) Nyquist Theory
b) Aliasing
c) Voltage quantization
d) Time quantization
e) Time jitter
f) Sampling rate

10.6 For type of noise give a definition in 16 words or less. Also give one way to reduce the effect of this noise
a) White noise
b) 1/f noise
c) Magnetic field noise
d) Electric field noise

10.7 Give a short answer in 16 words or less.
a) How do we choose the sampling rate?
b) How do we choose the gain of our amplifier?
c) When do we need a S/H module?
d) When does a data acquisition system have aliasing?

10.8 The following input/output data was collected for a distance transducer (0,5), (1,7), (2,8), (3,11), (4,14), where the first number is distance in cm, and the second number is resistance in kΩ. Calculate average linearity of reading (percent) and average linearity of full scale (percent). Calculate the sensitivity of this transducer. Is this transducer monotonic or nonmonotonic?

10.9 The following input/output data was collected for a distance transducer (0,10), (1,12), (2,14), (3,12), (4,10), where the first number is distance in cm, and the second number is resistance in kΩ. Calculate average linearity of reading (percent) and average linearity of full scale (percent). Calculate the sensitivity of this transducer. Is this transducer monotonic or nonmonotonic?

10.9b Explain how an ultrasonic distance transducer is temperature sensitive.

10.10 Let x(n) be a sampled data representing the angular position of a shaft. It is stored as a 0.01 radian decimal fixed-point number. E.g. x(n) ranges from 0 to 628. This data is sampled every 10 ms. Derive equations for angular velocity and angular acceleration.

10.11 Consider the nonlinear transducer $y = x^2$. The range of x is $1 \leq x \leq 2$. Thus, the range of y is $1 \leq y \leq 4$. Let the desired resolution be $\Delta_x = 0.01$. $n_x = r_x/\Delta_x = 100$ alternatives or about 7 bits. How many ADC bits will be needed?

10.12 Consider the linear transducer $y = 2x+1$. The range of x is $0 \leq x \leq 1$. Thus, the range of y is $1 \leq y \leq 3$. Let the desired resolution be $\Delta_x = 0.01$. $n_x = r_x/\Delta_x = 100$ alternatives or about 7 bits. How many ADC bits will be needed?

10.13 What is the resolution of a 13-bit ADC in dB_{FS}?

10.14 How much white noise in a 10 MΩ resistor if the system bandwidth is 100 kHz?

10.15 How much white noise in a 100 MΩ resistor if the system bandwidth is 10 MHz?

10.16 How can you experimentally differentiate between white noise and pink noise?

10.17 What do you see if you touch the measurement probe of an oscilloscope? Explain.

D10.18 Design a computer based data acquisition system that measures pressure. The pressure transducer is built with four resistive strain gauges placed in a DC bridge. When the pressure is zero, each gauge has a 120 Ω resistance making the bridge output, y, zero. When pressure is applied to the transducer two gauges are compressed (which lowers their resistance) and two are expanded (which increases their resistance.) At full scale pressure ($p = \pm 100$ dynes/cm^2), each resistor changes by ± 2 Ω. The transducer/bridge output impedance is therefore 120 Ω. You may assume the transducer is linear. The desired pressure resolution is 1 dyne/cm^2. The frequencies of interest are 0 to 100 Hz, and the 2 pole Butterworth analog low pass filter will have a cutoff (gain=0.707) frequency of 100 Hz (you will design it in b). In terms of choosing a sampling rate, you may assume the LPF removes all signals above 100 Hz.

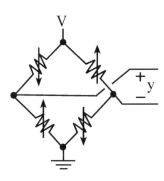

D10.19 Design a temperature acquisition system using a type-T thermocouple. The range is $1 \leq T \leq 200$ °C. The desired resolution is $\Delta T = 1$ °C. The frequencies of interest are DC to 5 Hz. Show the analog interface and data acquisition software.

D10.20 One of the difficulties in mass producing a transducer is to create multiple copies that can be used interchangeably. In this pressure measurement device, the R_3 node is the transducer that is sensitive to pressure. Assume R_3 equals $100\Omega(1+mP+\alpha T)$, where $100m$ is the pressure sensitivity, and p is the input pressure. The R_4 node is a dummy device but is not affected by pressure. R_4 equals $100\Omega(1+\alpha T)$. Both R_1 and R_4 have equal temperature sensitivity, $\partial R_3/\partial T = \partial R_4/\partial T = 100\alpha$. Assume the INA gain is 10. The resistors R_1 and R_2 are the other part of the bridge.

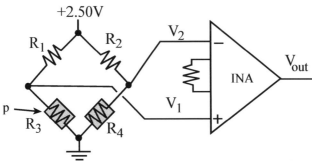

a) Assume R_1 and R_2 are not temperature dependent Show that the output is much less sensitive to temperature than if R_4 were a fixed value.
b) If R_2 is replaced with a potentiometer we can adjust the offset. Derive an equation of showing the output voltage as a function of p and R_2.
c) If R_1 is replaced with a potentiometer we can adjust the sensitivity. Derive an equation of showing the output voltage as a function of p and R_1.

D10.21 This problem deals with the classification and reduction of noise.
a) Describe a **single experimental procedure** (measurement) which could **identify** (or differentiate) the **type**(s) of **noise** existing on the circuit.
b) For each of these three types of noise (white noise, 60 Hz noise, or 1/f noise), give a typical outcome of the experimental procedure.
c) Give one approach (other than analog or digital filtering) which will reduce white noise.
d) Give one approach (other than analog or digital filtering) which will reduce 60 Hz noise.
e) Give one approach (other than analog or digital filtering) which will reduce 1/f noise.

D10.22 A temperature transducer has the following relationship
$R = 200+10T$ where R is in Ω and T is in °C.
The problem specifications are range is $30 \leq T \leq 50$°C. The resolution should be 0.01°C. The frequencies of interest are 0 to 100 Hz. The transducer dissipation constant is 20mW/°C. The ADC range is 0 to +3 volts and the sampling rate is 1000 Hz.
a) How many ADC bits are required?
b) What is the maximum allowable noise at the amp *output*?
c) Design the *analog amplifier/filter*.

D10.23 Design a wind direction measurement instrument. You are given a transducer that has a resistance that is linearly related to the wind direction. As the wind direction varies from 0 to 360 degrees, the transducer resistance varies from 0 to 1000 Ω. The frequencies of interest are 0 to 0.5 Hz, and the sampling rate will be 1 Hz.
a) Show the analog interface between the transducer and the ADC. Only the +3.3 V supply can be used. Show how the analog components are powered. Give chip numbers but not pin numbers. Specify the type and tolerance of resistors and capacitors.
b) Write the initialization and function that measures the wind direction and returns a 16-bit unsigned result with units of degrees. I.e., the value varies from 0 to 359. You do not have to write software that samples at 1 Hz, simply a function that measures wind direction once.

10.8. Lab Assignments

Lab 10.1. This experiment will use a thermistor and the ADC to construct a digital thermometer. The temperature range should be 20 to 40 °C. If the current temperature is above the upper limit in the specified range, a red LED should be turned on. You can test this feature by shorting the thermistor leads together (zero resistance.) If it is below the lower limit of the specified range, a yellow LED should be turned on. Similarly, you can test this feature by disconnecting one wire of the thermistor (infinite resistance.) Otherwise, a green LED will stay on indicating the temperature is within the specified range. The temperature measurements will be displayed as fixed-point numbers on an LCD or OLED. Design your system with the best possible resolution. The temperature component is 0 to 1 Hz. Experimentally verify noise level, time-constant, and accuracy of your system.

Lab 10.2. The objective of this lab is to build a digital sound recorder for human speech. You will first need to interface an external RAM to store the data. Next, you will need to design an analog circuit that interfaces a microphone to the ADC of the microcontroller. Investigate the frequency components of human speech and design your system to pass these frequencies. You will also need a mechanism to playback the recorded sound, so design an audio amplifier that interfaces a DAC to a speaker. Your human interface should include a buttons to trigger sound recording, stop recording, start playback, and stop playback.

Lab 10.3. Design a thermocouple-based thermometer with a range of 0 to 50 °C. You can use an ice bucket for the reference, or you could design the thermistor-based thermometer of Lab 10.1 and use it to compensate for the cold junction of the thermocouple. Design your system with the best possible resolution. The temperature measurements will be displayed as fixed-point numbers on an LCD or OLED. The temperature component is 0 to 1 Hz. Experimentally verify noise level, time-constant, and accuracy of your system.

Lab 10.4. Design a digital scale using a strain gauge force transducer. Select the range of the scale to match the linear range of your force transducer. You can build a force transducer using a slide pot and a spring. Design your system with the best possible resolution. The force measurements will be displayed as fixed-point numbers on an LCD or OLED. Experimentally verify noise level, time-constant, and accuracy of your system.

Lab 10.5. Design two digital position measurement systems using a slide potentiometer as the transducer. Select the range of the scale to match the linear range of your transducer. The position measurements will be displayed as fixed-point numbers on an LCD. The first system will use the ADC, and the second system will use an astable multivibrator (TLC555) and input capture. Design your systems with the best possible resolution. Experimentally verify noise level, time-constant, and accuracy of both systems.

Lab 10.6. Design an autoranging voltmeter. The three ranges are 0 to 2V, 0 to 0.2V, and 0 to 0.02V. The hardware/software system automatically adjusts the range providing the best possible measurement resolution. Display both the numerical and graphical results on an LCD or OLED display. Write a graphical device driver for the LCD or OLED and use it to graph the time-varying voltage in real time. Experimentally determine the measurement resolution for each range and compare it to the expected theoretical resolution. Analyze the various sources of measurement error in your system.

Lab 10.7. Design a data acquisition system using an IR distance sensor. See Figure 10.8. Design a system for static measurement of tilt and a dynamic measurement of motion.

Lab 10.8. Design a data acquisition system using an accelerometer. If the sensor is not moving, it response to the earth's gravity and can be used to determine the tilt angle. If the sensor is moving it responses to both gravity and the acceleration of the sensor. Design a system for static measurement of tilt and a dynamic measurement of motion.

Lab 10.9. Design a data acquisition system using a force sensing resistor (FSR). The sensor resistance depends on the pressure being applied to the sensing area. When no pressure is applied to the FSR its resistance will be larger than 1MΩ. Design a static measurement of pressure and test it with standard weights.

Figure 10.37. Force sensing resistors can measure pressure.

Figure 10.38. Multiple sensors, actuators, and displayed interfaces using modules from Grove Seeedstudio.

11. Introduction to Communication Systems

Chapter 11 objectives are to:
- Introduce basic concepts of networks
- Present master/slave, ring and multidrop networks based on UARTs
- Introduce the concept of wireless communications

The goal of this chapter is to provide a brief introduction to communication systems. Communication theory is a richly developed discipline, and much of the communication theory is beyond the scope of this book. Nevertheless, the trend in embedded systems is to employ multiple intelligent devices, therefore the interconnection will be a strategic factor in the performance of the system. These devices will be developed by different manufacturers, thus the interconnection network must be flexible, robust, and reliable. Consequently, this chapter focuses on implementing communication systems appropriate for embedded systems. The components of an embedded system typically combined to solve a common objective, thus the nodes on the communication network will cooperate towards that shared goal. In particular, requirements of an embedded system, in general, involve relatively low bandwidth, static configuration, and a low probability of corrupted data. We will introduce the internet of things (IoT) by designing a two systems that connect to the internet, one wired and one wireless.

11.1. Fundamentals

In the serial interfacing chapter, we considered the hardware interfaces between computers. In this chapter, we will build on those ideas and introduce the concepts of networks by investigating a couple of simple networks. A communication network includes both the physical channel (hardware) and the logical procedures (software) that allow users or software processes to communicate with each other. The network provides the transfer of information as well as the mechanisms for process synchronization. It is convenient to visualize the network in a hierarchical fashion as shown in Figure 11.1.

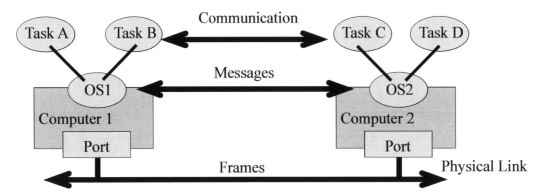

Figure 11.1. *A layered approach to communication systems.*

The International Standards Organization (ISO) defines a 7-layer model called the **Open Systems Interconnection** (OSI), described in Chapter 9 of Volume 3. Figure 11.1 shows a simple 3-layer model.

At the lowest level, frames are transferred between I/O ports of the two (or more) computers along the physical link or hardware channel. At the next logical level, the operating system (OS) of one computer sends messages or packets to the OS on the other computer. The message protocol will specify the types and formats of these messages. Typically, error detection and correction is handled at this level. Messages typically contain four fields:

1) Address information field
    physical address specifying the destination/source computers
    logical address specifying the destination/source processes
2) Synchronization or handshake field
    Physical synchronization like shared clock, start and stop bits
    OS synchronization like request connection or acknowledge
    Process synchronization like semaphores
3) Data field
    ASCII text (raw or compressed)
    Binary (raw or compressed)
4) Error detection and correction field
    Vertical and horizontal parity
    Checksum
    Block correction codes (BCC)

Observation: Communication systems often specify bandwidth in total bits/sec, but the important parameter is the data transfer rate.

Observation: Often the bandwidth is limited by the software and not the hardware channel.

At the highest level, we consider communication between users or tasks. A process is a complete software task that has a well-defined goal. For example, when a file is to be printed on a network printer, the OS creates a process that

1) establishes connection with the remote printer,
2) reads blocks from the hard disk drive and sends the data to the printer
    may have to manipulate graphics/colors for the specific printer
    network driver will break the data into message packets,
3) disconnect the printer.

Many embedded systems require the communication of command or data information to other modules at either a near or a remote location. Because the focus of this book is embedded systems, we will limit our discussion with communication with devices within the same room. A **full duplex** channel allows data to transfer in both directions at the same time. In a **half duplex** system, data can transfer in both directions but only in one direction at a time. Half duplex is popular because it is less expensive (2 wires) and allows the addition of more devices on the channel without change to the existing nodes.

Information, such as text, sound, pictures and movies, can be encoded in digital form and transmitted across a channel, as shown in Figure 11.2. The channel will have a maximum information per second it can transmit, called **channel capacity**. In order to improve the effective bandwidth many communication systems will compress the information at the source, transmit the compressed version, and then decompress the data at the destination. Compression essentially removes redundant information in such a way that the decompressed data is identical (**lossless**) or slightly altered but similar enough (**lossy**). For example, a 400 pixels/inch photo compressed using the JPEG algorithm will be 5 to 30 times smaller than the original. A **guided medium** focuses the transmission energy into a well-defined path, such as current flowing along copper wire of a twisted pair cable, or light traveling along a fiber optic cable. Conversely, an **unguided medium** has no focus, and the energy field diffuses as in propagates, such as sound or EM fields in air or water. In general, for communication to occur, the transmitter must encode information as **energy**, the channel must allow the energy to move from transmitter to receiver, and the receiver must decode the energy back into the information, see Figure 11.2. In an analog communication system, energy can vary continuously in amplitude and time. A digital communication signal exists at a finite number of energy levels for discrete amounts of time. Along the way, the energy may be lost due to **attenuation**. For example, a simple $V=I*R$ voltage drop is in actuality a loss of energy as electrical energy converted to thermal energy. A second example of attenuation is an RF cable splitter. For each splitter, there will be 50% attenuation, where half the energy goes left and the other half goes right through the splitter. Unguided media will have attenuation as the energy propagates in multiple directions. Attenuation causes the received energy to be lower in amplitude than the transmitted energy. A second problem is **distortion**. The transfer gain and phase in the channel may be function of frequency, time, or amplitude. Distortion causes the received energy to be different shape than the transmitted energy. A third problem is **noise**. The noise energy is combined with the information energy to create a new signal. White noise and EM field noise were discussed in Chapter 10. **Crosstalk** is a problem where energy in one wire causes noise in an adjacent wire. We quantify noise with **signal-to-noise ratio** (SNR), which is the ratio of the information signal power to noise power.

$$\text{SNR(dB)} = 10 \cdot \log_{10}\left(\frac{\text{Average signal power}}{\text{Average noise power}}\right)$$

Checkpoint 11.1: Why do we measure SNR as power and not voltage?

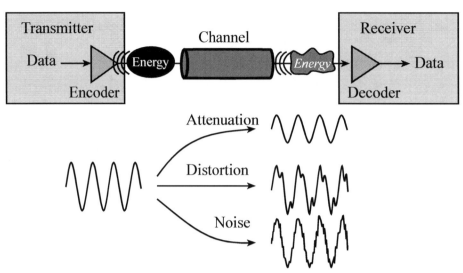

Figure 11.2. Information is encoded as energy, but errors can occur during transmission.

We can make an interesting analogy between time and space. A communication system allows us transfer information from position A to position B. A digital storage system allows us transfer information from time A to time B. Many of the concepts (encoding/decoding information as energy, noise, error detection/correction, security, and compression) apply in an analogous manner to both types of systems.

Checkpoint 11.2: We measure the performance of a communication system as bandwidth in bits/sec. What is the analogous performance measure of a digital storage system?

Errors can occur when communicating through a channel with attenuation, distortion and added noise. If the receiver detects an error, it can send a negative acknowledgement so the transmitter will retransmit the data. The CAN and ZigBee protocols handle this detection-retransmission process automatically. Networks based on the UART port will need to define and implement error detection. I.e., we can add an additional bit to the serial frame for the purpose of detecting errors. With **even parity**, the sum of the data bits plus the parity bit will be an even number. The framing error in the UART can also be used to signify the data may be corrupted. The CAN network sends a **longitudinal redundancy check**, LRC, which is the exclusive or of the bytes in the frame. The ZigBee network adds a checksum, which is the sum of all the data.

There are many ways to improve transmission in the channel, reducing the probability of errors. The first design choice is the selection of the interface driver. For example RS422 is less likely to exhibit errors than RS232. Of course having a driver will be more reliable than not having a driver. The second consideration is the cable. Proper shielding can improve SNR. For example, Cat6 Ethernet cables have a separator between the four pairs of twisted wire, which reduce the crosstalk between lines as compared to Cat5e cable. If we can separate or eliminate the source of added noise, the SNR will improve. Reducing the distance and reducing the bandwidth often will reduce the probability of error. If we must transmit long distances, we can use a repeater, which accepts the input and retransmits the data again.

Error correcting codes are beyond the scope of this book. However, we can present two simple error correcting codes. The first error correcting code involves sending three copies of each data. The receiver will compare the three versions received and majority vote will decide which value to use. A second error correcting code uses both parity and LRC. For example, assume we wished to send the message "Ciao". Encoded as ASCII characters the data are 0x43, 0x69, 0x61, and 0x6F. The first step is to display the binary data in 2-D.

	Byte 0	Byte 1	Byte 2	Byte 3
Bit 7	0	0	0	0
Bit 6	1	1	1	1
Bit 5	0	1	1	1
Bit 4	0	0	0	0
Bit 3	0	1	0	1
Bit 2	0	0	0	1
Bit 1	1	0	0	1
Bit 0	1	1	1	1

The second step is to add an even parity to each byte and add a LRC at the end. Notice that the even parity is the exclusive OR of each bit in the vertical column and the LRC is the exclusive OR of each bit in the horizontal row. The parity bit for the LRC (or the LRC bit for the parity) will be the exclusive OR of all the data bits.

	Byte 0	Byte 1	Byte 2	Byte 3	LRC
Parity	1	0	1	0	0
Bit 7	0	0	0	0	0
Bit 6	1	1	1	1	0
Bit 5	0	1	1	1	1
Bit 4	0	0	0	0	0
Bit 3	0	1	0	1	0
Bit 2	0	0	0	1	1
Bit 1	1	0	0	1	0
Bit 0	1	1	1	1	0

Now, if any one bit in this 9-row by 5-column matrix is flipped, we can determine which byte is in error by the parity and which bit is in error by the LRC. Rather than asking for retransmission, we simply correct the error. These are very simple error correcting codes, but they illustrate that we can send more bits than the minimum and use those extra bits in a creative way to either detect or correct errors.

11.2. Communication Systems Based on the UARTs

In this section, we will present three communication systems that utilize the UART port. If the distances are short, half duplex can be implemented with simple **open collector** or **open-drain** digital-level logic. Open drain logic has two output states: low and off. In the off state the output is not driven high or low, it just floats. The 10 kΩ pull-up resistor will passively make the signal high if none of the open drain outputs are low. The microcontroller can make its **TxD** serial outputs be open drain (**ODE** on the TM4C). This mode allows a half duplex network to be created without any external logic (although pull-up resistors are often used). Three factors will limit the implementation of this simple half duplex network: 1) the number nodes on the network, 2) the distance between nodes; and 3) presence of corrupting noise. In these situations a half duplex RS485 driver chip like the SP483 made by Sipex or Maxim can be used.

The first communication system is **master-slave** configuration, where the master transmit output is connected to all slave receive inputs, as shown in Figure 11.3. This provides for broadcast of commands from the master. All slave transmit outputs are connected together using wire-or open drain logic, allowing for the slaves to respond one at a time. The **ODE** (Open Drain Enable) in the slaves should be set to activate open drain mode on transmitters. The low-level device driver for this communication system is identical to the UART driver developed in Program 5.9. When the master performs UART output it is broadcast to all the slaves. There can be no conflict when the master transmits, because a single output is connected to multiple inputs. When a slave receives input, it knows it is a command from the master. In the other direction, however, a potential problem exists because multiple slave transmitters are connected to the same wire. If the slaves only transmit after specifically being triggered by the master, no collisions can occur.

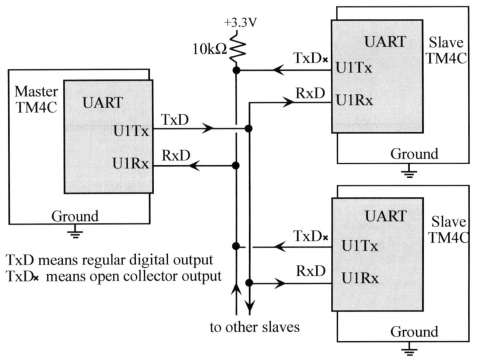

Figure 11.3. A master-slave network implemented with multiple microcontrollers.

Checkpoint 11.3: What voltage level will the master RxD observe if two slaves simultaneously transmit, one making it a logic high and the other a logic low?

The next communication system is a **ring network**. This is the simplest distributed system to design, because it can be constructed using standard serial ports. In fact, we can build a ring network simply by chaining the transmit and receive lines together in a circle, as shown in Figure 11.4. Building a ring network is a matter as simple as soldering a RS232 cable in a circle with one DB9 connector for each node. Messages will include source address, destination address and information. If computer A wishes to send information to computer C, it sends the message to B. The software in computer B receives the message, notices it is not for itself, and it resends the message to C. The software in computer C receives the message, notices it is for itself, and it keeps the message. Although simple to build, this system has slow performance (response time and bandwidth), and it is difficult to add/subtract nodes.

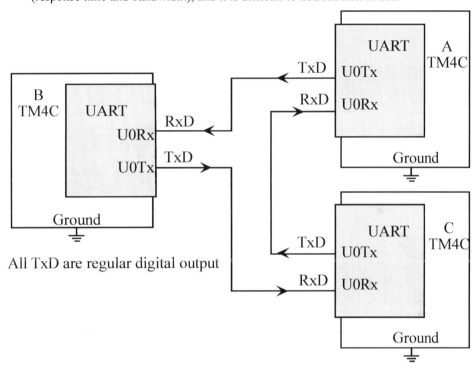

Figure 11.4. A ring network implemented with three microcontrollers.

Checkpoint 11.4: Assume the ring network has 10 nodes, the baud rate is 100,000 bits/sec, and there are 10 bits/frame. What is average time it takes to send a 10 byte message from one computer to another?

The third communication system is a very common approach to distributed embedded systems, called **multi-drop**, as shown in Figure 11.5. To transmit a byte to the other computers, the software activates the SP483 driver and outputs the frame. Since it is half duplex, the frame is also sent to the receiver of the computer that sent it. This echo can be checked to see if a collision occurred (two devices simultaneously outputting.) If more than two computers exist on the network, we usually send address information first, so that the proper device receives the data. Many collisions can be avoided looking to see if the receiver is active before transmitting.

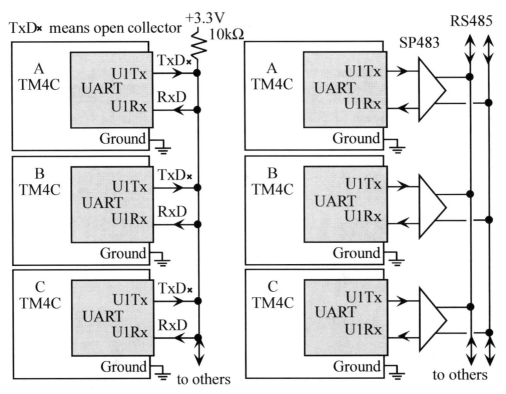

Figure 11.5. *A multidrop network is created using a half duplex serial channel implemented with open drain logic or with RS485 drivers.*

Checkpoint 11.5: How can the transmitter detect a collision had corrupted its output?

Checkpoint 11.6: How can the receiver detect a collision had corrupted its input?

There are many ways to check for transmission errors. You could use a **longitudinal redundancy check** (LRC) or horizontal even parity. The error check byte is simply the *exclusive-OR* of all the message bytes (except the LRC itself). The receiver also performs an *exclusive-OR* on the message as well as the error check byte. The result will equal zero if the block has been transmitted successfully. Another popular method is **checksum**, which is simply the modulo$_{256}$ (8-bit) or modulo$_{65536}$ (16-bit) sum of the data packet. In addition, each byte could have (but doesn't have to) include even parity.

There are two mechanisms that allow the transmission of variable amounts of data. Some protocols use start (STX=0x02) and stop (ETX=0x03) characters to surround a variable amount of data. The disadvantage of this "termination code" method is that binary data cannot be sent because a data byte might match the termination character (ETX). Therefore, this protocol is appropriate for sending ASCII characters. Another possibility is to use a byte count to specify the length of a message. Many protocols use a byte count. The ZigBee frames, for example, have a byte count in each frame.

11.3. Wireless Communication

The details of how wireless communication operates are beyond the scope of this book. Nevertheless, the interfacing techniques presented in this book are sufficient to implement wireless communication by selecting a wireless module and interfacing it to the microcontroller. In general, one considers bandwidth, distance, topology and security when designing a wireless link. Bandwidth is the fundamental performance measure for a communication system. In this book, we define bandwidth of the system as the information transfer rate. However, when characterizing the physical channel, bandwidth can have many definitions. In general, the bandwidth of a channel is the range of frequencies passed by the channel (<u>Communication Networks</u> by Leon-Garcia). Let $G_x(f)$ be the gain versus frequency of the channel. When considering EM fields transmitted across space, we can define **absolute bandwidth** as the frequency interval that contains all of the signal's frequencies. **Half-power bandwidth** is the interval between frequencies at which $G_x(f)$ has dropped to half power (-3dB). Let f_c be the carrier frequency, and P_x be the total signal power over all frequencies. The **equivalent rectangular bandwidth** is $P_x/G_x(f_c)$. The null-to-null bandwidth is the frequency interval between first two nulls of $G_x(f)$. The FCC defines **fractional power containment bandwidth** as the bandwidth with 0.5% of signal power above and below the band. The **bounded power spectral density** is the band defined so that everywhere outside $G_x(f)$ must have fallen to a given level. The purpose of this list is to demonstrate to the reader that, when quoting performance data, we must give both definition of the parameter and the data. If we know the channel bandwidth W in Hz and the *SNR*, we can use the **Shannon–Hartley Channel Capacity Theorem** to estimate the maximum data transfer rate C in bits/s:

$$C = W \cdot \log_2(1 + SNR)$$

For example, consider a telephone line with a bandwidth W of 3.4 kHz and *SNR* of 38 dB. The dimensionless ***SNR*** = $10^{(38/10)}$ = 6310. Using the Channel Capacity Theorem, we calculate C = 3.4 kHz * $\log_2(1 + 6310)$ = 43 kbits/s.

One of the simplest modules we can use for wireless embedded systems is the TI CC2500, which is a low-power 2.4-GHz RF transceiver with a SPI interface. The CC2500 is intended for 2400 to 2483.5 MHz Industrial, Scientific and Medical (ISM) applications. The computer interface uses a SPI protocol, the clock circuit is based on an external crystal, and the antenna circuit must be tuned for the 2.4 GHz frequency. The eZ430-RF2500 from Texas Instruments is a low-cost development tool based around the MSP430 and the TI CC2500 low-power transceiver. The CC430 microcontroller includes both an MSP430 and RF.

Example 11.1. Design a system that can communicate at 1000 bytes/sec between two microcontrollers within the same room without security.

Solution: This low bandwidth and low power can be solved with a radio-frequency (RF) link that supports *Bluetooth* Low Energy (BLE). This short distance is classified as a Short Range Device (SRD). There are many RF communication modules that could have been used. As illustrated in Figure 11.6, the CC2650 interface has three parts. The CC2650 is a complete Cortex M3 microcontroller with Bluetooth protocol in firmware. Basically, the master microcontroller on the left transmits data via its SSI, and the microcontroller on the right receives the data with its SSI. It is a transceiver, meaning data can flow across the link in both directions.

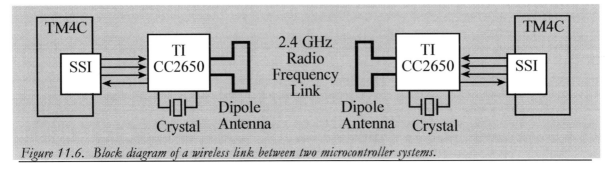

Figure 11.6. Block diagram of a wireless link between two microcontroller systems.

ZigBee is another low-cost wireless solution for embedded systems. The name is derived from the behavior of honey bees. Honey bees distributed across a large open field implement a mesh network in order to communicate information to their hive. They do this by message relaying. A bee distant from the hive will fly a particular zigzag pattern that represents the information. A second bee nearer the hive will repeat the pattern. The relay continues until the information reaches the hive. ZigBee is a standard that defines a set of communication protocols for low-data-rate, very low-power, short-range wireless networking. It can operate under battery power, and last for years because there are multiple types of low-power sleep modes. It is an appropriate solution for sensor networks, meter reading, industrial automation, security systems, and patient monitoring. ZigBee is an extension of the IEEE 802.15.4b standard. It is lower cost and lower performance than Bluetooth or IEEE 801.11b, as shown in Table 11.1. The range values in this table represent performance indoors–outdoors. ZigBee modules come in a variety of power versus performance models. In other words, you can run at lower power if you are willing to sacrifice distance and data rate.

	Data rate	Range	Wireless applications
ZigBee	20 to 250 Kbps	10–100m	Sensor Networks
Bluetooth	1 to 3 Mbps	2–10m	Headset, mouse
IEEE 802.11b	1 to 11 Mbps	38–140m	Internet connection
IEEE 802.11g	1 to 54 Mbps	38–140m	Internet connection
IEEE 802.11n	1 to 72 Mbps	70–250m	Internet connection

Table 11.1. Comparison of wireless protocols.

The ZigBee protocol is layered. The top layer is the application layer, implemented as the user program. At this layer software in one node sends messages to another node. This section will focus on this layer, because we will purchase a ZigBee module that performs the lower layers automatically. The second layer is the application support sublayer (APS). Below the APS is the network layer (NWK). Below the NWK is the media access control (MAC) layer. Below the MAC is the physical layer, which includes the RF transmitter and receiver. For more information about how ZigBee works, see http://www.zigbee.org/. You also could do a web search on XBee, which is a low-cost implementation of ZigBee.

ZigBee is a personal area network (PAN), as shown in Figure 11.7. There can be a coordinator, devices that support all functions, or devices that support some functions. A typical application of ZigBee is a remote sensor network. Consider each of the nodes in Figure 11.7 is capable of collecting sensor data. The sensor database can be centralized at the PAN coordinator or distributed across the entire system.

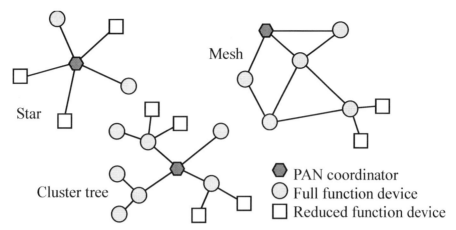

Figure 11.7. *ZigBee used to create a remote sensor network.*

One low-cost implementation of ZigBee is the XBee module from Digi (formally MaxStream). These modules take ZigBee and wrap it into a simple-to-use serial command set, called AT commands. These modules allow a very reliable and simple communication between microcontrollers with a serial port. Both point-to-point and multi-point networks are supported. The hardware involves interfacing 3.3 V full duplex serial channel to the XBee module, as shown in Figure 11.8.

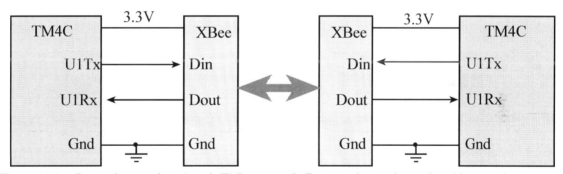

Figure 11.8. *Circuit diagram for a 2-node ZigBee network. Power and grounds are shared between the microcontroller and its XBee but not between XBees.*

The full software solution to this network will be left as Lab 11.2. In this simple configuration, there can be up to 256 nodes on the network. The initialization software must establish the baud rate at 9600 bits/sec. The system bandwidth therefore will be on the order of 960 bytes/sec. The <CR> symbol refers to the carriage return, character 13, existing as one 8-bit ASCII character. To place the XBee in AT command mode, we execute steps 1 to 5 over and over until we get an OK response.

1. Send a dummy character like 'X'.
2. Wait 1.1 second (greater than the one second guard time).
3. Output "+++".
4. Wait 1.1 second (greater than the one second guard time).
5. Wait for response, should be **OK<CR>.**

Each node must establish its address (called my) and the destination address. For example, if this computer is at address 0x01 and wishes to send packets to the computer at address 0x02, then this computer executes these AT commands. The software should wait 20 ms after each AT command. When an AT command is executed correctly, the XBee responds with **OK<CR>**. One of the simplest modes is **Application Programming Interface** (API) mode 1, which allows the nodes to send and receive packets:

ATDL02<CR>	Sets destination address to 2 (number given in hexadecimal)
ATDH0<CR>	Sets destination high address to 0
ATMY01<CR>	Sets my address to 1 (number given in hexadecimal)
ATAP1<CR>	Sets API mode 1 (packets)
ATCN<CR>	Ends command mode

The other node in the point-to-point connection performs a similar initialization, but obviously with the **my** and **destination** addresses reversed. Using the API mode simplifies the application software, because all message routing, error detection/retransmission, and message acknowledgement occurs at lower levels automatically. A data transmission frame has the following format:

0x7E, LengthHi, LengthLo, 0x01, ID, DestHi, DestLo, 0x00, b1, ..., bn, Chksum

All API mode 1 frames begin with 0x7E. The next two bytes are the length of the message, which will be the number of bytes after the length and before the checksum. The length does not include the four bytes comprising the 0x7E, which is the length itself and the checksum. The fourth byte 0x01 signifies this is a transmit data packet. The ID should be used as a message sequence number. I.e., as this computer sends packets to the destination computer, the **ID** is sequenced as 1, 2, ... 255, 1, 2, ... An ID of 0 is not used. This sequence number guarantees the packets arrive at the destination in the same order as they were sent.

The two bytes **DestHi,DestLo** specify the destination node address. The high byte should be zero for this configuration. The next byte provides options for the frame, which should be 0. Bytes **b1** through **bn** are the data to be transmitted. Because there is a frame length, this data can be formatted however you wish. The last byte of the frame will be a checksum. Let **sum** be the 8-bit addition of all bytes not including 0x7E delimiter and the length. We calculate **Chksum** as 0xFF-**sum**. In this way the receiver can add up all the bytes after the length and including the **Chksum** in order to get the result 0xFF.

For example, assume we wish to send the message "He" to node 2. Also assume this is the 254th frame sent, so the ID will be 0xFD. The message has a length of 7. The checksum is calculated as

0xFF-(0x01+0xFD+0x00+0x02+0x00+0x48+0x65) = 0x52.

The oscilloscope recording for this frame is shown in Figure 11.9.

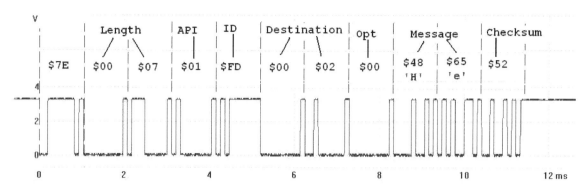

Figure 11.9. *API transmit frame measured on the Din pin of the XBee module.*

When the transmitted frame is properly delivered, the XBee sends an acknowledgment to the transmitter. The length is always 3 bytes. The API code is 0x89. The **ID** matches the corresponding value of the transmitted frame. The next byte is a status field, and 0x00 means success. A status of 0x01 means no acknowledgement received, which may mean the destination node does not exist or is turned off. A status of 0x02 means CCA failure and 0x03 means the message was purged. The **Chksum** byte is calculated in the same manner as all the API frames.

0x7E, 0x00, 0x03, 0x89, ID, 0x00, Chksum

Figure 11.10 shows a scope trace when an acknowledge frame was reported to the transmitter. The **ID** of this frame is 0xC1. The checksum for this frame is 0xFF-(0x89+0xC1+0x00) = 0xB5.

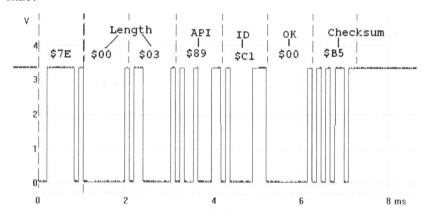

Figure 11.10. *API acknowledge frame measured on the Dout pin of the XBee module.*

The receiver with a **my** address matching the **destination** address of the transmitted frame will be given that frame as an API packet type 0x81. A data receive frame has the following format:

0x7E, LengthHi, LengthLo, 0x81, SourceHi, SourceLo, RSSI, 0x00, b1, ..., bn, Chksum

The source field identifies the node that sent the message. The **RSSI** (received signal strength indictor) is the decimal equivalent measure of the signal. For example 0x3A means the received signal strength is 58 dBm. The option field should be zero. Bytes **b1** through **bn** contains the data. Figure 11.11 shows a scope trace of the received message corresponding to the transmitted message in Figure 11.9

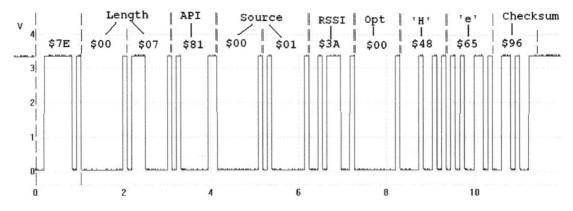

Figure 11.11. API receive frame measured on the Dout pin of the XBee module.

11.4. Internet of Things

11.4.1. Basic Concepts

With the proliferation of embedded systems and the pervasiveness of the internet, it is only natural to connect the two. The internet of things (IoT) is the combination of embedded systems, which have sensors to collect data and actuators to affect the surrounding, and the internet, which provides for ubiquitous remote and secure communication. This section will not describe how the internet works, but rather we will discuss both the general and specific approaches for connecting embedded systems to the internet. For an excellent description of the **TCP/IP** (Transmission Control Protocol/Internet Protocol) protocol the reader is referred to W. Richard Stevens, TCP/IP Illustrated, Volume 1: The Protocols. For a general description of the internet of things, see Vasseur and Dunkels, Interconnecting Smart Objects with IP. These two books provides good overviews of network technologies used for connecting devices.

Figure 11.12 illustrates the distributed approach taken with the internet of things. A **distributed** solution deploys multiple sensors and actuators connected by the internet. Another name given for an embedded system connected to the internet is **smart object**. Smart objects include sensors to collect data, processing to detect events and make decisions, and actuators to manipulate the local environment. There are many reasons to consider a distributed solution (network) over a centralized solution. Often multiple simple microcontrollers can provide a higher performance at lower cost compared to one computer powerful enough to run the entire system. Some embedded applications require input/output activities that are physically distributed. For real-time operation there may not be enough time to allow communication between a remote sensor and a central computer. Another advantage of distributed system is improved debugging. For example, we could use one node in a network to monitor and debug the others. Often, we do not know the level of complexity of our problem at design time. Similarly, over time the complexity may increase or decrease. A distributed system can often be deployed that can be scaled. For example, as the complexity increases more nodes can be added, and if the complexity were to decrease nodes could be removed. Table 11.2 lists some existing applications and the things they sense or control.

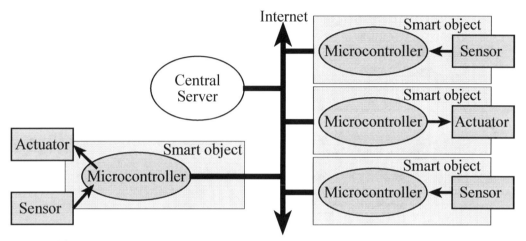

Figure 11.12. The internet of things places input, output and processing at multiple locations connected together over the internet.

Industrial Automation	Factories, machines, shipping
Environment	Weather, pollution, public safety
Smart Grid	Electric power, energy delivery
Smart Cities	Transportation, hazards, public services
Social Networks	Ideas, politics, sales, and communication
Home Networks	Lighting, heat, security, information
Building Networks	Energy, hazards, security, maintenance
Structural Monitors	Bridges, roads, building
Health Care	Heart function, medical data, remote care
Law enforcement	Crime, public safety

Table 11.2. Applications of smart objects.

Challenges. On a local scale, the design of smart objects faces the same challenges existing in all embedded systems: power, size, reliability, longevity, and cost. Luckily the deployment of billions of microcontrollers into the market has created a technology race to reduce power, size and cost while increasing the performance. At the microcontroller level things are getting smaller, but at the network level, complexity is increasing and protocols are constantly changing as the world's thirst for information and communication rapidly grows.

Standardization. The existence of standards allows for a wide variety of objects to communicate with each other. Adhering to a standard will increase the acceptance of our device by customers, and allow our customers to apply our device to solve problems we never envisioned. **uIP** is a light-weight implementation of the IP stack specifically designed to operate with the available memory resources of smart objects. In this section we will start with a microcontroller with the hardware and software to implement TCP/IP protocols, and build our application on top of this standard.

Interoperability means our device can function with a wide range of other devices made with different technologies, sold by different vendors, and produced by different companies.

Evolution is process of how new technologies are introduced into the market. If there is one constant in this world, it is that things will change. Every thousand years, one big discovery fundamentally changes how we operate (fire, language, metal tools). More frequently, change is introduced gradually such that those technologies that give us a competitive advantage survive. If we build our business model on the premise evolutionary change, then we can be nimble to deploy new technology when it provides lower cost and/or better performance.

Stability. Even though technology will advance, our customers demand products that work reliably, for a long time, and in a manner with which they are comfortable. Over the last 50 years automotive technology has drastically improved, but the driving experience, how we drive, has remained almost constant.

Abstraction. You will notice the approach in this section differs widely from the other examples in this book. The rest of the book deploys a bottom up approach. With bottom-up education, the details are first explained, so there is no magic, and then abstraction occurs by encapsulating that we fully understand. In this section we will purchase hardware and software with capabilities to communicate with the internet, and use this abstraction without fully understanding how some of the lower levels operate.

Scalability. ARM reports over 50 billion devices with an ARM core have been shipped from 1993 to 2013, and predicts another 50 billion before the end of this decade. In order to be effective and profitable, we need to develop systems that can scale.

Security. Because embedded systems are deployed in life-critical situations, and because the quality of service affect our profits, we must protect the system from a determined adversary. A chain is only as strong as its weakest link. Security cannot be obtained simply by operating in secret, because once the secret is out, the system will be extremely vulnerable. "Security by obscurity" is a very poor design method. Security involves more than encrypting the data. The first aspect of security is **confidentiality**. We must decide what it means to view/change the data and who has the right to read/write. Authentication is the means to ensure the identity of the sender is correct. Confidentiality will require both logical and physical measures to protect against an attack. Encryption makes it harder for an unauthorized party to view a message. The second aspect is data **integrity**. For most of the applications listed in Table 11.2 it is important that data reach the rightful recipient in an unaltered fashion. To support network integrity we need techniques that support both detection and prevention. The third aspect is **availability**. A secure communication not only requires the correct data arrive at the correct place, but also at the correct time. A Denial of Service (DoS) attack attempts to breach the availability of the network. For wired networks, we can reroute traffic along multiple paths. With wireless networks, we can channel hop by switching channels on a pseudorandom fashion, making it harder for an attacker to jam. For more information on security, see Frank Stajano, Security for Ubiquitous Computing.

11.4.2. Layered Model

Most networks provide an **abstraction** that hides low-level details from high-level operations. This abstraction is often described as layers. The International Standards Organization (ISO) defines a 7-layer model called the **Open Systems Interconnection** (OSI), as shown on the left of Figure 11.13. It provides a standard way to classify network components and operations. The **Physical** layer includes connectors, bit formats, and a means to transfer energy. Examples include RS232, controller area network (CAN), modem V.35, T1, 10BASE-T, 100BASE-TX, DSL, and 802.11a/b/g/n PHY. The **Data link** layer includes error detection and control across a single link (single hop). Examples include 802.3 (Ethernet), 802.11a/b/g/n MAC/LLC, PPP, and Token Ring. The **Network** layer defines end-to-end multi-hop data communication. The **Transport** layer provides connections and may optimize network resources. The **Session** layer provides services for end-user applications such as data grouping and check points. The **Presentation** layer includes data formats, transformation services. The **Application** layer provides an interface between network and end-user programs.

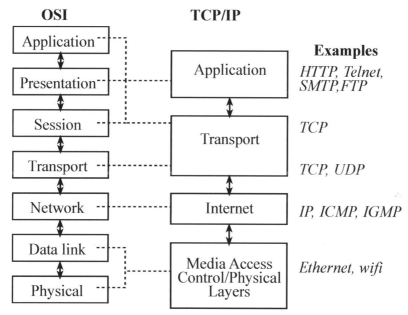

Figure 11.13. The TCP/IP model has four layers.

The **TCP/IP** model of the **Internet** does not adhere to such a strict layered structure, but does recognize four broad layers: scope of the software application; the end-to-end transport connection; the internetworking range; and the direct links as shown on the right of Figure 11.13. Examples of applications include Telnet, FTP (File Transfer Protocol), and SMTP (Simple Mail Transfer Protocol). Examples of transport include TCP (Transmission Control Protocol) and UDP (User Datagram Protocol). TCP provides reliable, ordered delivery of data from a software task on one computer to another software task running on another computer. For applications that do not require reliable data stream service UDP can be used. UDP provides a datagram service that emphasizes reduced latency over reliability. Examples of network include IP (Internet Protocol), ICMP (Internet Control Message Protocol) and IGMP (Internet Group Management Protocol). **Ethernet** is the physical link explored later in this section. In this section we will develop projects at the application layer. The communication of **bits** happen at the physical layer, **frames** at the data link layer, **packets** or **datagrams** at the network layer, **segments** at the transport layer, and **messages** at the application layer.

11.4.3. Message Protocols

The layered format can be seen in the message packet formats, as overviewed in Figure 11.14. At the lowest level are physical frames. Ethernet frames contain a header, 46 to 1500 bytes of payload, and a trailer. The header includes address, type and length information. If the there is less than 46 bytes of Ethernet data, zeros are added (padding) to make the Ethernet payload at least 46 bytes. The trailer includes error checking (CRC). At the IP level, packets include a header and payload. The header of an IP packet includes a 32-bit destination IP address, typically shown as four 8-bit numbers (e.g., 176.31.244.1). Some of these IP addresses are reserved for communicating within nodes on a local network. The Domain Name System (DNS) host can be used to translate domain names to IP addresses. Computers that communicate only with each other via TCP/IP, but are not connected to the Internet, need not have globally unique IP addresses. IP addresses for private networks are listed in Table 11.3. These IP addresses could be used for embedded systems that use TCP/IP to communicate, but are not connected to the internet.

Start	*End*	*Number of addresses*
10.0.0.0	10.255.255.255	2^{24}
172.16.0.0	172.31.255.255	2^{20}
192.168.0.0	192.168.255.255	2^{16}

Table 11.3. Private IP addresses.

Because of the growth of the internet, the 32-bit IP address (IPv4) is being replaced with a 128-bit address (IPv6), which will provide for about $3 \cdot 10^{38}$ addresses. The IP header is 20 to 40 bytes and contains the source IP address, destination IP address, and length.

The UDP header is 8 bytes and contains the source port, destination port, length, and checksum, see Table 11.4. The IP address specifies the node, and ports are addresses within the source and destination nodes.

> Source port: 16-bit number of the process that sent the packet, could be zero
> Destination port: 16-bit number of the process to receive the packet.
> Length: 16-bit number specifying the size in bytes of the data to follow
> Checksum: 16-bit modulo addition of all data, UDP header, and IP header

Table 11.4. UDP header format.

The TCP header is 20 bytes with the possibility of additional and optional information, see Table 11.5. The sequence and acknowledgment numbers allow the receiver to properly sort segments of data that were received out of order. The flags specify different modes of the TCP communication. The SYN flag means the first of a sequence of packets, and the FIN flag means the last. The RST flag terminates a connection. The URG flag means the urgent pointer specifies a piece of data the application urgently needs.

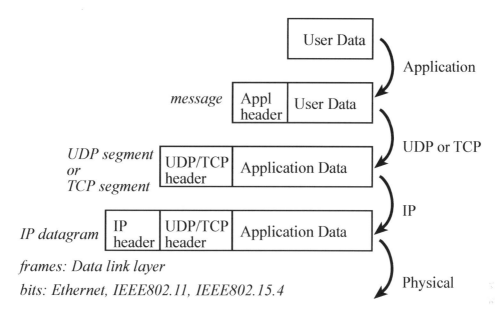

Figure 11.14. Overview of message packets used at various layers.

Source port: 16-bit number of the process that sent the packet, could be zero
Destination port: 16-bit number of the process to receive the packet.
Sequence number: 32-bit number defining the position of this data
Acknowledgement: 32-bit number of the next data expected to be received
Hlen: 4-bit field of the header size (including options) divided by 4
Flags: 6-bit field with FIN, SYN, RST, PSH, ACK, and URG
Window: 16-bit number specifying the number of bytes the receiver can accept
Checksum: 16-bit modulo addition of all data, TCP header, and IP header
Urgent pointer: 16-bit field pointing to a place in the stream urgently needed

Table 11.5. TCP header format.

11.4.4. Web server

This first application creates a web server that maintains a web page displaying local data, see Figures 11.15 and 11.16. The components of the system are a sensor and sensor interface (Figure 10.28 and Program 10.1), an EK-TM4C1294XL LaunchPad, Texas Instruments TivaWare, and a router connected to the Internet. The Dynamic Host Configuration Protocol server provides an IP address, and is typically initiated via a DHCP broadcast, when it connects. DHCP provided the address 192.168.0.107, a local address on its network. This example was built on top of the uIP stack delivered as part of TivaWare. First, you need to download TivaWare. I first ran the **enet_uip** example found in the **TivaWare_C_Series-2.1.0.12573\examples\boards\ek-tm4c1294xl\enet_uip** folder. I copied this example, and changed the web server as shown in Program 11.1.

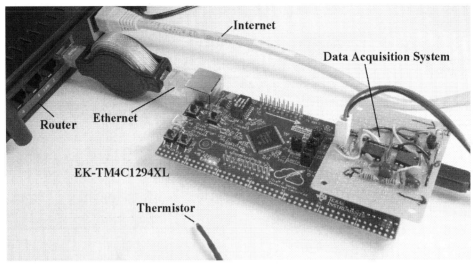

Figure 11.15. The thermistor measures temperature and the LaunchPad serves pages to the internet.

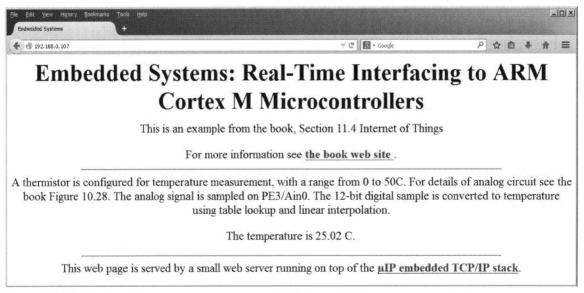

Figure 11.16. The thermistor measures temperature and the LaunchPad serves pages to the internet.

Program 11.1 shows the code you need to modify to create your own remote sensor smart object. When another node sends a request to this server, this node will respond with html code to render the page. The page is divided into three parts. The first part (**default_page_buf1of3**) and last part (**default_page_buf3of3**) are fixed. The application callback function, **httpd_appcall**, is invoked when the web page is requested. This callback function calls our application function **Board_Update** which collects sensor data from the thermistor and rebuilds the middle part of the html code (**default_page_buf2of3**). The meta code automatically refreshes every 5 seconds.

```c
const char default_page_buf1of3[] =
  "HTTP/1.0 200 OK\r\n"
  "Server: UIP/1.0 (http://www.sics.se/~adam/uip/)\r\n"
  "Content-type: text/html\r\n\r\n"
  "<!DOCTYPE HTML PUBLIC \"-//W3C//DTD HTML 4.01 Transitional//EN"
  "http://www.w3.org/TR/html4/loose.dtd\">"
  "<html> <head>"
  "<meta http-equiv=\"refresh\" content=\"5\">"
  "<title>Embedded Systems</title></head>"
  "<body> <center>"
  "<h1>Embedded Systems: Real-Time Interfacing"
  "to ARM Cortex M Microcontrollers</h1>"
  "<p>This is an example from the book, Section 11.4 Internet of Things</p>"
  "<p> For more information see "
  "<a href=\"http://users.ece.utexas.edu/~valvano/arm/outline.htm\">"
  "<b>the book web site</b> </a>."
  "<hr width=\"75%\">"
  "<p>A thermistor is configured for temperature measurement, "
  "with a range from 0 to 50C. "
  "For details of analog circuit see the book Figure 10.28. "
  "The analog signal is sampled on PE3/Ain0.  "
  "The 12-bit digital sample is converted to temperature using table lookup "
  "and linear interpolation.</p>  "
  "<p>The temperature is ";
uint32_t const buf1of3_Size = (sizeof(default_page_buf1of3) - 1);
char default_page_buf2of3[] = "12.01";
uint32_t buf2of3_Size = (sizeof(default_page_buf2of3) - 1);
const char default_page_buf3of3[] =
  " C.</p>"
  "<hr width=\"75%\">"
  "<p>This web page is served by a small web server running on top of "
  "the <a href=\"http://www.sics.se/~adam/uip/\"><b>&micro;IP embedded TCP/IP "
  "stack</b></a>.</center> </body> </html>";
uint32_t const buf3of3_Size = (sizeof(default_page_buf3of3) - 1);
void Board_Update(void){uint32_t data,temperature;
  data = ADC0_InSeq3();                     // 12-bit ADC, 0 to 4095
  temperature = ADC2Temperature(data);      // temperature, 0.01C
  Fix2Str(temperature,default_page_buf2of3); // 5 ASCII characters
  buf2of3_Size = 5; // in this case it is fixed size (but it could vary)
}
```

Program 11.1. The thermistor measures temperature and the LaunchPad serves pages to the internet.

To run the internet examples described in this section download and unzip the IoT examples into **examples\boards** so the directory path looks like this

TivaWare_C_Series-2.1.0.12573
    examples
        boards
            CC31xxxx
                ek-tm4c1294xl-enet_uip_temperature
                ek-tm4c123gxl-boost-cc3100_basic_wifi_UDP
                ek-tm4c123gxl-boost-cc3100_starter
                ek-tm4c1294xl-boost-cc3100_starter

11.4.5. UDP communication over wifi

The approach for implementing a smart object over wifi is to begin with a hardware/software platform that implements IEEE801.11 wifi. The CC3100BOOST is a boosterpack that can be used with the MSP432 LaunchPad, the TM4C123 LaunchPad, the TM4C1294 LaunchPad, the MSP432E4 LaunchPad, or with a CC31XXEMUBOOST emulation module, see Figure 11.17. The emulation module can be used early in a project to develop wireless applications using a "generic" microcontroller. After a prototype is configured, the project can select a microcontroller and design the actual smart object. In this design we will use either of the two TM4C LaunchPads and develop a solution that transmits UDP packets from one smart object to another. UDP is simpler than TCP and appropriate for applications requiring simplicity and speed. Furthermore, to use UDP the application must tolerate lost or out of order packets. UDP provides a best-effort datagram delivery service.

Figure 11.17. The CC3100 booster packet provides IEEE802.11 wireless connectivity.

The actual TCP/IP software stack resides in firmware on the booster pack itself. Therefore, when using any of the wireless booster packs the first step is to upgrade the firmware. One way to upgrade the firmware is to use the CC31XXEMUBOOST emulation module. The examples of this section ran on version 3.3 booster packs without needing to upgrade the firmware.

Program 11.2 shows the client software, which samples the ADC and sends UDP packets. Line 1 specifies the name of the access point (AP) to which the node will connect. There is a mechanism using SmartConfig to automate this discovery, but in this example I named the AP **Valvano** so I used a manual method to define the connection between the node and AP. The UDP payload will have a type field, which is defined in line 2. The destination IP address is hard-coded in line 3. For this application, the server was at IP address at 192.168.0.101, which in hex is C0.A8.00.65. The port number, which is a 16-bit value defining which process in the server should receive the data, is specified in line 4. There are a long list of registered port numbers that have special purposes, so I chose a port number larger than 1024 to avoid selecting any of these special purpose port numbers. Lines 5 and 6 define the payload for the UDP packet. Line 15 sets the bus clock to 50 MHz. The PLL needs to be active for the ADC to operate. Line 16 initializes the ADC channel 7 using PD0. Line 17 initializes the CC3100. After executing line 18 we will be connected and have IP address. Line 19 will return the network configuration. Lines 21-24 define the address and port to which the USP packet will be sent. Line 25 defines and opens a socket. In this example we leave the socket open, but it is ok to close the socket, go into low-power mode, and reopen the connection after sleeping. Lines 26-29 will sample the ADC and create a new message. Line 30 sends the UDP packet through the open socket. The wait in line 32 defines the rate at which packets are sent. Each of the wifi functions will return a success flag (error code). In this simple program we ignored the return values, assuming it was ok. In the version on the web, the process is restarted on error.

```
#define SSID_NAME      "Valvano"    // AP to connect to                    1
#define ATYPE          'a'          // analog data type                    2
#define IP_ADDR        0xC0A80065   // server IP                           3
#define PORT_NUM       5001         // Port number to be used              4
#define BUF_SIZE       12           //                                     5
UINT8 uBuf[BUF_SIZE];               // UDP packet payload                  6
int main(void){
  UINT8              IsDHCP = 0;
  _NetCfgIpV4Args_t  ipV4;
  SlSockAddrIn_t     Addr;
  UINT16             AddrSize = 0;
  INT16              SockID = 0;
  UINT32             data;
  unsigned char      len = sizeof(_NetCfgIpV4Args_t);
  initClk();             // PLL 50 MHz, ADC needs PPL active              15
  ADC0_InitSWTriggerSeq3(7);  // Ain7 is on PD0                           16
  sl_Start(0, 0, 0); // Initializing the CC3100 device                    17
  WlanConnect();         // connect to AP                                 18
  sl_NetCfgGet(SL_IPV4_STA_P2P_CL_GET_INFO,&IsDHCP,&len,                //19
              (unsigned char *)&ipV4);                                  //20
  Addr.sin_family = SL_AF_INET;                           //                21
  Addr.sin_port = sl_Htons((UINT16)PORT_NUM);             //                22
  Addr.sin_addr.s_addr = sl_Htonl((UINT32)IP_ADDR);       //                23
  AddrSize = sizeof(SlSockAddrIn_t);                      //                24
  SockID = sl_Socket(SL_AF_INET,SL_SOCK_DGRAM, 0);        //                25
  while(1){
    uBuf[0] = ATYPE;     // analog data type                              26
    uBuf[1] = '=';       //                                               27
    data = ADC0_InSeq3(); // 0 to 4095, Ain7 is on PD0                    28
    Int2Str(data,(char*)&uBuf[2]); // 6 digit number                      29
    sl_SendTo(SockID, uBuf, BUF_SIZE, 0,                  //                30
              (SlSockAddr_t *)&Addr, AddrSize);           //                31
    ROM_SysCtlDelay(ROM_SysCtlClockGet() / 25); // 40ms                   32
  }
}
```

Program 11.2. Client software that measures ADC data and sends UDP packets.

Program 11.3 shows the server software, which accepts UDP packets and plots the data on an ST7735 graphics LCD. Line 1 specifies the name of the access point (AP) to which the node will connect. The client and server use the same AP, which I named **Valvano**, so I used the manual method to define the connection between the node and AP. The UDP payload will have a type field, which is defined in line 2. Lines 16, 22-25 configure the wifi connection in a similar way as the client. Lines 17-20 initialize the ST7735 LCD and output a welcome message. Line 21 configures the LCD graphics routines specifying the range on the y-axis of the plot. Raw ADC data will be plotted versus time. Lines 26-29 define an IP address and port to use. Line 31 defines and opens a socket, and lines 32-33 bind the port to that socket. Lines 34-35 receive a UDP packet. Just like the client, we leave the socket open. If we wished to save power, we could close the socket, go into low-power mode, and reopen the connection after sleeping. Lines 36-51 decode the packet and plot the data on the LCD.

```
#define SSID_NAME      "Valvano"    // AP to connect to                 1
#define ATYPE          'a'          // analog data type                 2
#define IP_ADDR        0xC0A80065   // server IP                        3
#define PORT_NUM       5001         // Port number to be used           4
#define BUF_SIZE       12           //                                  5
UINT8 uBuf[BUF_SIZE];               // UDP packet payload               6
int main(void){
  UINT8              IsDHCP = 0;
  _NetCfgIpV4Args_t  ipV4;
  SlSockAddrIn_t     Addr, LocalAddr;
  UINT16             AddrSize = 0;
  INT16              SockID = 0;
  INT16              Status = 1;   // ok
  UINT32             data;
  unsigned char      len = sizeof(_NetCfgIpV4Args_t);
  initClk();         // PLL 50 MHz, ADC needs PPL active                16
  ST7735_InitR(INITR_REDTAB);                      // Initialize       17
  ST7735_OutString("Internet of Things\n");        //                  18
  ST7735_OutString("Embedded Systems\n");          //                  19
  ST7735_OutString("Vol. 2, Valvano");             //                  20
  ST7735_PlotClear(0,4095);   // range from 0 to 4095                  21
  sl_Start(0, 0, 0); // Initializing the CC3100 device                 22
  WlanConnect();     // connect to AP                                  23
  sl_NetCfgGet(SL_IPV4_STA_P2P_CL_GET_INFO,&IsDHCP,&len,      //       24
               (unsigned char *)&ipV4);                        //       25
  LocalAddr.sin_family = SL_AF_INET;                           //       26
  LocalAddr.sin_port = sl_Htons((UINT16)PORT_NUM);             //       27
  LocalAddr.sin_addr.s_addr = 0;                               //       28
  AddrSize = sizeof(SlSockAddrIn_t);                           //       29
  while(1){
    SockID = sl_Socket(SL_AF_INET,SL_SOCK_DGRAM, 0);           //       31
    Status = sl_Bind(SockID, (SlSockAddr_t *)&LocalAddr,       //       32
                     AddrSize);                                //       33
    Status = sl_RecvFrom(SockID, uBuf, BUF_SIZE, 0,            //       34
        (SlSockAddr_t *)&Addr, (SlSocklen_t*)&AddrSize );//              35
    if((uBuf[0]==ATYPE)&&(uBuf[1]== '=')){                     //       36
```

```
    int i,bOk; uint32_t place;                      //   37
    data = 0; bOk = 1;                              //   38
    i=4;   // ignore possible negative sign              39
    for(place = 1000; place; place = place/10){    //   40
      if((uBuf[i]&0xF0)==0x30){ // ignore spaces        41
        data += place*(uBuf[i]-0x30);              //   42
      }else{                                        //   43
        if((uBuf[i]&0xF0)!= ' '){                  //   44
          bOk = 0;                                  //   45
        }                                           //   46
      }                                             //   47
      i++;                                          //   48
    }                                               //   49
    if(bOk){                                        //   50
      ST7735_PlotLine(data);                        //   51
      ST7735_PlotNextErase();                       //   51
    }
   }
  }
 }
}
```
Program 11.3. Server software that receives UDP packets and plots results on the LCD.

Since UDP transmission is "best effort" we could lose packets or receive packets out of order. In this simple example we will not know if either of these errors were to occur. If we wished to have a more reliable transmission, we could have used TCP. Program 11.2 line 25 would have specified a socket stream instead of a datagram. To create a TCP communication, use the example software in the tcp_socket folder.

```
SockID = sl_Socket(SL_AF_INET,SL_SOCK_STREAM, 0);    // TCP socket
```

11.4.6. Client-Server Paradigm

The client-server paradigm is the dominant communication pattern over the Internet. The socket API supports two forms of this paradigm, a TCP-based connection-oriented form and a UDP-based connection-less form. Table 11.6 shows various popular applications and the underlying application and transport protocols they use.

Application	Application Protocol	Transport Protocol
email	SMTP	TCP
Remote terminal	Telnet	TCP
Web	HTTP	TCP
File transfer	FTP	TCP
Streaming media	HTTP, RTP	TCP or UDP
Telephony	SIP, RTP, Skype	TCP or UDP

Table 11.6. Applications their protocols and Transport protocol on which they depend.

The popular Hyper-Text Transfer Protocol (HTTP), which is used in the implementation of the World Wide Web (WWW) is TCP-based. A web browser like Firefox is a client (a HTTP client) and a web server like openweathermap.org is a server (a HTTP server). The paradigm can be seen to function as outlined in Figure 11.18.

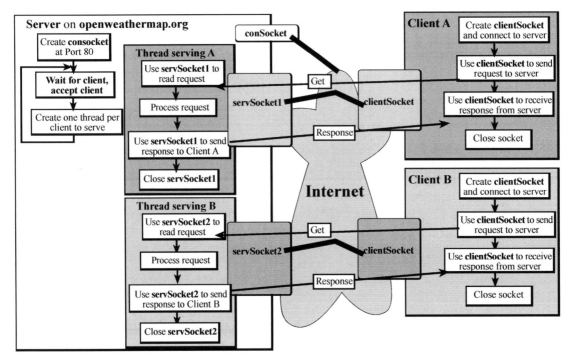

Figure 11.18. *Client-Server paradigm with a multi-threaded server, serving multiple clients.*

A server running at a machine (in the next section we will communicate with openweathermap.org) creates a socket bound to a port xyz. This port number is either a standard port or one chosen specifically for the service. For example, if the server is a **web server** that is publicly accessible, then the port number is 80. On the other hand, if it's a proprietary server, the service may choose an arbitrary number for xyz (say 6565) and let clients know about it so they can connect for service. This socket is called a connection socket (**conSocket**) where the server waits for connection requests from clients. A client that wants to access the service provided by the server creates a client socket (**clientSocket**). The client needs to know the hostname and the port at which the service is running to establish the connection using the client socket. Once the server sees the request, it accepts it on a separate socket (shown as **servSocket1** or **servSocket2** in the figure). It does not serve the client on the connection socket because, if it takes too long to serve the client then the server will be inaccessible to other clients seeking service. By accepting each client at a separate socket, the server can remain ready for connection requests from new clients.

Once the server accepts a client (say A) it creates a **thread** to service that client by passing to the thread the socket at which the client was accepted (server tells Client A about **servSocket1**). The main thread is now free to accept other clients. The exchange between client A and the thread serving it may involve multiple messages transferred back and forth or a single request/response (shown as **Get** and **Response** in the figure). A Web Browser like Firefox makes one connection for each object it requests from a Web Server. Note that each client connection is to port 80, which is a standard port at which Web Servers run.

11.4.7. Access a Weather Server using TCP

In this section, we will have the Launchpad connect to a web server running at **openweathermap.org**, a public server that gives weather information for any city in the world. So, the process is to create a socket and connect to the weather server running on openweathermap.org at port 80. Once connected, we send it a request using the HTTP GET command. The server responds with a well-formatted string containing weather details for the city whose weather information was sought in the GET request.

The **CC3100GetWeather_4C123** application creates a web client that fetches weather data from a server. To see the features of the server, go to openweathermap.org, enter Austin Texas into the search box, and click search. The goal of this smart object is to display weather data on the LCD. The approach for implementing a smart object over wifi is to begin with a hardware/software platform that implements IEEE801.11 wifi. The CC3100BOOST is a boosterpack that can be used with the TM4C123 LaunchPad as shown in Figure 11.19.

Figure 11.19. Client-Server paradigm with the LaunchPad+CC3100 as a client fetching weather data.

A version of this client is shown in Program 11.4. To run this application you will need an access point (AP). The first step will be to edit lines 97 – 99 to define connection to your access point, change these three lines in **main.c**

```
#define SSID_NAME    "Valvano"           // Access point name
#define SEC_TYPE     SL_SEC_TYPE_WPA
#define PASSKEY      "Calvin"            // Password
```

Notice the server web address and the HTTP GET request in lines 97 and 98. You can see/edit the GET request sent to the weather server. You could edit line 206 to see weather in a different city if you wish. The ASCII code %20 represents the space character in the name, rather than a delimiter in the GET request.

```
#define REQUEST "GET /data/2.5/weather?q=Austin%20Texas&
      units=metric HTTP/1.1\r\n
      User-Agent: Keil\r\nHost:api.openweathermap.org\r\n
      Accept: */*\r\n\r\n"
```

Notice the server web address and the HTTP GET request in lines 97 and 98. You can see/edit the GET request sent to the weather server. You could edit line 206 to see weather in a different city if you wish. Lines 209 – 212 initialize the LaunchPad allowing for output on the UART, switch input, and LED output. The function call on line 213 configures the policies on the SimpleLink software driver. Line 215 will start the CC3100 hardware. Lines 217 – 223 will connect to the access point. Programs 11.2 and 11.3 called the function **sl_NetCfgGet()** at this point to determine the IP address the AP assigned to this client. The local IP address will not be needed in this example. Lines 226 – 228 use the DNS to translate the logical IP address (**openweathermap.org**) to its physical address (**144.76.83.20**). Lines 230 – 236 open a socket in the weather server. As we saw in Figure 11.18, we send the socket request to port 80 of the public **conSocket** and the server response by giving us our own private **servSocket** socket. The SimpleLink driver keeps a data structure of the details of this connection and we can refer to it as a simple number (e.g., **SockID**). Lines 239 – 240 send the TCP packet containing the GET request to the server. Line 241 receives the response from the server. Embedded in the response are the weather details we requested. Line 242 closes the socket, and line 245 prints the response out UART0.

There are two ways to observe the weather data: 1) over the UART using a terminal program like PuTTY, or 2) looking at **Recvbuff** in the debugger. To understand how the client-server paradigm works, place breakpoints at these lines in **main.c**:

Line 220	Connect to access point (name, password, type)
Line 227	Domain Name System (web address to IP address)
Line 236	Create Socket, open connection (IP address, port 80)
Line 240	Send TCP (weather request)
Line 241	Receive TCP

There is also a main program file called **main2015.c** that parses the weather data and displays in on a Nokia5110 LCD, as shown in Figure 11.19.

```c
int main(void){int32_t retVal;  SlSecParams_t secParams;      //207
  char *pConfig = NULL; INT32 ASize = 0; SlSockAddrIn_t  Addr;
  initClk();        // PLL 50 MHz                              209
  UART_Init();      // Send data to PC, 115200 bps             210
  LED_Init();       // initialize LaunchPad I/O                211
  UARTprintf("Weather App\n");                                //212
  retVal = configureSimpleLinkToDefaultState(pConfig);        //213
  if(retVal < 0)Crash(4000000);
  retVal = sl_Start(0, pConfig, 0);                           //215
  if((retVal < 0) || (ROLE_STA != retVal) ) Crash(8000000);
  secParams.Key = PASSKEY;                                    //217
  secParams.KeyLen = strlen(PASSKEY);                         //218
  secParams.Type = SEC_TYPE; // OPEN, WPA, or WEP              219
  sl_WlanConnect(SSID_NAME, strlen(SSID_NAME),0,&secParams,0);//220
  while((0==(g_Status&CONNECTED))||(0== (g_Status&IP_AQUIRED))){
    _SlNonOsMainLoopTask();                                   //222
  }                                                           //223
  UARTprintf("Connected\n");
  while(1){
    strcpy(HostName,"openweathermap.org");                    //226
    retVal = sl_NetAppDnsGetHostByName(HostName,              //227
            strlen(HostName),&DestinationIP, SL_AF_INET);     //228
    if(retVal == 0){                                          //229
      Addr.sin_family = SL_AF_INET;                           //230
      Addr.sin_port = sl_Htons(80);                           //231
      Addr.sin_addr.s_addr = sl_Htonl(DestinationIP);         //232
      ASize = sizeof(SlSockAddrIn_t);                         //233
      SockID = sl_Socket(SL_AF_INET,SL_SOCK_STREAM, 0);       //234
      if(SockID >= 0){
        retVal=sl_Connect(SockID,(SlSockAddr_t *)&Addr,ASize);//236
      }
      if((SockID >= 0)&&(retVal >= 0)){
        strcpy(SendBuff, REQUEST);                            //239
        sl_Send(SockID, SendBuff, strlen(SendBuff), 0);       //240
        sl_Recv(SockID, Recvbuff, MAX_RECV_BUFF_SIZE, 0);     //241
        sl_Close(SockID);                                     //242
        LED_GreenOn();
        UARTprintf("\r\n\r\n");
        UARTprintf(Recvbuff);  UARTprintf("\r\n");            //245
      }
    }
    while(Board_Input()==0){}; // wait for touch
    LED_GreenOff();
  }
}
```

Program 11.4. Client software that fetches weather data from openweathermap.org (CC3100GetWeather_4C123).

11.4.8. Other CC3100 Applications

This section lists the sample applications are also provided for MSP430F5739, TM4C123GH6PM and SimpleLink Studio. The source code for these examples can be found in the examples directory after downloading CC3100SDK, the SimpleLink Wi-Fi CC3100 Software Development Kit (SDK) from the TI website. For more details on each example, see the docs folder included in the CC3100SDK download. The CC3100 comes preloaded with CC3100 BoosterPack comes preloaded with Out of Box HTML pages. Out of box demo highlights the following features: Simple WLAN Connection Using Smart Config, and easy access to CC3100 using mDNS and HTTP Server.

Antenna Selection. This is a reference implementation for antenna-selection scheme running on the host MCU, to enable improved radio performance inside buildings

Connection Policies. This application demonstrates the usage of the CC3100 profiles and connection-policies.

Send Email. This application sends an email using SMTP to a user-configurable email address at the push of a button.

Enterprise Network Connection. This application demonstrates the procedure for connecting the CC3100 to an enterprise network.

File Download. This application demonstrates file download from a cloud server to the on board serial Flash.

File System. This application demonstrates the use of the file system API to read and write files from the serial Flash.

Get Time. This application connects to an SNTP cloud server and receives the accurate time.

Get Weather. This application connects to 'Open Weather Map' cloud service and receives weather data.

Getting Started in AP Mode. This application configures the CC3100 in AP mode. It verifies the connection by pinging the connected client.

Getting Started in Station Mode. This application configures the CC3100 in STA mode. It verifies the connection by pinging the connected Access Point.

HTTP Server. This application demonstrates using the on-chip HTTP Server APIs to enable static and dynamic web page content.

IP Configuration. This application demonstrates how to enable static IP configuration instead of using DHCP.

mDNS. This application registers the mDNS service for broadcasting and attempts to get the service by the name broadcasted by another device.

Mode Configuration. This application demonstrates switching between STA and AP modes.

NWP Filters. This application demonstrates the configuration of Rx-filtering to reduce the amount of traffic transferred to the host, and to achieve lower power consumption.

NWP Power Policy. This application shows how to enable different power policies to reduce power consumption based on use case in the station mode.

P2P (Wi-Fi Direct). This application configures the device in P2P (Wi-Fi Direct) mode and demonstrates how to communicate with a remote peer device.

Provisioning AP. This application demonstrates the use of the on Chip HTTP server for Wi-Fi provisioning in AP Mode, building upon example application 7.8 above.

Provisioning with SmartConfig. This application demonstrates the usage of TI's SmartConfig™ Wi-Fi provisioning technology. The Wi-Fi Starter Application for iOS and Android is required to use this application. It can be downloaded from following link: http://www.ti.com/tool/wifistarter or from the Apple App store and Google Play.

Provisioning with WPS. This application demonstrates the usage of WPS Wi-Fi provisioning with CC3100.

Scan Policy. The application demonstrates the scan-policy settings in CC3100.

SPI Diagnostics Tool. This is a diagnostics application for troubleshooting the host SPI configuration.

SSL/TLS. The application demonstrates the usage of certificates with SSL/TLS for application traffic privacy and device or user authentication

TCP Socket. The application demonstrates simple connection with TCP traffic.

Transceiver Mode. The application demonstrates the CC3100 transceiver mode of operation.

UDP Socket. The application demonstrates simple connection with UDP traffic.

XMPP Client. The application demonstrates instant messaging using a cloud based XMPP server.

These were the steps I used to create the UDP communication example in Section 11.4.5. I began with the starter application, ek-tm4c123gxl-boost-cc3100_starter. I first changed `SSID_NAME` to match our access point

```
#define SSID_NAME    "Valvano"    // AP to connect to
```

Next, I compiled, downloaded and ran this application onto two LaunchPad+CC3100 systems, observing the operating on PuTTY. The interpreter output should show it has connected and shows the IP assigned to these two nodes by the AP. I could run the **ping** command to check the wifi connection to my AP.

Once I was sure my two LaunchPad+CC3100 systems could communicate with my AP, I made a copy of the starter application by copy-pasting the entire folder. I renamed this new folder to ek-tm4c123gxl-boost-cc3100_basic_wifi_UDP. I opened the new project in the compiler IDE and opened the **main.c** from the udp_socket example folder. I added and/or merged the source code from **main.c** of udp_socket into **starter.c** of the new project. The event handlers and the main project needed merging, but the `BsdUdpClient` and `BsdUdpServer` functions were simply added. I changed the IP address to match the address given to the server.

```
#define IP_ADDR       0xC0A80068
```

I then loaded a version that called the client (send UDP) on one system

```
while(1){ BsdUdpClient(PORT_NUM) };
```

and loaded a version that called the server (receive UDP) on the other system

```
while(1){ BsdUdpServer(PORT_NUM) };
```

I ran the two systems in the debugger to see that packets were being sent. I did not use SmartConfig, because I knew the name of the AP. The last step was to modify the client and server so the client collects data and the server displays it.

11.5. Exercises

11.1 For each term give a definition in 32 words or less.
a) Master-slave
b) Multidrop
c) Channel capacity
d) Ring network
e) Lossy compression
f) Crosstalk
g) Stuff bits
h) Frequency shift key
i) Phase encoding
j) Compression

11.2 For each pair of terms compare and contrast in 32 words or less.
a) Full duplex versus half duplex
b) Lossless versus lossy compression
c) Guided versus unguided media
d) LRC versus checksum
e) Originate versus answer

11.3 What fundamental property is used to transmit information through a channel?

11.4 Consider a telephone line with a channel bandwidth of 2 kHz and **SNR** of 40 dB. What is the maximum data rate possible?

11.5 Consider a telephone line with a channel bandwidth of 4 kHz and **SNR** of 30 dB. What is the maximum data rate possible?

11.6 Assume node 0x31 wishes to send the message "Hello" to node 0x32. Let ID = 0. Let RSSI = 0x3A.
a) Give the API mode 1 frame node 0x31 sends to its XBee module. Include the entire message from 0x7E up to and including the checksum.
b) Give the API mode 1 frame 0x32 will receive from its XBee module.

11.7 Assume node 0x41 wishes to send the message "Ciao" to node 0x42. Let ID = 0x12. Let RSSI = 0x3A.
a) Give the API mode 1 frame node 0x41 sends to its XBee module. Include the entire message from 0x7E up to and including the checksum.
b) Give the API mode 1 frame 0x42 will receive from its XBee module.

D11.8. The objective of this problem is to design a communication network using four single chip microcontrollers placed in a ring. Develop a message protocol that supports addressing and variable length data. Perform the I/O in the background using interrupt synchronization.

11.6. Lab Assignments

Lab 11.1. The overall goal of this lab is to design, implement and test a peer-to-peer communication system. Peer-to-peer means people on two computers communicate without the people on the other computers seeing the information. The system must use a ring-connected RS232 serial channel (Figure 11.3), must use interrupt-driven I/O, and must have a layered software configuration. The lowest-level software performs serial I/O. The middle-level software sends message packets from one computer to another. The highest level software interfaces with the human operator (keypad/LCD) and provides a mechanism to create a peer-to-peer connection. In a layered system, software in one layer can only call routines within that layer or the layer immediately below it. You need a way to see who is on the network, and a way to request/accept/terminate connection between two operators. Local operator input/output will occur via a keypad and LCD. You may assume all nodes on the system are willing to cooperate and not perform malicious activity. On the other hand it is possible that another computer on the network may not be plugged in, or the network connection may be broken.

The communication system between two or more microcontrollers will be designed in three layers. The first layer, the physical layer is implemented by the UART hardware and the interrupt-driven device driver. The second layer may consist of a simplified binary synchronous communication protocol (BSC). At this level message packets will be transmitted between the two machines. The highest level will be a keypad interpreter and an LCD display. The LCD should show interactive feedback to the operator creates messages to be sent and displays messages received.

Lab 11.2. The objective of this lab is to design a point-to-point ZigBee network using the XBee module. Each microcontroller will have a keypad and a two-line LCD display. When the operator types on the keypad, characters are displayed on one line of the LCD. When the operator hits the send key, those characters are sent and eventually displayed on the second line of the LCD of the other microcontroller. Debug the system in a bottom-up manner. Write simple main programs in the Rx and Tx systems to test the low-level functionality of the interface. Add appropriate debugging instruments, such as profiles and dumps. Connect unused pins from both devices to a single logic analyzer and record a thread profile (which program runs when) as a stream of data is passed from the Tx system to the Rx system. Write a main program to measure maximum bandwidth of your channel without using hardware flow control. Furthermore, you should determine which component limits bandwidth. Modify the transmitter so it outputs pseudo data as fast as possible. Modify the receiver so it does not display data on the LCD. Rather, the receiver will check for this pattern of characters and count the number of errors. Add minimally intrusive debugging instruments to determine if and where data is lost. Measure the maximum range of your system. In particular, find the maximum distance where the system performs reliable communication.

Appendix 1. Glossary

1/f noise A fundamental noise in resistive devices arising from fluctuating conductivity. Same as pink noise.
2's complement (see two's complement).
60 Hz noise An added noise from electromagnetic fields caused by either magnetic field induction or capacitive coupling.
accumulator High-speed memory located in the processor used to perform arithmetic or logical functions. Any of the registers R0 to R12 on the ARM Cortex-M processor can be used as accumulators.
accuracy A measure of how close our instrument measures the desired parameter referred to the NIST.
acknowledge Clearing the interrupt flag bit that requested the interrupt.
active thread A thread that is in the ready-to-run circular linked list. It is either running or is ready to run.
actuator Electro-mechanical or electro-chemical device that allows computer commands to affect the external world.
ADC Analog to digital converter, an electronic device that converts analog signals (e.g., voltage) into digital form (i.e., integers).
address bus A set of digital signals that connect the CPU, memory and I/O devices, specifying the location to read or write for each bus cycle. See also control bus and data bus.
address decoder A digital circuit having the address lines as input and a select line as output (see select signal)
aging A technique used in priority schedulers that temporarily increases the priority of low priority treads so they are run occasionally. (See starvation)
aliasing When digital values sampled at f_s contain frequency components above ½ f_s, then the apparent frequency of the data is shifted into the 0 to ½ f_s range. See Nyquist Theory.
alternatives The total number of possibilities. E.g., an 8-bit number scheme can represent 256 different numbers. An 8-bit digital to analog converter (DAC) can generate 256 different analog outputs.
anode The positive side of a diode. Current enters the anode side of a diode. Contrast with cathode.
answer modem The device that receives the telephone call.
anti-reset-windup Establishing an upper bound on the magnitude of the integral term in a PID controller, so this term will not dominate, when the errors are large.
arithmetic logic unit (ALU) Component of the processor that performs arithmetic and logic operations.
arm Activate so that interrupts are requested. Trigger flags that can request interrupts will have a corresponding arm bit to allow or disallow that flag to request interrupts. Contrast to enable.
armature The moving structure in a relay, the part that moves when the relay is activated. Contrast to frame.
ASCII American Standard Code for Information Interchange, a code for representing characters, symbols, and synchronization messages as 7 bit, 8-bit or 16-bit binary values.
assembler System software that converts an assembly language program (human readable format) into object code (machine readable format).
assembly directive Operations included in the program that are not executed by the computer at run time, but rather are interpreted by the assembler during the assembly process. Same as pseudo-op.
assembly listing Information generated by the assembler in human readable format, typically showing the object code, the original source code, assembly errors, and the symbol table.
asynchronous bus A communication protocol without a central clock where is the data is transferred using two or three control lines implementing a handshaked interaction between the memory and the computer.
asynchronous protocol A protocol where the two devices have separate and distinct clocks
atomic Software execution that cannot be divided or interrupted. Once started, an atomic operation will run to its completion without interruption. On most computers the assembly language instructions are atomic. All instructions on the ARM® Cortex-M processor are atomic except store and load multiple, `STM LDM`.

autoinitialization The process of automatically reloading the address registers and block size counters at the end of a previous block transfer, so that DMA transfer can occur indefinitely without software interaction.

availability The portion of the total time that the system is working. MTBF is the mean time between failures, MTTR is the mean time to repair, and availability is MTBF/(MTBF+MTTR).

bandwidth In communication systems, the information transfer rate, the amount of data transferred per second. Same as throughput. In analog circuits, the frequency at which the gain drops to 0.707 of the normal value. For a low pass system, the frequency response ranges from 0 to a maximum value. For a high pass system, the frequency response ranges from a minimum value to infinity. For a bandpass system, the frequency response ranges from a minimum to a maximum value. Compare to frequency response.

bandwidth coupling Module A is connected to Module B, because data flows from A to B.

bang-bang A control system where the actuator has only two states, and the system "bangs" all the way in one direction or "bangs" all the way in the other, same as binary controller.

bank-switched memory A memory module with two banks that interfaces to two separate address/data buses. At any given time one memory bank is attached to one address/data bus the other bank is attached to the other bus, but this attachment can be switched.

basis Subset from which linear combinations can be used to reconstruct the entire set.

baud rate In general, the baud rate is the total number of bits (information, overhead, and idle) per time that are transmitted. In a modem application it is the total number of sounds per time are transmitted

bi-directional Digital signals that can be either input or output.

biendian The ability to process numbers in both big and little-endian formats.

big endian Mechanism for storing multiple byte numbers such that the most significant byte exists first (in the smallest memory address). See also little endian.

binary A system that has two states, on and off.

binary controller Same as bang-bang.

binary recursion A recursive technique that makes two calls to itself during the execution of the function. See also recursion, linear recursion, and tail recursion.

binary semaphore A semaphore that can have two values. The value=1 means OK and the value=0 means busy. Compare to counting semaphore.

bipolar transistor Either a NPN or PNP transistor.

bipolar stepper motor A stepper motor where the current flows in both directions (in/out) along the interface wires; a stepper with four interface wires. Contrast to unipolar stepper motor.

bit Basic unit of digital information taking on the value of either 0 or 1.

bit rate The information transfer rate, given in bits per second. Same as bandwidth and throughput.

bit time The basic unit of time used in serial communication. With serial channel bit time is 1/baud rate.

blind-cycle A software/hardware synchronization method where the software waits a specified amount of time for the hardware operation to complete. The software has no direct information (blind) about the status of the hardware.

block correction code (BCC) A code (e.g., horizontal parity) attached to the end of a message used to detect and correct transmission errors.

blocked thread A thread that is not scheduled for running because it is waiting on an external event.

blocking semaphore A semaphore where the threads will block (so other threads can perform useful functions) when they execute wait on a busy semaphore. Contrast to spinlock semaphore.

Board Support Package (BSP) A set of software routines that abstract the I/O hardware such that the same high-level code can run on multiple computers.

borrow During subtraction, if the difference is too small, then we use a borrow to pass the excess information into the next higher place. For example, in decimal subtraction 36-27 requires a borrow from the ones to tens place because 6-7 is too small to fit into the 0 to 9 range of decimal numbers.

bounded waiting The condition where once a thread begins to wait on a resource, there are a finite number of threads that will be allowed to proceed before this thread is allowed to proceed.

break-before-make In a double-throw relay or double-throw switch, there is one common contact and two separate contacts. Break-before-make means as the common contact moves from one of separate contacts to another, it will break off (finish bouncing and no longer touch) the first contact before it makes (begins to bounce and starts to touch) the other contact. A *form C* relay has a *break-before-make* operation.

break or trap A break or a trap is a debugging instrument that halts the processor. With a resident debugger, the break is created by replacing specific op code with a software interrupt instruction. When encountered it will stop your program and jump into the debugger. Therefore, a break halts the software. The condition of being in this state is also referred to as a break.

breakdown A transducer that stops functioning when its input goes above a maximum value or below a minimum value. Contrast to dead zone.

breakpoint The place where a break is inserted, the time when a break is encountered, or the time period when a break is active.

brushed DC motor A motor where the current reversals are produced with brushes between the stator and rotor. They are less expensive than brushless DC motors.

brushless DC motor (BLDC) A motor where the current reversals are produced with shaft sensors and an electronic controller. They are faster and more reliable than brushed DC motors.

buffered I/O A FIFO queue is placed in between the hardware and software in an attempt to increase bandwidth by allowing both hardware and software to run in parallel.

burn The process of programming a ROM, PROM or EEPROM.

burst DMA An I/O synchronization scheme that transfers an entire block of data all at once directly from an input device into memory, or directly from memory to an output device.

bus A set of digital signals that connect the CPU, memory and I/O devices, consisting of address signals, data signals and control signals. See also address bus, control bus and data bus.

bus bandwidth The number of bytes transferred per second between the processor and memory.

bus interface unit (BIU) Component of the processor that reads and writes data from the bus. The BIU drives the address and control buses.

busy-wait synchronization A software/hardware synchronization method where the software continuously reads the hardware status waiting for the hardware operation to complete. The software usually performs no work while waiting for the hardware. Same as gadfly.

byte Digital information containing eight bits.

calibration drift A change occurring over time in the transfer function used to calculate the measured variable from the raw measurements. For example, assume we have a voltmeter that uses the equation $V = C1*n/1024+C2$ to calculate voltage V in mV from a raw ADC value n. C1 and C2 are calibration constants determined by measurements on known voltages. Calibration drift occurs when the correct values for C1 and C2 vary over time.

carrier frequency the average or midvalue sound frequency in the modem.

carry During addition, if the sum is too large, then we use a carry to pass the excess information into the next higher place. For example, in decimal addition 36+27 requires a carry from the ones to tens place because 6+7 is too big to fit into the 0 to 9 range of decimal numbers.

cathode The negative side of a diode. Current exits the cathode side of a diode. Contrast to anode.

causal The property where the output depends on the present and past inputs, but not on any future inputs.

ceiling Establishing an upper bound on the result of an operation. See also floor.

certification A process where a governing body (e.g., FDA, NASA, FCC, DOD etc.) gives approval for the use of the device. It usually involves demonstrating the device meets or exceeds safety and performance criteria.

channel The hardware that allows communication to occur.

checksum The simple sum of the data, usually in finite precision (e.g., 8, 16, 24 bits).

closed loop control system A control system that includes sensors to measure the current state variables. These inputs are used to drive the system to the desired state.
CMOS A digital logic system called complementary metal oxide semiconductor. It has properties of low power and small size. Its power is a function of the number of transitions per second. Its speed is often limited by capacitive loading.
cohesion A cohesive module is one such that all parts of the module are related to each other to satisfy a common objective.
common anode LED display A display with multiple LEDs, configured with all of the LED anodes connected together, there are separate connections to the cathodes (current flows in the common anode and out the individual cathodes).
common cathode LED display A display with multiple LEDs, configured with all of the LED cathodes connected together, there are separate connections to the anodes (current flows in the individual anodes and out the common cathode).
common mode For a system with differential inputs, the common mode properties are defined as signals applied to both inputs simultaneously. Contrast to differential mode.
common mode input impedance Common mode input voltage divided by common mode input current.
common mode rejection ratio For a differential amplifier, CMRR is the ratio of the common mode gain divided by the differential mode gain. A perfect CMRR would be zero.
compiler System software that converts a high-level language program (human readable format) into object code (machine readable format).
complex instruction set computer (CISC) A computer with many instructions, instructions that have varying lengths, instructions that execute in varying times, many instructions can access memory, one instruction may both read and write memory, fewer and more specialized registers, and many different types of addressing modes. Contrast to RISC.
compression ratio The ratio of the number of original bytes to the number of compressed bytes.
concurrent programming A software system that supports two tasks to be active at the same time. A computer with interrupts implements concurrent programming.
condition code register (CCR) Register in the processor that contains the status of the previous ALU operation, as well as some operating mode flags such as the interrupt enable bit.
control bus A set of digital signals that connect the CPU, memory and I/O devices, specifying when to read or write for each bus cycle. See also address bus and data bus.
control coupling Module A is connected to Module B, because actions in A affect the control path in B.
control unit (CU) Component of the processor that determines the sequence of operations.
cooperative multi-tasking A scheduler that cannot suspend execution of a thread without the thread's permission. The thread must cooperate and suspend itself. Same as nonpreemptive scheduler.
counting semaphore A semaphore that can have any signed integer value. The value>0 means OK and the value≤0 means busy. Compare to binary semaphore.
CPU bound A situation where the input or output device is faster than the software. In other words it takes less time for the I/O device to process data, than for the software to process data. Contrast to I/O bound.
CPU cycle A memory bus cycle where the address and R/W are controlled by the processor. On microcontrollers without DMA, all cycles are CPU cycles. Contrast to DMA cycle.
crisp input An input parameter to the fuzzy logic system, usually with units like cm, cm/sec, °C etc.
crisp output An output parameter from the fuzzy logic system, usually with units like dynes, watts etc.
critical section Locations within a software module, which if an interrupt were to occur at one of these locations, then an error could occur (e.g., data lost, corrupted data, program crash, etc.) Same as vulnerable window.
cross-assembler An assembler that runs on one computer but creates object code for a different computer.
cross-compiler A compiler that runs on one computer but creates object code for a different computer.

cycle steal DMA An I/O synchronization scheme that transfers data one byte at a time directly from an input device into memory, or directly from memory to an output device.

cycle stretch The action where some memory cycles are longer allowing time for communication with slower memories, sometimes the memory itself requests the additional time and sometimes the computer has a preprogrammed cycle stretch for certain memory addresses

DAC Digital to analog converter, an electronic device that converts digital signals (i.e., integers) to analog form (e.g., voltage).

data acquisition system A system that collects information, same as instrument.

data bus A set of digital signals that connect the CPU, memory and I/O devices, specifying the value that is being read or written for each bus cycle. See also address bus and control bus.

data communication equipment (DCE) A modem or printer connected a serial communication network.

data terminal equipment (DTE) A computer or a terminal connected a serial communication network.

dead zone A condition of a transducer when a large change in the input causes little or no change in the output. Contrast to breakdown.

deadlock A scenario that occurs when two or more threads are all blocked each waiting for the other with no hope of recovery.

decibel A measure of the relative amplitude of two voltages: $dB = 20 \log_{10}(V_1/V_2)$. It is also refers to the relative amplitude of two powers: $dB = 10 \log_{10}(P_1/P_2)$.

defuzzification Conversion from the fuzzy logic output variables to the crisp outputs.

desk checking or dry run We perform a desk check (or dry run) by determining in advance, either by analytical algorithm or explicit calculations, the expected outputs of strategic intermediate stages and final results for a set of typical inputs. We then run our program can compare the actual outputs with this template of expected results.

device driver A collection of software routines that perform I/O functions.

differential mode For a system with differential inputs, the differential mode properties are defined as signals applied as a difference between the two inputs. Contrast to common mode.

differential mode input impedance Differential mode input voltage divided by differential mode input current.

digital signal processing Processing of data with digital hardware or software after the signal has been sampled by the ADC, e.g., filters, detection and compression/decompression.

direct memory access (DMA) the ability to transfer data between two modules on the bus, this transfer is usually initiated by the hardware (device needs service) and the software configures the communication, but the data is transferred without explicit software action for each individual piece of data

direction register Specifies whether a bi-directional I/O pin is an input or an output. We set a direction register bit to 0 (or 1) to specify the corresponding I/O pin to be input (or output.)

disarm Deactivate so that interrupts are not requested, performed by clearing the arm bit.

Discrete Fourier Transform (DFT) A technique to convert data in the time domain to data in the frequency domain. N data points are sampled at f_s. The resulting frequency resolution is f_s/N.

DMA Direct Memory Access is a software/hardware synchronization method where the hardware itself causes a data transfer between the I/O device and memory at the appropriate time when data needs to be transferred. The software usually can perform other work while waiting for the hardware. No software action is required for each individual byte.

DMA cycle A memory bus cycle where the address and R/W are controlled by the DMA controller. Contrast to CPU cycle.

double byte Two bytes containing 16 bits. Same as halfword.

double-pole relay Two separate and complete relays, which are activated together. Contrast to single pole.

double-pole switch Two separate and complete switches. The two switches are electrically separate, but mechanically connected. Such that both switches are activated together. Contrast to single pole.

double-throw relay A relay with three contact connections, one common and two throws. The common will be connected to exactly of one the two throws (see single-throw).

double-throw switch A switch with three contact connections. The center contact will be connected exactly one of the other two contacts. Contrast with single-throw.

double word Two words containing 64 bits.

download The process of transferring object code from the host (e.g., the PC) to the target microcontroller.

drop-out An error that occurs after a right shift or a divide, and the consequence is that an intermediate result loses its ability to represent all of the values. E.g., I=100*(N/51) can only result in the values 0, 100, or 200, whereas I=(100*N)/51 properly calculates the desired result.

dual address DMA Direct memory access that requires two bus cycles to transfer data from an input device into memory, or from memory to an output device.

dual port memory A memory module that interfaces to two separate address/data buses, and allows both systems read/write access the data.

duty cycle For a periodic digital wave, it is the percentage of time the signal is high. When an LED display is scanned, it is the percentage of time each LED is active. A motor interfaced using pulse-width-modulation allows the computer to control delivered power by adjusting the duty cycle.

dynamic allocation Data structures like the TCB that are created at runtime by calling malloc() and exist until the software releases the memory block back to the heap by calling free(). Contrast to static allocation.

dynamic RAM Volatile read/write storage built from a capacitor and a single transistor having a low cost, but requiring refresh. Contrast with static RAM.

EEPROM Electrically erasable programmable read only memory that is nonvolatile and easy to reprogram. EEPROM can be erased and reprogrammed multiple times.

embedded computer system A system that performs a specific dedicated operation where the computer is hidden or embedded inside the machine.

emulator An in-circuit emulator is an expensive debugging hardware tool that mimics the processor pin outs. To debug with an emulator, you would remove the processor chip and attach the emulator cable into the processor socket. The emulator would sense the processor input signals and recreate the processor outputs signals on the socket as if an actual processor were actually there, running at full speed. Inside the emulator you have internal read/write access to the registers and processor state. Most emulators allow you to visualize/record strategic information in real-time without halting the program execution. You can also remove ROM chips and insert the connector of a ROM-emulator. This type of emulator is less expensive, and it allows you to debug ROM-based software systems.

EPROM Same as PROM. Electrically programmable read only memory that is nonvolatile and requires external devices to erase and reprogram. It is usually erased using UV light.

erase The process of clearing the information in a PROM or EEPROM, using electricity or UV light. The information bits are usually all set to logic 1.

EVB Evaluation Board, a board-level product used to develop microcontroller systems.

even parity A communication protocol where the number of ones in the data plus a parity bit is an even number. Contrast with odd parity.

external fragmentation A condition when the largest file or memory block that can be allocated is less than the total amount of free space on the disk or memory.

fan out The number of inputs that a single output can drive if the devices are all in the same logic family.

Fast Fourier Transform (FFT) A fast technique to convert data in the time domain to data in the frequency domain. N data points are sampled at f_s. The resulting frequency resolution is f_s/N. Mathematically, the FFT is the same as the DFT, just faster.

FET Field effect transistor, also JFET.

filter In the debugging context, a filter is a Boolean function or conditional test used to make run-time decisions. For example, if we print information only if two variables x,y are equal, then the conditional (x==y) is a filter. Filters can involve hardware status as well.

finite impulse response filter (FIR) A digital filter where the output is a function of a finite number of current and past data samples, but not a function of previous filter outputs.

Finite State Machine (FSM) An abstract design method to build a machine with inputs and outputs. The machine can be in one of a finite number of states. Which state the system is in represents memory of previous inputs. The output and next state are a function of the input. There may be time delays as well.

firm real-time A system that expects all critical tasks to complete on time. Once a deadline as passed, there is no value to completing the task. However, the consequence of missed deadlines is real but the overall system operates with reduced quality. Streaming audio and video are typical examples. Compare to hard real-time and soft real-time.

fixed-point A technique where calculations involving nonintegers are performed using a sequence of integer operations. E.g., 0.123*x is performed in decimal fixed-point as (123*x)/1000 or in binary fixed-point as (126*x)>>10.

flash EEPROM Electrically erasable programmable read only memory that is nonvolatile and easy to reprogram. Flash EEPROMs are typically larger than regular EEPROM.

floating A logic state where the output device does not drive high or pull low. The outputs of open collector and tristate devices can be in the floating state. Same as HiZ.

floor Establishing a lower bound on the result of an operation. See also ceiling.

follower An analog circuit with gain equal to 1, large input impedance and small output impedance. Same as voltage follower.

frame A complete and distinct packet of bits occurring in a serial communication channel.

frame The fixed structure in a relay or transducer. Contrast to armature.

framing error An error when the receiver expects a stop bit (1) and the input is 0.

frequency response The frequency at which the gain drops to 0.707 of the normal value. For a low pass system, the frequency response ranges from 0 to a maximum value. For a high pass system, the frequency response ranges from a minimum value to infinity. For a bandpass system, the frequency response ranges from a minimum to a maximum value. Same as bandwidth.

frequency shift key (FSK) A modem that modulates the digital signals into frequency encoded sine waves.

friendly Friendly software modifies just the bits that need to be modified, leaving the other bits unchanged, making to easier to combine modules.

full duplex channel Hardware that allows bits (information, error checking, synchronization or overhead) to transfer simultaneously in both directions. Contrast with simplex and half duplex channels.

full duplex communication A system that allows information (data, characters) to transfer simultaneously in both directions.

functional debugging The process of detecting, locating, or correcting functional and logical errors in a program, typically not involving time. The process of instrumenting a program for such purposes is called functional debugging or often simply debugging.

fuzzification Conversion from the crisp inputs to the fuzzy logic input variables.

fuzzy logic Boolean logic (true/false) that can take on a range of values from true (255) to false (0). Fuzzy logic **and** is calculated as the minimum. Fuzzy logic **or** is the maximum.

gadfly A software/hardware synchronization method where the software continuously reads the hardware status waiting for the hardware operation to complete. The software usually performs no work while waiting for the hardware. Same as busy wait.

gauge factor The sensitivity of a strain gauge transducer, i.e., slope of the resistance versus displacement response.

gibibyte (GiB) 2^{30} or 1,073,741,824 bytes. Compare to **gigabyte**, which is 1,000,000,000 bytes.

half duplex channel Hardware that allows bits (information, error checking, synchronization or overhead) to transfer in both directions, but in only one direction at a time. Contrast with simplex and full duplex channels.

half duplex communication A system that allows information to transfer in both directions, but in only one direction at a time.

halfword Two bytes containing 16 bits. Same as double byte.

handshake A software/hardware synchronization method where control and status signals go both directions between the transmitter and receiver. The communication is interlocked meaning each device will wait for the other.

hard real-time A system that can guarantee that a process will complete a critical task within a certain specified range. In data acquisition systems, hard real-time means there is an upper bound on the latency between when a sample is supposed to be taken (every $1/f_s$) and when the ADC is actually started. Hard real-time also implies that no ADC samples are missed. Compare to soft real-time and firm real-time.

Harvard architecture A computer architecture where instructions are fetched from a different bus from where data are fetched.

heartbeat A debugging monitor, such as a flashing LED, we add for the purpose of seeing if our program is running.

hexadecimal A number system that uses base 16.

HiZ A logic state where the output device does not drive high or pull low. The outputs of open collector and tristate devices can be in the floating state. Same as floating.

hold time When latching data into a device with a rising or falling edge of a clock, the hold time is the time after the active edge of the clock that the data must continue to be valid. See setup time.

hook An indirect function call added to a software system that allows the user to attach their programs to run at strategic times. These attachments are created at run time and do not require recompiling.

hysteresis A condition when the output of a system depends not only on the input, but also on the previous outputs, e.g., a transducer that follows a different response curve when the input is increasing than when the input is decreasing.

I/O bound A situation where the input or output device is slower than the software. In other words it takes longer for the I/O device to process data, than for the software to process data. Contrast to CPU bound.

I/O device Hardware and software components capable of bringing information from the external environment into the computer (input device), or sending data out from the computer to the external environment (output device).

I/O port A hardware device that connects the internal software with external hardware.

I_{IH} Input current when the signal is high.

I_{IL} Input current when the signal is low.

immediate An addressing mode where the operand is a fixed data or address value.

impedance loading A condition when the input of stage n+1 of an analog system affects the output of stage n, because the input impedance of stage n+1 is too small and the output impedance of stage n is too large.

impedance The ratio of the effort (voltage, force, pressure) divided by the flow (current, velocity, flow).

incremental control system A control system where the actuator has many possible states, and the system increments or decrements the actuator value depending on either in error is positive or negative.

indexed An addressing mode where the data or address value for the instruction is located in memory pointed to by an index register.

infinite impulse response filter (IIR) A digital filter where the output is a function of an infinite number of past data samples, usually by making the filter output a function of previous filter outputs.

input bias current Difference between currents of the two op amp inputs.

input capture A mechanism to set a flag and capture the current time (TCNT value) on the rising, falling or rising&falling edge of an external signal. The input capture event can also request an interrupt.

input impedance Input voltage divided by input current.

input noise current Current noise refereed to the op amp inputs.

input noise voltage Voltage noise refereed to the op amp inputs.

input offset current Average current into the two op amp inputs.

input offset voltage Voltage difference between the two op amp inputs that makes the output zero.

instruction register (IR) Register in the control unit that contains the op code for the current instruction.

instrument An instrument is the code injected into a program for debugging or profiling. This code is usually extraneous to the normal function of a program and may be temporary or permanent. A print statement added to your source code is an example of a permanent instrument, because removal requires editing and recompiling.

instrument A embedded system that collects information, same as data acquisition system.

instrumentation The debugging process of injecting or inserting an instrument.

instrumentation amp A differential amplifier analog circuit, which can have large gain, large input impedance, small output impedance, and a good common mode rejection ration.

internal fragmentation Storage that is allocated for the convenience of the operating system but contains no information. This space is wasted.

interrupt A software/hardware synchronization method where the hardware causes a special software program (interrupt handler) to execute when its operation to complete. The software usually can perform other work while waiting for the hardware.

interrupt flag A status bit that is set by the timer hardware to signify an external event has occurred.

interrupt mask A control bit that, if programmed to 1, will cause an interrupt request when the associated flag is set. Same as **arm**.

interrupt service routine (ISR) Program that runs as a result of an interrupt.

interrupt vector 32-bit values in ROM specifying where the software should execute after an interrupt request. There is a unique interrupt vector for each type of interrupt including reset.

intrusive The debugger itself affects the program being tested. See nonintrusive.

Inverse Discrete Fourier Transform (IDFT) A technique to convert data in the frequency domain to data in the time domain. There are N data points and the sampling rate is f_s. The resulting frequency resolution is f_s/N.

invocation coupling Module A is connected to Module B, because A calls B.

I_{OH} Output current when the signal is high. This is the maximum current that has a voltage above V_{OH}.

I_{OL} Output current when the signal is low. This is the maximum current that has a voltage below V_{OL}.

isolated I/O A configuration where the I/O devices are interfaced to the computer in a manner different than the way memories are connected.

jerk The change in acceleration; the derivative of the acceleration. Third derivative of shaft angle.

Johnson noise A fundamental noise in resistive devices arising from the uncertainty about the position and velocity of individual molecules. Same as thermal noise and white noise.

Karnaugh map tabular representation of the input/output relationship for a combinational digital function, the inputs possibilities are placed in the row and column labels, and the output values are placed inside the table

kibibyte (KiB) 2^{10} or 1024 bytes. Compare to **kilobyte**, which is 1000 bytes.

latch As a noun, it means a register. As a verb, it means to store data into the register.

latched input port An input port where the signals are latched (saved) on an edge of an associated strobe signal.

latency In this book latency usually refers to the response time of the computer to external events. For example, the time between new input becoming available and the time the input is read by the computer. For example, the time between an output device becoming idle and the time the input is the computer writes new data to it. There can also be a latency for an I/O device, which is the response time of the external I/O device hardware to a software command.

LCD Liquid Crystal Display, where the computer controls the reflectance or transmittance of the liquid crystal, characterized by its flexible display patterns, low power, and slow speed.

LED Light Emitting Diode, where the computer controls the electrical power to the diode, characterized by its simple display patterns, medium power, and high speed.

light-weight process Same as a thread.

linear filter A filter where the output is a linear combination of its inputs.

linear recursion A recursive technique that makes only one call to itself during the execution of the function. Linear recursive functions are easier to implement iteratively. We draw the execution pattern as a straight or linear path. See also recursion, binary recursion, and tail recursion.

linear variable differential transformer (LVDT) A transducer that converts position into electric voltage.

little endian Mechanism for storing multiple byte numbers such that the least significant byte exists first (in the smallest memory address). Contrast with big endian.

loader System software that places the object code into the microcontroller's memory. If the object code is stored in EPROM, the loader is also called a EPROM programmer.

Local Area Network (LAN) A connection between computers confined to a small space, such as a room or a building.

logic analyzer A hardware debugging tool that allows you to visualize many digital logic signals versus time. Real logic analyzers have at least 32 channels and can have up to 200 channels, with sophisticated techniques for triggering, saving and analyzing the real-time data.

LSB The least significant bit in a number system is the bit with the smallest significance, usually the right-most bit. With signed or unsigned integers the significance of the LSB is 1.

maintenance Process of verifying, changing, correcting, enhancing, and extending a system.

make before break in a double-throw relay or double-throw switch, there is one common contact and two separate contacts. Make before break means as the common contact moves from one of separate contacts to another, it will make (finishing bouncing) the second contact before it breaks off (start bouncing) the first contact. A *form D* relay has a *make before break* operation.

mailbox A formal communication structure, similar to a FIFO queue, where the source task puts data into the mailbox and the sink task gets data from the mailbox. The mailbox can hold at most one piece of data at a time, and has two states: mailbox has valid data or mailbox is empty.

mark A digital value of true or logic 1. Contrast with space.

mask As a verb, mask is the operation that selects certain bits out of many bits, using the logical and operation. The bits that are not being selected will be cleared to zero. When used as a noun, mask refers to the specific bits that are being selected.

Mealy FSM A FSM where the both the output and next state are a function of the input and state

measurand A signal measured by a data acquisition system.

mebibyte (MiB) 2^{20} or 1,048,576 bytes. Compare to **megabyte**, which is 1,000,000 bytes.

membership sets Fuzzy logic variables that can take on a range of values from true (255) to false (0).

memory A computer component capable of storing and recalling information.

memory-mapped I/O A configuration where the I/O devices are interfaced to the computer in a manner identical to the way memories are connected.

microcomputer A small electronic device capable of performing input/output functions containing a microprocessor, memory, and I/O devices, where small means you can carry it.

microcontroller A single chip microcomputer like the TI MSP430, Freescale 9S12, Intel 8051, PIC16, or the Texas Instruments TM4C123.

mnemonic The symbolic name of an operation code, like `mov str push`.

modem An electronic device that MOdulates and DEModulates a communication signal. Used in serial communication across telephone lines.

monitor A monitor is a debugger feature that allows us to passively view strategic software parameters during the real-time execution of our program. An effective monitor is one that has minimal effect on the performance of the system. When debugging software on a windows-based machine, we can often set up a debugger window that displays the current value of certain software variables.

Moore FSM A FSM where the both the output is only a function of the state and the next state is a function of the input and state

MOSFET Metal oxide semiconductor field effect transistor.

MSB The most significant bit in a number system is the bit with the greatest significance, usually the left-most bit. If the number system is signed, then the MSB signifies positive (0) or negative (1).

multiple access circular queue MACQ A data structure used in data acquisition systems to hold the current sample and a finite number of previous samples. It is always full and new data overwrites the oldest.

multi-threaded A system with multiple threads (e.g., main program and interrupt service routines) that cooperate towards a common overall goal.

mutual exclusion or **mutex** Thread synchronization where at most one thread at a time is allowed to enter.

negative feedback An analog system with negative gain feedback paths. These systems are often stable.

negative logic A signal where the true value has a lower voltage than the false value, in digital logic true is 0 and false is 1, in TTL logic true is less than 0.7 volts and false is greater than 2 volts, in RS232 protocol true is -12 volts and false is +12 volts. Contrast with positive logic.

nibble 4 binary bits or 1 hexadecimal digit.

nonatomic Software execution that can be divided or interrupted. Most lines of C code require multiple assembly language instructions to execute, therefore an interrupt may occur in the middle of a line of C code. The instructions store and load multiple, `STM LDM`, are nonatomic.

nonintrusive A characteristic when the presence of the collection of information itself does not affect the parameters being measured. Nonintrusiveness is the characteristic or quality of a debugger that allows the software/hardware system to operate normally as if the debugger did not exist. Intrusiveness is used as a measure of the degree of perturbation caused in program performance by an instrument. For example, a print statement added to your source code and single-stepping are very intrusive because they significantly affect the real-time interaction of the hardware and software. When a program interacts with real-time events, the performance is significantly altered. On the other hand, an instrument that toggles an LED on and off (requiring just 1 μs to execute) is much less intrusive. A logic analyzer that passively monitors the address and data by is completely nonintrusive. An in-circuit emulator is also non-intrusive because the software input/output relationships will be the same with and without the debugging tool.

nonlinear filter A filter where the output is not a linear combination of its inputs. E.g., median, minimum, maximum are examples of nonlinear filters. Contrast to linear filter.

nonpreemptive scheduler A scheduler that cannot suspend execution of a thread without the thread's permission. The thread must cooperate and suspend itself. Same as cooperative multi-tasking.

nonreentrant A software module which once started by one thread, should not be interrupted and executed by a second thread. A nonreentrant modules usually involve nonatomic accesses to global variables or I/O ports: read modify write, write followed by read, or a multistep write.

nonvolatile A condition where information is not lost when power is removed. When power is restored, then the information is in the state that occurred when the power was removed.

Nyquist Theorem If a input signal is captured by an ADC at the regular rate of fs samples/sec, then the digital sequence can accurately represent the 0 to ½ fs frequency components of the original signal.

object code Programs in machine readable format created by the compiler or assembler. The S19 records contain object code.

odd parity A communication protocol where the number of ones in the data plus a parity bit is an odd number. Contrast with even parity.

op amp An integrated analog component with two inputs, (V2,V1) and an output (Vout), where Vout=K•(V2-V1). The amp has a very large gain, K. Same as operational amplifier.

op code, opcode, or **operation code** A specific instruction executed by the computer. The op code along with the operand completely specifies the function to be performed. In assembly language programming, the op code is represented by its mnemonic, like `LDR`. During execution, the op code is stored as a machine code loaded in memory.

open collector A digital logic output that has two states low and HiZ. Same as open drain and wire-or-mode.

open drain A digital logic output that has two states low and HiZ. Same as open collector and wire-or-mode.

open loop control system A control system that does not include sensors to measure the current state variables. An analog system with no feedback paths.

operand The second part of an instruction that specifies either the data or the address for that instruction. An assembly instruction typically has an op code and an operand (e.g., #55). Instructions that use inherent addressing mode have no operand field.

operating system System software for managing computer resources and facilitating common functions like input/output, memory management, and file system.

originate modem the device that places the telephone call.

oscilloscope A hardware debugging tool that allows you to visualize one or two analog signals versus time.

output compare A mechanism to cause a flag to be set and an output pin to change when the timer matches a preset value. The output compare event can also request an interrupt.

output impedance Open circuit output voltage divided by short circuit output current.

overflow An error that occurs when the result of a calculation exceeds the range of the number system. For example, with 8-bit unsigned integers, 200+57 will yield the incorrect result of 1.

overrun error An error that occurs when the receiver gets a new frame but the receive FIFO and shift register already have information.

parallel port A port where all signals are available simultaneously. E.g., Port A is an 8-bit parallel port.

parallel programming A software system that supports two or more programs being executed at the same time. A computer with multiple cores implements parallel programming.

partially asynchronous bus a communication protocol that has a central clock but the memory module can dynamically extend the length of a bus cycle (cycle stretch) if it needs more time

path expression A software technique to guarantee subfunctions within a module are executed in a proper sequence. For example, it forces the user to initialize I/O device before attempting to perform I/O.

PC-relative addressing An addressing mode where the effective address is calculated by its position relative to the current value of the program counter.

performance debugging or profiling The process of acquiring or modifying timing characteristics and execution patterns of a program and the process of instrumenting a program for such purposes is called performance debugging or profiling.

periodic polling A software/hardware synchronization method that is a combination of interrupts and busy wait. An interrupt occurs at a regular rate (periodic) independent of the hardware status. The interrupt handler checks the hardware device (polls) to determine if its operation is complete. The software usually can perform other work while waiting for the hardware.

Personal Area Network (PAN) A connection between computers controlled by a single person or all working toward for a well-defined single task.

phase shift key (PSK) a protocol that encodes the information as phase changes between the sounds.

photosensor A transducer that converts reflected or transmitted light into electric current.

physical plant The physical device being controlled.

PID controller A control system where the actuator output depends on a linear combination of the current error (P), the integral of the error (I) and the derivative of the error (D).

pink noise A fundamental noise in resistive devices arising from fluctuating conductivity. Same as 1/f noise.

pole A place in the frequency domain where the filter gain is infinite.

polling A software function to look and see which of the potential sources requested the interrupt.

port External pins through which the microcontroller can perform input/output. Same as I/O port.
port A port is a software construct. It is part of communications endpoint. A port together with an IP address and the protocol type constitutes a socket. A port is a 16-bit number, commonly known as the port number. Port number 80 is commonly used on servers that provide HTTP web pages.
positive feedback An analog system with positive gain feedback paths. These systems will saturate.
positive logic a signal where the true value has a higher voltage than the false value, in digital logic true is 1 and false is 0, in TTL logic true is greater than 2 volts and false is less than 0.7 volts, in RS232 protocol true is +12 volts and false is -12 volts. Contrast with negative logic.
potentiometer A transducer that converts position into electric resistance.
precision A term specifying the degrees of freedom from random errors. For an input signal, it is the number of distinguishable input signals that can be reliably detected by the measurement. For an output signal, it is the number of different output parameters that can be produced by the system. For a number system, precision is the number of distinct or different values of a number system in units of "alternatives". The precision of a number system is also the number of binary digits required to represent all its numbers in units of "bits".
preemptive scheduler A scheduler that has the power to suspend execution of a thread without the thread's permission.
priority When two requests for service are made simultaneously, priority determines which order to process them.
private Can be accessed only by software modules in that local group.
private variable A variable that is used by a single thread, and not shared with other threads.
process The execution of software that does not necessarily cooperate with other processes.
producer-consumer A multithreaded system where the producers generate new data, and the consumers process or output the data.
profiling See performance debugging.
program counter (PC) A register in the processor that points to the memory containing the instruction to execute next.
PROM Same as EPROM. Programmable read only memory that is nonvolatile and requires external devices to erase and reprogram. It is usually erased using UV light.
promotion Increasing the precision of a number for convenience or to avoid overflow errors during calculations.
pseudo interrupt vector A secondary place for the interrupt vectors for the convenience of the debugger, because the debugger cannot or does not want the user to modify the real interrupt vectors. They provide flexibility for debugging but incur a run time delay during execution.
pseudo op Operations included in the program that are not executed by the computer at run time, but rather are interpreted by the assembler during the assembly process. Same as assembly directive.
pseudo-code A shorthand for describing a software algorithm. The exact format is not defined, but many programmers use their favorite high-level language syntax (like C) without paying rigorous attention to the punctuation.
public Can be accessed by any software module.
public variable A global variable that is shared by multiple programs or threads.
pulse width modulation A technique to deliver a variable signal (voltage, power, energy) using an on/off signal with a variable percentage of time the signal is on (duty cycle). Same as **variable duty cycle**.
Q The Q of a bandpass filter (passes f_{min} to f_{max}) is the center pass frequency ($f_o=(f_{max}+f_{min})/2$) divided by the width of the pass region, $Q=f_o/(f_{max}-f_{min})$. The Q of a bandreject filter (rejects f_{min} to f_{max}) is the center reject frequency ($f_o=(f_{max}+f_{min})/2$) divided by the width of the reject region, $Q=f_o/(f_{max}-f_{min})$.
quadrature amplitude modem (QAM) a protocol that used both the phase and amplitude to encode up to 6 bits onto each baud.

qualitative DAS A DAS that collects information not in the form of numerical values, but rather in the form of the qualitative senses, e.g., sight, hearing, smell, taste and touch. A qualitative DAS may also detect the presence or absence of conditions.

quantitative DAS A DAS that collects information in the form of numerical values.

RAM Random Access Memory, a type of memory where is the information can be stored and retrieved easily and quickly. Since it is volatile the information is lost when power is removed.

range Includes both the smallest possible and the largest possible signal (input or output). The difference between the largest and smallest input that can be measured by the instrument. The units are in the units of the measurand. When precision is in alternatives, range=precision•resolution. Same as span

read cycle data flows from the memory or input device to the processor, the address bus specifies the memory or input device location and the data bus contains the information at that address

read data available The time interval (start,end) during which the data will be valid during a read cycle, determined by the memory module

real-time A characteristic of a system that can guarantee an upper bound (worst case) on latency.

real-time system A system where time-critical operations occur when needed.

recursion A programming technique where a function calls itself.

reduced instruction set computer (RISC) A computer with a few instructions, instructions with fixed lengths, instructions that execute in 1 or 2 bus cycles, only load and store can access memory, no one instruction can both read and write memory, many identical general purpose registers, and a limited number of addressing modes. Contrast to CISC.

reentrant A software module that can be started by one thread, interrupted and executed by a second thread. A reentrant module allow both threads to properly execute the desired function. Contrast with non-reentrant.

registers High-speed memory located in the processor. The registers in the ARM® Cortex-M include R0 through R15.

relay A mechanical switch that can be turned on and off by the computer.

reliability The ability of a system to operate within specified parameters for a stated period of time. Given in terms of mean time between failures (MTBF).

reproducibility (or **repeatability**) A parameter specifying how consistent over time the measurement is when the input remains fixed.

requirements document A formal description of what the system will do in a very complete way, but not including how it will be done. It should be unambiguous, complete, verifiable, and modifiable.

reset vector The 32-bit value at memory locations 0x0000.0004 specifying where the software should start after power is turned on or after a hardware reset.

resistance temperature device (RTD) A linear transducer that converts temperature into electric resistance.

resolution For an input signal, it is the smallest change in the input parameter that can be reliably detected by the measurement. For an output signal, it is the smallest change in the output parameter that can be produced by the system, range equals precision times resolution. The units are in the units of the measurand. When precision is in alternatives, range=precision•resolution.

response time Similar to latency, it is the delay between when the time an event occurs and the time the software responds to the event.

ritual Software, usually executed once at the beginning of the program, that defines the operational modes of the I/O ports.

ROM Read Only Memory, a type of memory where is the information is programmed into the device once, but can be accessed quickly. It is low cost, must be purchased in high volume and can be programmed only once. See also EPROM, EEPROM, and flash EEPROM.

rotor The part of a motor that rotates.

round robin scheduler A scheduler that runs each active thread equally.

roundoff The error that occurs in a fixed-point or floating-point calculation when the least significant bits of an intermediate calculation are discarded so the result can fit into the finite precision.

RTD Resistance temperature device, a sensor used to measure temperature, usually made from platinum

sample and hold A circuit used to latch a rapidly changing analog signal, capturing its input value and holding its output constant.

sampling rate The rate at which data is collected in a data acquisition system.

saturation A device that is no longer sensitive to its inputs when its input goes above a maximum value or below a minimum value.

scan or scanpoint Any instrument used to produce a side effect without causing a break (halt) is a scan. Therefore, a scan may be used to gather data passively or to modify functions of a program. Examples include software added to your source code that simply outputs or modifies a global variable without halting. A scanpoint is triggered in a manner similar to a breakpoint but a scanpoint simply records data at that time without halting execution.

scheduler System software that suspends and launches threads.

Schmitt Trigger A digital interface with hysteresis making it less susceptible to noise.

scope A logic analyzer or an oscilloscope, hardware debugging tools that allows you to visualize multiple digital or analog signals versus time.

select signal The output of the address decoder (each module on the bus has a separate address decoder); a Boolean (true/false) signal specifying whether or not the current address of the bus matches the device address

semaphore A system function with two operations (wait and signal) that provide for thread synchronization and resource sharing.

sensitivity The sensitivity of a transducer is the slope of the output versus input response. The sensitivity of a qualitative DAS that detects events is the percentage of actual events that are properly recognized by the system.

serial communication A process where information is transmitted one bit at a time.

serial communications interface (SCI) A Freescale term for a device to transmit data with asynchronous serial communication protocol (same as UART and ACIA.)

serial peripheral interface (SPI) A device to transmit data with synchronous serial communication protocol. Same as SSI.

serial port An I/O port with which the bits are input or output one at a time.

servo A DC motor with built in controller. The microcontroller specifies desired position and the servo adds/subtracts power to move the shaft to that position.

setup time When latching data into a register with a clock, it is the time before an edge the input must be valid. Contrast with hold time.

shot noise A fundamental noise that occurs in devices that count discrete events.

signed two's complement binary A mechanism to represent signed integers where 1 followed by all 0's is the most negative number, all 1's represents the value -1, all 0's represents the value 0, and 0 followed by all 1's is the largest positive number.

sign-magnitude binary A mechanism to represent signed integers where the most significant bit is set if the number is negative, and the remaining bits represent the magnitude as an unsigned binary.

simplex channel Hardware that allows bits (information, error checking, synchronization or overhead) to transfer only in one direction. Contrast with half duplex and full duplex channels.

simplex communication A system that allows information to transfer only in one direction.

simulator A simulator is a software application that simulates or mimics the operation of a processor or computer system. Most simulators recreate only simple I/O ports and often do not effectively duplicate the real-time interactions of the software/hardware interface. On the other hand, they do provide a simple and interactive mechanism to test software.

single address DMA Direct memory access that requires only one bus cycle to transfer data from an input device into memory, or from memory to an output device.

single-pole relay A simple relay with only one copy of the switch mechanism. Contrast with double pole.

single-pole switch A simple switch with only one copy of the switch mechanism. One switch that acts independent from other switches in the system. Contrast with double-pole.

single-throw switch A switch with two contact connections. The two contacts may be connected or disconnected. Contrast with double-throw.

slew rate The maximum slope of a signal. If the time-varying signal V(t) is in volts, the slew rate is the maximum dV/dt in volts/s.

socket A socket is an application endpoint for communication that encapsulates IP address, the transport protocol and the port.

soft real-time A system that implements best effort to execute critical tasks on time, typically using a priority scheduler. Once a deadline as passed, the value of completing the task diminishes over time. Compare to hard real-time and firm real-time.

software interrupt See trap.

software maintenance Process of verifying, changing, correcting, enhancing, and extending software.

solenoid A discrete motion device (on/off) that can be controlled by the computer usually by activating an electromagnet. For example, electronic door locks on automobiles.

source code Programs in human readable format created with an editor.

space A digital value of false or logic 0. Contrast with mark.

span Same as range.

spatial resolution The volume over which the DAS collects information about the measurand.

specificity The specificity of a transducer is the relative sensitivity of the device to the signal of interest versus the sensitivity of the device to other unwanted signals. The sensitivity of a qualitative DAS that detects events is the percentage of events detected by the system that are actually true events.

spinlock semaphore A semaphore where the threads will spin (run but do no useful function) when they execute wait on a busy semaphore. Contrast to blocking semaphore.

stabilize The debugging process of stabilizing a software system involves specifying all its inputs. When a system is stabilized, the output results are consistently repeatable. Stabilizing a system with multiple real-time events, like input devices and time-dependent conditions, can be difficult to accomplish. It often involves replacing input hardware with sequential reads from an array or disk file.

stack Last in first out data structure located in RAM and used to temporarily save information.

stack pointer (SP) A register in the processor that points to the RAM location of the stack.

start bit An overhead bit(s) specifying the beginning of the frame, used in serial communication to synchronize the receiver shift register with the transmitter clock. See also stop bit, even parity and odd parity.

starvation A condition that occurs with a priority scheduler where low priority threads are never run.

static allocation Data structures such as an FSM or TCB that are defined at assembly or compile time and exist throughout the life of the software. Contrast to dynamic allocation.

static RAM Volatile read/write storage built from three transistors having fast speed, and not requiring refresh. Contrast with dynamic RAM.

stator The part of a motor that remains stationary. Same as frame.

stepper motor A motor that moves in discrete steps.

stop bit An overhead bit(s) specifying the end of the frame, used in serial communication to separate one frame from the next. See also start bit, even parity and odd parity.

strain gauge A transducer that converts displacement into electric resistance, force or pressure.

string A sequence of ASCII characters, usually terminated with a zero.

symbol table A mapping from a symbolic name to its corresponding 16-bit address, generated by the assembler in pass one and displayed in the listing file.

synchronous bus a communication protocol that has a central clock; there is no feedback from the memory to the processor, so every memory cycle takes exactly the same time; data transfers (put data on bus, take data off bus) are synchronized to the central clock

synchronous protocol a system where the two devices share the same clock.

synchronous serial interface (SSI) A device to transmit data with synchronous serial communication protocol. Same as SPI.

tachometer a sensor that measures the revolutions per second of a rotating shaft.

tail recursion A technique where the recursive call occurs as the last action taken by the function. See also recursion, binary recursion, and linear recursion.

thermal noise A fundamental noise in resistive devices arising from the uncertainty about the position and velocity of individual molecules. Same as Johnson noise and white noise.

thermistor A nonlinear transducer that converts temperature into electric resistance.

thermocouple A transducer that converts temperature into electric voltage.

thread The execution of software that cooperates with other threads. A thread embodies the action of the software. One concept describes a thread as the sequence of operations including the input and output data.

thread control block TCB Information about each thread.

three-pole relay Three separate and complete relays, which are activated together (see single pole).

three-pole switch Three separate and complete switches. The switches are electrically separate, but mechanically connected. The three switches turned on and off together (see single pole).

throughput The information transfer rate, the amount of data transferred per second. Same as bandwidth.

time constant The time to reach 63.2% of the final output after the input is instantaneously increased.

time profile and execution profile Time profile refers to the timing characteristic of a program and execution profile refers to the execution pattern of a program.

tolerance The maximum deviation of a parameter from a specified value.

total harmonic distortion (THD) A measure of the harmonic distortion present and is defined as the ratio of the sum of the powers of all harmonic components to the power of the fundamental frequency.

transducer A device that converts one type of signal into another type.

trap A trap is similar to a regular or hardware interrupt: there is a trigger that invokes the execution of an ISR. On the Cortex-M, there are two software interrupts: supervisor call and PendSV. Hardware interrupts are triggered by hardware events, while software interrupts are triggered explicitly by software. To invoke a PendSV, the software sets bit 28 of the NVIC_INT_CTRL_R register. Same as software interrupt.

tristate The state of a tristate logic output when off or not driven.

tristate logic A digital logic device that has three output states low, high, and off (HiZ).

truncation The act of discarding bits as a number is converted from one format to another.

two's complement A number system used to define signed integers. The MSB defines whether the number is negative (1) or positive (0). To negate a two's complement number, one first complements (flip from 0 to 1 or from 1 to 0) each bit, then add 1 to the number.

two-pole relay two separate and complete relays, which are activated together (same as double pole).

two-pole switch Two separate and complete switches. The switches are electrically separate, but mechanically connected. The two switches turned on and off together which are activated together, same as double-pole.

ultrasound A sound with a frequency too high to be heard by humans, typically 40 kHz to 100 MHz.

unbuffered I/O The hardware and software are tightly coupled so that both wait for each other during the transmission of data.

unipolar stepper motor A stepper motor where the current flows in only one direction (on/off) along the interface wires; a stepper with 5 or 6 interface wires.

universal asynchronous receiver/transmitter (UART) A device to transmit data with asynchronous serial communication protocol.

unsigned binary A mechanism to represent unsigned integers where all 0's represents the value 0, and all 1's represents is the largest positive number.

vector A 32-bit address in ROM containing the location of the interrupt service routines. See also reset vector and interrupt vector.

velocity factor (VF) The ratio of the speed at which information travels relative to the speed of light.
vertical parity the normal parity bit calculated on each individual frame, can be even or odd parity
V_{OH} The smallest possible output voltage when the signal is high, and the current is less than I_{OH}.
V_{OL} The largest possible output voltage when the signal is low, and the current is less than I_{OL}.
volatile A condition where information is lost when power is removed.
volatile A property of a variable in C, such that the value of the variable can change outside the immediate scope of the software accessing the variable.
voltage follower An analog circuit with gain equal to 1, large input impedance and small output impedance. Same as follower.
von Neumann architecture A computer architecture where instructions are fetched from the same bus as data are fetched.
vulnerable window Locations within a software module, which if an interrupt were to occur at one of these locations, then an error could occur (e.g., data lost, corrupted data, program crash, etc.) Same as critical section.
white noise A fundamental noise in resistive devices arising from the uncertainty about the position and velocity of individual molecules. Same as Johnson noise and thermal noise.
wire-or-mode A digital logic output that has two states low and HiZ. Same as open collector.
word Four bytes containing 32 bits.
workstation A powerful general purpose computer system having a price in the $3K to 50K range and used for handling large amounts of data and performing many calculations.
write cycle data is sent from the processor to the memory or output device, the address bus specifies the memory or output device location and the data bus contains the information
write data available time interval (start,end) during which the data will be valid during a write cycle, determined by the processor
write data required time interval (start,end) during which the data should be valid during a write cycle, determined by the memory module
XON/XOFF A protocol used by printers to feedback the printer status to the computer. XOFF is sent from the printer to the computer in order to stop data transfer, and XON is sent from the printer to the computer in order to resume data transfer.
Z Transform A transform equation converting a digital time-domain sequence into the frequency domain. In both the time and frequency domain it is assumed the signal is band limited to 0 to ½fs.
zero A place in the frequency domain where the filter gain is zero.

Appendix 2. Solutions to Checkpoints

Checkpoint 1.1: A microcomputer is a small computer that includes a processor, memory, and input/output. A microprocessor is a small processor that includes registers, ALU, control unit and a bus interface unit. A microcontroller is a single chip microcomputer.
Checkpoint 1.2: Flash ROM is higher density because it requires fewer transistors per bit.
Checkpoint 1.3: An input port is hardware that is part of the computer, and it is the channel through which information enters into the computer. An input interface includes hardware components external to the computer, the input port, and software, which all together perform the input function.
Checkpoint 1.4: Typical input devices include the keys on the keyboard, mouse and its buttons, touch pad, DVD reader, and microphone. USB drives, Ethernet, and wireless can be used for input and output.
Checkpoint 1.5: Typical output devices include the LEDs on the keyboard, monitor, speaker, printer, CD burner, and speaker. The floppy disk can be used for input and output.
Checkpoint 1.6: CU (control unit) BIU (bus interface unit) and ALU (arithmetic logic unit) are all part of the processor. DMA stands for direct memory access, which is a high-speed mechanism to move data directly from input to memory, or move data directly from memory to output.

Checkpoint 1.7: An embedded system is a microcomputer with mechanical, chemical and electrical devices attached to it, programmed for a specific dedicated purpose, and packaged up as a complete system.
Checkpoint 1.8: We store temporary information like pictures, sound recordings, and text messages.
Checkpoint 1.9: We store nonvolatile information like programs, address book, music, and calendar.
Checkpoint 1.10: The software in the alarm clock must maintain time using a real-time clock, output the current time on the display, respond to button pushes updating parameters as required, check and see if the current time matches the alarm time.
Checkpoint 1.11: A requirement is a detailed performance parameter that the system must satisfy, generally derived from the overall objective of the system. A constraint is a condition defining how the system will be developed, generally restricting the range of solutions from which the system will be built.
Checkpoint 1.12: If two modules output to the same port, then the second module will undo the function of the first one. For example if one module says "go" and the other one says "stop", then the order of execution determines the resulting function. A similar error can occur for input ports.
Checkpoint 1.13: When I start the wait, I observe the current time. Next, I add t to this time, calling it *endTime*. Lastly, I wait for the time on the wait to pass *endTime*.
Checkpoint 1.14: 0 and 1V are considered low because they are less than V_{IL}. 2V is not defined. 3, 4, and 5V are considered high because they are more than V_{IH}. TM4C inputs are 5V tolerant.
Checkpoint 1.15: We must check four inequalities. Yes, V_{OL}(0.5V) is less than V_{IL}(0.8V), V_{OH}(4.4V) is greater than V_{IH}(2V), I_{OH}(4mA) is greater than I_{IH}(20μA), and I_{OL}(4mA) is greater than I_{IL}(0.4mA).
Checkpoint 1.16: No, only three of the four inequalities are true. V_{OL}(0.4V) is less than V_{IL}(1.5V), V_{OH}(2.4V) is *NOT* greater than V_{IH}(3.5V), I_{OH}(0.4mA) is greater than I_{IH}(1μA), and I_{OL}(4mA) is greater than I_{IL}(1μA).
Checkpoint 1.17: The correct resistor value is (3.3-2-0.5V)/20mA = 40Ω.
Checkpoint 1.18: 2½ decimal digits is 200 alternatives, which is about 8 bits.
Checkpoint 1.19: 10 binary bits is 1024 alternatives, which is about 3 decimal digits.
Checkpoint 1.20: 6½ decimal digits is 2,000,000 alternatives, which is about 21 bits.
Checkpoint 1.21: The rule of thumb says 2^{60} is about 10^{18}, which is 18 decimal digits. 2^4 is 16, which is about 1½ decimal digits. Together, we have 19½ decimal digits.
Checkpoint 1.22: First, break into nibbles 1110, 1110, 1011, then convert each, 0xEEB.
Checkpoint 1.23: First, convert each hex digit one at a time 0011 1000 0000 0000, then combine to get 2_0011100000000000.
Checkpoint 1.24: Each hex digit needs 4 bits, so a total of 20 bits will be required
Checkpoint 1.25: 2_01101010 equals 64+32+8+2=106.
Checkpoint 1.26: 0x23 equals 2*16+3 = 35.
Checkpoint 1.27: 37 equals 32+4+1 = 2_00100101 = 0x25.
Checkpoint 1.28: 202 equals 128+64+8+2 = 2_11001010 = 0xCA.
Checkpoint 1.29: They are the same, both equally 53 (3*16+5=.48+5)
Checkpoint 1.30: -31 equals –128+64+32+1 = 2_11100001 = 0xE1.
Checkpoint 1.31: The character '0' is represented in ASCII as 0x30.
Checkpoint 1.32: Converting each character to ASCII yields "48656C6C6F20576F726C6400"
Checkpoint 1.33: The integer part ranges from -32768 to +32767, so the value of the fixed-point number ranges from -32.768 to +32.767.
Checkpoint 1.34: π is about 3142, with a resolution of 0.001.
Checkpoint 1.35: π is about 804, with a resolution of 1/256.
Checkpoint 1.36: Using a calculator, we find √5-(161/72) is -0.000043. √5-(682/305) is 0.000002.
Checkpoint 1.37: y = (1000•x-53•x_1+1000•x_2+51•y_1-903•y_2)/1000.
Checkpoint 2.1: 0x2200.0000 + 32*n + 4*b=0x2200.0000 + 32*0x1003 +4*5 =0x2200.0000 +0x20060 + 0x14 = 0x2202.0074.
Checkpoint 2.2: 0x2200.0000 + 32*n + 4*b=0x2200.0000 + 32*0x1000+4*20 =0x2200.0000 +0x20000 + 0x50 = 0x2202.0050.

Checkpoint 2.3: 0x4200.0000 + 32*n + 4*b=0x4200.0000 + 32*3 +4*2 =0x4200.0000 +0x00060 + 0x08 = 0x4200.0068.
Checkpoint 2.4: The addressing mode specifies where the instruction will read or write data.
Checkpoint 2.5: It will access the 32-bit value at 0x2000.0008. R3 is not updated.
Checkpoint 2.6: It will access the 32-bit value at 0x2000.0000, then R3 is updated to 0x2000.0008.
Checkpoint 2.7: In C, change the `int32_t` to `int16_t`. In assembly, change 40 to 20, and `LSL #2` to `LSL #1`.
Checkpoint 2.8: Current flows into the output across the transistor to ground. In other words, the open collector gate can sink current to ground but not source current from the power supply.
Checkpoint 2.9: The sum of two unsigned 32-bit numbers is a 33-bit unsigned number. Consider smallest and largest values. 0+0=0, and $(2^{32}-1)+(2^{32}-1) = (2^{33}-2)$.
Checkpoint 2.10: The sum of two signed 32-bit numbers is a 33-bit signed number. Consider smallest and largest values. $(-2^{31})+(-2^{31}) = (-2^{32})$, and $(2^{31}-1)+(2^{31}-1) = (2^{32}-2)$.
Checkpoint 2.11: The discontinuity is between -128 and +127, which are adjacent on the wheel.
Checkpoint 2.12: The product of two unsigned 32-bit numbers is a 64-bit unsigned number. Consider smallest and largest values. 0*0=0, and $(2^{32}-1)*(2^{32}-1) = (2^{64}-2^{33}+1)$.
Checkpoint 2.13: The product of two signed 32-bit numbers is a 64-bit signed number. Consider smallest and largest products. $(-2^{31})* (2^{31}-1) = (-2^{62}-2^{32})$, and $(-2^{31})*(-2^{31}) = (2^{62})$. (2^{62}) will not fit into a signed 63-bit number.
Checkpoint 2.14: No, because the product goes into 64 bits. See Checkpoints 2.12 and 2.13.
Checkpoint 2.15: The conditional **BHS** is used with unsigned comparisons and **BGE** with signed.
Checkpoint 2.16: Nothing happens if the software writes to an input port.
Checkpoint 2.17: If the software reads this output port it gets the values on the pins. For example, if the user mistakenly grounded the output pin (very bad thing to do), and the software writes a '1'; when it reads it will get '0'.
Checkpoint 2.18: Since there are 8 bits in a port and 8 bits in the direction register, each bit can be individually programmed as input or output.
Checkpoint 2.19: See Table 2.4 for the TM4C123, with PCTL = 0001, PD5 is UART6 transmitter. See Table 2.5 for the MSP432E401Y, with PCTL = 0001, PD5 is UART2 transmitter.
Checkpoint 2.20: Nothing happens. Since none of the address bits are selected, none of the port bits are affected.
Checkpoint 2.21: The base address for Port D is 0x4000.7000.
```
#define PD72   (*((volatile uint32_t *)0x40007210))
     PD72 = 0x84; // sets PD7 PD2, other 6 bits are not affected
```
Checkpoint 2.22: The base address for Port B is 0x4000.5000.
```
#define PB650 (*((volatile uint32_t *)0x40005184))
     PB650 = 0x61; // sets PB6 PB5 PB0, other 5 bits are not affected
```
Checkpoint 2.23: It will still operate according to specifications, but it may be more expensive to build or it may be harder to order components to build it.
Checkpoint 2.24: It will no longer operate according to specifications.
Checkpoint 2.25: Change the specification from 6 MHz to 8 MHz. Change the line
```
   SYSCTL_RCC_R +=   0x00000540;    // 10101, configure for 16 MHz crystal
to SYSCTL_RCC_R +=   0x00000380;    // 01110, configure for 8 MHz crystal
```
Change the line (400/8 is 50 MHz, so SYSDIV2 is 7)
```
   SYSCTL_RCC2_R += (4<<22);        // configure for 80 MHz clock
to SYSCTL_RCC2_R += (7<<22);        // configure for 50 MHz clock
```
Checkpoint 2.26: Change 60000 to 80000.
Checkpoint 3.1: The policy rewards bad behavior. A better policy would be to reward good behavior.

Checkpoint 3.2: A private function has no underline in the name (e.g., `OutNibble`). A public function has the module name, underline, then operation in the name (e.g., `FIFO_Put`).

Checkpoint 3.3: A local variable begins with a lower case letter. A global variable begins with an upper case letter.

Checkpoint 3.4: A local variable is created by defining it after an open brace { but before any executable code. E.g., this routine as three local variables `littleVar mediumVar largeVar`
```
void Process(void){
  int8_t littleVar;
  int16_t mediumVar;
  int32_t largeVar;
```

Checkpoint 3.5: A global variable is created by defining it outside the function. E.g., this program as three global variables `LittleVar MediumVar LargeVar`
```
  int8_t LittleVar;
  int16_t MediumVar;
  int32_t LargeVar;
void Process(void){
```
You can also define global variables inside the function using `static`. The scope of static variables is restricted to the function. E.g.,
```
void Process(void){
  static int8_t LittleVar;
  static int16_t MediumVar;
  static int32_t LargeVar;
```

Checkpoint 3.6: `static` changes a local variable from temporary to permanent allocation. `static` does not change the scope of a local variable. `static` changes a global variable from public to private scope (this file only). `static` does not change the allocation of a global variable. `static` changes a function from public to private scope (this file only). It does not change how the function runs or how it is stored.

Checkpoint 3.7: `const` changes a global variable from RAM allocation to ROM allocation. `const` does not change the scope of a variable.

Checkpoint 3.8: In a Moore FSM, the output depends only on the state. In a Mealy FSM, the output depends on the state and the input.

Checkpoint 3.9: Redefine these parameters
```
#define BSP_InPort          GPIO_PORTA_DATA_R
#define BSP_InPort_DIR      GPIO_PORTA_DIR_R
#define BSP_InPort_DEN      GPIO_PORTA_DEN_R
#define BSP_OutPort         GPIO_PORTB_DATA_R
#define BSP_OutPort_DIR     GPIO_PORTB_DIR_R
#define BSP_OutPort_DEN     GPIO_PORTB_DEN_R
#define BSP_GPIO_EN         SYSCTL_RCGCGPIO_R
#define BSP_InPort_Mask     0x00000001  // bit mask for Port A
#define BSP_In_M            0x00000030  // bit mask for pins 5,4
#define BSP_In_Shift        0x00000004  // shift value for Input pins
#define BSP_OutPort_Mask    0x00000002  // bit mask for Port B
#define BSP_Out_M           0x0000006E  // bit mask for pins 6-1
#define BSP_Out_Shift       0x00000001  // shift value for Output pins
```

Checkpoint 3.10: It will first sit up, then it will stand up.

Checkpoint 3.11:
```
            AddIndexFifo(CAN1, 100, char, 0, 1)
            AddIndexFifo(CAN2, 100, char, 0, 1)
            AddIndexFifo(CAN3, 100, char, 0, 1)
```

Checkpoint 3.12:   `AddIndexFifo(F1, 256, int16_t, 0, 1)`
    `AddIndexFifo(F2, 256, int16_t, 0, 1)`
Checkpoint 3.13: Search the free list to see if the address `&Heap[SIZE*i]` is free.
Checkpoint 3.14: Ignore size parameter, return 100 bytes regardless of the request.
Checkpoint 3.15: A runtime flag can be turned on during a checkup procedure while the device is in service. A compile time flag removes all debugging so the software system is faster and smaller.
Checkpoint 3.16: There are three advantages of leaving the instruments. First, the system was tested with the instruments and works to specification with the instruments. There is no guarantee the system will still work if the instruments are removed. Second, the instruments could provide run time checks to catch failures during operation. Third, the instruments could be used during system checkup (recalibration, diagnostic checkup etc.)
Checkpoint 3.17: The existence of the instrument has a small but inconsequential effect on the system performance. The time to execute the instrument is small compared to the time between executions of the instrument.
Checkpoint 3.18: The base address of Port A is 0x4000.0000. $4*2^3 = 0x20$.
`#define GPIO_PORTA3 (*((volatile uint32_t *)0x40000020))`
`#define Debug_HeartBeat() (GPIO_PORTA3 ^= 0x08)`
Checkpoint 3.19: Each output is about 3 instructions, so I expect it to be 4 to 6 bus cycles wide. This time delay explains why Program 3.18 had a measurement value 6 cycles larger than Program 3.17 (244-238).
Checkpoint 4.1: Each output is about 3 instructions, so I expect it to be 4 to 6 bus cycles wide.
Checkpoint 4.2: Negative logic means when we touch the switch the voltage goes to 0 (low). Formally, negative logic means the true voltage is lower than the false voltage. Positive logic means when we touch the switch the voltage goes to +3.3 (high). Formally, positive logic means the true voltage is higher than the false voltage.
Checkpoint 4.3: For PA2, we need input with pull-up. DIR bit 2 is low (input), AFSEL bit 2 is low (not alternate), PUE bit 2 high (pull-up) and PDE bit 2 low (not pull-down). For PA3, we need input with pull-down. DIR bit 3 is low (input), AFSEL bit 3 is low (not alternate), PUE bit 3 low (no pull-up) and PDE bit 3 high (pull-down).
Checkpoint 4.4: Increasing voltage will increase torque. To increase current you would have to replace the motor with a motor having lower resistance.
Checkpoint 4.5: No, if you know the initial position, you step slow enough, and the system does not load the motor. Yes if the you do not know initial position, you step too fast, or if you load the motor.
Checkpoint 4.6: If Out goes first, Out goes 1, 2, 3, and waits on 5, In goes 2, 3. Now both are running (Out runs 6, and In runs 4, 5, and waits on 6). After Out runs 6, In finishes 6.
If In goes first, In waits on 2. Then the steps are the same as above.
Checkpoint 4.7: There is 1 byte of data per 10 bits of transmission. So, there are 960 bytes/sec.
Checkpoint 4.8: divider = 1010.010100_2. or 10 and 20/64 = 10.3125. The baud rate is 6MHz/10.3125/16 which is 36.36 kHz.
Checkpoint 4.9: 50,000,000/38400/16 is 81.3802, which is similar to 81 and 24/64. `UART0_IBRD_R` is 81 `UART0_FBRD_R` is 24. The baud rate is 50MHz/(81+24/64)/16 which is 38402 bits/sec.
Checkpoint 4.10: RXFE is set and cleared by hardware. It means receive FIFO empty. To make it 0 means to put data into the FIFO. Software cannot clear this flag. An incoming UART frame will clear RXFE.
Checkpoint 4.11: TXFF is set and cleared by hardware. It means transmit FIFO full. To make it 0 means to get data from the FIFO. Software cannot clear this flag. An outgoing UART frame will clear TXFF.
Checkpoint 4.12: The data will be received in error (values will not be correct). The receiver could appear to get two input frames for every one frame transmitted. It will probably cause framing errors (FE). It would cause parity errors if active.

Checkpoint 4.13: The data will be received in error (values will not be correct). The receiver will appear to get one input frame for every one frame transmitted. It will probably not cause framing errors (FE). It would cause parity errors if active.

Checkpoint 5.1: Trigger flag set by hardware; the device is armed by software; the device is enabled for interrupts in the NVIC; the processor is enabled for interrupts (PRIMASK I bit is clear).

Checkpoint 5.2: The processor is enabled for interrupts by clearing the I bit in the PRIMASK. Execute
```
      CPSIE  I
```
Checkpoint 5.3: Instruction is finished; registers R0–R3, R12, LR, PC, and PSR are pushed; PC is set with interrupt vector address; LR = 0xFFFFFFF9. The last two steps can occur in either order.

Checkpoint 5.4: No, if the average producer rate exceeds the average consumer rate, the FIFO will always fill regardless of size. However, if the average producer rate is less than the average consumer rate, the FIFO full errors can be eliminated by increasing size.

Checkpoint 5.5: At the end of the inner nested program, interrupts would be enabled. So, the last part of the outer section would be running with interrupts enabled. In this example `Stuff2B` runs with interrupts enabled.

```
Critical1                      Critical2
  Disable                        Disable
  Stuff1A                        Stuff2A
  Call Critical2                 Enable
  Stuff1B                        return
  Enable
  return
```

Checkpoint 5.6: The RxFifo is empty when there is no input data. Software is waiting for hardware.
Checkpoint 5.7: The TxFifo is empty when there is no output data. Hardware is waiting for software.
Checkpoint 6.1: An input capture event occurs on the selected edge on an input pin.
Checkpoint 6.2: Timer value is copied into the input capture latch, the flag is set, and if armed an interrupt is requested.
Checkpoint 6.3: Switching pins from PB6 to PB2 means switching from Timer0 to Timer3.

```
void TimerCapture_Init(void){
  SYSCTL_RCGCTIMER_R |= 0x08;        // activate timer3
  SYSCTL_RCGCGPIO_R  |= 0x00000002;  // activate port B
  Count = 0;                         // allow time to finish activating
  GPIO_PORTB_DEN_R |= 0x04;          // enable digital I/O on PB2
  GPIO_PORTB_AFSEL_R |= 0x04;        // enable alt funct on PB2
  GPIO_PORTB_PCTL_R = (GPIO_PORTB_PCTL_R&0xFFFFF0FF)+0x00000700;
  TIMER3_CTL_R &= ~0x00000001;       // disable timer3A during setup
  TIMER3_CFG_R = 0x00000004;         // configure for 16-bit timer mode
  TIMER3_TAMR_R = 0x00000007;        // configure for input capture mode
  TIMER3_CTL_R &= ~(0x000C);         // TAEVENT is rising edge
  TIMER3_TAILR_R = 0x0000FFFF;       // start value
  TIMER3_IMR_R |= 0x00000004;        // enable capture match interrupt
  TIMER3_ICR_R = 0x00000004;         // clear timer0A capture flag
  TIMER3_CTL_R |= 0x00000001;        // enable timer0A
  NVIC_PRI8_R =(NVIC_PRI8_R&0x00FFFFFF)|0x40000000; // Timer3A=priority 2
  NVIC_EN1_R = 0x00000008;           // enable interrupt 35 in NVIC
  EnableInterrupts();}
void Timer3A_Handler(void){
  TIMER3_ICR_R = 0x00000004;         // acknowledge timer3A capture match
  Count = Count + 1;}
```

Checkpoint 6.4: Switching microcontrollers means changing the I/O pin. On the MSP432E401Y, one of the T0CCP0 choices is PD0 (other choices would have been PA0 or PL4). See Table 2.5.

```
void TimerCapture_Init(void){
  SYSCTL_RCGCTIMER_R |= 0x01;       // activate timer0
  SYSCTL_RCGCGPIO_R |= 0x00000008;  // activate port D
  Count = 0;                         // allow time to finish activating
  GPIO_PORTD_DEN_R |= 0x01;          // enable digital I/O on PD0
  GPIO_PORTD_AFSEL_R |= 0x01;        // enable alt funct on PD0
  GPIO_PORTD_PCTL_R = (GPIO_PORTD_PCTL_R&0xF0FFFFFF)+0x00000003;
  TIMER0_CTL_R &= ~0x00000001;       // disable timer0A during setup
  TIMER0_CFG_R = 0x00000004;         // configure for 16-bit timer mode
  TIMER0_TAMR_R = 0x00000007;        // configure for input capture mode
  TIMER0_CTL_R &= ~(0x000C);         // TAEVENT is rising edge
  TIMER0_TAILR_R = 0x0000FFFF;       // start value
  TIMER0_IMR_R |= 0x00000004;        // enable capture match interrupt
  TIMER0_ICR_R = 0x00000004;         // clear timer0A capture flag
  TIMER0_CTL_R |= 0x00000001;        // enable timer0A
  NVIC_PRI4_R =(NVIC_PRI4_R&0x00FFFFFF)|0x40000000; // Timer0A=priority 2
  NVIC_EN0_R = 0x00080000;           // enable interrupt 19 in NVIC
  EnableInterrupts();}
void Timer0A_Handler(void){
  TIMER0_ICR_R = 0x00000004;         // acknowledge timer0A capture match
  Count = Count + 1;}
```

Checkpoint 6.5: `TIMER1_ICR_R = TIMER_ICR_CBECINT; // bit10 =0x0400`

Checkpoint 6.6: One-shot/periodic: on the cycle when the timer counts down to 0.

Checkpoint 6.7: One-shot/periodic: TnTORIS bit is set, interrupt may be requested, the output pin is toggled, and ADC conversion may be triggered.

Checkpoint 6.8: PWM: on the cycle when the timer equals the value in the Match Register or the Interval Load Register.

Checkpoint 6.9: PWM: output pin cleared (set if inverting mode) on match or set (cleared if inverting mode) on reload.

Checkpoint 6.10: The V_{OL} of the 7406 at 40 mA will be 0.7V, so there will be 4.3V across the coil.

Checkpoint 7.1: There is always 1 start bit. So with 8 data bits and one stop, there will be 10 total bits.

Checkpoint 7.2: Full duplex is both directions at the same time, while half duplex is both directions, but only one direction at a time.

Checkpoint 7.3: With synchronous serial (SSI), the transmitter and receiver operate on the same clock, which is included in the cable. With asynchronous serial (UART), the transmitter and receiver operate with clocks of similar frequencies, and the receiver uses transitions in the data to synchronize with the transmitter.

Checkpoint 7.4: The MAX3232 allows the signal to travel farther and be less sensitive to noise.

Checkpoint 7.5: Capacitance reduces the slew rate of the signals causing a reduction in the maximum possible baud rate. The end result is capacitance reduces the bandwidth of the channel.

Checkpoint 7.6: Setup time is the time before a clock edge the input data must be valid. Hold time is the time after a clock edge the data must continue to be valid.

Checkpoint 7.7: 80/4 is 20 MHz<25 MHz. Change `SSI0_CPSR_R` from a 2 to a 4.

Checkpoint 7.8: The lower the resistance the higher the I_{OH}. The higher the current the faster the rise time. A faster rise time means the bus operates faster.

Checkpoint 7.9: Open drain is the same as open collector. There are two output states float and low. Formally, open drain means the drain pin of the N-channel MOS transistor is open. See Figure 1.20.

Checkpoint 7.10: If no device sends an acknowledgement, the SDA signal will float high, generating a negative acknowledgement.

Checkpoint 7.11: Half duplex because communication goes both directions, but only one way at a time.

Checkpoint 7.12: Master 2 wins because the 0 in address bit 4 dominates over the 1.

Checkpoint 7.13: If they both send the same address and the same sequence of data bits, both will finish without getting a lost-arbitration error. If they both send the same address and but different data values, an arbitration will occur during the data transfer, and the master with the smaller data value will win arbitration.

Checkpoint 8.1: Speed requires power. The power goes up as bandwidth increases.

Checkpoint 8.2: Output impedance is open circuit voltage divided by short circuit current. 3.3V/30mA is 110Ω.

Checkpoint 8.3: The offset voltage is 0.5 mV. The error will be 0.5 mV times 100, which will be 50 mV.

Checkpoint 8.4: The offset voltage is 3 mV. The error is 3 mV times 1000, which would be 3 V.

Checkpoint 8.5. The gain-bandwidth product of the MAX494 is 500 kHz. The bandwidth of the circuit will be 500 kHz divided by the gain, which will be 5 kHz. The noise density of the MAX494 is 25 nV/√Hz. The noise will be 25 nV*sqrt(5000), which is about 1.8 µV.

Checkpoint 8.6. The input impedance of the op amp determines the input impedance of a noninverting amplifier, which for the OPA227 is 10 MΩ.

Checkpoint 8.7. 1*4096/3 = 1365 or 1*4095/3 = 1365.

Checkpoint 9.1: This is a one dimensional search of at most 24 possibilities. Choose C = 1.6nF, so $1/(2\pi*100e3*1.6e-9) = 995$ Hz..

Checkpoint 9.2: This is a two dimensional search of at most ½ 24^2 possibilities. Choose R_1 = 47 kΩ and R2 = 82 kΩ.

Checkpoint 9.3: A switching or buck-boost regulator can increase voltage, e.g., TPS63002.

Checkpoint 9.4: Average current is 500mA-hr/5years*(1day/24hour)*(1year/365days) = 20 µA.

Checkpoint 9.5: Energy storage is 20 µA*1year*(24hour/1day)* (365days/1year) = 175 mA-hr. So, choose the GM041842 that has 250 mA-hr.

Checkpoint 9.6: There is a JTAG debugger interface, a logic analyzer interface, and there are test points.

Checkpoint 10.1: Set 10=1000/2k, so k=50. For every ADC sample, calculate the average of the previous 50 samples.

Checkpoint 10.2: If the system is linear, the precisions are equal.

Checkpoint 10.3: Since the response in Figure 10.29 is approximately linear, the resolution will be 10°C/256, which is about 0.04 °C.

Checkpoint 10.4: At 60 BPM, the `Rcount` will be 4 and the difference `Rlast-Rfirst` will be 480. If the heart rate increases a little, then the difference will reduce to 479, giving a measurement of 60.13. Thus the inherent measurement resolution is about 0.12 BPM. Because integer math is used, the system will have a resolution of 1 BPM.

Checkpoint 11.1: Since information is encoded as energy, and data is transferred at a fixed rate, each energy packet will exist for a finite time. Energy per time is power.

Checkpoint 11.2: The performance measure for a storage system is information density in bits/cm^3.

Checkpoint 11.3: With open collector outputs, the low will dominate over HiZ. The signal will be low.

Checkpoint 11.4: On average, it will take N/2 transmissions for the message to go from one computer to another. There are 10 bits/frame, so there are 10,000 bytes/sec. Because there are 10 bytes/message, it takes 1ms to transmit a message. Because it has to be sent 5 times, it takes 5ms on average.

Checkpoint 11.5: The frame sent by a transmitter is echoed to its own receiver. If the data does not match, or if there are any framing or noise errors then a collision occurred.

Checkpoint 11.6: Parity could be used to detect collisions. Also the message could have checksum added. Framing or noise errors can also indicate a collision.

Index

`#define` 128
`#endif` 131
`#ifndef` 131
& operation 88
^ operation 88
| operation 88
<< operation 91
>> operation 91
1/f noise 390, 488, 546
100BASE-TX 529
10BASE-T 529
16-bit number 46, 47, 48
2N2222 .... 38, 108, 316, 323, 584
2N2907 584
2's complement binary 424
555 hardware timer 300, 505
5V-tolerant 31
60 Hz noise 546
64-bit 96
6N139 407
7405 89
7406 89
7407 352
74HC04 .... 34, 203, 205
74HC05 35
74HC123 302
74HC244 .... 90, 202, 205
74HC374 .... 90, 202, 206
74HC595 362
74LS04 33
74LS05 35
8-bit number 43
AAPCS 65, 100
ABI 65
abstraction 20, 142, 150, 151, 155, 529
AC coupling 490
acceleration 221, 478, 479
accelerometer 360
accuracy 464, 546
acknowledge 546
acknowledgement 253
acquisition time 428
active 546
active stack pointer selection 70
actuator 329, 546
AD627 .... 406, 502
ADC 49, 72, 423, 463, 546
ADC linear 424
ADC precision 423

ADC range 423
ADC resolution 423
ADC sampling 430
ADC speed 425
`ADC_EMUX_R` 429
`ADC_ISC_R` 430
`ADC_SSMUX3_R` 430
`ADC3_Handler` 432, 507
`ADD` 82, 84, 94
adder circuit 95
addition 93
address decoder 546
addressing mode 83
ADXL362 360
`AFSEL` 211, 214
aliasing 412, 482, 485, 546
`ALIGN` 101
aligned access 85
alkaline battery 448
allocate local variable 100
allocation 128, 143
alternatives 40
ALU 8
amplifier 481
analog 4
analog filter 481, 484
analog latch 428
analog low pass filter 409
analog multiplexer 481
analog signal 411
analog to digital converter 49
`AND` 88
and operation 88
Anti-Reset-Windup 546
aperture time 428
API 151
Application Binary Interface 65
application program interface 151
application programmer interface 15
application specific integrated circuit 201
approximate 51
APSR 65
`AREA` 101
arithmetic logic unit 8
arithmetic shift right 91
arm 252, 546
ARM Architecture Procedure Call Standard 65
ARM® Cortex-M processor 64
armature 320, 475, 546

ASCII 45
ASIC 201
ASPSEL 70
ASR 92
assembly language 81
assembly line 208
`assert()` 130
astable multivibrator 300, 505
asynchronous bus 546
asynchronous communication system 341
asynchronous protocol 546
atomic 254, 264, 546
attenuation 515
autoinitialization 547
average accuracy 464
average accuracy of reading 464
average accuracy or full scale 464
B instruction 96
back emf 316
background 168
background thread 253
bandwidth 6, 196, 227, 232, 339, 396, 514, 521, 547
bandwidth coupling 148
bang-bang 471, 547
bank-switched memory 547
BASEPRI 66, 214, 252
basis 44
battery 447
battery charging circuits 449
baud rate 232, 238, 342
BCC 547
bias current 395
BIC 88
bi-directional 547
biendian 46
big-endian 46
binary 40, 43
binary fixed-point 49
binary full adder 95
binary operation 88
binary semaphore 547
biopotentials 501
BIOS 152
bipolar junction transistor 36, 316
bipolar stepper motor 222
bit time 232, 339
bit-banding 67
    Ports 106, 108
BIU 7
BJT 36, 316
BL instruction 96
BLDC 326
blind-cycle 198, 215, 221, 547
block correction code 547
blocked 547
blocking semaphore 547
BLX instruction 96
board support package 150
Boolean monitor 187
Booster Packs 76
bottom-up design 28, 147
bounce 281, 283
break 548
break before make 548
break error 235
breakdown 471, 548
breakpoint 185
    conditional 185
bridge 495
Brownian motion 486
brushed DC motor 325
brushless DC motor 326
BSP 150, 160
BSS138 352
buck-boost regulator 443
buffered I/O 260, 261, 548
burn 548
bus 4, 64, 206
bus arbitration 204
bus fault 69, 105
bus interface unit 7
BUSY protocol 344
busy wait 196, 199, 227, 239
Butterworth high-pass filter 410
Butterworth low-pass filter 410
BX instruction 82, 96
bypass capacitors 391
byte 43
C bit 65, 94
C0G 391, 410
CAERIS 295
calibration drift 464, 497
call graph 21
CamelCase 139
CAN 72, 76
capacitance 283
capacitive coupling 490
capacitive loading 30
capacitor 389
carbon composition 390
carry 94
causal 17, 548
causal dependency 263
CC2500 521
CC3100 534
CCP0 292
Central Limit Theorem 483
ceramic capacitors 391
channel 513
char 44

checksum ... 525
CISC ... 9, 549
Clear to Send ... 344
clients ... 124
closed loop control system ... 329, 549
CMN ... 94
CMOS ... 30, 34
CMP ... 94
CMRR ... 398, 475
code space ... 84
coefficient of variation ... 465
cohesion
    coincidental ... 149
    communicational ... 149
    functional ... 149
    logical ... 149
    procedural ... 149
    sequential ... 149
    temporal ... 149
cohesion ... 149
coil interface ... 316
coincidental cohesion ... 149
comment field ... 82
comments ... 140
common anode ... 549
common cathode ... 549
common mode input impedance ... 394, 398
common mode rejection ratio ... 398, 405
communication ... 513
communicational cohesion ... 149
compiler ... 549
complement ... 88
complementary binary ... 424
complementary metal-oxide semiconductor ... 30
complex instruction set computer ... 9
complexity abstraction ... 142
concurrent ... 228, 264
concurrent programming ... 26
condition code ... 94
conditional branch ... 96
const ... 128
const global ... 145
constant ... 128
constraint ... 19
context switch ... 253
CONTROL ... 70
control coupling ... 148
control system ... 329, 463
control unit ... 7
cooperative multi-tasking ... 549
corner case ... 184
counting semaphore ... 256, 549
coupling ... 146
    bandwidth ... 148
    control ... 148

invocation ... 148
covenant not to compete ... 57
coverage ... 184
coworkers ... 124
CPS ... 10
CPU bound ... 197, 260, 261, 549
crest factor ... 487
crisp input ... 549
crisp output ... 549
critical section ... 263, 549
cross-assembler ... 549
cross-compiler ... 22, 549
CSIC ... 201
CTS ... 344
CU ... 7
current sense amplifier ... 453
customer specific integrated circuit ... 201
cyber-physical system ... 10
cycle stretch ... 550
D flip-flop ... 89
DAC ... 412, 463, 550
DAC precision ... 412
DAC PWM ... 422
DAC range ... 412
DAC resolution ... 412
Darlington transistor ... 316
data acquisition system ... 550
data available interval ... 203
Data Carrier Detect ... 344
data communication equipment ... 233, 336
data flow graph ... 20
data flow problem ... 169, 275
data register ... 103
data required interval ... 203
Data Set Ready ... 344
data space ... 84, 85
data terminal equipment ... 233, 336
Data Terminal Ready ... 344
DB25 ... 233, 342
DB9 ... 233
dB_{FS} ... 485
DC motor ... 157, 322, 325
    selection ... 323
    speed ... 323
    torque ... 323
DCB ... 101
DCD ... 101, 268, 344
DCE ... 336
DCode bus ... 4, 64
DCW ... 101
dead zone ... 471, 550
deadlock ... 550
deallocate local variable ... 100
debounce ... 281, 283
debugger window ... 556

debugging ... 182
  functional ... 185
debugging instrument ... 182
decibel ... 550
decibels full scale ... 485
decimal digits ... 41
decimal fixed-point ... 49
deep sleep ... 451
defuzzification ... 550
DEN ... 211, 214
denormalized ... 54
derivative ... 434, 479
design ... 21
design for test ... 456
desk check ... 550
desk-checking ... 183
deterministic ... 210
deterministic model ... 208
device driver ... 15, 151, 153, 550
device latency ... 253
differential ... 466
differential input impedance ... 394
differential mode input impedance ... 398
digital filter ... 434, 480, 481
digital logic ... 29
digital low pass filter ... 435
digital output ... 214
digital signal processing ... 550
digital to analog converter ... 358
DIR ... 211
direct interface ... 241
direct memory access ... 6, 200, 550
direction bit ... 102
direction register ... 102, 103, 550
disable ... 252
`DisableInterrupts` ... 266
disarm ... 252, 550
Discrete Fourier Transform ... 484
distance ... 473
distributed ... 526
distributed programming ... 26
distributed systems ... 515
division ... 93
DMA ... 6, 200, 550
DNS ... 530
Domain Name System ... 530
double-pole switch ... 550
double-throw ... 320
double-throw switch ... 551
download ... 551
droop rate ... 428
drop-out ... 50
DRV2, **DRV4**, **DRV8** ... 214
dry run ... 550
DS1307 ... 388

DS1631A ... 388
DSR ... 344
DTE ... 336
DTR ... 344
dual port memory ... 551
dumps ... 183, 456
duty cycle ... 322, 551
dynamic allocation ... 551
dynamic efficiency ... 126
dynamic RAM ... 551
dynamic range ... 425
E12 ... 440
E192 ... 440
E24 ... 440
EAR ... 7
edge-triggered interfacing ... 211
EEPROM ... 3, 164, 551
effective address ... 87
effective address register ... 7
effective number of bits ... 425
efficiency of a regulator ... 442
EIA-561 ... 342
EIA-574 ... 233, 342
EKG ... 461, 501
electret microphone ... 474, 506
electric field induced noise ... 490
electric field interference ... 342
electrocardiogram ... 501
electrolytic ... 391
electromagnetic field induction ... 489
email ... 542
embedded computer system ... 551
embedded system ... 10, 164
empirical model ... 18
emulator ... 181, 551
`EnableInterrupts` ... 266
enclosure ... 457
EndCritical ... 266
endpoint ... 380
energy storage ... 447
ENOB ... 425
entry point ... 145
enum ... 136
EOR ... 88
EPROM ... 551
EPSR ... 65
EQU ... 101
erase ... 551
Ethernet ... 72, 529
ethics ... 56
EVB ... 551
even parity ... 516
exceptions ... 267
exclusive or ... 88
execution profile ... 562

exponent	53
false negative	480
false positive	480
fan in	149
fan out	30, 149, 551
Faraday's law of induction	17
fault injection	456
FAULTMASK	66
ferrite bead	391
FIFO	169, 257, 260, 262, 275
filter	552
FilterPro	411
Finite State Machine	155, 162, 289
firm real time	196, 552
first in first out circular queue	169
fixed-point	48, 413, 424
flash ADC	427
flash EEPROM	552
flash ROM	3, 58
flicker noise	488
flip flop	
D latch	89
gated D	89
set-reset	89
floating state	90
floating-point	53
flow control	344
flowchart	23
font	215
force	475
force sensing resistor	476, 512
foreground	168
foreground thread	166, 253
fork	167
fork and join	25
FQP27P06	318
fractional sensitivity	467
frame	232, 320, 338, 475, 514
framing error	235, 236, 516
`free`	177
frequency measurement	312
frequency resolution	312
frequency response	396, 470
friendly	106
FSM	155
FSR	476
full duplex	234, 340, 515, 552
functional abstraction	142
functional cohesion	149
functional debugging	185, 552
functions	23, 96
fuzzy logic	552
gadfly	196, 552, 553
gain error	413
galvanic noise	489

gate	151
gauge factor	475, 552
general-purpose computer	15
global	128, 129, 143, 144
GP2Y0A21YK	472
ground	342
guided medium	515
HAL	150, 152
half duplex	340, 351, 515, 553
half-cell potential	501
half-stepping	221, 223
halfword	46
handler mode	70
Handler mode	270
handshake	231
hard real time	196, 553
hardware abstraction layer	150, 152
Harvard architecture	5, 64
H-bridge	222, 319
HD44780	215
header file	130
device driver	153
included only once	131
including	128
heap	177
heart rate	501
heartbeat	187
heat conduction	17
hexadecimal	4, 42
h_{fe} current gain	36, 37
hiberation	451
hierarchical programming	146
high impedance	90
HiZ	90, 553
hold time	204, 217, 553
holding torque	226
hook	308
human-computer interface	12
hysteresis	406, 471, 553
I/O bound	197, 227, 259, 261, 553
I/O device	553
I/O ports	103, 553
I/O-mapped I/O	6
I^2C	72, 364, 368, 370
I^2S	72
IBE	212
I_{CC} supply current	33
ICE	180
ICI/IT	66
ICode bus	4, 64
ideal op amp	393
IEEE	42
IEEE Code of Ethics	57
IEEE488	232
IEV	212

IFOV 465
I_{IH} current 29, 214
I_{IL} current 29, 214
IME 212
immediate addressing 83
immediate addressing mode 553
impedance 553
impedance loading 553
INA122 406
in-circuit emulator 551
incremental control system 553
indexed addressing 83
indexed addressing mode 553
information hiding 148
input capture 292, 302, 554
input impedance 398, 399, 402, 468
input parameters 65
input port 4, 101
input voltage noise 395
instruction register 7
instrument 554
instrumentation 182, 462
instrumentation amp 405
integration 479
interface 4
interface latency 229, 253
interlocked 231
intermittent polling 279
International Standards Organization 514, 529
Internet 529
internet of things 526, 533, 534
interrupt 26, 64, 146, 199, 252, 554
interrupt flag 554
interrupt mask 554
interrupt numbers 270
interrupt polling 557
interrupt vectors 267
intrusive 185, 556
intrusiveness 183
inverting amplifier 399
invocation coupling 148
I_{OH} current 29, 214
I_{OL} current 29, 214
IP address 530
IPSR 65
IR distance sensor 472, 473
IR instruction register 7
IRF540 .... 316, 324
IRF9540 318
IS Interrupt Sense bit 212
ISO 514, 529
isolated I/O 554
isolation 407
ISR 268
ISR_NUMBER 70

IT conditional 99
jerk 221, 554
Johnson noise 554
join 167
Joint Test Action Group 71
JTAG 71, 182
Kahn Process Networks 209
Karnaugh map 554
KCL 29
keyboard
  matrix 243
KiB 42
kibibit 42
kibibyte 42, 554
Kibit 42
Kirchhoff's Current Law 29
Kirchhoff's Voltage Law 29
KPN 209, 263
KVL 29
L293 .... 222, 319
L6203 222
label field 82
last-in-first-out 68
latch 554
latched input 554
late arrival interrupt 271
latency 196, 236, 253, 297, 339, 555
LaunchPad 72, 73, 76, 77, 78, 81
layered software system 151
LCD 215, 555
LDO regulator 442
LDR 83, 86, 87
lead acid battery 448
least significant bit 43
least squares fit 467
LED 187, 555
level shifter 352
light-weight process 555
Li-Ion 448
linear 17
linear regulator 441
linear time-invariant 17
linear variable differential transformer 555
linearity 413, 467
line-trigger 491
link register 64
linked list 156
linked structure 156
linker 152
liquid crystal display 215
Lithium-ion battery 448
Little's Theorem 174
little-endian 46
LM1117 444
LM3S3748 383

LM3Sxxx...ix
LM4041...506
LM4F120H5QR......................................72
local...129, 143
local variable
    allocate..100
    deallocate....................................100
logging..456
logic analyzer....................................180
logic level conversion.......................352
logic probe...180
logical
    and..88
    complement..................................88
    exclusive or..................................88
    or operation..................................88
    shift left.......................................91
    shift right.....................................91
logical cohesion.................................149
long..47
low pass filter...................................408
low power..283
low-power design..............................445
LP2981..442
LR..64
LRC..517
LSL..92
LSR..92
LTI...17
LVDT..555
MACQ................................434, 435, 556
magnetic field pickup.......................489
mailbox..256
maintenance................................22, 145
make before break...........................555
`malloc`..177
Manchester encoding........................337
man-machine interface......................12
mantissa...53
mark...555
mask...88
master-slave......................................518
matrix keyboard................................243
MAX3232.....................234, 338, 343, 569
MAX4783..437
MAX483..351
MAX494......................................393, 446
MAX5353................358, 419, 445, 506
MAX604..446
maximum accuracy of reading........464
maximum accuracy or full scale.....464
maximum error.................................464
Maxwell–Faraday equation...............17
Mealy FSM..................................156, 162
measurand..................................461, 555

measure supply current..................450
mebibyte..43
membership Sets..............................555
memory...............................4, 66, 85, 555
memory effect...................................447
memory leak.....................................178
memory manager.............................177
memory mapped I/O............................4
memory read cycle................................6
memory write cycle..............................6
memory-mapped I/O........................555
message...514
message pipe....................................381
metal film..390
method...97
MiB..43
microcomputer............................2, 555
microcontroller........................2, 7, 555
microphone.......................................474
microprocessor.....................................2
minimally intrusive....................183, 456
mitered corner.................................457
mixed-signal design.........................400
MLA..96
MLS..96
modular debugging..........................184
modular programming....................145
monitor.................................187, 456, 556
monostable.......................................302
monotonic..........................210, 413, 424
Moore FSM..................................156, 157
morals..56
MOSFET..3, 316
most significant bit.....................43, 44
motion artifact.................................489
motor controller................................13
MOV..82
MSP main stack pointer....................64
MSP432E401Y....................................76
MUL..96
multi-drop..................................351, 519
multiple access circular queue...434, 556
multiplexed interface......................242
multiplication....................................93
multiply..96
    with accumulate..........................96
    with subtract................................96
multithreaded..................................252
multi-threaded.................................556
mutual exclusion......................266, 556
N bit..65
N-channel MOSFET.........................318
negative feedback.....................394, 397
negative logic..................39, 40, 202, 556
negative number................................45

negative predictive value 481
nested interrupt 271
Nested Vectored Interrupt Controller 64, 270
network 513
Newton's second law 17
NiCad battery 448
Nickel-Cadmium battery 448
Nickel-metal hydride battery 448
NiMH battery 448
NIST 413, 464
noise 395, 486, 491
Non Return to Zero Invert 378
nonatomic 264, 556
non-blocking 210
non-compete clause 57
nonintrusive 556
nonintrusiveness 183
noninverting amplifier 402
nonlinear 413
nonlinear transducer 472
nonmonotonic 413, 472
nonpolarizable electrode 501
nonpreemptive scheduler 556
nonreentrant 263, 556
non-return-to-zero 336
non-return-to-zero-inverted 336
nonvolatile 3, 556
normalized mean response time 174
NOT gate 88
NPN 36
NPN transistor 36
NRZ 336
NRZI 336, 378
null 177
null-termination 46
NVIC 64, 270
NVIC_EN0_R 270
NVIC_PRI0_R 270
NVIC_ST_CTRL_R 114, 280
NVIC_ST_CURRENT_R 114, 280
NVIC_ST_RELOAD_R 114, 280
NVIC_SYS_PRI3_R 280
Nyquist Theorem 412, 482
object code 556
ODE 214, 518
offset binary 424
offset current 395
offset error 413
offset voltage 395
ohmmeter 300
OLED 187
one shot mode 305
one's complement 44
op amps 393
OPA2330 393

OPA2350 .... 393, 396
opcode field 82
open collector 557
open collector NOT 89
open drain 35, 214
open loop 219
open loop control system 557
Open Systems Interconnection 514, 529
open-loop output impedance 394
operand field 82
operand2 84
operating system 557
optical sensor 295
ORN 88
ORR 88
oscilloscope 180
output compare 305, 557
output impedance 394, 398, 399, 400, 403
output port 4, 102
overflow 50, 93
overrun 235, 236
oversampling 425, 483
pacemaker 12, 24
packet 514
parallel 4
parallel I/O port 101
parallel port 557
parallel programming 25, 167
parity 338, 516, 556
parity error 235
partially asynchronous bus 557
PC program counter 64
PCB 457
P-channel MOSFET 318
PC-relative addressing 83, 557
PCTL 80
PDE 213
Peltier effect 477
performance debugging 557
performance map 18
performance maps 472
period measurement 298
periodic interrupt 307
periodic interrupts 279
periodic polling 200, 279, 557
personal area network 522
Petri Net 206
photosensor 557
physical plant 329
physical Plant 557
pi filter 391
PicoScope 180
PID controller 557
PID USB Packet ID 380
Ping))) 473

pink noise 488, 557
pipe 381
pipeline 7
place-value 44
PLL 111, 283
PN2222 .... 38, 108, 316, 323, 455
PN2907 317
PNP 318
PNP transistor 37
pointer 83, 156
Polarizable electrode 502
pole 320, 557
polling 227
pop 68
POP 68, 100, 120
port 558
portability 117
position transducer 472
positive feedback 394
positive logic 39, 40, 202
positive predictive value 481
potentiometer 505, 558
power block 441
power brick 441
power budget 447
power gain 396
power metric 174
power up 71
PPB 64
precision 40, 48, 298, 411, 465, 481, 483
precision of the period measurement 298
preemptive scheduler 558
pressure 509
prevalence 480
primary battery 447
PRIMASK 66, 252, 270
primitive data 148
printer 344
printf 155
priority 196, 272
private 129, 138, 142
private global 153
private peripheral bus 4, 64
procedural cohesion 149
procedure 23, 96
process 252, 514, 558
processor 2
producer/consumer 169, 257
profiling 182, 557
program counter 8, 64
program module 145
program status register 8, 65
PROM 558
pseudo-code 558
PSP process stack pointer 64

PSR 8, 65
public 129, 138, 143, 558
public global 143
PUE 213
pull-down 213
pull-up 38, 39, 213
pulse width measurement 302
pulse width modulation 322, 558
PUSH 68, 100
PWM 72, 322
PWM DAC 422
Q bit 65
QEI 72
QRB1134 .... 157, 295
qualitative DAS 480, 559
quantitative DAS 480, 559
quiescent current 442
R-2R ladder 415
race condition 208, 267
rail-to-rail 393, 399
RAM 2, 3, 58, 559
random access memory 2
range 49, 411, 465, 559
range of the period measurement 298
read cycle 6, 559
read data available 559
read only memory 2
readable output port 102
real time clock (RTC) 451
real-time 196, 236, 253, 559
real-time system 11, 559
reduced instruction set computer 8
reed relay 320
reentrant 168, 263, 559
reference voltage 401
registers 7, 64
regulator 441
relay 320
repeatability 559
reproducibility 466, 559
Request to Send 344
requirement 19
reset 71
    LR 71
resistance bridge 495
resistance measurement 300, 302
resistance temperature device 559
resistor 389
resolution 48, 49, 298, 302, 312, 411, 465, 498, 559
resolution of a period measurement 298
response time 559
ring network 519
ripple 390
RIS 212
RISC 8, 559

ritual 559
RJ45 342
rollover 246
ROM 2, 3, 559
ROM-emulator 551
`ROR` 92
rotate 92
rotor 219, 325
round robin scheduler 559
roundoff 55
RS232 .... 233, 342, 343
RS422 347
RS423A 347
RS485 .... 351, 518
`RSB` 94
`RSSI` 526
RTD 559, 560
RTS 344
`RXE` 237
`RXFE` 240
`RXX` 92
S/H 428, 463
S/N 468
Salen-Key 410
sample and hold 428, 481, 560
sampling rate 279, 483, 560
saturation 471, 560
scan 560
scanned interface 241
ScanPoint 560
scheduler 560
SCI 560
SCL 364
scope 129, 135, 142
SCSI 232
SDA 364
`SDIV` 91, 96
secondary battery 447
security 528
Seebeck effect 477
select signal 560
semaphore 560
sensitivity 475, 481, 560
sensitivity drift 469
sensor 329
separation of policies from mechanisms 148, 151
sequential cohesion 149
serial 4
serial communication 336, 560
serial port 560
serial transmission 232
set-reset flip flop 89
settling time 418
setup time 204, 217, 560
Shannon–Hartley Channel Capacity Theorem 521
Sharp GP2Y0A21YK 472
shielding 494
shift 91
short 46
shot noise 488, 560
sigma delta ADC 425
sign extension 87
signal processing 462
signal to noise ratio 418
signal-to-noise and distortion ratio 424
signal-to-noise ratio 468
signed byte 44
signed long 47
signed number 44
signed short 47
simplex communication system 341, 560
simulation 22
simulator 560
SINAD 424, 485
single-pole switch 561
single-precision floating-point 53
single-throw 320
single-throw switch 561
sleep mode 447
slew rate 30, 214, 396
slide potentiometer 511
slip 224
Small Computer Systems Interface 232
smart object 526, 534
`SMLAL` 96
smoke detector 13, 27
`SMULL` 96
SNR 418, 515
snubber diode 316
socket 535, 536
soft real time 196, 561
software 2
software abstraction 155
solenoids 321
solid state relay 108, 320
sound 506
SP stack pointer 64, 68
SP483 .... 351, 519
space 561
`SPACE` 101
span 561
spatial resolution 465, 561
speaker 506
specification 19
specificity 468, 481, 561
spectrum analyzer 492
SPI 71, 357, 560
spinlock semaphore 561
square wave 309
SRAM 3

SRL	214
SSI	71, 341, 353, 562
SSR	108, 320
ST7066U	215
stability	18
stabilize	183, 185
stabilizing	561
stack	68, 100, 561
empty	69
next	69
overflow	69
`pop`	100
pop data	68
`push`	100
push data	68
rules	69
top	69
underflow	69
stack pointer	8, 64, 68
stamp	148
standard deviation	499
standard error	464
standard resistance	440
standard values	439
start bit	232
`StartCritical`	266
`Startup.s`	268
starvation	561
state estimator	329
`static`	129
static allocation	561
static efficiency	126
static local	144
static RAM	3, 561
static sensitivity	467
statistical control	466
stator	219, 325
stepper motor	219
stepper motor selection	226
stop bit	232
STR	86, 87
straight binary	424
strain gauge	475, 509, 561
stream pipe	381
string	46
structured programming	24
student's t test	498
SUB	94
subroutine	96
subtraction	93
successive approximation ADC	425
supply current	33, 34
switch	320
switch bounce	281, 283
switching regulator	443
synchronous bus	561
synchronous communication system	341
synchronous protocol	561
system bus	4, 64
system capacity	174
SysTick	114, 187, 188, 279
T bit	66, 71
table data structure	421
TAEVENT	294
tail chaining	271
tantalum	391
TATOIM	307
TATORIS	307
TCP	529
TCP/IP	526, 529
TCR	454
temperature	495
Temperature Coefficient of Resistance	454
temporal cohesion	149
test points	456
test statistic	500
testing	22
testing interface	456
THD	424
thermal noise	487, 562
thermistor	476, 495, 562
thermocouple	476, 562
thick film	390
thin film	389
thread	24, 166, 252, 562
thread control block	562
thread mode	70
Thread mode	270
thread mode privilege level	70
thread scheduler	167
threshold detector	394, 396, 406
throughput	196, 562
Thumb	7, 81
THUMB	101
Thumb-2 technology	7, 81
time	4
time constant	470, 562
time delay	307
time interval	201
time profile	562
time quantizing	412, 482
timing diagrams	204
timing equations	201
TIP120	316, 323, 325, 584
TIP125	317, 584
TLC2274	393
TLC555	300
TM4C1294	76
tolerance	389, 454
top	69

top-down design 18, 147
torque 226, 323
total harmonic distortion 424
TPA731 .... 445, 506
t_{pd} propagation delay 33
TPL 70
TPS63002 443
TPS78233 .... 442, 444
track and hold 428
traffic controller 13
transducer 462, 466, 481, 562
transient response 470
transistor-transistor logic 30
transition time 30
trap 548, 562
trigger 252
tristate 562
tristate driver 90
tristate logic 32, 562
true positive 480
truncation 55
TTL 30
two's complement 44
two-key rollover 282
TXE 237
TXFF 240
`typedef` 135
UART 71, 232, 275, 336, 562
UART0_DR_R 234
UDIV 91, 96
UDP 529, 534, 536
ultrasonic distance sensor 473
ultrasound 562
UMLAL 96
UMULL 96
unary operation 88
unbuffered I/O 562
unfriendly 106
unguided medium 515
unipolar stepper motor 223
Universal Asynchronous Receiver/Transmitter 232
Universal Serial Bus 377
unnormalized 55
unsigned char 43
unsigned long 47

unsigned number 43
unsigned short 46
USB 72, 377
utilization factor 174
V bit 65, 94
Valvano Postulate 483
variance 499
$V_{be(SAT)}$ 36
vectors 267
velocity 221, 478, 479
verbose 456
verification 185
V_{IH} voltage 31
V_{IL} voltage 31
V_{OH} voltage 31
V_{OL} voltage 31, 214
volatile 3, 563
voltage comparator 406
voltage dropout 442
voltmeter 511
von Neumann architecture 2, 5
VRLA 448
vulnerable window 263, 563
wait for interrupt 452
wall wart 441
waveform generation 419
web server 531
WFI 283, 452
white noise 390, 487, 563
wirewound resistor 390
word 47
write cycle 6, 563
write data available 563
write data required 563
X7R 391
XBee 523
XON/XOFF 341
Z bit 65
Z Transform 563
Z5U 391
zero 563
zero drift 469
zero pad 86
ZigBee 522
β current gain 36, 37

Reference Material

Vector address	Number	IRQ	ISR name in **Startup.s**	NVIC	Priority bits
0x00000038	14	-2	PendSV_Handler		
0x0000003C	15	-1	SysTick_Handler		
0x00000040	16	0	GPIOPortA_Handler	NVIC_PRI0_R	7 – 5
0x00000044	17	1	GPIOPortB_Handler	NVIC_PRI0_R	15 – 13
0x00000048	18	2	GPIOPortC_Handler	NVIC_PRI0_R	23 – 21
0x0000004C	19	3	GPIOPortD_Handler	NVIC_PRI0_R	31 – 29
0x00000050	20	4	GPIOPortE_Handler	NVIC_PRI1_R	7 – 5
0x00000054	21	5	UART0_Handler	NVIC_PRI1_R	15 – 13
0x00000058	22	6	UART1_Handler	NVIC_PRI1_R	23 – 21
0x0000005C	23	7	SSI0_Handler	NVIC_PRI1_R	31 – 29
0x00000060	24	8	I2C0_Handler	NVIC_PRI2_R	7 – 5
0x00000064	25	9	PWMFault_Handler	NVIC_PRI2_R	15 – 13
0x00000068	26	10	PWM0_Handler	NVIC_PRI2_R	23 – 21
0x0000006C	27	11	PWM1_Handler	NVIC_PRI2_R	31 – 29
0x00000070	28	12	PWM2_Handler	NVIC_PRI3_R	7 – 5
0x00000074	29	13	Quadrature0_Handler	NVIC_PRI3_R	15 – 13
0x00000078	30	14	ADC0_Handler	NVIC_PRI3_R	23 – 21
0x0000007C	31	15	ADC1_Handler	NVIC_PRI3_R	31 – 29
0x00000080	32	16	ADC2_Handler	NVIC_PRI4_R	7 – 5
0x00000084	33	17	ADC3_Handler	NVIC_PRI4_R	15 – 13
0x00000088	34	18	WDT_Handler	NVIC_PRI4_R	23 – 21
0x0000008C	35	19	Timer0A_Handler	NVIC_PRI4_R	31 – 29
0x00000090	36	20	Timer0B_Handler	NVIC_PRI5_R	7 – 5
0x00000094	37	21	Timer1A_Handler	NVIC_PRI5_R	15 – 13
0x00000098	38	22	Timer1B_Handler	NVIC_PRI5_R	23 – 21
0x0000009C	39	23	Timer2A_Handler	NVIC_PRI5_R	31 – 29
0x000000A0	40	24	Timer2B_Handler	NVIC_PRI6_R	7 – 5
0x000000A4	41	25	Comp0_Handler	NVIC_PRI6_R	15 – 13
0x000000A8	42	26	Comp1_Handler	NVIC_PRI6_R	23 – 21
0x000000AC	43	27	Comp2_Handler	NVIC_PRI6_R	31 – 29
0x000000B0	44	28	SysCtl_Handler	NVIC_PRI7_R	7 – 5
0x000000B4	45	29	FlashCtl_Handler	NVIC_PRI7_R	15 – 13
0x000000B8	46	30	GPIOPortF_Handler	NVIC_PRI7_R	23 – 21
0x000000BC	47	31	GPIOPortG_Handler	NVIC_PRI7_R	31 – 29
0x000000C0	48	32	GPIOPortH_Handler	NVIC_PRI8_R	7 – 5
0x000000C4	49	33	UART2_Handler	NVIC_PRI8_R	15 – 13
0x000000C8	50	34	SSI1_Handler	NVIC_PRI8_R	23 – 21
0x000000CC	51	35	Timer3A_Handler	NVIC_PRI8_R	31 – 29
0x000000D0	52	36	Timer3B_Handler	NVIC_PRI9_R	7 – 5
0x000000D4	53	37	I2C1_Handler	NVIC_PRI9_R	15 – 13
0x000000D8	54	38	Quadrature1_Handler	NVIC_PRI9_R	23 – 21
0x000000DC	55	39	CAN0_Handler	NVIC_PRI9_R	31 – 29
0x000000E0	56	40	CAN1_Handler	NVIC_PRI10_R	7 – 5
0x000000E4	57	41	CAN2_Handler	NVIC_PRI10_R	15 – 13
0x000000E8	58	42	Ethernet_Handler	NVIC_PRI10_R	23 – 21
0x000000EC	59	43	Hibernate_Handler	NVIC_PRI10_R	31 – 29
0x000000F0	60	44	USB0_Handler	NVIC_PRI11_R	7 – 5
0x000000F4	61	45	PWM3_Handler	NVIC_PRI11_R	15 – 13
0x000000F8	62	46	uDMA_Handler	NVIC_PRI11_R	23 – 21
0x000000FC	63	47	uDMA_Error	NVIC_PRI11_R	31 – 29

Table 5.1. Some of the interrupt vectors for the TM4C/MSP432E4.

C Data type	C99 Data type	Precision	Range
`unsigned char`	`uint8_t`	8-bit unsigned	0 to +255
`signed char`	`int8_t`	8-bit signed	-128 to +127
`char`	`char`	8-bit	ASCII characters
`unsigned int`	`unsigned int`	compiler-dependent	
`int`	`int`	compiler-dependent	
`unsigned short`	`uint16_t`	16-bit unsigned	0 to +65535
`short`	`int16_t`	16-bit signed	-32768 to +32767
`unsigned long`	`uint32_t`	unsigned 32-bit	0 to 4294967295L
`long`	`int32_t`	signed 32-bit	-2147483648L to 2147483647L
`float`	`float`	32-bit float	$\pm 10^{-38}$ to $\pm 10^{+38}$
`double`	`double`	64-bit float	$\pm 10^{-308}$ to $\pm 10^{+308}$

Data types in C ([1]some compilers treat char as unsigned, some treat char as signed)

		BITS 4 to 6							
		0	1	2	3	4	5	6	7
	0	NUL	DLE	SP	0	@	P	`	p
B	1	SOH	XON	!	1	A	Q	a	q
I	2	STX	DC2	"	2	B	R	b	r
T	3	ETX	XOFF	#	3	C	S	c	s
S	4	EOT	DC4	$	4	D	T	d	t
	5	ENQ	NAK	%	5	E	U	e	u
0	6	ACK	SYN	&	6	F	V	f	v
	7	BEL	ETB	'	7	G	W	g	w
T	8	BS	CAN	(	8	H	X	h	x
O	9	HT	EM	)	9	I	Y	i	y
	A	LF	SUB	*	:	J	Z	j	z
3	B	VT	ESC	+	;	K	[	k	{
	C	FF	FS	,	<	L	\	l	|
	D	CR	GS	-	=	M	]	m	}
	E	SO	RS	.	>	N	^	n	~
	F	SI	US	/	?	O	_	o	DEL

Table 1.10. Standard 7-bit ASCII.

Parameter	PN2222 (I_C=150mA) PN2907 (I_C=150mA)	2N2222 (I_C=500mA) 2N2907 (I_C=500mA)	TIP120 (I_C=3A) TIP125 (I_C=3A)
h_{fe}	100	40	1000
V_{BEsat}	0.6	2	2.5 V
V_{CE} at saturation	0.3	1	2 V

Design parameters for the 2N2222 and TIP120.

Chip	Current	Comment
MC3479	0.35 A	Stepper driver
L293D	0.6 A	Dual, diodes
L293	1 A	Dual
TPIC0107	3 A	Direction, fault status
L6203	5 A	Dual

H-bridge drivers

Family	Example	I_{OH}	I_{OL}	I_{IH}	I_{IL}	fan out
Standard TTL	7404	0.4 mA	16 mA	40 µA	1.6 mA	10
Schottky TTL	74S04	1 mA	20 mA	50 µA	2 mA	10
Low Power Schottky	74LS04	0.4 mA	4 mA	20 µA	0.4 mA	10
High speed CMOS	74HC04	4 mA	4 mA	1 µA	1 µA	
TM4C 2mA-drive	TM4C	2 mA	2 mA	2 µA	2 µA	
TM4C 4mA-drive	TM4C	4 mA	4 mA	2 µA	2 µA	
TM4C 8mA-drive	TM4C	8 mA	8 mA	2 µA	2 µA	

Table 1.4. The input and output currents of various digital logic families and microcontrollers.

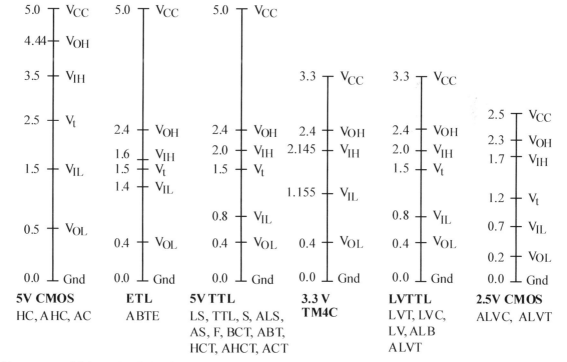

Figure 1.15. Voltage thresholds for various digital logic families.

Memory access instructions

```
LDR     Rd, [Rn]           ; load 32-bit number at [Rn] to Rd
LDR     Rd, [Rn,#off]      ; load 32-bit number at [Rn+off] to Rd
LDR     Rd, [Rn,#off]!     ; load 32-bit number at [Rn+off] to Rd, preindex
LDR     Rd, [Rn],#off      ; load 32-bit number at [Rn] to Rd, postindex
LDRT    Rd, [Rn,#off]      ; load 32-bit number unprivileged
LDR     Rd, =value         ; set Rd equal to any 32-bit value (PC rel)
LDRH    Rd, [Rn]           ; load unsigned 16-bit at [Rn] to Rd
LDRH    Rd, [Rn,#off]      ; load unsigned 16-bit at [Rn+off] to Rd
LDRH    Rd, [Rn,#off]!     ; load unsigned 16-bit at [Rn+off] to Rd, pre
LDRH    Rd, [Rn],#off      ; load unsigned 16-bit at [Rn] to Rd, postindex
LDRHT   Rd, [Rn,#off]      ; load unsigned 16-bit unprivileged
LDRSH   Rd, [Rn]           ; load signed 16-bit at [Rn] to Rd
LDRSH   Rd, [Rn,#off]      ; load signed 16-bit at [Rn+off] to Rd
LDRSH   Rd, [Rn,#off]!     ; load signed 16-bit at [Rn+off] to Rd, pre
LDRSH   Rd, [Rn],#off      ; load signed 16-bit at [Rn] to Rd, postindex
LDRSHT  Rd, [Rn,#off]      ; load signed 16-bit unprivileged
LDRB    Rd, [Rn]           ; load unsigned 8-bit at [Rn] to Rd
LDRB    Rd, [Rn,#off]      ; load unsigned 8-bit at [Rn+off] to Rd
LDRB    Rd, [Rn,#off]!     ; load unsigned 8-bit at [Rn+off] to Rd, pre
LDRB    Rd, [Rn],#off      ; load unsigned 8-bit at [Rn] to Rd, postindex
LDRBT   Rd, [Rn,#off]      ; load unsigned 8-bit unprivileged
LDRSB   Rd, [Rn]           ; load signed 8-bit at [Rn] to Rd
LDRSB   Rd, [Rn,#off]      ; load signed 8-bit at [Rn+off] to Rd
LDRSB   Rd, [Rn,#off]!     ; load signed 8-bit at [Rn+off] to Rd, pre
LDRSB   Rd, [Rn],#off      ; load signed 8-bit at [Rn] to Rd, postindex
LDRSBT  Rd, [Rn,#off]      ; load signed 8-bit unprivileged
LDRD    Rd,Rd2,[Rn,#off]   ; load 64-bit at [Rn+off] to Rd,Rd2
LDRD    Rd,Rd2,[Rn,#off]!  ; load 64-bit at [Rn+off] to Rd,Rd2,pre
LDRD    Rd,Rd2,[Rn],#off   ; load 64-bit at [Rn] to Rd,Rd2, postindex
LDMFD   Rn{!}, Reglist     ; load reg from list at Rn(inc), !update Rn
LDMIA   Rn{!}, Reglist     ; load reg from list at Rn(inc), !update Rn
LDMDB   Rn{!}, Reglist     ; load reg from list at Rn(dec), !update Rn
STMIA   Rn{!}, Reglist     ; store reg from list to Rn(inc), !update Rn
STMFD   Rn{!}, Reglist     ; store reg from list to Rn(dec), !update Rn
STMDB   Rn{!}, Reglist     ; store reg from list to Rn(dec), !update Rn
STR     Rt, [Rn]           ; store 32-bit Rt to [Rn]
STR     Rt, [Rn,#off]      ; store 32-bit Rt to [Rn+off]
STR     Rt, [Rn,#off]!     ; store 32-bit Rt to [Rn+off], pre
STR     Rt, [Rn],#off      ; store 32-bit Rt to [Rn], postindex
STRT    Rt, [Rn,#off]      ; store 32-bit Rt to [Rn+off] unprivileged
STRH    Rt, [Rn]           ; store least sig. 16-bit Rt to [Rn]
STRH    Rt, [Rn,#off]      ; store least sig. 16-bit Rt to [Rn+off]
STRH    Rt, [Rn,#off]!     ; store least sig. 16-bit Rt to [Rn+off], pre
STRH    Rt, [Rn],#off      ; store least sig. 16-bit Rt to [Rn], postindex
STRHT   Rt, [Rn,#off]      ; store least sig. 16-bit unprivileged
STRB    Rt, [Rn]           ; store least sig. 8-bit Rt to [Rn]
STRB    Rt, [Rn,#off]      ; store least sig. 8-bit Rt to [Rn+off]
STRB    Rt, [Rn,#off]!     ; store least sig. 8-bit Rt to [Rn+off],pre
```

```
STRB    Rt, [Rn],#off       ; store least sig. 8-bit Rt to [Rn], postindex
STRBT   Rt, [Rn,#off]       ; store least sig. unprivileged
STRD Rd,Rd2,[Rn,#off]       ; store 64-bit Rd,Rd2 to [Rn+off]
STRD Rd,Rd2,[Rn,#off]!      ; store 64-bit Rd,Rd2 to [Rn+off], pre
STRD Rd,Rd2,[Rn],#off       ; store 64-bit Rd,Rd2 to [Rn], postindex
PUSH    Reglist             ; push 32-bit registers onto stack
POP     Reglist             ; pop 32-bit numbers from stack into registers
ADR     Rd, label           ; set Rd equal to the address at label
MOV{S}  Rd, <op2>           ; set Rd equal to op2
MOV     Rd, #im16           ; set Rd equal to im16, im16 is 0 to 65535
MOVT    Rd, #im16           ; set Rd bits 31-16 equal to im16
MVN{S}  Rd, <op2>           ; set Rd equal to -op2
```

Branch instructions
```
B    label    ; branch to label       Always
BEQ  label    ; branch if Z == 1      Equal
BNE  label    ; branch if Z == 0      Not equal
BCS  label    ; branch if C == 1      Higher or same, unsigned ≥
BHS  label    ; branch if C == 1      Higher or same, unsigned ≥
BCC  label    ; branch if C == 0      Lower, unsigned <
BLO  label    ; branch if C == 0      Lower, unsigned <
BMI  label    ; branch if N == 1      Negative
BPL  label    ; branch if N == 0      Positive or zero
BVS  label    ; branch if V == 1      Overflow
BVC  label    ; branch if V == 0      No overflow
BHI  label    ; branch if C==1 and Z==0   Higher, unsigned >
BLS  label    ; branch if C==0 or  Z==1   Lower or same, unsigned ≤
BGE  label    ; branch if N == V      Greater than or equal, signed ≥
BLT  label    ; branch if N != V      Less than, signed <
BGT  label    ; branch if Z==0 and N==V   Greater than, signed >
BLE  label    ; branch if Z==1 or N!=V    Less than or equal, signed ≤
BX   Rm       ; branch indirect to location specified by Rm
BL   label    ; branch to subroutine at label
BLX  Rm       ; branch to subroutine indirect specified by Rm
CBNZ Rn,label ; branch if Rn not zero
CBZ  Rn,label ; branch if Rn zero
IT{x{y{z}}}cond ; if then block with x,y,z T(true) or F(false)
TBB  [Rn, Rm]           ; table branch byte
TBH  [Rn, Rm, LSL #1]   ; table branch halfword
```

Mutual exclusive instructions
```
CLREX                                ; clear exclusive
LDREX{cond}  Rt,[Rn{,#offset}]       ; load 32-bit exclusive
STREX{cond}  Rd,Rt,[Rn{,#offset}]    ; store 32-bit exclusive
LDREXB{cond} Rt,[Rn]                 ; load 8-bit exclusive
STREXB{cond} Rd,Rt,[Rn]              ; store 8-bit exclusive
LDREXH{cond} Rt,[Rn]                 ; load 16-bit exclusive
STREXH{cond} Rd,Rt,[Rn]              ; store 16-bit exclusive
```

Miscellaneous instructions

```
BKPT    #imm            ; execute breakpoint, debug state 0 to 255
CPSIE   F               ; clear faultmask F=0
CPSIE   I               ; enable interrupts   (I=0)
CPSID   F               ; set faultmask F=1
CPSID   I               ; disable interrupts (I=1)
DMB                     ; data memory barrier, memory access to finish
DSB                     ; data synchronization barrier, instructions to finish
ISB                     ; instruction synchronization barrier, finish pipeline
MRS Rd,SpecReg          ; move special register to Rd
MSR Rd,SpecReg          ; move Rd to special register
NOP                     ; no operation
SEV                     ; Send Event
SVC #im8                ; supervisor call (0 to 255)
WFE                     ; wait for event
WFI                     ; wait for interrupt
```

Logical instructions

```
AND{S} {Rd,} Rn, <op2>  ; Rd=Rn&op2     (op2 is 32 bits)
BFC  Rd,#lsb,#width     ; clear bits in Rn
BFI  Rd,Rn,#lsb,#width  ; bit field insert, Rn into Rd
ORR{S} {Rd,} Rn, <op2>  ; Rd=Rn|op2     (op2 is 32 bits)
EOR{S} {Rd,} Rn, <op2>  ; Rd=Rn^op2     (op2 is 32 bits)
BIC{S} {Rd,} Rn, <op2>  ; Rd=Rn&(~op2)  (op2 is 32 bits)
ORN{S} {Rd,} Rn, <op2>  ; Rd=Rn|(~op2)  (op2 is 32 bits)
TST     Rn, <op2>       ; Rn&op2        (op2 is 32 bits)
TEQ     Rn, <op2>       ; Rn^op2        (op2 is 32 bits)
LSR{S} Rd, Rm, Rs       ; logical shift right Rd=Rm>>Rs   (unsigned)
LSR{S} Rd, Rm, #n       ; logical shift right Rd=Rm>>n    (unsigned)
ASR{S} Rd, Rm, Rs       ; arithmetic shift right Rd=Rm>>Rs (signed)
ASR{S} Rd, Rm, #n       ; arithmetic shift right Rd=Rm>>n  (signed)
LSL{S} Rd, Rm, Rs       ; shift left Rd=Rm<<Rs (signed, unsigned)
LSL{S} Rd, Rm, #n       ; shift left Rd=Rm<<n  (signed, unsigned)
REV     Rd, Rn          ; Reverse byte order in a word
REV16   Rd, Rn          ; Reverse byte order in each halfword
REVSH   Rd, Rn          ; Reverse byte order in the bottom halfword,
                        ;   and sign extends to 32 bits
RBIT    Rd, Rn          ; Reverse the bit order in a 32-bit word
SBFX Rd,Rn,#lsb,#width  ; signed bit field and extract
UBFX Rd,Rn,#lsb,#width  ; unsigned bit field and extract
SXTB {Rd,}Rm{,ROR #n}   ; Sign extend byte
SXTH {Rd,}Rm{,ROR #n}   ; Sign extend halfword
UXTB {Rd,}Rm{,ROR #n}   ; Zero extend byte
UXTH {Rd,}Rm{,ROR #n}   ; Zero extend halfword
```

Arithmetic instructions
```
    ADD{S}  {Rd,} Rn, <op2>      ; Rd = Rn + op2
    ADD{S}  {Rd,} Rn, #im12      ; Rd = Rn + im12, im12 is 0 to 4095
    CLZ     Rd, Rm               ; Rd = number of leading zeros in Rm
    SUB{S}  {Rd,} Rn, <op2>      ; Rd = Rn - op2
    SUB{S}  {Rd,} Rn, #im12      ; Rd = Rn - im12, im12 is 0 to 4095
    RSB{S}  {Rd,} Rn, <op2>      ; Rd = op2 - Rn
    RSB{S}  {Rd,} Rn, #im12      ; Rd = im12 - Rn
    CMP     Rn, <op2>            ; Rn - op2      sets the NZVC bits
    CMN     Rn, <op2>            ; Rn - (-op2)   sets the NZVC bits
    MUL{S}  {Rd,} Rn, Rm         ; Rd = Rn * Rm     signed or unsigned
    MLA     Rd, Rn, Rm, Ra       ; Rd = Ra + Rn*Rm  signed or unsigned
    MLS     Rd, Rn, Rm, Ra       ; Rd = Ra - Rn*Rm  signed or unsigned
    UDIV    {Rd,} Rn, Rm         ; Rd = Rn/Rm       unsigned
    SDIV    {Rd,} Rn, Rm         ; Rd = Rn/Rm       signed
    UMULL   RdLo,RdHi,Rn,Rm      ; Unsigned long multiply 32by32 into 64
    UMLAL   RdLo,RdHi,Rn,Rm      ; Unsigned long multiply, with accumulate
    SMULL   RdLo,RdHi,Rn,Rm      ; Signed long multiply 32by32 into 64
    SMLAL   RdLo,RdHi,Rn,Rm      ; Signed long multiply, with accumulate
    SSAT    Rd,#n,Rm{,shift #s}  ; signed saturation to n bits
    USAT    Rd,#n,Rm{,shift #s}  ; unsigned saturation to n bits
```
Notes Ra Rd Rm Rn Rt represent 32-bit registers
        value   any 32-bit value: signed, unsigned, or address
        {S}   if S is present, instruction will set condition codes
        #im8   any value from 0 to 255
        #im12   any value from 0 to 4095
        #im16   any value from 0 to 65535
        {Rd,}   if Rd is present Rd is destination, otherwise Rn
        #n   any value from 0 to 31
        #off   any value from -255 to 4095
        label   any address within the ROM of the microcontroller
        SpecReg APSR,IPSR,EPSR,IEPSR,IAPSR,EAPSR,PSR,MSP,PSP,
                 PRIMASK,BASEPRI,BASEPRI_MAX,FAULTMASK, or CONTROL.
        Reglist is a list of registers. E.g., {R1,R3,R12}
        op2   the value generated by <op2>

Examples of flexible operand <op2> creating the 32-bit number. E.g., **Rd = Rn+op2**
```
    ADD Rd, Rn, Rm              ; op2 = Rm
    ADD Rd, Rn, Rm, LSL #n      ; op2 = Rm<<n   Rm is signed, unsigned
    ADD Rd, Rn, Rm, LSR #n      ; op2 = Rm>>n   Rm is unsigned
    ADD Rd, Rn, Rm, ASR #n      ; op2 = Rm>>n   Rm is signed
    ADD Rd, Rn, #constant       ; op2 = constant, where X and Y are hexadecimal digits:
```
- produced by shifting an 8-bit unsigned value left by any number of bits
- in the form 0x00XY00XY
- in the form 0xXY00XY00
- in the form 0xXYXYXYXY

Made in the USA
San Bernardino, CA
06 November 2019